T0311852

This translation of a classic Russian work on geocryology makes available for the first time in English a wide ranging and up-to-date review of permafrost science, unique in presenting the Russian viewpoint. This revealing account demonstrates how the field developed in the former USSR (largely in isolation from related studies elsewhere), and provides a fascinating insight into the extent of Russian scientific involvement and input. The fundamental physics of frozen ground, geotechnical procedures for construction problems, distribution of permafrost in terms of geological history, and planetary geocryology are all considered. This English edition brings the work to a larger readership, allowing the value of the knowledge and concepts developed to be realized more widely.

General Geocryology

Studies in Polar Research

This interdisciplinary series, aimed at all scientists with an interest in the world's cold regions, reflects the growth of research activity in the polar lands and oceans and provides a means of synthesizing the results. Each book in the series covers the present state of knowledge in a given subject area, resulting in a series which provides polar scientists with an invaluable, broad-ranging library.

Series Editor

D. W. Walton, British Antarctic Survey, Cambridge, UK

Advisory Editorial Board

G. di Prisco, Institute of Protein Biochemistry and Enzymology, Naples, Italy
D. Elliot, University of Ohio, USA
C. Parkinson, Goddard Space Flight Centre, Maryland, USA
M. Sonneson, University of Lund, Sweden
W. Vincent, Université Laval, Quebec, Canada
P. J. Williams, Carleton University, Ottawa, Canada

Other titles in this series

The Antarctic Circumpolar Ocean
Sir George Deacon

Canada's Arctic Waters in
International Law
Donat Pharand

Vegetation of the Soviet Polar Deserts
V. D. Aleksandrova, transl. D. Love

The Frozen Earth
P. J. Williams and M. W. Smith

The Biology of Polar Bryophytes and
Lichens
Royce Longton

Microbial Ecosystems of Antarctica
Warwick Vincent

Chronological List of Antarctica
Expeditions
Robert K. Headland

The Age of the Arctic
G. Osherenko and O. R. Young

Subantarctic Macquarie Island
*P. M. Selkirk, R. D. Seppelt and
D. R. Selkirk*

The Growth and Decay of Ice
G. H. S. Lock

Antarctic Fish and Fisheries
Karl-Hermann Kock

A History of Antarctic Science
G. E. Fogg

The Biology of the Southern Ocean
G. A. Knox

Skua and Penguin
E. Young

General Geocryology

E. D. YERSHOV
Moscow State University

Technical Editor Peter J. Williams
Carleton University, Ottawa, Canada

PUBLISHED BY THE PRESS SYNDICATE OF THE UNIVERSITY OF CAMBRIDGE
The Pitt Building, Trumpington Street, Cambridge, United Kingdom

CAMBRIDGE UNIVERSITY PRESS
The Edinburgh Building, Cambridge CB2 2RU, UK
40 West 20th Street, New York NY 10011–4211, USA
477 Williamstown Road, Port Melbourne, VIC 3207, Australia
Ruiz de Alarcón 13, 28014 Madrid, Spain
Dock House, The Waterfront, Cape Town 8001, South Africa

http://www.cambridge.org

First published in Russian as
Obshchaya Geokriologiya
by Nedra, © E. D. Yershov 1990

First published in English by Cambridge University Press 1998 as
General Geocryology

English edition © Cambridge University Press 1998

First published 1998
First paperback edition 2004

Typeset in 11/14 pt Times [VN]

A catalogue record for this book is available from the British Library

Library of Congress Cataloguing in Publication data

Yershov, E. D. (Eduard Dmitrievich)
  General geocryology / E. D. Yershov.
    p. cm. – (Studies in polar research)
  ISBN 0 521 47334 9 (hardback)
  1. Frozen ground. I. Title. II. Series
GB641.E75 1998
551.3′8–dc21   97-6004 CIP

ISBN 0 521 47334 9 hardback
ISBN 0 521 60757 4 paperback

Contents

Editor's note

It is fitting that the holder of the only Professorial Chair of Geocryology (or permafrost studies as it used to be called), that at Moscow State University, is the compiler of a standard work on the subject. The book is the first of its type to be translated into English. Professor Eduard D. Yershov, who assumed the Chair in 1982, demonstrates the scope of Russian science and geotechnology for cold regions. The fundamental physics of frozen ground, geotechnical procedures for construction problems, distribution of permafrost in terms of geological history and even extraterrestial (planetary) geocryology are all considered.

It is timely that such a book be made available in English because of the extent of interest and activity in the Russian permafrost regions on the part of companies and institutions internationally. This interest, which relates to the oil and gas industry especially but extends to other sectors, has come with the momentous changes since the end of the Soviet era. Although earlier textbooks could be obtained outside the Soviet Union, there was little of the exchanges otherwise characteristic of science and research. The Siberian permafrost areas and oil and gas fields were of course, particularly difficult of access by foreigners.

With the lack of personal, international contacts, the standards of translation of the papers available were not high. This further limited the chances that Russian progress, scientifically and technologically, could be appraised correctly by those outside the country. The Russian scientists and engineers working on the problems of foundation construction in areas prone to permafrost, have far outnumbered those elsewhere. Moscow State University's Department of Geocryology produces tens of graduates in the subject each year. The extent of the Russian studies and their potential for geotechnical procedures and environmental remediation has not been fully understood.

In preparing the English version of Professor Yershov's book it was decided to invest much effort in ensuring a readable text, accurate in translation especially of technical and scientific terms, with an overall presentation of a standard equal to that accorded by Cambridge University Press to any comparable work by an English language author. Working with a translation provided from Moscow, the text was first substantially reworked into correct and readable English while attempting to preserve as far as possible the author's surely typically Russian style. In the interests of scientific accuracy a number of the chapters were further reviewed by specialists in the topics who generously gave of their time. The responsibility for shortcomings is not, of course, theirs, whose-ever it may be. Suffice it to say that the flaws that remain are in spite of the best efforts of many people. We would welcome readers' comments on the translation of specific items. Important, too, was the help of the bibliographer of Russian materials at the Scott Polar Research Institute, Cambridge. Her work and that of several Russian colleagues is reflected in the *Notes on translations and definitions*. All these individuals are recorded in the acknowledgements below but it is appropriate to note here the willingness of Cambridge University Press to make the necessary budgetary provision for this diverse team effort. Dr Maria Murphy of the Press has dealt very imaginatively with the complexities of the project from its conception.

It is hoped that the translation of Professor Yershov's work will not merely provide details of interest to scientists in geocryology. It is important, too, that the extent of the Russian scientific involvement, as demonstrated by this book, be recognized by those working for international projects for oil and gas and other industrial or environmental projects within Russia. Quite generally in cold regions engineering (in Russia these regions are not limited to the Arctic or the North) there has been a failure to involve scientific research in the most cost-effective manner. The economic implications of the permafrost for the Siberian oil and gas industry are profound as indeed they are for the development of Canadian and Alaskan oil and gas. By opening Professor Yershov's book to a wider readership the value of the knowledge and concepts developed in Russia should be more fully realized worldwide.

Acknowledgements

The editor and publishers are indebted to those who have assisted in various ways in preparing the English version.

Isabella M.T. Warren, Russian Bibliographer, Scott Polar Research Institute, Cambridge, has been a constant source of help in refining the translation. The following gave expert advice in interpreting the finer points of particular technical terms in the original Russian: Evgenny Chuvilin, Evgenny Aksenov and Olga Toutoubalina, Moscow State University; and Vlad Roujansky, EBA Engineering, Edmonton.

Scientists and engineers, all in some degree geocryologists, have given advice on English terminology in their own areas of scientific and engineering expertise: R. van Everdingen, Arctic Institute of North America, Calgary; P.D. Groenevelt, Guelph University; D.W. Hayley, EBA Engineering, Edmonton; A. Judge, Geological Survey of Canada; J.A. Heginbottom, International Permafrost Association; B. Ladanyi, Ecole Polytechnique, Montreal; W.G. Rees, Scott Polar Research Institute, Cambridge; E. Jetchik, D.W. Riseborough, J. K. Torrance and T.L. White, Geotechnical Science Laboratories, Carleton University.

The initial translation was carried out in Moscow by N.B. Guseva (Introduction, Chapters 1 and 12-19); A.D. Anikin (Chapter 2); and L.V. Kholdobaeva and A.K. Stroganov (Chapters 3-11).

The editor adds his indebtedness to Kari Williams, for word processing, including the setting-up of the many equations and tables, and to Inger E. Williams, for preparing the English version of all the figures.

Notes on translations and definitions

In transliterations, the system used by the Scott Polar Research Institute's bibliographic services has been followed. The following dictionaries, glossaries and similar items have been used:

Challinor's Dictionary of Geology, 6th edn (Ed. Wyatt). 1986. University of Wales Press, Cardiff. 374 pp.

Slovar' Sokrashcheniy Russkogo Yazyka, 1977 [Dictionary of Abbreviations of the Russian Language].

Elsevier's Dictionary of Earth Sciences: Russian–English (compilers: K.P. Bhatnagar amd S.K. Battacharya) Elsevier, 1991. 1023pp.

Everdingen, R. O. van. 1994. *Multi-language Glossary of Permafrost and Related Ground-ice Terms*. International Permafrost Assn., Permafrost Terminology Working Group. 311 pp.

Glossary of Permafrost and Related Ground-ice Terms. Permafrost Subcommittee, Associate Committee for Geotechnical Research, National Research Council Canada, Technical Memorandum 142. 1988. 156 pp.

Oxford English Dictionary, 1953. 12 vols. (incl. supplement of 1973 and 1986, 4 vols.). Clarendon Press, Oxford.

Oswell, J.M., Doorduyn, A., Costin A. and A. Hanna. 1995. A comparison of CIS and ASTM soil classification systems. *Proc. 48th. Canadian Geotechnical Society Conference*. 573–582.

Poppe, V. and R. J. E. Brown. 1976. *Russian-English Glossary of Permafrost Terms*. Associate Committee for Geotechnical Research, National Research Council Canada. 25 pp.

Romanovsky, N. N., Konishchev V. N. and G. E. Rosenbaum. 1992. *Russian–English Glossary of Geocryology Terms*. International Permafrost Assn., Permafrost Terminology Working Group. 59 pp. (pre-publication copy).

Sanger, F.J. 1973. Editor's Note on Russian Terms for Soils. *2nd International Conference on Permafrost. USSR Contribution*. National Academy of Sciences, Washington. xi–xii.

The Concise Dictionary of Earth Sciences (Eds. Allaby A. and Allaby, M.) Oxford University Press. 1990. 410 pp

The Oxford Russian Dictionary (Eds.: Falla, Unbegaun and Howlett) 1993. Oxford Univ. Press. Oxford, New York. 1340 pp.

Russian words accepted in literal translation, the meanings of which may be unfamiliar or unusual

(Unless otherwise noted, the references are to items listed above.)

Aleurite: unconsolidated deposits, grain size 0.05 to 0.005 mm. Aleurolite is similar, but consolidated (in the geological, not geotechnical, sense, – van Everdingen, pers. comm.)

Ataxite, -itic: disorder, -disordered, irregular

BAM: Baikal–Amur Mainline

ChPI: Chitinskiy Politekhnicheskiy Institut

Cryolithozone, Cryolithic zone: usually synonymous with permafrost (regions) – including extensive areas where permafrost may be discontinuous or even rare. Indeed phenomena of seasonally freezing soils found far distant from permafrost are referred to as typical of the cryolithozone

Cryolite: ice considered as a mineral in the ground (not to be confused with the mineral cryolite, $Na_3 Al F_6$)

Cryopeg: ground which is unfrozen because of dissolved salts (usually below permafrost)

Crystallisation heat: latent heat of fusion

Dell: this transliteration, according to Romanovsky *et al.* (1992), means indeed, a small wooded valley (Oxford English Dictionary)

Deserption: thermal or cryogenic fracture or breakup

Deserptium: surficial creep deposit resulting from volume change (Poppe and Brown, 1976)

Dispersion, dispersed: 1. fineness of particles 2. dispersed (loose) nature of material

Dust, dusty: a grain size fraction (0.05–0.01). Corresponds to coarse silt in English.

Eluvium: residual deposit following washing-out of fine material

Golets ice: 'bold mountain ice' (Romanovsky *et al.* 1992)., 'ice accumulated under a coarse surface layer from meltwater in spring' (p. 267 this book)

Gosstroy: Gosudarstvennyy Komitet po Delam Stroitel'stva (State Committee for Construction Affairs)

Illuvial: material moved into stratum by percolating water

Intercalated: inserted after deposition of bed (used, e.g. of ice layers or lenses)

Khasyrey: thermokarst depression in epigenetic permafrost (contr. *alas* formed in syngenetic permafrost, with melting of syngenetic wedges). According to Romanovsky *et al.* 1992, a local, West Siberian term

Laida: coastal plain covered by tidal water (Evgenny Chuvilin, pers. comm.)

LGI: Leningradskiy Gornyy Institut (Leningrad Mining Institute)

Massive: (1) having ice in pores only; thus no segregation ice, and the ice masses are essentially no larger than pore size. (2) large (of a body of ground)

MISI Moskovskiy Inzhenerno-Stroitel'nyy Institut (Moscow Engineering and Construction Institute)

MSU: Moscow State University (this is an English abreviation, the transliterated Russian being MGU)

NIIOSP: Nauchno-Issledovatel'skiy Institut Osnovaniy i Podzemnykh Sooruzheniy (Scientific Research Institute for Foundations and Underground Structures)

Oligo-mictic: consisting of one to two dominant minerals (*Oxford English Dictionary* -'word of Russian origin')

Paludification: swampiness, conversion to swamp or marsh

Paludial: pertaining to marshes

PNIIIS: Proizvodstvennyy i Nauchno-Issledovatel'skiy Institut po Inzhenernym Izyskaniyam v Stroitel'stve (Industrial and Scientific Research Institute for Constructing Engineering)

Polynya: an area of open water in an otherwise extensive ice cover (on river, sea etc.)

RSFSR: Rossiyskaya Sovetskaya Federativnaya Sotsialisticheskaya Respublika (Russian Soviet Federated Socialist Republic)

Riphaean: Russian term for a period of Mid-proterozoic, from 1650 Ma to 680 Ma.

Rudaceous: rubbly (*Oxford English Dictionary*). Grains coarser than sand size.

Schlieren: repeated small ice lenses (German origin, here apparently often synonymous with 'streaks')

Suffosion: a spreading out into a sub-stratum (e.g. filling of a solution cavity or a cavity from melting ice)

Tripoli: non-diatomaceous sandstone, heavily-weathered in situ, therefore friable. (Metcalf, Robert W. 1946. *Tripoli. Information Circular*. U. S. Dept. of Interior, Bureau of Mines, I. C. 7371.)

Tuffolava: extrusive rocks intermediate between tuffs and lavas (*Challinor's Dictionary of Geology*).

USSR: Union of Soviet Socialist Republics

VSEGINGEO: Vsesoyuznyy Nauchno-Issledovatel'skiy Institut Gidrogeologii i Inzhenernoy Geologi (All-Union Scientific Research Institute for Hydrology and Engineering Geology)

Yedoma: remnants of ice-rich pleistocene plain, appear as hills following erosion. (Potapov, Eugene. 1993. *Ecology and Energetics of Rough-legged Buzzard in the Kolyma River Lowlands*. Thesis, D. Phil., Oxford.) See also pp. 312–15.

Frequently misinterpreted words

Readers acquainted with the Russian language may be interested in the following words which often cause confusion in translations. A direct transliteration or a literal translation gives an incorrect or, at least, misleading meaning. Some erroneous translations have recurred in so many publications as to have gained a certain currency. As a consequence they have often caused difficulties for the proper understanding and assessment of the Russian literature of geocryology. The correct translations, so far as could be established for the present work, are given below (**bold**). They are arranged alphabetically, according to the transliterated Russian word.

dispersnost = disperse: this word is widely used in the Russian, where English would require '**fine-grained**' ('greater dispersion' = '**more fine-grained**'); also used, however, in the sense of '**widely dispersed**' (i.e. widely spaced or loosely packed).

kriolitozona = cryolithic zone = cryolithozone: usually synonymous with **permafrost regions** – including extensive areas where permafrost may be discontinuous or even rare. Indeed phenomena of seasonally freezing soils found far distant from permafrost, are referred to as typical of the cryolithozone (for example, in Chapter 3). A term covering all ground affected by freezing (regardless of permafrost) would be useful. Cryolithosphere is occasionally used in this sense.

plotnost': – **density,** in the normal sense, but is also used in the sense of 'strength' or '**intensity**' for example of a flow of water (and would then usually be omitted in English).

suglinok: silty soil with more sand than clay (See quantitative limits in: Sanger, F. J. in *Second Intern. Conf. Permafrost, USSR Contribution*, National Acad. Sci., Washington. 1973 pp. xi, xii). Russian translators often use 'silty loam' for these soils but 'loam' in English is normally restricted to an agricultural or soil science context. Generally shown in the present work as **sandy-silt** or similar.

SNiP: – although 'snip' is an appealing acronym, sometimes used directly by English-speaking engineers, the translation **Building Norms and Regulations** seems preferable.

supes: silty soil with more clay than sand generally translated in the present work as '**clay-rich sandy silt**' or similar. See also comments under suglinok.

poroda: **ground, rock** or **soil**, according to context. Skal'naya poroda and gornaya poroda: **bedrock**

talyy: literally, '**thawed**' but often used for **unfrozen** part of ground, of soil sample etc. without reference to any previously frozen state. Thus, 'unfrozen soil', '**unfrozen** ground'. Where this meaning is intended rather than using the symbols t, th, or tha (used as subscripts) in equations and figures have been replaced by unf in the English.

migratsionno-segregatsionnyy led: literally, **migrational-segregation ice**: ice accumulating by cryosuction, normally in the form of **ice lenses** or **layers** (also called —**schlieren**), and characteristic of frost heaving soils.

ob'yemno-gradientnoye napryazheniye: literally, **volume-gradient stress**: an '**all-round' stress**, esp. stress inducing shrinkage (volumetric strain). The term embodies the concept of gradients of such stress, which leads to cracking and shearing as a result of the differential volume changes.

zhila: '**wedge**' or 'vein'. **Ice wedge** in the English sense, is povtorno zhil'nyy led = 'recurring ice wedge' although the 'recurring' (povtorno) may often be omitted. Thus confusion may exist between ice wedge, and **ice vein** -usually the initial (first year) ice infilling of the tensile fracture crack. See also van Everdingen, R. O. (1994).

usadka: **shrinkage** (as in English, a distinction may be made between 'shrinkage' and 'consolidation' (soil mechanics sense) – although the process is similar)

Place names and personal names

Geographic names are given in anglicized form where such exists and is generally recognized (e.g. St. Petersburg). Otherwise a transliterated form (e.g. River Irtysh) is given following the guidelines of the British Permanent Commission on Geographic Names. The transliterated Russian term is used for geographical features ('strait', 'bay', 'island', 'mountain range' etc.) *except for* 'sea' and 'lake'. The guidelines normally correspond to those advocated by the United States Board of Geographical Names and are those used in The Times Atlas of the World and in current maps of the National Geographic Society, Washington.

Personal names are transliterated in the normal fashion except where the person is known by a different form in English.

Peter J. Williams
Geotechnical Science Laboratories   Scott Polar Research Institute
Carleton University,   University of Cambridge
Ottawa   Cambridge

May 1997

Abstract

Theoretical concepts of the science dealing with frozen ground and cryogenic-geological processes are given. The principles of formation and development of frozen ground and seasonal freezing-thawing are elucidated. The particular features of the composition, structure and properties of frozen ground are described together with the conditions of their formation. The principles and methods of geocryological investigations and the classification of perennially frozen ground are considered. Also presented are the principles of engineering geology for design, construction and operation of engineering structures in the permafrost zone.

This work is particularly relevant for students studying in the specialization 'hydrogeology and engineering geology'.

Preface

Geocryology (the study of frozen soils) is a natural and historical science and a branch of geology, concerned with the laws of the formation and the evolution in time and space of frozen ground, its composition, cryogenic structure and properties, and with cryogeological processes and phenomena. The frozen ground may be hundreds of meters thick (up to 1500 m) in the region comprising the freezing zone of the lithosphere characterized by freezing temperatures (to $-15\,°C$) and inclusions of ice or ice crystals.

As any other branch of knowledge, geocryology has resulted from practical needs, and its coming into being has reflected the economic development of huge permafrost tracts, which include currently 25% of all land on our planet and some 50% of the territory of the former USSR.

The subject of geocryology is now well-defined, as are a range of its basic problems, its practical and scientific significance; techniques and procedures for special geocryological researches have evolved; with its major fields and trends established, the prospects of geocryology gaining both in science and in application have proved very promising. Vladimir I. Vernadskiy has noted earlier that it is the limits of cooling below the ground surface which define a task relating to the solution of problems which are all of great scientific and practical importance.

The topic of geocryology is frozen ground, including underground ice and snow accumulations. According to the views of A.B. Dobrovolskiy, V.I. Vernadskiy and P.I. Koloskov, frozen ground occurs in the cryosphere, which is a thermodynamic envelope of the Earth where ice, water and vapour can exist simultaneously under negative temperatures. Permafrost is a natural geological formation noted for its distinctive laws of genesis, existence, evolution and distribution on the planet. Looking to outer space, most planets of the Solar system and other celestial bodies appear to be cryogenic, i.e. to be noted for permafrost developed on them. In other words, our perspective extends from the cryology of the Earth to that of the planets or Universe, cosmic cryology.

Several works of generalization have been written on the fundamentals of geocryology. *General Frozen Ground Studies* by M.I. Sumgin, S.P. Kachurin, N.I. Tolstikhin and V.P. Tumel, came in 1940, and in 1959 *Fundamentals of Geocryology*, the magnum opus, put out by the Institute of Frozen Ground Studies. The first edition of the textbook *General Frozen Ground Studies* was printed in 1967, its second edition in 1978, and in 1981 came *Frozen Ground Studies* (*Concise Course*) prepared and edited by V.A. Kudryavtsev. *General Frozen Ground Studies* edited by P.I. Mel'nikov and N.I. Tolstikhin appeared in 1974. All these books have proved very important in laying the foundations of geocryology; now they are rarities, and what is more, many of their chapters are in need of considerable reworking due to substantial advances made recently in formulating the theoretical and applied (engineering) fundamentals of geocryology.

The present textbook, based on the latest advances in science and practice, expounds in condensed form the fundamentals of dynamic, lithogenic, regional, historical and engineering geocryology. It is designed for students and lecturers in geocryology at universities (the 'Engineering geology and groundwater hydrogeology' specialization), and in geological exploration, mining, oil and gas, building and transport, at institutions of higher education. There is no doubt it will be useful for a wide range of geologists in research and commercial bodies, and for many technicians in design and survey, building and mining establishments engaged in exploration and economic development of the permafrost regions.

Colleagues from the Department of Geocryology in the Geology Faculty at Moscow State University participated in compiling the textbook. Some chapters and subsections were written in collaboration with V.Ye. Afanasenko (Ch. 13), L.S. Garaguliya (Introduction, #5; Ch. 16, #1–3; Ch. 19, #2), I.D. Danilov (Ch.9), K.A. Kondrat'yeva (Ch. 14, #3; Ch. 15, #1, 2; Ch. 16, #1 and 4), S.Yu. Parmuzin (Ch. 17–19). Participating in writing individual subsections were Ye.N. Dunayeva (Ch. 15, #3), V.Ye. Romanovskiy (Ch. 1 #4 and 5; Ch. 10, #1), S.F. Khrutskiy (Ch. 10, #3; Ch. 12, #3); Ye.M. Chuvilin (Ch. 8, #4; Ch. 9, #3).

Ye. M. Chuvilin was responsible for preparing the manuscript for publication. Assisting him at all stages of its preparation were T.N. Kosatkova, L.A. Nikulicheva, O.N. Patrik.

The author would like to thank all these contributors, especially K.A. Kondrat'yeva who has read the final manuscript, for their invaluable help.

Chapter 2 was translated by A.D. Anikin. The Preface and Chapters 3 –
11 were translated by L.A. Kholdobayeva and A.K. Stroganov. The Introduction, Chapter 1 and Chapters 12 – 19 were translated by N.B. Guseva.

Introduction

1 Geocryology as part of planetary cryology

Geocryology is a branch of a more general science - cryology of the planets. Frozen ground as a natural-historical geological formation is not unique to and typical of the Earth only. It is widely developed on other planets of the Solar system such as Mars and Pluto as well as on the satellites of Jupiter, Saturn, Uranus, Neptune and Pluto. This becomes clear and evident if we take into account the fact that the void of the Universe with its temperature being close to absolute zero is 'a kingdom of cold'. Planets and their satellites, asteroids and other solid bodies within this space, must naturally be in thermodynamic equilibrium with the space environment. Consequently the temperature of at least the near-surface rock units on the overwhelming majority of planets far removed from the Sun must be below $0 °C$. There is always a chemical compound (matter) dominant in the atmosphere of a planet, which exists simultaneously in three states: solid, liquid and gaseous under a given temperature, depending on the temperature of each planet. The solid phases of this matter (ice) on the planet can form either separate large 'ice' aggregates and monomineral rocks or enter into the composition of frozen rocks in the form of the particular mineral component. Solid celestial bodies of the Universe must be cryogenic (Table 1) in the overwhelming majority of cases, i.e. be characterized by development and existence of frozen rocks.

On Earth H_2O is the widely distributed and chemically active matter penetrating into other compounds, minerals and rocks and is able to exist in three phases simultaneously, changing its state of aggregation by phase transitions. Processes of energy- and mass-exchange responsible for the phase transitions 'water – ice', 'ice – vapour' and 'water – vapour' are the essence of the cryogenic processes of, in this case, the water type. Under the conditions of other planets and celestial bodies carbon dioxide, methane, ammonia, hydrogen and other matters can be the material basis for

Table 1. *The proposed examples of space bodies of the cryogenic type*

Category	Planets and other celestial bodies	Ice varieties (established or proposed)
Completely cryogenic	Rings of Saturn Rings of Uranus	H_2O (I, amorphous?) Altered CH_4
	Miranda, Ariel, Umbriel, Titania, Oberon (satellites of Uranus)	H_2O (I)
	Pluto and its satellite Charon,	CH_4
	Triton (satellite of Neptune)	CH_4, H_2O ?
	Phobos and Deimos (satellites of Mars)*	—
	Amalthea, Metis, Adrastea Thebe (satellites of Jupiter)**	—
Continuous surface cryogenic	Io (satellite of Jupiter)	SO_2
	Europa (satellite of Jupiter)	H_2O (I)
	Ganymede (satellite of Jupiter)	H_2O (I, II, V, VI, VII, amorphous within the polar caps?)
	Callisto (satellite of Jupiter)	H_2O (I, II, V, VI, VII)
	Mimas, Tethys, Dione, Hyperion, Iapetus (satellites of Saturn)	H_2O (I)
	Mars	H_2O, CO_2
	Titan (satellite of Saturn)	H_2O, (I, II, V, VI, VII) clathrate hydrates
	Enceladus (satellite of Saturn)	H_2O (I, previously amorphous ?); clathrate hydrates (?), NH_3 hydrates (?)
Incompletely cryogenic and deeply cryogenic	Periodically stable varieties of comets and asteroids (with strongly extended elliptical orbits) and meteorites (going near the Sun) can be assigned to these cryogenic communities at the instant of being at their perihelion by convention	H_2O in cometary nucleus
Discontinuous surface cryogenic	The Earth, the Moon*	H_2O

*Cryotic bedrock and regolith
**Cryotic rock

cryogenic phenomena in addition to water, according to I.Ya. Baranov. Thus for example comparison of the spectra of the giant planets shows that the spectral lines of ammonia are gradually weakening while the lines of methane are intensifying in going from the spectrum of Jupiter to the spectrum of Neptune, resulting from the ammonia freezing out with the decrease of temperature. It is evident that glacial and other ice covers must be formed on planets in the same sequence (in accordance with gaseous-phase crystallization at lower temperatures). Actually while H_2O ice prevails in Earth's cryolithosphere, H_2O ice and CO_2 ice prevail in the cryolithosphere of Mars, so one must also follow a change to formation of ice from carbon dioxide and ammonia with water, and then from methane, proceeding towards the most distant planets of the Jupiter group. Consequently one might distinguish between the planets and their satellites by the development of the cryogenic processes of water (Earth), carbon dioxide-water (Mars), water-methane (Triton - satellite of Neptune), methane (Pluto) and of other types (Table 1) within the Solar system. For example it is common knowledge that both the poles of Mars are covered with thick ice 'caps'. Thus the northern polar cap consists of H_2O ice and contains a great amount of dust. The upper layer of the southern polar cap consists mainly of CO_2 ice and does not contain any dust. According to R.O. Kuz'min's opinion (1983) the Martian cryolithosphere has a three-layered structure with permafrost thickness varying in the range from 1 or 2 to 4 or 5 km (Fig. 1). Surface deposits (a few kilometers thick) representing glacial formations of a peculiar kind consisting of CO_2 ice and H_2O ice stratifications, gas hydrates and loose (dusty) sedimentary rocks (Fig. 2), were found within the polar regions of that planet with the help of images obtained by the 'Mariner 9' space vehicle.

Investigation of the Solar system's celestial bodies has allowed American researchers to publish in recent years interesting data on the peculiar features typical of the planets and their satellites. Thus the Galilean satellite of Jupiter, Io, is the unique celestial body of the Solar system. It is supposed that the greater part of the surface of this satellite is composed of solidified sulphur and solid SO_2 or SO_2 ice. The surfaces of the other Galilean satellites such as Europa, Ganymede and Callisto consist mainly of H_2O ice which, probably, is polluted by fine foreign material from fragments of meteorites. Study of the spectral reflecting power of Tethys, Dione, Rhea, Iapetus and Hyperion (satellites of Saturn) allowed the scientists to conclude that the surface of these satellites, except for the dark side of Iapetus, consists of all but pure H_2O ice with a less than 1% admixture of mineral dust. It is not inconceivable that the ice can include a great amount of gas.

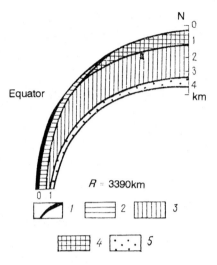

Fig. 1. Diagram of section (north pole – equator) of the Martian cryolithosphere along a meridian (after R.O. Koz'min): 1 – H_2O ice; 2 – H_2O ice and gas hydrate; 3 – H_2O ice and CO_2 liquid; 4 – CO_2 ice; 5 – supercooled water and gas hydrate.

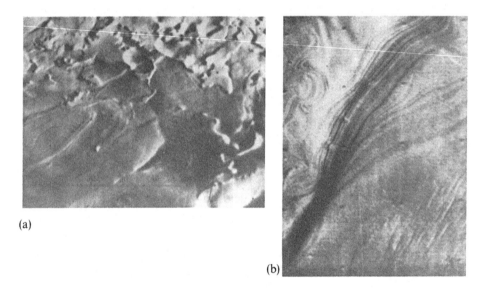

(a)

(b)

Fig. 2. Flat layered terrain on Mars lacking craters: *a* – an oval 'layered' plateau, 40 × 20 km in size; *b* – alternating glacial formations (with separate layers from 50 to 100 m in thickness).

Titan (a satellite of Saturn) is likely to have an ethane-methane ocean and its surface consists mainly of 'volatile' ices such as NH_3 and CH_4. It is supposed that there is a combination of H_2O ice and NH_3 hydrates in the composition of the surface of Enceladus (a satellite of Jupiter). This satellite has the

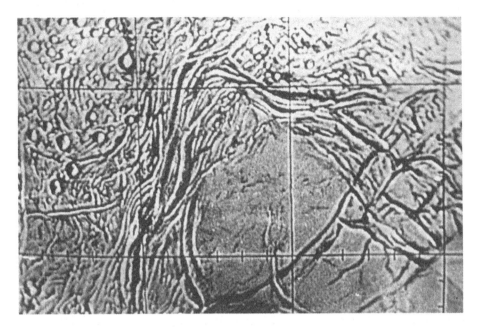

Fig. 3. The surface of Enceladus (from pictures taken from the spacecraft '*Voyager* 1' and '*Voyager* 2'). Scale 1:5 000 000.

highest reflecting power in the Solar system. It is possible that its surface is covered at all times with H_2O ice and $NH_3.2H_2O$ ice formed as a result of liquid water flowing out of the satellite interior. The largest part of the surface of the satellite is a plain with furrows (Fig. 3). Enceladus is the most active and the youngest satellite of Saturn from the geological standpoint with the density of craters being less than that on the other satellites even within the areas with the greatest crater concentration.

As in the case of Jupiter's and Saturn's satellites, H_2O ice was identified on the surface of all five satellites of Uranus (Miranda, Ariel, Umbriel, Titania, Oberon) (see Table 1). In this case the images from the 'Voyager 2' vehicle travelling between the orbits of Miranda and Ariel in 1986, show highly cratered areas as well as traces of tectonic activity in the form of extended faults and furrows on the surface of all the satellites.

American scientists proposed on the basis of experimental research that Neptune's satellite Triton is similar to Titan. It is thought that this satellite is shrouded in a mainly nitrogen atmosphere while there are oceans of liquid nitrogen with floating fragments of methane ice. It is possible that there is water hoar frost on the satellite's surface. It is supposed also that the planet most distant from the Sun in the Solar system, Pluto, contains a great amount of methane ice.

One can recognize a number of categories, classes, groups and varieties of planets, their satellites, asteroids and other celestial bodies (comets, meteorites, cosmic dust accumulations) with respect to conditions of development of frozen ground (granular or cemented with H_2O, CO_2, NH_3, CH_4 ice etc. or materials containing these varieties of 'ices') and with respect to their distribution over the planets.

Heat flux arrival at the surface of a celestial body from the interior q and from outside Q is of fundamental importance in such a classification. At the first approximation one can recognize the planets without ($q = 0$) and with ($q > 0$) heat flux to the surface of the planet from the interior (Table 2) in accordance with the state of the core of the planet (cold or hot) and physical and chemical exo- and endogenous thermodynamic processes and reactions. With respect to the heat arrival at the solid celestial body surface from a source of radiation (the Sun or from another celestial body) one can distinguish the planets without heat flux arrival at their surface from the outside ($Q = 0$ -when the planet is situated far away from the sources of thermal radiation) and with small ($Q > 0$) or great external heat flux ($Q \gg 0$ -when the planet is situated near the external thermal source). Various combinations of heat balance parameters cause a number of variations in the permafrost distribution over the planets and celestial bodies. In this respect completely cryogenic, incompletely cryogenic, deeply cryogenic, continuous-surface-cryogenic, discontinuous-surface-cryogenic and completely uncryogenic categories of planets and solid celestial bodies can be recognised (see Tables 1 and 2).

When the relationships between the periods of a planet's rotation around its axis τ_{rot} and of its revolution around the Sun τ_{rev} are also considered, one can recognize classes of planets and their satellites with symmetric ($\tau_{rot} > \tau_{rev}$) permafrost distribution and with asymmetric distribution ($\tau_{rot} = \tau_{rev}$), with permafrost being developed over the night side only. Daily temperature fluctuations with the period being less than the annual revolution of the celestial body around the external source of radiation are typical of the planets of the symmetric type. Thus consideration of the nature of the heat balance (of relations between the values of internal and external thermal fluxes and of the specific nature of the heating of the planet for the period of its rotation) allows us to recognize nine classes of planets which vary greatly in cryogenic structure i.e. in the particular permafrost occurrence (in depth) and distribution (over the surface) within the planets and celestial bodies (see Table 2).

The particular nature of the water balance of a planet (or of the carbon dioxide, hydrogen, ammonia, etc. balance), with the presence or absence of

Table 2. *Conditions of development and arrangement of planets of cryogenic type*

	With respect to the character of thermal and water balance		With respect to cryogenic stability and presence of seasonal freezing (thawing)
Community	class	Group	Variety
$q = 0$,	symmetric		
$Q \to 0$			
Completely cryogenic $q > 0$ $Q \to 0$	$\tau_{rot} \gtreqless \tau_{rev}$ symmetric		
continuous surface cryogenic	$\tau_{rot} \gtreqless \tau_{rev}$ asymmetric		
$q = 0$ $Q > 0$ Incompletely cryogenic	$\tau_{rot} = \tau_{rev}$ symmetric		
$q = 0$ $Q >> 0$ Deeply cryogenic	$\tau_{rot} > \tau_{rev}$ *asymmetric*		
	$\tau_{rot} = \tau_{rev}$ symmetric		
$q > 0$ $Q > 0$ Discontinuous surface cryogenic	$\tau_{rot} \gtrless \tau_{rot}$ asymmetric		
	$\tau_{rot} = \tau_{rev}$ symmetric		
$q > 0$ $Q >> 0$ uncryogenic	$\tau_{rot} > \tau_{rev}$		

Every class of planets is subdivided into atmospheric and atmosphere free groups with regard to particular features of the balances of water, carbon dioxide and of other types.

Every group of planets and celestial bodies is subdivided into varieties: cryogenically stable (with a circular orbit of movement near the heat source) and cryogenically unstable or periodically stable (with a strongly extended elliptical orbit).

Every group of planets is subdivided into cryogenic varieties with presence of seasonal freezing and thawing (with the value of angle between the axis of rotation of a planet and the plane of its orbit $\alpha \neq 90°$) and absence of seasonal freezing (thawing) or cooling (heating) (with $\alpha = 90°$)

Seasonal freezing (thawing) and cooling (heating) of rocks is absent or weakly expressed the liquid phase being responsible for physico-chemical transformations of the parent rocks and sedimentary materials, the prevailing kind of phase transitions, etc. are also essential criteria for the classification of lithogenesis on planets and celestial bodies. The planets having an atmosphere are characterized by the equilibrium and quasi-equilibrium nature of the balance of the prevailing three-phase matter making a closed cycle of transitions from the liquid or solid state into the gaseous one and vice versa from the gaseous state into the liquid or solid. Planets without an atmosphere are characterized by absence of these intrinsic balances.

The stability of permafrost on planets and celestial bodies depends essentially on the character of their orbits and periodic remoteness from a source of radiation. In the case of a circular orbit the celestial body will be characterized by a cryogenically stable type of permafrost development while in the case of a highly elliptical orbit it will be characterized by a periodically stable cryogenic type. As the celestial body approaches a source of radiation, decrease of thickness and area of the permafrost will be observed, while the further the celestial body is removed from it, the more severe will be the permafrost conditions. The periodically stable permafrost regime must manifest itself most clearly, for example, on such planetoids (asteroids) as Adonis and Apollo (approaching the Sun closer than Venus and moving a great distance from it, nearly reaching Jupiter's orbit at aphelion), as well as on comets with large eccentricities of orbit and with periods of revolution as long as hundreds or thousands of years. For celestial bodies with highly elliptical orbits (comets) and hyperbolic orbits (meteorites), with periodically stable regimes of permafrost, we cannot speak about the cryogenic varieties but about the possibility of moving them from one cryogenic category to another within the limits of the classification scheme given (for example, from completely cryogenic to deeply cryogenic or to discontinuous surface cryogenic character of permafrost distribution).

And finally one can recognize varieties of cryogenic celestial bodies with respect to presence or absence of seasonally frozen (or seasonally thawed) ground depending on the angle α of tilt of the planet or satellite axis with respect to the orbital plane. Seasonal freezing and thawing (or cooling and heating) of the surface rocks of the lithosphere will be absent in this case if the angle $\alpha = 90°$. If $\alpha \neq 90°$ a layer of seasonal thawing (freezing) and a layer of annual temperature fluctuations will be formed as typically, they are, for example, on Earth.

There are examples in Table 1 of the supposed correspondence between the planets, their satellites, asteroids, comets and meteorites well known today and the recognized classifications of categories, classes and groups of celestial bodies. The summary of the classification (which as yet is rather sketchy) provides support for the view that celestial bodies of the cryogenic type prevail among the known solid celestial bodies, and are hence typical of the Universe.

Thus the planet Earth is one of many planets of the cold type. It is of the cryogenically stable type, of the kind with a water atmosphere, of the symmetric class, with a discontinuous cryogenic surface and the presence of a marked process of seasonal freezing and thawing (cooling and heating) of the ground.

2   Frozen rocks as natural-historical geological formations

A few periods of active formation and existence of frozen ground alternating with periods of its disappearance or sharp decrease in extent can be discerned in the history of this planet's development. The distribution of frozen ground over the Globe in ancient epochs is associated with regions where ancient continental glaciations occurred and their moraine deposits are found. Therefore during the first half of the Earth's geological history (2.5 billion years ago) permafrost is unlikely to have been widespread because the platform massifs and continental areas were poorly developed while radioactive heating from the interior was still very great. However by the Early Proterozoic (2.1 – 2.5 billion years ago) the permafrost already existed on the North American continent and in Southern Africa. In the Late Proterozoic (600 – 1000 million years ago) it existed within Northern and Southern America, Greenland, Australia, Central and Southern Africa, the Russian Platform, Ural, Kazakhstan, Southern China and Korea. In the Palaeozoic (240 – 400 million years ago) the permafrost occupied (with interruptions) Cental and Southern Africa, Brazil, Southern America, Antarctica, mountain regions of India, Australia and the Arabian peninsula. In the Mesozoic and Early Cenozoic the frozen ground is unlikely to have been widespread. In the early Late Cenozoic (25 million years ago) cooling had occurred again and a set of glaciations took place while in the Pliocene and Pleistocene the perennial freezing of ground began which is still in progress. The regions of the widest permafrost development in this case were North America, Europe, Asia, Antarctica, Greenland. The most ancient traces of the permafrost in the deposits of the Early Pleistocene (more than 700 thousand years ago) and in those of the Late Pliocene (more than 1.8 million years ago) have been reported within the Kolyma lowland, Alaska and Canada.

Thus the data on known glacial events in the Earth's geological history point to a few large time intervals when the permafrost would have been widespread. The reasons for the irregular periodicity of the great glacial epochs and of permafrost development over the planet are still under discussion and poorly known. The development of permafrost was a possibility only when the ground was subjected to negative temperatures (below $0\,°C$ or $273.1\,K$). This caused the transition of ground water into ice and hence the transition of soils into the qualitatively new (frozen) state. The possibility or impossibility of negative temperatures in the surface layer of the lithosphere is determined by the relationships of the components of the Earth's energy (thermal) balance. It can be supposed that the periods of glacial epochs and permafrost originated as a result of the appropriate climate changes on the planet being predetermined first of all by astronomic factors and associated with particular tectonic developments as well. Thus using plate tectonics as a basis the largest periods of cooling on the Earth can be associated with the formation of gigantic continents (Laurasia, Gondwana, Pangaea, etc.) situated at one of the planet's poles.

Therefore frozen ground should be considered not as an exceptional phenomenon in the Earth's history, but as a natural-historical formation occurring repeatedly in the course of geological development in various parts of the planet. At the same time composition, structure, particular textural features and the other characteristics of frozen ground existing in ancient epochs are unlikely to have been identical to the present ones, being subjected to essential evolutional change in accordance with the irreversibility of evolution and types of lithogenesis. Such irreversibility in the Earth's history manifested itself in progressive replacement of the vulcanogenic sedimentary forms at first by humid, then by arid forms, and finally we can see a trend toward prevalence of the cryogenic type of lithogenesis over the others. At the same time the change from the predominantly chemogenic formation of basin sediments to chemogenic terrigenous and, beginning from the Cenozoic, to terrigenous biogenic ones is clearly followed. This is associated with a steady increase of the total area of platforms as well as with the migration of biota to dry land and with a sharp increase of total biomass.

The perennially frozen materials being rather specific formations, vary as far as their composition, cryogenic structure, type of cryogenesis, cryogenic age, temperature regime, thickness, ice content and other characteristics are concerned.

By *frozen* ground proper is usually meant geological formations characterized by negative temperature, by moisture contents exceeding that of unfrozen (pellicular bound) water W_{unf} under a given temperature and by ice cementing mineral particles and filling cavities, pores and fissures. Soils (clastics, sand, clay and peat) as well as faulted or weathered magmatic, metamorphic and cemented-sedimentary rocks can constitute the frozen ground. Surface ice (river, lake, marine, glacial and other types) and underground ice (buried, wedge, segregated, sheet etc.) and snow accumulations are considered in this case as monomineral rocks while ice is considered as a specific mineral. Rocks with negative temperature, with moisture content less than W_{unf} under a given temperature and without ice (monolithic magmatic, metamorphic, cemented-sedimentary rocks) are termed *cryotic*.

Among the variety of frozen and cryotic materials soils represent multicomponent polyphase capillary, porous and often colloidal ground systems which are most complex subjects of investigation. Water, H_2O, usually occurs in three states of aggregation: in the form of ice, vapour and unfrozen water. The unfrozen water represents a portion of the bound water which has not frozen out, the content of which decreases as the negative temperature is lowered. Ice and unfrozen water are in steady dynamic equilibrium. Thus with increase in temperature the ice begins to melt and replenishes the unfrozen water content while with the lowering of temperature the ice content increases in a soil at the expense of the unfrozen water. Consequently the frozen ground is a highly dynamic system responding to any change in the external thermodynamic conditions. Frozen soils differ from unfrozen ones first of all by their solid nature, i.e. by ice cementation of mineral particles, and presence of particular (cryogenic) texture and structure. All this is conditioned by phase transition of ground water into ice at freezing and is accompanied by a large variety of complex physico-chemical processes. There is an associated movement (*migration*) of film water from the unfrozen to the freezing part of the ground, with coagulation and aggregation of the soil particles with shrinkage and loss of water content, below the freezing front and with dispersion, breaking, swelling and heaving of the ground in the frozen part by the wedging effect of the 9% volume increase on freezing of the moisture migrating to this part in thin films. Freezing of soils with moisture migration causes ground differentiation (segregation) with the massive frozen (mineral skeleton portion) and the visually observed migration-segregation ice layers (layered, reticulate, porphyritic-like, lenticular, etc.) forming a specific *cryogenic structure*. Cryogenic structures of hard rocks of various genesis and composition depend mainly on the nature of voids and their distribution (fissures, pores, cavities, etc.). Ice streaks and inclusions occupy those voids which were completely or partly filled with water before freezing. In the event that ice in the form of visual interlayers and separate inclusions is absent in frozen soils and occupies pore space in the form of ice cement only, the uniform cryogenic structure formed is termed *massive*. In essence, in this case, we are dealing with a cryogenic structure.

The essential differences between frozen and unfrozen soils are noted in their chemical mineral composition and degree of dispersion. This is associated with the specific character of geochemical and weathering processes proceeding within the permafrost zone. Thus the soils cemented with ice are characterized by a higher than usual carbon dioxide content, with reducing conditions and acidity clearly evident. This establishes favourable conditions for decomposition of silicates and migration of chemical elements, and for the formation of lower ion oxides and gley horizons. In these conditions hydrous mica and montmorillonite are formed, and such minerals as vivianite, pyrite, marcasite, siderite, etc. are accumulated. The specific geochemical situation of the permafrost zone (low temperatures, shortage of exchange and oxidizing processes, paludification, etc.) contributes to the preservation of vegetable and animal remains, formation of various hydrogenous compounds (methane, hydrogen sulphide) and completion of the process of humus formation at a less mature stage causing wide development not of humic acids but rather of mobile and aggressive fulvic acids and their organic mineral compounds (fulvates, chelates etc.). Sharp intensification of such processes as carbonation and sulphonation (precipitation of poorly soluble salts and formation of calcite, mirabilite and gypsum under decreasing temperature), occurs, with cryogenic 'desalination' of the permafrost pore solution and cryogenic concentration (especially of highly soluble salts) in subpermafrost waters, the formation of horizons containing salt water and brine having negative temperature (*cryopegs*) and with the reverse hydrochemical ground water redistribution, the formation of gas hydrates, etc.

The manifest domination of physical weathering (cryoeluvium formation) over the chemical within the permafrost region causes wide distribution of poorly sorted clastic materials, cemented with ice and characterized by the clearly defined poorly sorted nature and heteroporosity of the mineral skeleton (cryogenic conglomerate, ice breccia, etc.). The fine-grained portion of the granulometric spectrum (particles smaller than 1 mm) within the permafrost zone has the distinctive feature of being high in dust content (up to 60% and higher). This is connected solely with the specific nature of the cryogenic *weathering* processes (with their repeated freezing and thawing) being accompanied by fracturing of sand particles and agglomeration (coagulation) of clay (colloidal) particles. It is precisely this process which provides an explanation for the wide development of loess-like deposits within the permafrost zone, characterized by predominance of particles of the dust size fraction (0.05 – 0.01 mm). As a consequence of the particular, unusual features of composition and cryogenic structure of frozen sediments they also have distinctly different properties as a whole, compared with unfrozen materials. In association with the strong manifestation of physical weathering (of the cryohydration type) and intensification of slope processes (of the freezing type), placer deposits of eluvial, solifluction, alluvial and other (mainly continental) genesis prevail among the sedimentary formations within the permafrost zone.

Ground ice occurs in frozen strata in the form of independent ice formations – the monomineral material (ice) is widely developed within the permafrost zone. Wedge ice is formed along frost (temperature) cracks, and ice bodies in various frost mounds, sheet ice deposits, etc. are widely distributed.

Numerous classifications of perennially frozen ground based on the subdivision of frozen materials with respect to any one or a few features, for example, with respect to ice content, cryogenic structure, permafrost genesis and age, temperature, permafrost thickness, etc., have been devised recently for the multicomponent composition of frozen ground and the inherent complex linkages responsible for its existence and development.

In a general approach permafrost is subdivided into epicryogenic and syncryogenic forms with respect to the type of freezing. Materials transformed into the perennially frozen state after completion of the accumulation of the sediments and their diagenetic modification (the process of transformation of sediment into rock), are epicryogenic. *Epicryogenic* rock units are formed mainly in the course of one-sided freezing from above in connection with global or regional cooling and build up their thickness with a deepening of the permafrost base. *Syncryogenic* rocks are formed as a rule from sedimentary (basin and continental) deposits on the existing frozen substratum when sediment accumulation and transition into the frozen state take place practically synchronously (simultaneously on a geological time scale). They are always underlain by epigenetically frozen materials and build up their thickness by the rising of the permafrost table as a result of sediments being progressively accumulated and freezing simultaneously into the perennially frozen state. Some researchers have described *diacryogenic* (parasyncryogenic) rock units formed recently in the course of freezing (from above and from the sides) of moisture-supersaturated unlithified ground (newly deposited sediments and silts). Complex, diagenetic physical chemical processes in them preceding the freezing process are far from completion. Taliks in basin deposits freezing in the bottom conditions of shallowing water bodies are a good example. Various combinations of epicryogenetic, syncryogenetic and diagenetic strata in vertical section form polycryogenic strata, very widely distributed in the permafrost zone.

It should be noted that below the permafrost base in hard rocks as well as in sediments, materials having negative temperature but not containing ice often occur. This is usually associated with squeezing out of highly soluble salts (chlorides of calcium, magnesium, sodium, etc.) from pore solutions of the freezing rocks and their concentration in deeper horizons below the permafrost base. As a consequence highly mineralized sub-, intra- and suprapermafrost waters (up to 200 g/l and higher) occur (in 'cryopegs') the freezing temperature of which depends on the concentration of the salts in the solution and is always well below 0 °C. The thickness of such negative temperature strata with cryopegs varies in the range from the first ten to 500 or 700 m and greater and can be 1000 – 1500 m and greater.

Ground freezing is shown to have occurred at various times in the territory of the former USSR. It is supposed that the permafrost of Pliocene-Eopleistocene age and formed 1 – 2 million years ago, is preserved in thick strata composing the lowlands of the Far North while within the remainder of the territory such strata are usually of Pleistocene age. During the thawing of permafrost in the Holocene climatic optimum (8 – 4.5 thousand years ago) the permafrost, termed *relict* (Fig. 4), was conserved at various depths from the surface. Later on, during the period of Late Holocene cooling new permafrost formed and a joining of the relict and the Late Holocene permafrost took place, mostly in the territory of Siberia and the Far North. There was no such joining of these permafrost horizons south of Western Siberia and in the European North, where therefore two layers of permafrost occur.

Frozen ground is usually subdivided with respect to lifetime into three varieties: 1) short-term frozen ground existing for hours, or days, and which extends from a few centimetres to the first ten centimetres in thickness; 2) seasonally frozen ground existing for a few months, which is from a few tens of centimetres to a metre or two in thickness; 3) perennially frozen ground existing for years, or hundreds and thousands of years, and which extends from the first meters to many hundred metres in depth.

The periodic change of ground surface temperature observed during the period of a year causes various thermal effects in near-surface ground, soils and rock strata. In places where there is no perennial freezing (the mean annual temperature of the ground is positive i.e. $t_{mean} > 0 °C$) the *seasonally freezing ground* develops from the surface. The layer of seasonal freezing is underlain by thawed ground (or, in the far south) by ground which has not been frozen. And, vice versa, within the regions with permafrost, that is,

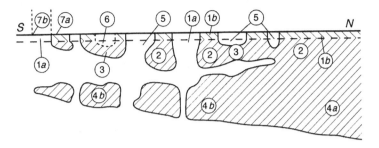

Fig. 4. Change of character of the permafrost from south to north: 1 – layer of
seasonal ground freezing (*a*) and of seasonal thawing (*b*); 2 and 3 – contemporary
continuous and discontinuous permafrost, respectively; 4 – relict continuous (*a*)
and discontinuous permafrost (*b*); 5 and 6 – open and closed taliks, respectively;
7 – southern limits of present (*a*) and relict permafrost (*b*).

when t_{mean} is below $0\,°C$ seasonal thawing occurs, i.e. the uppermost layer of
frozen ground is thawed during the warm period of a year. The layer of
seasonal thawing is always underlain by frozen ground. The depth of seasonal freezing m_{fr} and seasonal thawing m_{tha} varies usually from the first ten
centimetres to a few metres. The maximal deviation of mean monthly
ground surface temperature from the mean annual temperature t_{mean} is
termed the *amplitude of temperature fluctuation* A_o. The value of A_o (physical
value) is numerically equal to one half of the difference in mean temperature
for the coldest and warmest months. Temperature fluctuations are diminished with depth z, i.e. damping of the amplitude of the annual temperature
fluctuation takes place. The maximum depth at which the annual fluctuations are perceptible i.e. where $A_z = 0$, is termed *the depth of zero annual
amplitude or the depth of penetration of annual temperature fluctuations*, H_{an}.
The temperature of the ground over the period of a year is constant here (i.e.
$t_z(\tau) = $ const) and is termed *the mean annual temperature of the ground* t_{mean}.
The depth of the layer of annual temperature fluctuations in the territory of
the former USSR varies in the range from 5 to $20\,m$. Below this layer the
ground temperature varies in accordance with the geothermal gradient.
Within the area of development of permafrost the mean annual temperatures are in the range from $0\,°C$ to $-15\,°C$ and lower.

Perennially frozen ground is widely distributed now on the Globe and
occupies about 25% of the Earth's continents and nearly 50% of the
territory of the former USSR. Allowing for the seasonally frozen ground, the
area occupied by frozen ground is as large as 50% of the Earth's continents
and nearly 100% of the territory of the former USSR (Fig. 5). Great changes
and discontinuity in time as well as space are typical of frozen ground. Thus

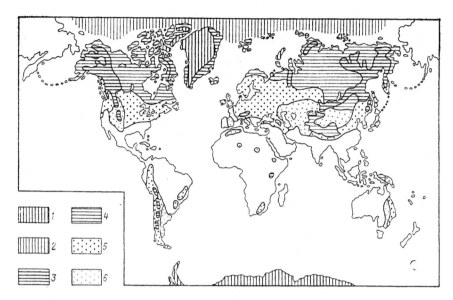

Fig. 5. Map of contemporary distribution of frozen ground on the Earth:
1–4 – permafrost (of types: 1 – glacier; 2 – ice; 3 – mountain; 4 – plain);
5–6 – seasonally frozen ground (5 – of humid; 6 – of arid type).

the short-term frozen and seasonally frozen layers are continuous, having the same upper boundary, the ground surface, while the lower boundary is situated at a not very great depth (from centimetres to the first metres). The structure of the perennially frozen ground is more complex because the upper boundary is situated at various depths from the ground surface on account of the processes of seasonal and perennial thawing (Fig. 4). In this connection the permafrost is subdivided into 'joining' permafrost, when the upper boundary coincides with the base of the layer of seasonal thawing, and 'non-joining' when there exists a layer of thawed ground between the table and the base of the seasonally freezing layer. When a number of permafrost layers one above the other are separated by thawed layers, 'layered' frozen strata are formed.

Continuous permafrost from the surface is usually developed in the northern regions only. However closed and open taliks often exist below large water bodies and, also in these regions, within the areas of intensive ground water circulation (see Fig. 4). The number and area of such taliks increases from the north southward where their formation depends in the majority of cases on the peculiar features of the ground surface radiation thermal balance. Regions or zones of practically continuous, discontinuous and island permafrost (Fig. 6) are recognized with respect to the nature of the

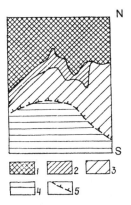

Fig. 6. Permafrost distribution from north to south: 1 – continuous permafrost; 2 – discontinuous permafrost; 3 – permafrost islands; 4 – area of previously frozen ground; 5 – southern limit of occurrence of permafrost.

spatial distribution of permafrost. A line termed *the southern limit of permafrost distribution* is used on maps to outline the southern border of the region where permafrost may occur today. The orientation and spherical shape of the Earth's surface results in the minimal amount of radiation arriving at the surface near the poles while the maximum is observed in the equatorial zone. This situation causes the formation of latitudinal zonation in distribution of the equilibrium temperature values of the surface layers of the lithosphere. Consequently the lowest negative surface temperatures (during the periods of permafrost existence) will be noted in high latitudes while the highest ones will be noted in low latitudes. In this connection usually wide and universal distribution of permafrost is observed in subpolar regions, a permafrost island distribution in middle latitudes and complete absence of permafrost in subequatorial regions of the Earth (see Fig. 5). The thickness of the permafrost must decrease with latitude (when moving from poles to equator) in accordance with change of negative temperature values. Frozen formations are represented not only by perennially frozen ground but also by perennial ice covers on the waters of northern seas as well as by ice caps and mountain glaciers on the Earth's continental areas (see Fig. 5).

As with the latitudinal zonation of permafrost (negative) temperatures and the change of permafrost thickness, the altitudinal zonation, associated with the fact that the negative temperature of the surface of the lithosphere decreases with increase of altitude, is clearly followed. Therefore instead of the term *southern limit* the notion *of the lower altitudinal limit* of permafrost distribution, above which mainly perennially frozen ground occurs, is used for mountain regions.

3   History of research of the zone of permafrost and the frozen materials composing this zone

We can recognize several stages in the history of research into seasonally and perennially frozen ground.

The first stage (from the sixteenth century to the first half of eighteenth century) is that of the first information concerning discoveries of the permafrost and the first attempts to explain the reasons for its existence. There is no doubt, though, that natives of the northern regions and Siberia always knew of its existence. The first data on permafrost (which was termed the 'Russian sphinx' in the West) probably began to appear in the literature from the beginning of the sixteenth century only when the Russians intensified the search for northern shipping lanes from Russia to China. A work from 1598, *A Description why there is no Way to sail from Archangel to the Chinese State and then to East India*, already contains clear views not only about the presence but also about the general reasons for the existence of perennial ice in the Arctic, connected with the small amount of Solar radiation arriving at high latitudes. Approximately at the same time information on the occurrence of permafrost in North America appeared. In the seventeenth century written reports on the discovery of permafrost in Siberia and within Novaya Zemlya began to arrive at Moscow from Russian provincial governors and travellers. Thus for example provincial governors of the Lena region P. Golovin and M. Glebov reported to Moscow in 1640 that 'the ground does not thaw completely even during the summer'. In Peter I's period some travellers as well as expeditions visited the north and east of Siberia.

At the time when the scientists of Western Europe were still disagreeing and the majority of them denying the possibility of the existence of permafrost, M.V. Lomonosov had not only recognized it but had offered the scientific explanation of the permafrost and its widespread nature. Thus Lomonosov wrote in his work *Birth of Metals as a Result of the Earth's Shaking* in 1757,

> ... heat and fire exist uninterruptedly in the Earth's interior. And we should see further if there is cold and ice there in contrast to this. True enough that vast Siberian territories, especially those lying near the Arctic sea, as well as spacious areas composing a ridge of the high mountain separating Siberia from the Chinese state have ground about two or three feet in thickness frozen throughout a summer at depth... It can be associated with winter cold overcoming summer heat and with the fact that these areas, one of them because of cold climate closeness, another one because of occurring high near the cold layer of atmosphere, are deprived of the atmosphere effect....

Thus an opinion on the reason for the existence of permafrost had been stated for the first time in the middle of the eighteenth century, i.e. it was pointed out that the permafrost is a result of two mutually opposing processes: summer heating and winter cooling. Lomonosov established the beginning of the doctrine of heat exchange between ground and the surrounding environment as the main factor responsible for the thermal state of the upper layer of the lithosphere.

The second stage (the second half of the eighteenth century to the first half of nineteenth century) of the history of the study of permafrost is that of accumulation of factual material by scientists and recognition of the widespread distribution of permafrost in the territories of Siberia, the Far East and the Far North. Two large expeditions studying the north-eastern region of Siberia (F.P. Vrangel, F.F. Matyushkin, N.M. Koz'min and others) and Novosibirskiye Ostrova [New Siberian Islands] (P.F. Anzhu, A.Y. Figurin and others) were working during the period from 1820 to 1828 and paying great attention to frozen ground and ground ice deposits. The first measurements of permafrost temperature were carried out by A. Erman in 1825 in Berezovo settlement in a well which was 18 m in depth. In 1828 Fedor Shergin, an employee of the Russian-American Company, had taken charge of work on sinking the deep well for water supply in Yakutsk and which took 9 years. In 1837 the sinking was ended at the depth of 116 m because there were no water-bearing horizons there. Thus was the famous Sherginskaya shaft created which is a historical memorial now and has played a rather important role in permafrost studies.

Academician A.F. Middendorf's expedition in 1842 – 1845 investigated the territory extending from the Sea of Okhotsk coast to the Yenisei river and was of great importance for the study of permafrost. Publication of this expedition's work had put an end to all the doubts about the existence of permafrost and was an important step in the development of geocryology. At the same time Middendorf established the relationship between the depth of seasonal (summer) thawing of ground and its lithological composition and thermal conductivity and had measured the temperature of the frozen ground in the Sheginskaya shaft, to a depth of 116 m – the first time such a depth was reached. These measurements allowed him to make the first hypothesis about the thickness of permafrost.

The third stage (the second half of nineteenth century to the beginning of the twentieth century) of the study of permafrost was marked by the appearance of geocryology as an applied (engineering) field of knowledge and is associated with the beginning of the industrial development of Siberia (construction of the Great Trans-Siberian main railroad, mining development, intensification of agricultural development and migratory movements to Siberia). Active frozen ground studies in that period were associated with the necessity of protecting many engineering constructions from deformations and failures as well as with severe problems with the water supply of the Zabaykal'ye and Amur rail roads.

The beginning of this stage (the second half of the nineteenth century) is marked by permafrost investigations at many sites within the territory of Siberia (G. Maydel', L.A. Yachevskiy, V.A. Obruchev, N.M. Koz'min, A.I. Voyeykov, *et al.*) and in North America (E.K. Leffingwell, K. Brooks, A.E. Porsild, etc.) as well as within the islands of the Arctic Ocean (I.A. Lopatin, A. Bunge, E. Toll, etc.). The problems of the formation of ground (wedge) ice, gigantic icings, groundwater and permafrost geological processes and phenomena were of prime importance in construction in the North and were a major preoccupation of the scientists.

L.A. Yachevskiy, in presenting the schematic map of distribution of permafrost and its southern limit in his report '*About the Perennially Frozen Soil of Siberia*' in 1889, made a great contribution to geocryology. He had pointed out the great importance of negative air temperature, snow cover thickness, geological structure, rock composition, water saturation and the orientation of slopes as well, for the formation of permafrost. A.I. Voyeykov had considered in 1889 the problem of the effect of climatic conditions on the process of freezing of the Earth's crust, in a published report on frozen ground situated along the rail roads in Siberia. Subsequently V.A. Obruchev and N.M. Koz'min presented the first scientific views of the intimate connection between hydrogeological and permafrost conditions and their mutual effect on each other.

In the 1890s a Committee consisting of A.I. Voyeykov, V.A. Obruchev, M.A. Rykachev and K.I. Bogdanovich was set up at the request of the Department of Siberian Rail Road Construction attached to the Russian Geographical Society under the chairmanship of I.V. Mushketov, for the study of frozen ground. In 1895 this Committee published '*The instructions for studying the soil permafrost of Siberia*', in essence the first published work on the problems of engineering geocryology. During the first quarter of the twentieth century permafrost studies continued in the same direction. In 1903 S.A. Pod'yakonov published a work '*Icings of Eastern Siberia and the reasons for their formation*'. In 1912 N.S. Bogdanov demonstrated various forms of building construction in the conditions of Zabaykal'ye, in his monograph '*Permafrost and construction upon it*'. V.A. Lvov's work '*Search for and testing of water supply sources in the western section of the Amur rail road in conditions of eternally frozen soil'* was published in 1916 and was the first great report on the distribution of hydrogeological problems within the permafrost regions.

The beginning of the twentieth century was marked by investigations in the field of climatology, soil science and agriculture made by researchers from the Department of Migration and from the Meteorological bureau. These investigations were by N.I. Prokhorov, P.I. Koloskov, M.I Sumgin, L.I. Prasolov, B.B. Polynov, V.N. Sukachev and others having close involvement with the permafrost conditions.

The fourth stage (beginning to middle of the twentieth century) is the stage of formation of geocryology as a science within the history of the study of freezing ground, by the work after the Great October socialist revolution. It is marked by intensive studies of permafrost and by the organization of special institutions for training specialists and for studying seasonally and perennially frozen ground.

In the 1920s and 1930s institutions began to appear which could not dispense with knowledge of ground conditions when constructing structures in the cold regions (State Institute of Metallurgical Works Design, the Road Research Office of the Central Department of Local Transport, Institute of Railroad Transport Engineers, etc.). In these organizations groups of researchers studying permafrost began to be formed. The necessity of generalizing the knowledge of geocryology accumulated by that time was reflected in the fundamental work by M.I. Sumgin *'Soil Permafrost within the USSR'* published in 1927. Publication of this work was a great event and a turning point in the history of permafrost research.

At the end of 1929, a permanent Commission for the Study of Permafrost, under the chairmanship of Academician V.A. Obruchev, was organized in the USSR Academy of Sciences on M.I. Sumgin's and Academician V.I. Vernadskiy's initiative. During the period from 1927 to 1936 the Commission organized regional scientific research stations for permafrost in Skovorodino, Petrovsko-Zabaykalskoye, Anadyr, Igarka, Yakutsk, Vorkuta, and Norilsk. In addition to purposeful permafrost studies in various regions of the country, this Commission was an organizing centre for all the problems of geocryology. Six all-Union conferences on geocryology were convened by the Commission from 1930 to 1939 bringing together numerous groups of geocryologists and coordinating investigations in the permafrost regions of the former USSR. It should be particularly noted that the permafrost investigations in connection with surveying and designing of the Baikal-Amur Mainline were carried out in 1932 – 1935, that is, in a very short time. In 1936 the Commission for Study of Permafrost was reorganized by the Academy of Sciences into the Committee on Permafrost in connection with a widening of the research.

In 1939 the Obruchev Institute of Frozen Ground Studies of the Academy of Sciences was organized in Moscow on the basis of the Committee on Permafrost. A number of scientific permafrost research stations were placed under this Institute. In addition a special permafrost laboratory was set up in Moscow and a permanent station was established near Zagorsk city as well to study seasonal ground freezing. The study of particular regional features of permafrost as well as laboratory investigations of composition, structure and properties of frozen ground were intensified in this period. Monographs and works of the Institute of Frozen Ground Studies (about 20 items) began to be published and four issues of *Frozen Ground Studies* appeared. In 1937 N.A. Tsytovich's and M.I. Sumgin's work *Basic Mechanics of Frozen Ground* was published. Results of the study of the physical and mechanical properties of frozen soils and their interaction with engineering constructions as well as principles of calculations and designs for foundations on frozen soils, were presented for the first time in this work. In 1940 the Institute of Frozen Ground Studies of the USSR Academy of Sciences published for the first time a general text *General Frozen Ground Studies* edited by V.A. Obruchev, the authors of which were M.I. Sumgin, S.P. Kachurin, N.I. Tolstikhin, V.F. Tumel'. This book remained for two to three decades the only textbook for postgraduate students and specialists of various types studying and developing the permafrost. In 1940 B.P. Veynberg's book *Ice* was published in which conditions of ice formation, its properties and texture, as well as ice in frozen ground, are considered. In 1955 P.A. Shumskiy's monograph *Principles of the Science of the Structure of Ice* summarizing all the results of the study of ice as a mineral and as a rock came out.

As early as the 1930s – 1940s a problem of training young specialists in permafrost studies arose. This problem led to M.I. Sumgin giving lectures on permafrost at the Leningrad Mining Institute and at Leningrad University and beginning from 1940 at the Moscow Geological Prospecting Institute. His followers continued lecturing on general frozen ground studies subsequently. From 1935 the same course was given at MSU at first by A.Ye. Fedosov, then from 1941 by S.S. Morozov and then by N.I. Bykovskiy and from 1947 by N.F. Poltev. However only in 1953 was the first Department of Frozen Ground Studies (now called geocryology) in the world organized at the Faculty of Geology, MSU, on Professor A.N. Mazarovich's initiative. This Department has become a centre for the training of specialists in geocryology. Professor V.A. Kudryavtsev became head of the Department. In 1961 the Department of Polar Lands (subsequently the Department of Cryolithology and Glaciology) was organized at the Faculty of Geography, MSU. By that time special courses on geocryology were given at a number of educational institutions while at the Obruchev Institute and at MSU a postgraduate course for training highly skilled specialists in geocryology was organized.

In the same period techniques for thermal engineering calculations and the physical and mathematical simulation of the soil freezing (thawing) processes were made (by P.A. Bogoslovsky, L.S. Leybenzon, D.V. Redozubov, M.M. Krylov, V.S. Lukyanov, M.D. Golovko, V.A. Koudryavtsev, G.B. Porkhayev, Kh.R. Khakimov, A.G. Kolesnikov and others) in connection with the need to predict ground temperature regimes and their control for the purposes of construction and agriculture. These investigations called for broadening and extending knowledge in the field of physics, mechanics and petrography of frozen, freezing and thawing soils and for studying the processes of water migration, phase transition of soil moisture, ice formation, cryogenic structure formation, ground heaving and settlement (M.I. Sumgin, A.Ye. Fedosov, N.A. Tsytovich, A.N. Gol'dshteyn, I.A. Tyutyunov, A.M. Pchelintsev, A.A. Ananyan, V.F. Zhukov, P.A. Shumskiy, S.S. Vyalov, B.I. Dalmatov and others). Extensive investigations at many sites were carried out during that period, to develop methods for construction on the permafrost associated with the mining industry, agriculture and water supply in the permafrost regions (N.A. Tsytovich, N.I. Saltykov, P.I. Koloskov, P.I. Mel'nikov, N.A. Vel'mina, V.P. Bakakin, A.I. Yefimov, V.G. Gol'dtman, V.M. Ponomarev, A.P. Tyrtikov, V.P. Dadykin and others). Knowledge of composition, structure, temperature regime, distribution of frozen strata, hydrogeological conditions, permafrost-geological processes and phenomena was compiled for a number of regions with permafrost (V.F. Tumel', I.Ya. Baranov, S.P. Kachurin, P.F. Shvetsov, V.A. Kudryavtsev, N.A. Grave, P.I. Mel'nikov, V.K. Yanovskiy, P.A. Solov'yev, B.N. Dostovalov, A.T. Akimov, A.I. Popov, L.A. Meyster, A.I. Kalabin, N.I. Tolstikhin, V.M. Maksimov, V.M. Ponomarev and others). The completion of that fruitful period in the history of permafrost investigations was marked by the publication of a two volume work *Basic Geocryology* (*Frozen Ground Studies*) in 1959 edited by P.F. Shvetsov, B.N. Dostovalov, N.I. Saltykov, and of the guide to methods *Field Geocryological* (*Frozen Ground*) *Investigations* in 1961.

The fifth stage (the second half of the twentieth century) is recent in the history of geocryological research and is characterized by the further development of the science and its differentiation into a number of scientific directions, with some of them pretending to a right to be independent disciplines.

In 1961 the Obruchev Institute of Frozen Ground Studies in the Academy of Sciences was reorganized and its structural subdivisions were placed under PNIIIS and NIIOSP of the USSR Gosstroy where large departments including a few laboratories had been formed. At the same time the Permafrost Institute of the Siberian section of the Academy of Sciences, USSR, led by P.I. Mel'nikov, was set up in Yakutsk, based on the North-East section of the Obruchev Institute. The scientific and technical laboratory base was organized at that institute and investigations into various field aspects of geocryology including those in many regions of new construction in Siberia were carried out. In 1966 the VIIIth All-Union Interdepartmental Conference on Geocryology (Frozen Ground Studies) was based at the Institute of Frozen Ground Studies of the North Section of the Academy of Sciences, USSR. The conference proceedings were published in eight issues. In the same period large scale geocryological investigations were carried out at PNIIIS and NIIOSP of the USSR Gosstroy, at VSEGINGEO of the USSR Ministry of Geology and at a number of departments of educational institutions such as MSU, MISI, LGI, ChPI, etc. as well as at a number of scientific research offices of the Ministries of Power, Gas and Oil Industry, Transport Engineering, etc..

Thus the methods of integrated permafrost survey and permafrost prediction were developed and permafrost hydrogeological and engineering-geological research in the construction of hydrotechnical structures and timber industry systems, atomic power plants, gas and oil pipelines, rail roads, etc. was carried out using these methods, at the Department of Frozen Ground Studies under the direction of Prof. V.A. Kudryavtsev. Scientific achievements of the Department were reflected in 25 volumes of the work *Frozen Ground Investigations*, in the textbooks *General Frozen Ground Studies* (1967, 1978, 1981), *Basics of Permafrost Prediction in Engineering Geological Investigations* (1974), *Methods of Geocryological Survey* (1979), *Ground Waters of the Permafrost Zone* (1983), *Physics, Chemistry and Mechanics of Frozen Ground* (1986), *Petrography of Frozen Ground* (1987) and others as well as in numerous monographs, technical guides and integrated geocryological maps.

During the same period wide scale scientific work on the permafrost zone was carried out in the various regions of the country by the Department of Cryolithology and Glaciology under Professor Popov's direction, and on the engineering-geological study of Western Siberia by the Department of

Soil Science and Engineering Geology under Academician Ye.M. Sergeyev's and Professor V.T. Trofimov's direction. By and large, typical for this period is the extended development of geocryology along such main scientific aspects as:

1. general, regional and historical geocryology (V.A. Kudryavtsev, I.Ya. Baranov, A.I. Popov, P.E. Shvetsov, V.V. Baulin, K.A. Kondrat'yeva, N.N. Romanovskiy, N.A. Grave, I.A. Nekrasov, P.A. Solov'yev, G.I. Dubikov and others);
2. thermal physics, physical chemistry and mechanics of frozen ground (N.A. Tsytovich, S.S. Vyalov, B.N. Dostovalov, V.A. Kudryavtsev, G.V. Porkhayev, V.G. Balobayev, B.A. Savel'yev, A.A. Ananyan, N.S. Ivanov, Z.A. Nersesova, I.A. Tyutyunov, S.E. Grechishchev, G.M. Fel'dman, E.D. Yershov, V.O. Orlov, Ye.P. Shusherina, M.K. Pekarskaya, K.F. Voytkovskiy and others);
3. engineering geocryology (N.A. Tsytovich, V.P. Ushkalov, N.I. Saltykov, V.F. Zhukov, G.V. Porkhayev, S.S. Vyalov, L.N. Khrustalyov, L.T. Roman, A.M. Pchelintsev, V.O. Orlov, B.I. Dalmatov and others);
4. methods of permafrost investigations and geocryological forecasting (V.A. Kudryavtsev, I.Ya. Baranov, L.S. Garagulya, K.A. Kondrat'yeva, A.V. Pavlov, G.M. Fel'dman and others);
5. cryolithology (A.I. Popov, E.M. Katasonov, N.N. Romanovskiy. Sh.Sh. Gasanov, N.A. Shilo, E.D. Yershov, I.D. Danilov, B.I. Vtyurin, T.N. Kaplina, S.V. Tomirdiaro, G.I. Dubikov and others);
6. ground water in the permafrost zone (N.I. Tolstkhin, A.I. Yefimov, A.I. Kalabin, P.F. Shvetsov, N.A Vel'mina, N.N. Romanovskiy, S.M. Fotiyev, V.E. Afanasenko, and others).

The *Geocryological Map of the USSR* at 1 : 2 5000 000 scale and the five volume monograph *Geocryology of the USSR* (5) are notable compilations of regional integrated geocryological material accumulated to date.

In 1970 the Scientific Council for Cryolithology of the Earth attached to the Academy of Sciences, of the former USSR, was organized in Moscow. The Council successfully coordinated all of the geocryological investigations in the country. Our geocryologists have taken active part in all the International Conferences on Permafrost. The International Association of Permafrost and the National Committee on Permafrost had, as their first President, Academician P.I. Mel'nikov.

At the present stage geocryology as a full branch of geology interacts with a number of interdisciplinary sciences. A few units are set up in the

system of the Academy of Sciences (the sections of lithogenesis and engineering geology in the permafrost zone are attached to the Joint Lithology
Committee of the USSR Academy of Sciences and the Engineering Geology and Hydrogeology Scientific Council attached to the USSR Academy
of Sciences), in Gosstroy of the former USSR and former RSFSR, as well
as in other Ministries and Departments.

4   Structure, problems and scientific themes of geocryology

At the present time further accelerated development and differentiation of permafrost science is taking place. A number of scientific directions
in geocryology have pretensions to being independent scientific disciplines
(cryolithology, engineering geocryology, etc.). However, as is obvious from
the foregoing, geocryology has to investigate practically all the aspects of
formation and development of the ground in the qualitatively new (frozen)
state. Therefore it is natural that the structure of this science is essentially the
same as that of geology in miniature. Dynamic, lithogenetic, regional,
historical and engineering geocryology are thus the constituents of geocryology (Fig. 7). Reasoning from this structure of geocryology it is clear
how important and essential are its connections with all the geological
disciplines as well as with the fundamental sciences (physics, chemistry,
mathematics, mechanics) and with geographical and biological sciences
(climatology, paleogeography, geobotany, soil science, etc.).

Dynamic geocryology deals with the processes of freezing and thawing,
cooling and heating which control the formation and development of seasonally frozen as well as perennially frozen ground. Consideration of these
and other processes proceeding in freezing, frozen and thawing ground
amounts to a thermodynamic approach in terms of energy. Development of
the thermodynamic and thermal-physical basis of formation of frozen
ground is based on the study of heat exchange in the system 'atmosphere –
lithosphere', of radiation-thermal and water-thermal balances, temperature
regime and moisture phase transitions in the ground as well as of thermal
processes in the upper layers of the Earth's crust. It is possible to study the
dynamics (to reconstruct the history of formation and development or to
forecast change) of perennially and seasonally frozen strata in time, in area
of distribution and in depth, given the close correlation of thermal aspects of
the ground freezing and thawing processes with the geologic-geographical
conditions of the environment in which these processes proceed. Predictions
are made in the context of natural historical evolution of the natural
environment as well as in the context of technological disturbance of the
natural environment at a global, regional or local scale. Such questions are

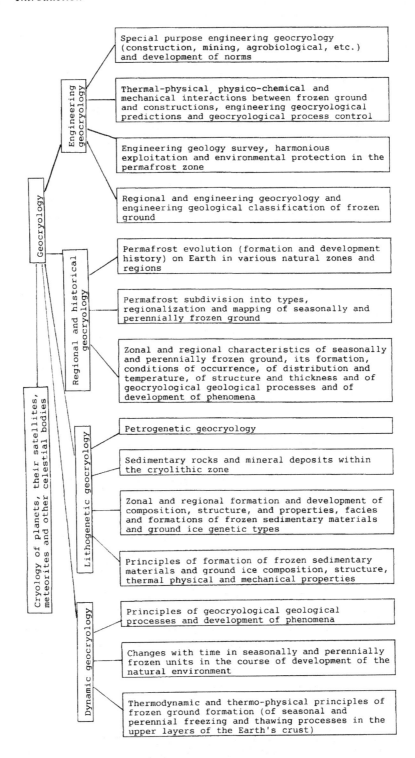

Fig. 7. Structure and content of the scientific thrusts in geocryology.

solved on the basis of physical and mathematical simulation and use of approximate and exact solutions to the problems of ground freezing and thawing (using computers), with regard to heat and mass exchange and internal and external sources of heat and ice formation in the range of negative temperatures. As a result the trend and character of the freezing process (degradational, stable, aggradational) are revealed, as well as the types of permafrost strata: short-, medium- and long-period degradation across the whole thickness; degrading in the upper part and aggrading in the lower part of the unit (and vice versa) with various periods of degradation and aggradation; aggrading across the whole thickness, etc. All this takes place against the background of development of the complex global crustal tectonic movements of various signs and amplitudes.

Another important part of dynamic geocryology is the study of, in essence, the thermal-physical, physico-chemical and mechanical principles of formation and development, and as well the forecasting of the permafrost geological processes resulting in geological phenomena (the specific permafrost topographic forms). All these processes and phenomena can be divided in a first approximation into three groups. The first group refers to permafrost geological processes and phenomena resulting from seasonal and perennial ground freezing and thawing, or cooling and heating (frost weathering, heaving, thaw settlement, frost fracturing, formation of ground and surface ice, thermokarst and polygonal topographic forms, cemetery mounds, frost mounds, etc.). The second group includes slope processes resulting from the effect of gravitational forces and causing various phenomena and topographic forms (sheet and differential solifluction, thermogenic and cryogenic detritus, subsidences, screes, rock streams, mountain terraces, stone fields and sorted stripes, mud- and solifluction flows, torrents, snow avalanches etc.). And finally, the freezing-geological processes and phenomena resulting from the effect of surface and ground waters, glaciers, snow patches, wind and other exogenous factors (wash-outs, transfer and accumulation of material in persistent water courses, glacial transport of material, nivation, thermal abrasion, thermal erosion, icing formation, sublimation and material transfer by wind, frozen ground erosion with help of thermal abrasion and collapse, bank retreat, etc.) belong in the third group. The study of the principles of permafrost geological processes, phenomena and topographic forms is not only of great general geological and lithological importance but also of practical interest as far as the engineering geology aspects are concerned.

Lithogenetic geocryology (cryolithology) reveals the general and specific principles of formation of frozen sedimentary materials and ice, their granular and chemical-mineral composition, texture and particular structural features, on the basis of the chemical, physico-chemical and mechanical processes in sediments in the permafrost regions, in the course of their stage by stage transformation. The main scientific thrusts in this part of geocryology are the following: research of material composition, textural-structural features and properties of frozen sedimentary material and ice (petrography and ice structures); investigation of the various genetic types, facies and composition, structure and properties with the aim of determining the genesis as well as mechanisms and the geomorphological and geological conditions of sedimentary deposition, the study of the formation and history of sedimentary materials within the permafrost regions. This is the study of the particular features of sedimentogenesis and transformation of a sediment into rock in the course of weathering, transfer, continental and basin accumulation and further diagenesis (sedimentary rock formation in the permafrost regions) on the basis of analysis of the stage of permafrost development. At the same time, investigating the principles of formation and development of frozen and cryotic magmatic, metamorphic and cemented sedimentary rocks, their structure and properties (petrogenetic geocryology, a subdivision for the future) should evidently be assigned to lithogenetic geocryology.

Since the formation and properties of frozen sedimentary materials are closely associated with the manner and conditions of their freezing as well as with the character and rate of sedimentation and tectonogenesis, lithogenetic geocryology attaches particular significance to the study of frozen bodies of various genesis. At the same time the synchronously and asynchronously epicryogenic and paleocryoeluvial continental masses, the syncryogenic masses of continental subaqueous and continental subaerial deposits, etc. can be recognized. The process of formation of sedimentary rocks within the permafrost zone and its results have proved to be so unique that researchers have recently begun to recognize the *cryogenic* type of lithogenesis (as a specific type). This type of cryolithogenesis is the youngest, appearing definitively only in the Proterozoic, becoming more important towards present time. It is characterized by a well-defined irreversibility of its evolution.

Regional and historical geocryology investigates zonal and regional features of the formation and development of seasonally and perennially frozen ground, their distribution, conditions of occurrence and temperature, structure and thickness of the frozen masses, and the permafrost geological processes and phenomena. Frozen ground and ice are classified with respect to their composition, cryogenic structure, genesis, age, and heat exchange conditions, and regionalization and mapping of the permafrost is conducted on this basis. Reconstruction of the history of frozen ground and its development within regions, continents and the Earth as a whole is an important aspect of this part of geocryology. The data on permafrost evolution and development now available are adequate mainly for the Quaternary period. The problems of distribution of frozen ground and its development (evolution) on the planet in the more ancient epochs still remain to be solved.

Engineering geocryology represents a particularly practical part of geocryology and deals with engineering and geological support for design, construction and operation of various engineering structures within the permafrost regions, for the carrying out and selection of the most reliable and economical ways for development of territory within the regions of seasonally and perennially frozen ground. All the specific types of engineering geological survey are thus conducted on the basis of regional engineering geocryology and should be carried out at the stage of predesign investigations. Compilation of engineering geocryological maps for various types of construction, prediction of engineering geocryological conditions in the course of economic development of a territory, as well as the development of general plans for harmonious exploitation of the environment within the permafrost zone recognizing changes of natural conditions, are the functional result of regional engineering geocryology.

Engineering geocryology surveys and studies (in the laboratory and field) of composition, cryogenic structure and physical- mechanical properties of frozen, freezing and thawing soils and ice as foundations, as materials and as the environment for engineering structures, are essential at the stages of technical project and shop drawings and in the effort to select the particular regions, sections and construction sites for economic development. Predicting the permafrost conditions and technological changes at the selected site as well as recommendations on the rational use of the geocryological environment and its protection are obligatory elements of an engineering geocryological survey.

At all the design stages and especially at the stage of working drawings it is necessary that primarily, attention should be given to the processes of interaction between frozen ground and constructions. The problems of thermal-physical and mechanical interactions between engineering constructions and frozen, freezing and thawing ground are solved on the basis of thermal physics and mechanics of frozen ground. Recently investigations of processes of physical and chemical interaction between subgrade and foundation have begun to receive attention and a number of questions have already been posed (of moisture redistribution, change in ice content, cryogenic structure and mechanical properties of soils adjacent to foundations, of the chemical reactions and processes, corrosion of foundations, etc.). Specific engineering geocryological predictions for the behaviour of subgrade and foundation for the periods of engineering construction and operation of the structure are made on the basis of qualitative and quantitative investigations of interaction between the frozen ground and the construction. In case of unfavourable predictions, measures are developed which are intended to deliberately change (transform) the freezing-geological situation (temperature, composition, ice content, cryogenic structure and soil properties) and the frozen ground processes and phenomena, that is, measures to control the effects of permafrost are drawn up.

Specialized engineering geocryology is focused on the solution of problems of an engineering geocryological nature for reliable and economical construction and operation of particular engineered structures on frozen, freezing and thawing soils. One can recognize the following aspects of specialized engineering geocryology: construction (industrial, municipal, hydrotechnical, highway, pipeline etc.), mining (underground constructions, oil-gas fields, shafts, subways, quarries, tunnels, etc.) and agrobiological (forestry and agriculture, etc.). The experience being accumulated in the field of engineering geocryological surveying is recognized in the development of various kinds of rules, instructions, standards, regulations, methods and other normative documents (at Union, republic and departmental level) setting out the requirements for the conduction of investigations at various stages of design of engineering constructions on freezing soils.

And finally, taking into consideration the modern trends in the development of geocryology there is no escape from the conclusion that the more general science – *planetary cryology* – has begun to develop steadily using successively the knowledge and principles of investigation of the frozen Earth. The cryology of Mars is progressing rapidly, the foundations of the Moon's cryology are developed, the data on the cryology of the Jupiterian group of planets are being accumulated, etc. In other words we can speak about development of geocryology into planetary or cosmic cryology in this perspective.

5   Methodological basis of geocryology

The laws of dialectical materialism provide the starting point for the construction of scientific concepts in geocryology*. Thus it is the most general dialectical law on the evolutional character of quantitative changes

---

*The initiation of the development of the methodological basis as far as geocryology is concerned is associated mainly with V.A. Kudryavtsev's work.

and the spasmodic character of transition (at the particular stage) of quantity into quality, which has predetermined in essence the appearance and existence of geocryology itself. Actually the change of ground temperature in positive or negative direction (without going through $0\,^{\circ}C$) causes transformations only of the quantitative properties of the earth materials, with these materials continuing to belong to the same class of earth material. When the temperature passes through $0\,^{\circ}C$ the phase transition of water into ice and of the earth material from the unfrozen state into the frozen one (and vice versa) occurs. With this there is a deep qualitative material transformation which manifests itself first of all in an essential difference of the soil or rock composition, structure, texture and most properties compared to those of the unfrozen and thawed materials. As this takes place a mineral (ice) is formed and specific new (cryogenic) materials appear, i.e. a quantitatively new geological object – *the frozen soil or rock* – is formed.

The correct understanding and interpretation of the negation of negation law, as applied to the process of the transition of the unfrozen sedimentary materials into the frozen state and then the subsequent thawing and transition into the thawed state, is of great interest for geocryology. The first negation of the unfrozen ground in this case consists not only in the transition of the sedimentary formations into the frozen state but also in an essential change of their granulometric, microaggregate and chemical mineral composition, structure, texture and properties. Therefore the second negation, i.e. the return of the frozen ground into the unfrozen state ('negation of negation' proper), will never lead this frozen ground to its original quality. That is the reason why researchers should distinguish not only the unfrozen (that is, never frozen) soils and rocks and the frozen ones, but also those which have been frozen in the past.

By and large the sedimentary formations in the permafrost follow complex evolutional paths never returning to the starting position, in accordance with the principle of indissolubility of unity and struggle of opposites. In this manner the law of the continuous development of the cryolithosphere must work in nature, i.e. of continuous progressive movement in a spiral in accordance with the general history of development of the lithosphere and of the whole Earth. From this follows the irreversibility of lithogenetic development in the Earth's history and the necessity to use principles of actualism and historical method simultaneously in the course of geocryological analysis.

The frozen earth materials are complex and rather dynamic natural objects formed in specific geosystems. The latter represent the multitude of natural components connected with each other in a certain order (typically, structure and properties) and making up the integral unit in the form of natural terrain complexes. It is natural that the approach to permafrost investigations should be systematic, complex, based on laws of universal and causal relationships between processes and phenomena, on continuity of development, on transformation of quantity into quality and other laws of dialectic materialism. The methodological importance of such an approach lies in the fact that the perennially frozen rocks are studied in relation to each element of the complex (system) individually and in their particular combination as a single whole. At the same time dependences are determined and particular and general principles are established. By the first is meant two way dependence, between two geocryological characteristics one of which is considered as a causative factor, the other one as an effect.

By general principle is meant the net result of the effect of all the natural factors typical of the given natural complex, on the formation of one of the geocryological characteristics expressed either by temperature regime of ground or by thickness or cryogenic structure of permafrost or by the particular type of seasonally and perennially frozen ground with its distinctive features.

Analysis and synthesis are the basis for the investigations of principles. The study of permafrost can be divided into a number of problems associated with the study of particular characteristics such as composition, structure, properties, thickness, temperature regime, etc. Each geocryological characteristic reflects the particular quality of an object and its association with other characteristics (properties, manifestations). By and large all the characteristics of the natural environment form the individual natural complex or *geosystem*. This association is complex and is individually inherent in each natural (including geocryological) characteristic. At the same time all geocryological characteristics are closely associated with each other. That is the reason why the dialectical approach (from complex to simple and then to complex again, from particular to general through analysis and then to general through synthesis, etc.) has become firmly established in geocryological investigations.

The process of cognition of natural geocryological systems is possible in the course of integrated geocryological surveying, revealing the dependencies and establishing experimentally principles of distribution and dynamics of various types of seasonally and perennially frozen ground and their inherent characteristics, by way of logic and mathematical modelling (based on experiment). A definition of survey was given by V.A. Kudryavtsev in 1961. *Permafrost (geocryological) survey* represents the complex field, laboratory and office work aimed at the study of seasonally and perennially frozen ground and permafrost geological processes and phenomena, and the establishment of the principles in accordance with the existing natural conditions, their natural change during the Pleistocene – Holocene and under technological effects as well as the compilation of permafrost maps and predictions of permafrost occurrence. The main method of permafrost surveying is the key landscape method, the essence of which is the following: at the first stage the typological landscape regionalization (microregionalization) of a territory is carried out with respect to factors and conditions associated with the formation and existence of particular types of seasonally and perennially frozen ground. Then the key areas typical of the recognized types of landscape are selected as the basis of the landscape regionalization map. Particular and general geocryological principles, hydrogeocryological and engineering geocryological conditions, are studied within these key areas during the field periods by various methods and then the geocryological maps are compiled by way of extending the obtained principles over the types of landscape analogous to those studied within the key area and traced on the map of landscape regionalisation.

The study of the freezing conditions of a territory should be carried out taking into account the dynamics of the whole natural environment. Thus, for example, the formation, spatial and temporal change of seasonally and perennially frozen ground should be considered in connection with the general geological history of development of the territory in Late Cenozoic, i.e. with the history of the sedimentary strata and their particular facies with the history of development of topographic, hydrogeological and hydrological conditions, vegetation, climate, etc. The necessity for consideration of the dynamics of permafrost is one of the items of the geocryological survey. Reconstruction of the history of development of frozen ground in the Neogene and Pleistocene within the region under study and adjacent territories, is a logical completion of the study of dynamics.

The permafrost conditions change drastically in connection with inevitable disturbance of the natural conditions as a result of extensive economic development of an area. Therefore special work on prediction (i.e. expected change) of the permafrost conditions in the course of natural development as well as because of anthropogenic effects on the natural environment, are carried out in the course of geocryological surveys. Geocryological prediction and development of measures intended to initiate change of permafrost conditions, are carried out on the basis of the revealed particular and general principles of formation and development of the frozen ground as well as permafrost geological processes and phenomena.

A large complex of field, laboratory and office methods is used to study geocryological conditions and principles of their formation, to determine classificational features of frozen rocks, taliks, geocryological processes and phenomena. Along with general scientific methods being used in other sciences, geocryology has its own, specific methods allowing for the study of the principles of seasonally and perennially frozen ground and freezing-geological processes and phenomena and their development as well as their physical nature, at various levels of knowledge (from microstructure to macrostructure).

One of the main aspects of any science is correct relations between theory and practice in the course of cognition: from practice to theory and from theory to practice again. Geocryology was initiated by practice, i.e. in the practical needs of human engineering activity in the regions of frozen ground and in connection with prospecting and exploitation of deposits within the permafrost regions. Consideration of practical questions inevitably demanded a temporary distraction from practice, for the establishment of relationships and general laws of ground freezing, i.e. for the development of the theoretical basis of the study of geocryology. Theoretical constructions based on practice are verified, supported and gain new momentum for further development.

I

Thermal-physical, physico-chemical and mechanical processes in freezing, frozen and thawing ground and their manifestation in the permafrost regions

1

Thermal-physical processes in freezing and thawing ground

1.1   Heat transfer and temperature field in ground

The temperature regime of the upper part of the lithosphere is a result of thermal interaction with the environment (atmosphere, space, etc.) and with the underlying strata. The amount of heat arriving at the Earth from the atmosphere (mainly from the Sun) is approximately three orders of magnitude larger than that arriving from the interior. The process of heat transfer in ground proceeds generally through radiation, convection and conduction.

Thermal radiation represents the process of emission of electromagnetic waves (radiant energy) by a heated body into the environment. The wave length corresponding to the highest value of emission by an absolutely black body is inversely related to its absolute temperature. The portion of heat transferred by radiation within the ground usually comprises less than a few percent of the value of the total heat flux.

Heat transfer by *convection*, q_{con}, is carried out by liquid and gas flowing through pores, cavities and fissures in rocks, with the portion of the heat being transferred in the course of migration (diffusion) of moisture in soils usually being small compared with that being transferred by conduction. In the case of forced water and gas convection (infiltration) in faulted hard or coarse clastic rocks and sands the convective mechanism of heat transfer plays a significant role and the neglect of this mechanism may cause fundamental errors in the analysis of the temperature field. Account is taken of the convective component of thermal flux either through the effective thermal conductivity factor λ_{ef}, covering the total heat transfer including convection and conduction or through the direct calculation of the thermal flux density q associated, for example, with the infiltration of water or pore solution: $q_{con} = \gamma_w I_{fil} v_w$ where γ_w is the density of water; v_w is its filtering rate and I_{fil} is the specific relative heat content of the infiltrating water (solution).

The main and the most basic mechanism of heat transfer in ground is *conduction*. Heat is conducted through the medium as a result of atomic and molecular vibrations in the crystal lattice, the intensity of which increase with temperature. This heat transfer occurs on account of the thermal conductivity of the material. From the mathematical point of view the process of stationary conductive heat transfer is depicted by Fourier's law: $q_{cond} = -\lambda \operatorname{grad} t$, where q_{cond} is the density of the conductive component of thermal flux; λ is the coefficient of heat conduction for the medium and is numerically equal to the amount of thermal energy passing through unit area per unit time with unit gradient of temperature.

The main parameter characterizing the thermal state of a rock is temperature. Its distribution in the ground is termed the temperature field. Generally the ground temperature varies from one point to another and in time, i.e. the temperature field is depicted by a three-dimensional non-stationary function $t(x, y, z, \tau)$. In nature the temperature field often varies markedly in two or even one of the space coordinates, while being practically invariable in the other. In keeping with this, the term 'two-dimensional $t(x, y, \tau)$' or 'one-dimensional $t(x, \tau)$' temperature field is usually used. Temperature fields are displayed graphically as isotherms (lines or surfaces) connecting equal temperature values (Fig. 1. 1). If the temperature field does not vary with time at any point this field is termed *stationary* (see Fig. 1.1a); otherwise it is characterized as *non-stationary* (see Fig. 1.1b). Fourier's heat conduction equation supplemented with the appropriate boundary (initial) conditions is used for describing the process of non-stationary conductive heat transfer mathematically and for calculating the temperature distribution in time and space. In the case of a one-dimensional temperature field it takes the form:

$$C_{vol}(z)\frac{\partial t}{\partial \tau} = \frac{\partial}{\partial z}\left(\lambda(z)\frac{\partial t}{\partial z}\right) + f(z) \tag{1.1}$$

where $f(z)$ is the intensity of the distributed heat sources or sinks; $C_{vol}(z)$ is the coefficient of thermal capacity of the medium, which is equal numerically to the amount of thermal energy required to heat unit volume of the substance by $1°C$. In the simplest case when the medium has constant thermal-physical characteristics and when the heat sources are absent ($\lambda = \text{const}, C_{vol} = \text{const}, f(z) = 0$) the equation of the non-stationary heat conduction takes the form:

$$\frac{\partial t}{\partial \tau} = a\frac{\partial^2 t}{\partial z^2} \tag{1.2}$$

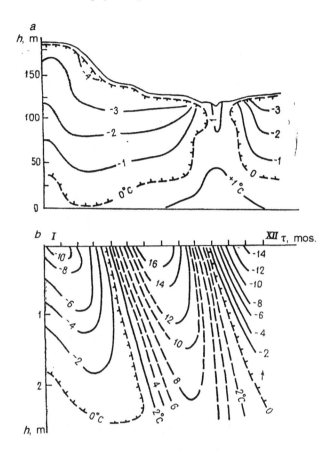

Fig. 1.1. Examples of graphical representation of temperature field:
a – two-dimensional stationary; b – one-dimensional non-stationary. The 0°C
isotherm is the boundary between unfrozen and frozen ground.

where $a = \lambda/C_{vol}$ the thermal diffusivity serving as an index of thermal inertia of the medium.

The changes in the temperature field in time can be rather varied (from monotonic to spasmodic) in the case of a non-stationary thermal regime. Fields which are periodic steady-state represent the most important case of non-stationary temperature fields under natural conditions. Their special feature is that the effect of the initial temperature distribution in the medium is virtually lost with the repeated temperature fluctuations on the surface. This distribution is so to say 'erased from the memory' of the medium and can be ignored beginning from some instant of time. It is this instant which points to the periodically stable regime of the process.

The most simple and at the same time in nature most widespread case of the periodic stationary regime is the sine-wave of the daily, annual and perennial air temperature fluctuations and the same sine-wave (harmonic) fluctuation of temperature caused by them in the ground. In this case we can speak of the propagation of temperature waves through the ground. We can represent these temperature changes formally by the simple relation:

$$t(0, \tau) = t^0_{mean} + A_0 \sin \frac{2\pi}{T} \tau \qquad (1.3)$$

where $t(0, q)$ is the temperature at the ground surface changing in time; A_0 is the physical amplitude of the fluctuation of this temperature (being equal to a half of the meteorological), in the annual (or daily or perennial) cycle; t^0_{mean} is the mean annual (mean daily or mean perennial) temperature at the ground surface around which the fluctuation with period T proceeds. Thus, the temperature fluctuations on the surface are uniquely determined by three values: t^0_{mean}, A_0 and T.

It is obvious that the actual temperatures under natural conditions, for example the annual temperature pattern on the surface, differ essentially from the sinusoidal because there are irregular warmings and coolings on the general background of temperature changes throughout a year (Fig 1.2a). However if the temperature is averaged over a ten-day period and especially over a period of months the temperature curve will increasingly approximate to the sinusoidal form. In the general case this curve can always be brought into the 'ideal' (perfect) sinusoidal form through the experimental determination of the areas S_1 and S_2 situated between the axis of zero temperature and the actual curve of temperature change for the period T equal to one year. The main parameters of such 'ideal' sinusoidal waves, i.e. the mean annual surface temperature t^0_{mean} and the physical amplitude of temperature fluctuation A_0 can be calculated in this case in the following way:

$$t^0_{mean} = \frac{S_1 - S_2}{T}, \qquad A_0 \approx \frac{S_1 + S_2}{T} \cdot \frac{\pi}{2}$$

Temperature fluctuations on the surface $t(0,\tau)$ cause temperature fluctuations in the ground $t(z,q)$ that is, the temperature wave formed on the surface propagates into the ground exciting a similar wave in the ground. At the same time when moving downward the wave meets a thermal resistance (the value being inversely proportional to the value of thermal conductivity, λ) as well as expends its power in the processes associated with the thermal capacity, C_{vol}, of the medium. Therefore the wave is damped out with depth, with A_0 decreasing gradually. The maximum depth of penetration into the

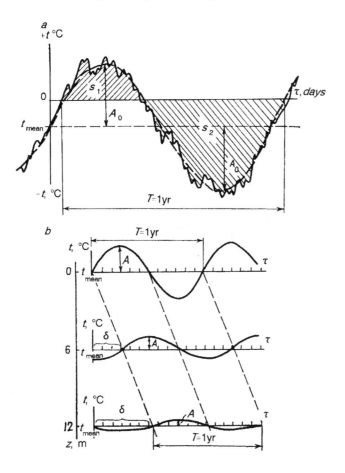

Fig. 1.2. Harmonic temperature fluctuations on the surface and in the underlying ground: a – temperatures observed on the surface reduced to the 'ideal' sinusoidal fluctuation; b – temperature wave propagation in the ground: A – the amplitude of the fluctuation; δ – the time shift of the fluctuation S_1 and S_2 – areas of temperature, degree-hours, in warm and cold periods, respectively.

ground of the surface temperature wave (perennial, annual and daily) is termed the *depth of propagation of temperature fluctuation or the depth of zero (perennial, annual or daily) temperature amplitude*. The near-surface ground delimited by these values of depth is termed the *layer of perennial (n-year)*, *annual or daily temperature fluctuations*, respectively.

The propagation of temperature waves through media without phase transitions is described by Fourier's laws, which follow from the solving of the so-called Fourier's problem of the propagation of temperature waves through a homogeneous medium of the unbounded half-space without phase transitions. Mathematical formulation of the problem looks as follows:

$$\frac{\partial t(z,\tau)}{\partial \tau} = a\frac{\partial^2 t(z,\tau)}{\partial z^2}, \quad z > 0$$

$$t(0,\tau) = t^0_{\text{mean}} + A_0 \sin\frac{2\pi}{T}\tau \tag{1.4}$$

The initial conditions are absent in the formulation of the problem because it is assumed that the process of heat transfer is already in the stable periodic regime. The solution of this problem is the following function:

$$t(z,\tau) = t^0_{\text{mean}} + A_0 e^{-z\frac{\sqrt{\pi C_{\text{vol}}}}{\lambda T}} \sin\left(\frac{\pi}{T}\tau - z\sqrt{\frac{\pi C_{\text{vol}}}{\lambda T}}\right) \tag{1.5}$$

This function determines the value of ground temperature at any depth z and in any instant of time τ under the corresponding ground surface temperature change. It describes completely the steady-state periodic temperature field formed in ground as a result of sine-wave temperature fluctuations on the surface. On the basis of this solution we can conclude that the fluctuations around the same mean temperature t^0_{mean} and of the same period T as on the surface are established in the ground as well (see Fig. 1.2b). The more sophisticated treatment of the solution given allows us to state the following relationships known as Fourier's laws which are fundamental for the process of propagation of temperature waves into the ground.

Fourier's first law states that the amplitude of temperature fluctuations decreases exponentially with depth:

$$A(z) = A_0 e^{-z\frac{\sqrt{\pi C_{\text{vol}}}}{\lambda T}} \tag{1.6}$$

It follows from this relationship that the rate of damping out of temperature fluctuations with depth depends on the properties of the medium: the higher the heat conduction (the lower its thermal resistance) and the lower the heat capacity, the slower is the damping out and the greater the depth to which the temperature fluctuations penetrate (all other things being the same).

The curves bounding the maximum and minimum values of the sine-wave temperature fluctuation at each particular depth are exponential curves defining the character of the damping of the amplitude of temperature fluctuation with depth (Fig. 1.3). In other words, the solution of Fourier's problem for the propagation of temperature waves through a homogeneous medium shows that all the possible temperature changes with

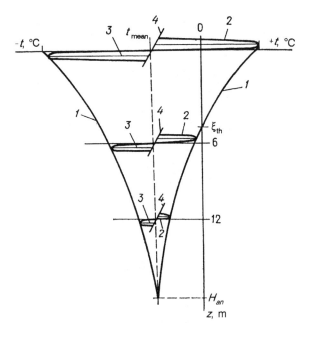

Fig. 1.3. Nature of damping of temperature fluctuations with depth in ground.
1 – envelope of temperature fluctuations; 2 – temperature fluctuations on the
surface and at depths of 6 and 12 m; 3–4 – amplitudes of fluctuations and time
axes at depths of 0, 6 and 12 m; m_{th} is the depth of seasonal ground thawing; H_{an}
is the depth of annual temperature propagation.

depth are confined to particular boundaries for each particular depth. These
boundaries are segments of the exponential curves $t^0_{mean} \pm A(z)$ where z
varies in the range from 0 to the depth of zero annual amplitudes. Thus if
actual data on temperature measurements in holes throughout a year are
entered in one figure (Fig. 1.4), the synchronous temperature curves will so
to speak 'fill' the area enveloped by the segments of the exponential curves.
To show temperature changes with depth geocryologists often use these
envelopes without presenting the particular temperature curves for the
particular instant of time (see Fig. 1.3).

Fourier's second law points to the fact that temperature fluctuations in
ground proceed with a phase shift proportional to the depth (see Fig. 1.2 b).
The time of the delay of the fluctuation can be calculated by the following
equation:

$$\tau(z) = \frac{1}{2}z\sqrt{\frac{TC_{vol}}{\pi\lambda}} \qquad (1.7)$$

Phase shift of the fluctuations with depth occurs as a consequence of the

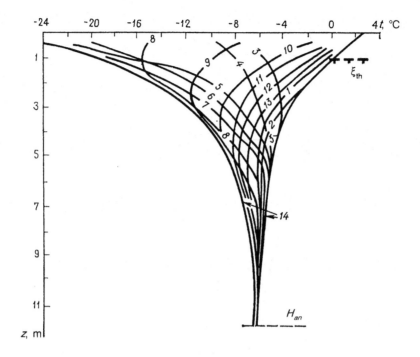

Fig. 1.4. Nature of change of ground temperature (in a bore-hole) for various moments of time: 1–13 – temperature curves (1 – 30/IX, 2 – 15/X, 3 – 24/XI, 4 – 17/XII, 5 – 20/I, 63/II, 7 – 20/II, 8 – 24/III, 9 – 26/ IV, 10 – 2/VI, 11 – 26/VI, 12 – 27/VII, 13 – 29/VIII; 14 – the curves of the envelopes. See also Fig. 1.3.

limitation of the velocity of propagation of the wave through the ground. This velocity depends on the thermal-physical properties of the ground and on the period of the fluctuations; therefore it is constant with depth and with time for homogeneous ground. In line with this law of Fourier, the minimal winter temperatures at some depth z, for example, will be observed not at the moment of their occurrence on the surface but after the time necessary for the wave to reach this depth z. Therefore under natural conditions the following pattern is observed: when, for example, a period of cooling or warming begins in the near-surface portion of the ground the period of warming (or cooling, respectively) is still in progress in the middle part of the layer of the annual temperature fluctuation, while the period of cooling (or warming) of the previous temperature wave is observed in the lower portion of the layer (see Fig. 1.4, curves 8 and 10).

Fourier's third law relates the depth of penetration of temperature fluctuations h with the period and amplitude of the temperature wave on the surface:

$$h = \sqrt{\frac{\lambda T}{\pi C_{vol}} \ln \frac{A_0}{A_h}} \tag{1.8}$$

Starting from this relation the depth of penetration of fluctuations is proved to increase with greater amplitude A_0 and period of fluctuation T. The maximal depth of penetration of the temperature fluctuation (of damping) is taken to be such a depth at which the amplitude A_h becomes less than 0.1 °C, i.e. a value equivalent to the accuracy of temperature measurements under natural conditions. Substituting the value $A_h = 0.1$ °C and various numerical values of period T of the fluctuations into the expression (1.8) we can show that the fluctuations with larger period penetrate to the greatest depth, while short-period fluctuations are damped near the surface, all other things being the same. Actually, as calculations and natural observations show, daily temperature fluctuations can penetrate to a depth of 1–2 m, annual fluctuations to 15–25 m, and 300 year fluctuations to 150–200 m and deeper. According to Fourier's third law, given two fluctuations of unequal periods T_1 and T_2 ($T_1 \neq T_2$), the depths z_1 and z_2 at which the same damping of the fluctuations occurs are connected by the following relationship:

$$z_2 = \sqrt{\frac{T_2}{T_1}} z_1 \tag{1.9}$$

Thus, given two temperature fluctuations at the ground surface of periods 1 year and 100 years, if it is known that the annual amplitude decreases by a factor of 10 at the depth of 10 m, it follows from expression (1.9) that the analogous damping of amplitude (by a factor of 10) of the 100-year fluctuation will occur at the depth of 100 m.

By and large, by virtue of the fact that the periodic (harmonic) temperature fluctuations (temperature waves) are observed in the near-surface ground, the annual (daily or perennial) period of the temperature change in ground is divided into two parts: a half-period of heating and a half-period of cooling. The amount of heat arriving at the ground for the half-period of heating and outgoing from it for the half-period of cooling is often termed the ground *heat turnover* in geocryology. In the case of a stable periodic temperature regime the heat turnover for each of the half-periods is equal in magnitude but opposite in sign. As it was shown by V.A. Kudryavtsev (17), when heat propagates through dry homogeneous soil where phase transitions are absent the expression for heat turnover through the ground surface in the layer z takes the form:

$$Q_{surf} = \sqrt{2}(A_0 - A_z) \sqrt{\frac{\lambda TC}{\pi}} \tag{1.10}$$

It should be stressed that all the principles discussed above of the formation of the temperature field in ground do not take into account the effect of moisture phase transitions and the presence of horizontal as well as certain vertical heat fluxes. Thus, for example, when the existence of the geothermal flux q is taken into account the mean annual temperature of ground within the depth of the layer of annual temperature fluctuations t_{mean} is not constant but increases in accordance with the geothermal gradient g, i.e:

$$t_{mean}(z) = t_{mean}^0 + (q/\lambda)z \qquad (1.11)$$

The geothermal gradient ($g = q/\lambda$) shows, for a given location, how many degrees the temperature changes with a change of depth of a unit length. The inverse of the geothermal gradient is termed the *geothermic depth*. It shows at what distance along the vertical the ground temperature changes by $1\,°C$. The geothermic depth is often considered to be $33\,m$ on the average within the continents, when approximate calculations are performed.

In addition to vertical heat fluxes from the Earth's interior and the associated geothermal temperature gradients in the upper part of the lithosphere, there are also the lateral heat fluxes and consequently the lateral temperature gradients. Extremely high horizontal temperature gradients in rocks are observed at the boundaries between one topographic element and another having a different orientation relative to the Sun, higher or lower degree of slope and moisture content, between land and sea and lake basins as well as large water courses. In other words, there often exist not one-dimensional but two- and three-dimensional temperature fields in nature.

The principles of formation of temperature fields are far more complex in moist soils and ground systems when their temperature passes through $0\,°C$. This is associated with the fact that the processes of water freezing or ice thawing (water phase transitions) are accompanied by release or absorption of a great amount of heat which changes the temperature pattern fundamentally. By its nature it is not simply a process of cooling or heating of ground in this case but the far more complex process termed the process of freezing or thawing of moist ground. One should consider the very nature of moisture phase transitions in moist soil systems to understand and to analyze this process correctly.

1.2   Freezing (crystallization) of water and melting of ice in the ground

Moist soils are extremely complex in composition, structure, properties and interaction between their components and phases when the temperature passes through $0\,°C$. All the natural waters are effectively solutions, i.e. consist of water molecules bound to a variable degree by solute ions. In soils this binding of water by ions is accompanied by an interaction with the active surfaces of mineral particles (adsorbed water) as well. Therefore when studying the structure and phase transitions of bound water in soils it is necessary to consider the model of free, bulk water texture serving as a primary standard for comparison.

In line with A.A. Ananyan's, B.N.Dostovalov's, O.Ya. Samoylov's, Ya. I. Frenkel's and other researchers' developments the processes of ice – free water – vapour phase transition (the process of *ice breaking down*) can be represented in the following way (11). With the temperature of the ice close to 0 °C each molecule situated at an ice lattice point has the minimal kinetic energy E_K and the maximal activation energy, i.e. combined bond energy E_B or the energy of interaction with the surrounding molecules. The E_K value proves to be considerably less than the E_B value in this case. Increase of the velocity of molecular movement occurs with heat input to the ice.

The incoming energy can be expended to overcome the bonds between the particular molecules in the ice texture and to make molecules execute translational jumps when $E_K > E_B$. As the energy action continues, there comes a time when more than 9–13% of the hydrogen bonds are broken and the process of isothermal melting begins. The proportion of the broken hydrogen bonds is determined from the relation between the heats of melting ($5.9 \, kJ \, mol^{-1}$) and of sublimation ($48.1 \, kJ \, mol^{-1}$).

The breaking of such an amount of hydrogen bonds of H_2O molecules occurs as a result of the formation of a sufficient number of vacancies in ice crystal lattice points resulting from the increasing number of the translational jumps of molecules. As a consequence the symmetric and tracery ice structure with long-range order breaks, as it were, into pieces. At the same time the groups of molecules or associations no longer adjoining each other in such an orderly fashion as they were in the ice are separated or isolated in the crystal body. Two conditions should be fulfilled for the formation of such separated associations: 1) the obligatory breaking of a number of hydrogen bonds between the neighbouring molecules and along the contours of the water associations; 2) setting of the separated molecular groups ('pieces of ice') into vibration (Brownian motion) around their equilibrium centres. The latter occurs on account of the transmission of impulses of the H_2O molecules. The increase of energy of the H_2O rotation-vibration movement of molecules at points in the crystal lattice practically stops in this case. In connection with this there is no further increase of ice temperature with the continuing arrival of heat.

In the course of ice melting the largest and the slowest associations of H_2O molecules can be broken down and split into smaller ones under the effects of momentum of the individual translating molecules as well as of their groups. The total volume of ice and water decreases in this case while the density increases as a result of breaking of some of the tetrahedrons in the hexagonal ice texture and of penetration of their molecules into voids still persisting in the unbroken portions of the crystal lattice. The coordination number of molecules in such a system increases becoming more than 4, which is to say that the number of hydrogen bonds of each molecule increases, i.e. the activation energy increases. In other words, the further breaking of the associations stops, having reached on the average a certain critical mass, while the arriving heat is expended mainly in increasing the kinetic energy of the molecules in associations and of the associations themselves. The temperature of the system begins to rise, while melting and further breaking of hydrogen bonds stops.

Based on the foregoing it is simple to consider the reverse course of the process, i.e. the transformation of water into ice (the process of water *freezing*). Thus, the number of translational jumps and the velocity of the diffusing H_2O molecules and of the small water associations decreases with the extraction of heat from the water. The moment being imparted by the translating particles weakens. Vibrations of large molecular associations composing the skeleton of the structure of the water around the equilibrium centres slows down. The water temperature decreases as the natural result. The increasing sluggishness of the associations means the hydrogen bonds between H_2O molecules are able to hold the associations and groups of molecules together. Brownian motion of the molecular associates slows down and then stops totally. Enlargement of the associations, their unification and their mean mass increase to the critical point and higher, i.e. transformation of the associations occurs, into the holocrystalline nuclei of ice which grow quickly and release heat of crystallization. Smaller associations as well as isolated diffusing H_2O molecules begin to add themselves to these enlarged ones. Because the coordination number of H_2O molecules is less in an ice lattice than in water and is equal to 4, as well as because ice crystallizes as a hexagonal crystal system characterized by the presence of free passages (holes), the volume of the system increases with the transition of water into ice while the density decreases. An expansion of water by 9% occurs on its transition into ice.

The process of unification and enlargement of H_2O associations and molecules discussed above is accompanied by the necessary formation of new hydrogen bonds between H_2O molecules, naturally with the release of heat which must be removed from the system.

It can be assumed that the process of freezing proceeds abruptly. For

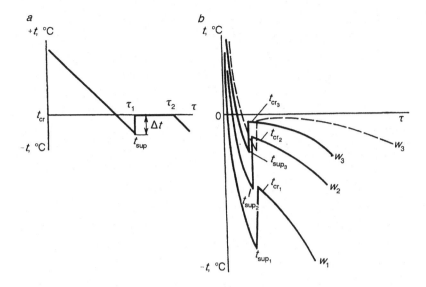

Fig. 1.5. Temperature change with time (τ) during supercooling t_{sup} and crystallization t_{cr} in (*a*) water and (*b*) freezing clay soil for various initial moisture contents $W_1 < W_2 < W_3$ and constant heat removal rates, Q'_1 (solid line) and Q'_2 (dotted line), $Q'_1 > Q'_2$.

example, as soon as the temperature of ice crystals decreases, equalizing of the temperature in the system begins and the temperature of water in contact with the ice crystals must decrease. However the water cannot decrease its temperature in this case, because it will be transformed into ice (i.e. the binding of associations of H_2O molecules can occur). In other words, any local lowering of temperature of the ice crystals or of the associations of H_2O molecules will cause crystallization and transition of some portion of the water into ice. Consequently, the temperature of the system cannot decrease until the transition of all the water into ice occurs (Fig. 1.5a). That is the reason why the freezing process is isothermic. It is accompanied by intensive enlargement of the associations, characterized by Brownian motion and by the release of heat of crystallization. It can be assumed, however, that there can be conditions when freezing proceeds essentially instantaneously because of very fast heat removal from the freezing system. The heat of crystallization will not be released in this case because there will be no transformation of water texture into ice texture. Thus the texture of H_2O is 'fixed' and a so-called *amorphous* (non-crystalline) ice texture is formed.

One of the most complex and unclear problems in the theory of *water crystallization* is the problem of the establishment of the critical mass for the associations of H_2O molecules when they become the crystal nuclei of ice, i.e. are transformed into full scale microcrystals as far as their texture is concerned. Investigations in recent years allow the assumption that this takes place at critical mass equal to $472\,H_2O$. The size of this unit ice cell is considered to be of the order of $10^{-20}\,cm^3$. When such a mass is reached, these associations can grow intensively and are the basis for the future crystal. The crystals usually grow radially from the centre of the nucleus. From the thermodynamic standpoint the additional forces binding the H_2O molecules and associations into ice texture, i.e. formation of a centre or centres of crystallization, are necessary for the initiation of the crystallization process.

There are two theories on the nature of ice microcrystal development. One of them (*homogeneous*) considers the formation of the crystallization centres as a regular process, with initiation of sufficient mass of the associations of H_2O molecules resulting from the homogeneous temperature fall in the medium slowing down the molecular motion velocities with the joining together of molecules through hydrogen bonding. The *heterogeneous* theory of the initiation of ice crystals is based on the inevitable presence of foreign particles (impurities) in the melt (water) serving as natural seed crystals, i.e. as the crystallization centres. The *heterogeneous* theory does not rule out the theoretical possibility of homogeneous initiation of crystals. However as it is practically impossible to free water of the impurities it is these foreign particles which give rise to the crystallization in practice.

Transition into a metastable state, reached when a liquid is supercooled to a temperature below its freezing point, is essential for the initiation of the homogeneous ice crystal. Such a unit crystal is formed spontaneously and its future depends on the degree of supercooling of the solution. The crystal will grow under sufficient supercooling of the system. Calculations show that the water can be in the metastable state in the temperature range between 0 and $-55°C$. It was possible to supercool distilled water to -12 to $-16°C$ and small drops of water to $-72°C$ in experiments. At such a temperature the crystallization is initiated in the whole volume of the drop at once. The supercooling of the volume of water does not usually exceed a few degrees and depends on the composition of the water, the background of its formation and the cooling rate.

According to the heterogeneous theory the forced crystallization is caused by seed crystals, given the small oversaturation or supercooling of the melt. The best seed crystals are particles (impurities) of the growing crystal (ice) itself. The next best are particles of isomorphous substances able to give solid solutions with the crystal, i.e. the so-called gas hydrates and crystalline hydrates. Atoms or ions of such an impurity occupy the corners of the water crystal lattice. Salt crystals not soluble in water, various mechanical irregularities and substances giving regular aggregations with the growing crystal under certain conditions, can serve as seed crystals too. On the whole, all solid bodies having a structure similar to that of ice, must cause crystallization.

The further growth of an ice crystal (after initiation of the nucleus) depends, on the one hand, on the particular features of the crystal-chemical structure and, on the other hand, on the conditions of its feeding. Thus, given the constant temperature gradient and the continuous increase of the growing crystal surface and, consequently, the increasing expenditures of energy on phase transition of water into ice, the water temperature around the crystal will rise slightly in the beginning, thereby slowing down its growth. Subsequently, lowering of the temperature with removal to outside the system of heat released in the course of water crystallization begins. In other words, the growth of ice crystals in water is likely to be spasmodic and cyclic in character. In response to the slight heating of water around the growing crystal and its further cooling, the differences in water density are able to cause temperature diffusion as well as diffusion of H_2O molecules with a consequent inflow of mass to the various portions of the crystal causing its growth. The rate of growth around the crystal edges and vertices (corners) is proved to be higher than that near its planes, i.e. growth anisotropy is observed. Ice needles and stars are formed because the crystal vertices elongate more rapidly. Later on they grow together into a continuous mass holding their skeleton shape along the edges only (19). Under the conditions of one-sided uniform cooling this leads to selection and to primary growth of the crystals which are oriented in the most favourable direction. The crystals which are oriented in the less favourable direction are wedged out. At the same time if a growing crystal plane meets any foreign body it begins to push this body away with a certain force termed the *crystallization* force.

In natural water containing dissolved salts differentiation (separation) of components is usually observed in the course of crystallization. In the course of cooling such a solution, molecules of salts as well as water molecules are organized into the elementary lattice. Separate formation of lattices of ice and of each salt occurs. Such a stratification is governed by the distinction between the crystallographic parameters of the ice and salt lattices. In the course of freezing the elementary lattices of salts are rejected to the boundary of the force field in which the ice crystal nucleus acts. Ice crystals usually contain a very small amount of salts, this being determined by the conditions of ice lattice formation. At the same time, the greater the crystallization rate, the greater the amount of salts captured by the crystal. The temperature of freezing (crystallization) $t^0{}_{cr}$ of salt solution is lower than $0°C$.

The water possesses a store of energy of $636 \, J \, g^{-1}$ while the ice possesses a store of energy of $302 \, J \, g^{-1}$ at the temperature of $0°C$. The difference in their stores of energy ($334 \, J \, g^{-1}$) represents the latent heat of fusion of ice. This heat, which is supplied from outside, is expended for the formation of the water lattice, i.e. for the change of the mean statistical distance between the H_2O molecules in ice and in water.

The problem of *phase transition of bound water into ice* and, first of all, the interpretation of the effect of the lowering of the freezing temperature of water in soils, is more complex. It is obvious in the most general molecular kinetic context that the process of phase transition of bound water into ice must be in agreement with the pattern of free (bulk) water. However one should take into account that the water which is bound by the mineral surface is characterized by a texture different from that of the free water. One can anticipate that the skeleton lattice consisting of associations of H_2O molecules, typical of free water, is absent in the bound water texture. All the H_2O molecules in bound water are likely to take part in the formation of a specific textural lattice corresponding to the textural pattern of the crystalline lattice of the mineral surface (substrate). This results from the epitaxial mechanism binding water to the surface, with the textural pattern of the bound water lattice being increasingly in agreement with the texture of the hydrating mineral crystalline lattice as the solid substrate is approached. At the same time the bonding energy will grow progressively. In other words the closer the mineral surface, the more distorted is the texture of the bound water as compared with free water. Therefore while it may be sufficient for the formation of the first holocrystalline ice nuclei in the free water that the vibrations of large associations of molecules composing the textural skeleton of water should basically decrease, with the hydrogen bonds between them able to hold them together under the falling temperature, this condition is not sufficient for the formation of the first ice nuclei in bound water. For the bound water, in which the large H_2O associations composing the textural lattice of the bulk water are mainly absent, it is necessary first of all to organize such associations. For this purpose sufficient lowering of temperature will be required to allow for a reduction of kinetic and vibrational energy of the individual molecules of bound water to an extent that the bond energy between molecules E_b, or the activation energy, is greater than the sum of the kinetic energy and the bond energy between these molecules and the mineral substrate E_m. In this case the hydrogen bonds of the H_2O

molecules will be able to overcome the effect of the surface forces of the mineral particles and to group the molecules into ice-like associations. The condition for the formation of the first holocrystalline ice nuclei in bound water is written as:

$$E_b > E_k + E_m \tag{1.12}$$

where E_m is a variable value increasing exponentially as the mineral surface is approached and as the thickness of the bound water film h decreases, i.e $E_m = f(1/h^n)$, where n is an exponent the value of which depends on the type of interaction between the water and substrate. Lowering of the bound water temperature down to t_{sup}, equal to the supercooling temperature necessary to fulfil this condition, will depend on the value of the energy of bonding of the water by the mineral substrate E_m. The more energetically bound the water is, the lower the temperature to which it must be super-cooled. Accordingly the supercooling temperature is characterized by in-creasingly lower negative values with decreasing initial moisture content of the soil ($W_1 < W_2 < W_3$) and, consequently, with decreasing thickness of the bound water films and associated increased energy of bonding between the H_2O molecules and the mineral surface (see Fig. 1.5b). Thus bound water supercooling in the course of cooling of the soil, depends first of all on the necessity of removing this portion of water from the action of the mineral particle surface forces, whereas supercooling of the free water is required merely to form the first holocrystalline ice nuclei. It follows from this also that the surface energy depends on the temperature, i.e. that distance over which it is effective decreases with temperature.

Thus once the ground system has reached a temperature equal to the temperature of the supercooled bound water ($t_{sup} < t_{cr}^{in}$), the energetically less bound part is outside the effect of the surface forces of the ground particles resulting in the formation of the associations of individual H_2O molecules and in their enlargement. The subsequent cessation of Brownian (vibratory) motion of the associations (as the heat is being removed from the system) will cause their transformation into the holocrystalline ice nuclei as well as the grouping of individual H_2O molecules and small associations around these, or foreign, crystallization nuclei. At the same time additional supercooling is not needed for this process to proceed in bound soil water, because, on the one hand, the fulfilment of condition (1.12) provides for the necessary fall of temperature (to $t_{sup} < t_{cr}^{in}$), and, on the other hand, the pore solution is characterized by the presence of a great number of foreign crystallization nuclei accelerating and enhancing the regrouping of water molecules into the ice texture. The process of new ice nuclei formation in bound water of the soil and their transformation into the ice crystals of full value as far as their texture is concerned will take place, naturally, with the release of the heat of phase transition. This will cause the spasmodic raising of temperature of the soil system and, consequently, of the bound water up to the temperature of the beginning of its freezing or crystallization t_{cr}^{in}.

The temperature at the beginning of bound water crystallization in soils depends on the value of bond energy between the soil moisture and the mineral surface E_m. The experimental investigations show that the higher this energy is and the thinner the bound water film, the lower is the temperature value of the beginning of crystallization. Thus, the temperature of the beginning of water freezing in sands is determined by a high content of rather weakly bound (nearly free) water and it approaches the freezing temperature of the bulk (pure) water ($t_{cr}^{in} \approx 0\,°C$). At this temperature all the free water proceeds to freeze in sands, and the duration of freezing depends on the moisture content of the sand and on the intensity of cooling. If the heat removal from the freezing sand sample is continued, its temperature will decrease only slowly to approximately -0.5 to $-1\,°C$ (Fig. 1.6) pointing to freezing out of the progressively more bound water, this being confirmed by the liberation of the crystallization heat in small amounts. The temperature of the beginning of bound water crystallization in clay-rich soils turns out to be essentially lower than that in sands and can reach, for example, values of the order of -2.5 to $-3.5\,°C$ in the montmorillonite clays with moisture content of 41–45%. One should take into account in this case that the presence of solute salts and exchange cations in the bound water of soils will also reduce the value t_{cr}^{in}. The further removal of the heat of phase transition from the clay-rich system causes a gradual lowering of the negative temperature as the freezing out of the increasingly more energetically bound layers of water proceeds (see Fig. 1.6). Thus the moisture phase transitions and the liberation of the heat of crystallization proceed not under a particular, constant temperature but in a range of negative temperatures in sands (although little pronounced) as well as in clay-containing soils. It is not unreasonable in this case to highlight only the temperature of the beginning of bound water crystallization (t_{cr}^{in}). The duration of crystallization of bound water as well as of free water will depend on the rate of heat removal from the system. A higher rate of heat removal will cause more rapid freezing out of bound water which is shown by a sharper knee in the curve of the freezing soil temperature change (see Fig. 1.5b).

Why is it necessary for the temperature to fall in the course of successive freezing out of the increasingly more bound water layers, after the first holocrystalline ice nuclei have developed and the temperature of the begin-

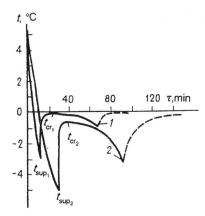

Fig. 1.6. Temperature curves, through supercooling (t_{sup}) and initial crystallization (t_{cr}) (solid line), and warming (dotted line), of sand (1) and clay (2).

ning of crystallization (t_{cr}^{in}) has been reached by the system? Let us consider broadly the process of the gradual freezing of bound water as heat removal continues, to explain the nature of this phase transition over a range of negative temperature.

The fact that the soil has reached the temperature of the beginning of crystallization ($t = t_{cr}^{in} < 0\,°C$) and that the first holocrystalline ice nuclei have been formed means that a new complex physico-chemical system with such interfaces as mineral surface-unfrozen (bound) water-ice has been formed. The mutual effect of the surface forces of mineral particles and ice will be minimal in this case, as the curve for the unfrozen water bond energy due to the mineral surface with respect to the thickness of the liquid-like film h, ($E_m(t) = f(1/h^n)$), and the curve for ice ($E_{IS}(t) = f(1/h^n)$) cross each other at a not very high energy level (Fig. 1.7, section I-I). Section I-I represents in this case a conventional surface dividing the unfrozen water layer into two parts. One, the left part, situated between this section and the mineral surface (MS) will be affected mainly by the surface forces of the mineral substrate. The other (right part), which is smaller in thickness, is the part of the water layer between the section I-I and the ice surface (IS) and is confined mainly by the surface forces of the ice substrate. It will have practically the same texture as the ice. In other words, this part of the unfrozen water represents a quasi-liquid film on the ice surface, the texture of which has already undergone modification. The surface forces of the mineral as well as of the ice substrate in this case cause distortion of the hydrogen bonds in adjacent layers of the unfrozen water modifying the water texture towards the texture of the crystalline lattices of the substrate. Such a distortion is transmitted to the

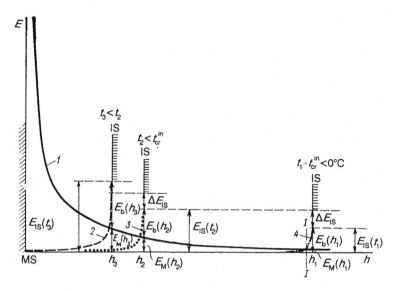

Fig.1.7. Relation of the unfrozen water film thickness h to the energy of adsorption (E) under the negative temperatures $t_1 > t_2 > t_3$: 1 – with the mineral surface MS; 2–4 – with the ice surface IS (2 – at h_3 and t_3, 3 – at h_2 and t_2, 4 – at h_1 and t_1).

adjacent water layers by the epitaxial mechanism. As the difference between the textural parameters of the ice and the water lattice is less than that between the textural parameters of mineral particles and water lattices, the hydrogen bond distortion caused by the ice surface will be weaker as compared with the mineral surface and must propagate over a shorter distance in the water. Therefore the thickness of the unfrozen water film bound by the mineral surface turns out to be essentially greater than that of the film subjected to the ice surface forces (see Fig.1.7). Note that the lowering of the negative temperature of such an ice-soil system by an equal value ($\Delta t_1 = \Delta t_2 = \Delta t_3 \ldots$) must cause an equal and essential increase of the free surface ice energy ($\Delta E_{IS}^1 = \Delta E_{IS}^2 = \Delta E_{IS}^3 = \ldots$). The increase of free surface energy of the ice will be an order of magnitude more than that of the mineral surface. In support of this fact it is noted that the coefficient of thermal expansion of ice (α -change of linear size of a body with change of its temperature by a unit) is more than one order of magnitude greater than that of minerals and soils ($\alpha = 40 - 60 \times 10^{-6}$ K^{-1} for ice, $0.4 - 8 \times 10^{-6}$ K^{-1} for the minerals of the skeleton of clay-rich soils, and $3 - 10 \times 10^{-6}$ K^{-1} for granite, sandstones and shales). At the same time the contraction of the solid bodies at cooling is associated with the reduction in the textural parameters of their crystal lattices. A greater change of these textural parameters causes appropriate (larger in magnitude) distortion (strain) of the hydrogen bonds in the surrounding unfrozen water and will manifest itself over a greater distance from the distorting surface. In other words, the energy of binding of water by an ice surface E_{IS} and the thickness of the layer of unfrozen water held by this surface will grow by an order of magnitude faster on lowering of the temperature than we can observe for the mineral surfaces of clay particles. In this connection the change of the value $E_m = f(1/h^n)$ with lowering of temperature can be neglected. At the same time the change in the curve $E_{IS}(t) = f(1/h^n)$ on account, for example, of equal increase of ice surface energy ΔE_{IS} on lowering of the temperature by an equal value Δt, must occur and is shown graphically in Fig. 1.7. In this case:

$$E_{IS}(t_1) < E_{IS}(t_2) < E_{IS}(t_3), \tag{1.13}$$

where $t_1 > t_2 > t_3$ while $|t_1 - t_2| = |t_2 - t_3| = \Delta t$ and $E_{IS}(t_2) - E_{IS}(t_1) = E_{IS}(t_3) - E_{IS}(t_2) = \Delta E_{IS}$. In other words, the ice surface forces will grow so profoundly on lowering the temperature that a portion of the unfrozen water previously bound by the mineral surface will experience the ice surface forces, i.e. *the phase transition of a portion of unfrozen water* into ice will occur. At the same time the thickness of the layer of water bound by the ice surface must increase.

On this basis the possibility exists of explaining the reason for further phase transitions of bound (unfrozen) water into ice (after the formation of the first holocrystalline ice nuclei at the temperature of the beginning of crystallization t_{cr}^{in}) not under a constant temperature t_{cr}^{in} but over a range of negative temperatures.

From the moment of the formation of the holocrystalline ice nuclei in freezing soil at $t_1 = t_{cr}^{in}$ it can be assumed that in the case of no further heat removal from the unfrozen soil the system reaches a quasi-equilibrium state between the unfrozen water and ice and there will be no further freezing of bound water. In this case, given the thickness h_1 of the bound water film being responsible for the distance from the mineral surface h_1 determining the temperature t_1, the following condition should be fulfilled:

$$E_{IS}(t_1) = E_b(t_1) + E_m(t_1) \tag{1.14}$$

where $E_{IS}(t_1)$ is the surface energy of ice expended in the interaction (bond) with the adjacent molecules of the unfrozen water. $E_b(t_1)$ and $E_m(t_1)$ are, respectively, the energy of the bond between H_2O molecules and the energy of the bond between unfrozen water molecules and the mineral surface situated at a distance h_1 from the surface at the temperature t_1. Such a quasi-equilibrium state can be disturbed only by additional heat input or removal from the system. The process of ice nuclei disappearing and their transition into the water texture, i.e. the process of soil thawing, will begin if there is an input of heat. At the same time the removal of a quantity of heat from the freezing ground system will cause the temperature of the ice, bound water and mineral particles to fall (for example, by $\Delta t = t_1 - t_2$), where $t_2 < t_{cr}^{in}$). Such temperature lowering will cause an increase in the surface energy of the ice by the value $\Delta E_{IS} = E_{IS}(t_2) - E_{IS}(t_1)$ and growth of energy of bonding between H_2O molecules in the bound water by ΔE_b (as $E_b(t_1) < E_b(t_2)$). The value of the energy of bonding of unfrozen water molecules with the mineral surface, as is obvious from the foregoing, will not be changed essentially ($E_m(t_1) \approx E_m(t_2)$), in contrast to that of the ice, which results in violating the condition (1.14) in such a manner that:

$$E_{IS}(t_1) > E_b(t_1) + E_m(t_1) \tag{1.15}$$

Consequently, on account of the greater bond energy between the unfrozen water molecules and the ice surface some of them will be released from the effect of mineral surface forces and experience the effect of ice surface forces, i.e. phase transition of unfrozen water into ice will occur, provided the liberated heat of crystallization is removed from the system. Freezing out of some portion of the unfrozen water and, consequently, a movement of the ice-unfrozen water interface from the h_1 position to the new equilibrium interface position h_2 should occur, fulfilling the condition:

$$E_{IS}(t_2) = E_b(t_2) + E_m(t_2).$$

Further heat removal from the freezing soil will cause renewed lowering of the negative temperature (for example by $\Delta t = t_2 - t_3 = t_1 - t_2$, where $t_3 < t_2$). The temperature lowering by the Δt value (as in the first case under consideration) will provide for the surface energy of ice increasing by the value $\Delta E_{IS} = E_{IS}(t_2) - E_{IS}(t_1) = E_{IS}(t_3) - E_{IS}(t_2)$ and for the energy of the bonds between H_2O molecules in the unfrozen water to increase, such that $E_b(t_3) > E_b(t_2) > E_b(t_1)$. This will violate the steady-state quasi-equilibrium condition resulting in freezing-out of an additional amount of unfrozen water, provided there is the necessary removal of latent heat of phase transition from the soil. It is easy (as in the previous case) to find the new h_3 position of the ice surface corresponding to the negative temperature $t_3 < t_2$ in Fig. 1.7.

Fig. 1.8 presents a curve well-known from experiments, showing the dependence of the unfrozen water content on the negative temperature.

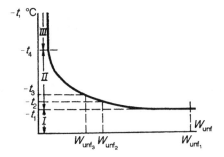

Fig. 1.8. Dependence of the unfrozen water content (W_{unf}) on the negative temperature, showing the intensive (*I*), slight (*II*) and very little (*III*) phase transitions regions.

From this figure we note that successive changes of ground negative temperature by an equal value Δt causes freezing out of smaller and smaller amounts of unfrozen water and the decrease by smaller and smaller amounts in the thickness of the bound water film: $h_1 - h_2 > h_2 - h_3$, resulting from a sharp increase in the value of energy of bonding between water molecules and mineral surface, with the unfrozen water film thinning in accordance with the curve $E_m = f(1/h^n)$. Thus, it is this curve which determines essentially the trend of the curve $W_{unf} = f(t)$. Really, under high negative temperatures there is intensive freezing-out of the slightly bound water in soil systems for a drop in temperature of $1\,°C$, while in the region of low negative temperatures such a temperature fall is marked by an essential slowing down or, in effect, cessation of the transition of unfrozen water into ice (see Fig. 1.7). With allowance made for this and in line with the curve $W_{unf} = f(t)$, N.A. Tsytovich in 1973 proposed dividing the process of water phase transitions in soils into three categories or regions (see Fig. 1.8): I – the region of intensive phase transitions, when the unfrozen water content changes by 10% with a change of $1\,°C$ in temperature, i.e. $\Delta W_{unf} > 10\%$; II – the region of less intensive transitions, when ΔW_{unf} changes in the range from 10 to 1%, with $\Delta t = 1\,°C$; III – the region of virtual absence of phase transitions, when $\Delta W_{unf} < 1\%$, , with $\Delta t = 1\,°C$.

The described patterns refer to chemically 'pure', bound water in soils. However under natural conditions the bound soil water is characterized by the presence of a significant amount of solute salts bonding the H_2O molecules – and in so doing retarding the transition into ice of this osmotically bound water. In line with this the condition of phase transition of bound water containing solute salts into ice (freezing-out of such unfrozen water in the range of negative temperature), given the film thickness h at its surface and the freezing temperature t, is described by:

$$E_{IS}(t) > E_b(t) + E_0(t) + E_m(t) \qquad (1.16)$$

where $E_0(t)$ is the energy of the bond between H_2O molecules in the unfrozen water and the ions of solute salts at distance h from the mineral surface. It turns out in accordance with this condition (1.16) that the higher the concentration of the unfrozen water solution and, consequently, the energy of interaction between H_2O molecules and ions, the lower is the negative temperature required to perform the phase transition of equal amounts of unfrozen water into ice.

The problem of the texture of the ice being formed may be of great interest when considering the phase transitions of unfrozen water in soils. It is assumed at the moment that the first type of ice (ice I) is formed in the soil when the greater part of the unfrozen water is freezing out. However it cannot be denied that the texture of the ice under formation must experience the effect of the mineral substrate surface forces. Such an effect can probably manifest itself (though very slightly) over a distance of the order of a few hundred nanometres, that is to say, at high negative temperatures. The effect will manifest itself most tangibly at a distance of tens and units of nanometers between the ice and the mineral surface, i.e. at low negative temperatures when the unfrozen water film thickness is rather small and completely under the effect of surface forces of the mineral particles (see Fig. 1.7, $h \ll h_3$). Really, the energy of interaction between the unfrozen water molecules and the mineral surface is proved in this case to be essentially higher than that between the unfrozen water molecules and the ice surface, i.e. E_{IS} ($t \ll t_3$) $\ll E_m$ ($h \ll h_3$). At the same time distortions of hydrogen bonds in the unfrozen water by the mineral substrate surface forces turn out to be so strong that they are able to strain the hydrogen bonds in the ice texture, i.e. in essence to distort this texture. Thus the texture of the ice crystal accomodates itself to the texture of the mineral substrate under the effect of the mineral forces of this substrate (the phenomenon of pseudomorphism). According to the theory of pseudomorphism, propagation of stresses (on account of the distortion of hydrogen bonds in this case) inside the texture of the intergrowing bodies can extend tens of nanometers and depends on the size of the gap between the textures of the contacting bodies, on the energy of interaction between them and on the 'rigidity' or elasticity of their lattices. As this takes place, of two intergrowing bodies the crystals with the lower value of crystal lattice energy experience the greatest changes of volume under pseudomorphism. It is natural that the surface ice crystals will be more strained in conformity with the frozen soil system under consideration.

Experiments conducted by a number of researchers on the building up of ice on various mineral surfaces represent a certain experimental verification (if not a strict one, because the ice build-up was in the direction away from the mineral substrate, and not the reverse as would apply for the freezing of the unfrozen water). These experiments have shown that in the vicinity of the mineral surface the ice has a particular texture, different from the texture of the ice formed at a substantial distance from the surface. The thickness of such a contact layer of ice and the degree of its change of texture turn out to be essentially dependent on the particular texture of the crystal lattice of the mineral surface. Thus smaller ice crystals were formed on a polycrystalline base in the course of ice build-up. The extremely small crystals of the ice coating were formed on materials which were amorphous in structure. The mineral substrate of gypsum, which has a fibrous structure, caused the development of an oriented crystal lamination with the long axes of the crystals oriented transversely to the fibres of the base. It was found experimentally that there is a dependence between the orientation of the optical axes of the ice crystals in the contact ice layer and the value and direction of cooling. It is obvious that the effects of ice crystal texture distortion will manifest themselves less in the case of the freezing out of the unfrozen water because the mineral surface effect on the ice does not directly manifest itself through the film of bound water.

Using all the above-cited ideas on the freezing of unfrozen water in soils in the range of negative temperatures we can show that the reverse phase transition (ice melting and increase of the unfrozen water content on this account) will also not proceed at once and not at a constant temperature but gradually with the progressive raising of negative temperature in the thawing soil system. All the above is supported strongly by the results of experimental investigations (see Fig. 1.6).

Thus, the moisture phase transitions proceed in fine-grained soils not at 0°C and not at a particular constant negative temperature but in a range of negative temperatures. Freezing or frozen fine-grained soils always contains some amount of unfrozen water. Thus it has been established experimentally that even at a temperature of −100°C there exists a thin film of unfrozen water of an order of 0.3 nm in thickness around the fine mineral particles. It must be stressed in this case that the unfrozen water should not be identified with bound water because the film of bound water of any thickness experiences the energy effect of mineral surface force only while the film of unfrozen water, being the same in thickness, experiences the effect of ice surface forces in addition. In other words, such a film of unfrozen water will be energetically bound and must be less mobile. At the same time the term 'unfrozen water in soils' includes practically all the types of ground moisture, as the situation may be: free water (freezing usually at 0°C and at temperatures much lower than 0°C in the conditions of high concentrations of pore solution); capillary water (freezing in the region of high negative temperatures, i.e. near 0°C); pellicular and adsorption water (freezing in a wide range of negative temperatures to -100°C and lower).

It should be stressed also that each negative temperature of a particular soil system in an equilibrium state must correspond to a strictly specified unfrozen water content or strictly specified thickness h of a water film. If this condition is violated, the thermodynamic state of the system will not be in equilibrium and the process of either freezing out of the excess (above the equilibrium value) moisture or of ice melting will proceed resulting in the unfrozen water content increasing up to its equilibrium value.

1.3 Sublimation and desublimation of moisture in frozen rocks

According to the thermodynamic viewpoint, *ice sublimation* represents the phase transition of a substance from the solid to the gaseous state without first forming a liquid and which occurs with heat absorption (the heat of sublimation is 2.83 kJ^{-1}g).

From the molecular-kinetic theoretical standpoint the mechanism of the sublimation and ablimation (desublimation) processes consists of the following. H_2O molecules residing in the lattice structure of the ice surface and having the highest velocity of heat (kinetic, vibratory) motion ($E_K > E_b$) escape into the gaseous surroundings, breaking free of the interaction with the rest of the molecules. Simultaneously with this process, there is the possibility of the process of the trapping of those individual molecules and chaotically moving groups in the gaseous surroundings which collide with the ice surface. These molecules lose the kinetic energy of their translational motion in the course of collision and are integrated into the crystal lattice of the ice surface. The process of ice sublimation proceeds in this case on condition that the process of abstraction of H_2O molecules from the ice surface dominates over the process of their trapping from the gaseous surroundings, otherwise the process of ablimation (desublimation) of water vapour on the crystal ice surface occurs.

Based on the hypothesis of the presence of a quasi-liquid film on the ice surface, by the process of *ice sublimation* is meant the process of evaporation of this film with its restoration by ice melting. Microphotographic investigations show that the sublimation process proceeds not uniformly from the whole ice surface but from the individual most energetically unstable portions. These are various faults in the ice crystal texture as well as ice crystal edges and vertices on its surface.

The ice surface proves to be strewn with a great number of shallow recesses of a cup-like shape (cones) at the beginning of sublimation. The sublimation cones are characterized in this period by small cross-section and depth, forming a 'honey-comb' pattern on the ice surface. Furthermore, they become deeper, wider and coalesce together partially, resulting in the formation of oriented recesses and grooves.

Evidently, there exists a limitation on the development of concave microrelief elements after which the sublimation of protuberances begins with, at this moment, high energy instability. Large smooth microrelief forms appear and further development of small sublimation cones occurs again on this gentle wavy ice surface. Subsequently the ice surface is likely to be subject to cycles of simple microrelief evolution similar to the foregoing one, on the ice surface inherited from the previous cycle. Thus one can refer to the varied cyclic nature of the changes within the various parts of the sublimating ice surface, the areas of which can increase and decrease around a particular mean value. It is this fact which explains the scatter in the values of ice sublimation intensity I_s, representing the ice mass loss in unit time for a constant cross-sectional area of the ice sample, determined by experimentation for various moments of time. However, by and large the experimental points are satisfactorily distributed about a constant (mean) value of the ice sublimation intensity (Fig. 1.9a). As the ice sublimation is developed only within the same energetically unstable portions on the sample surface initially, the minimum values of ice sublimation intensity occur then, increasing at first essentially with time and then stabilizing in response to the covering of the whole area of the ice sublimation surface by the processes of external heat-mass exchange. The sublimation process in samples with large- and small-crystal ice proceeds at different rates during the same time intervals. Thus this process proceeds faster in the samples of small-crystal ice on account of a great number of textural faults and high values of volumetric strain gradients as a result of small-crystal ice formation under conditions of low negative temperatures. Typically with such ice there is also less scatter about the mean value in the experimental data on the sublimation intensity.

The pattern of the ice sublimation process discussed applies also to ice occurring in macropores of soil systems. However it is complicated there by the presence of some amount of unfrozen water in dynamic equilibrium with ice. According to the principle of the equilibrium state of unfrozen water and ice in frozen soils, its amount must remain constant at a given temperature. Therefore by ice sublimation in soils can be understood not only the transition of ice from solid to gaseous state but also the evaporation of

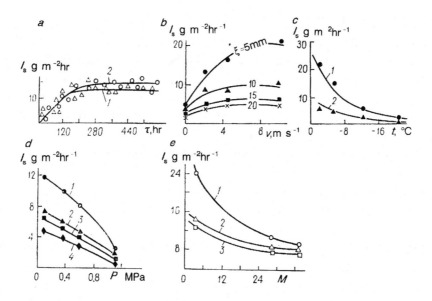

Fig. 1.9. Dependence of the ice sublimation intensity I_s: a – on duration of the process for samples of coarse-crystalline (1) and fine-crystalline (2) ice; b – on the air flow velocity v at $t = -5°C$ and various thickness of dry zones ξ_{dry} in samples of polymineral clay; c – on the temperatures of air t from $0°C$ moving with velocity $4.5\,\mathrm{m\,s}^{-1}$ (1) and stationary (2) for samples of polymineral clay at $\xi = 5\,\mathrm{mm}$; d – on pressure of stationary air above samples of ice (1), polymineral clay (2), kaolin (3), loam (4) (the values I_s given here are for the fixed instant of time $\tau = 220\,\mathrm{h}$ in the process); e – on the molecular composition of the stationary gas medium for samples of ice (1), polymineral clay (2) and kaolin (3); M = molecular weight.

unfrozen water with its progressive restoration by the melting of ice. It is known that the process of ice sublimation in soils is associated with the process of phase transitions and desorption proper as well as with the processes of external and internal heat and moisture transfer. Therefore it amounts to a process of ground drying at negative temperatures, or *frost drying*. At the same time the processes of the internal heat and mass transfer exert the predominant effect on the intensity of the frost drying because the intensity of ice sublimation (phase transitions) under natural conditions is limited by the rate of water vapour removal from the system, that is, by the resistance to moisture transfer rather than the provision of the process with energy.

The intensity of ice sublimation in the soil is defined experimentally as the density of moisture flux through the open surface of a sample, i.e. as a moisture mass Δm being lost by the soil from a unit cross-sectional area F during a time interval $\Delta\tau$: $I_s = \Delta m / F\Delta\tau$.

The intensity of ice sublimation from the soil is damped out and decreases with time as the thickness of the sublimated (dried) zone increases. Because the process of soil drying proceeds as an interaction with the vapour-gaseous medium, the thermodynamic conditions of this medium exert considerable effect on the intensity of the heat- and mass-exchange process with the exterior. The experimental data show that an increase of I_s with increasing air flux velocity is possible up to a particular limit ($v = 4-7\,\mathrm{m\,s}^{-1}$). The effect decreases with the increasing thickness of ground dried zone (see Fig. 1.9b). Lowering of the temperature causes decreasing sublimation intensity on account of the decrease in the moisture transfer coefficient and gradients of transfer potential (see Fig. 1.9c). Experimental investigation of the effect of the overall pressure of the air (in the range from 1×10^5 to $1.2 \times 10^{10}\,\mathrm{Pa}$) on I_s at a temperature of $-8\,°\mathrm{C}$ has shown that the sublimation intensity decreases as the pressure is raised (see Fig. 1.9d). This depends mainly on the decrease in the diffusion coefficient. The process of ice sublimation in soils in relation to the vapour-gaseous medium depends fundamentally on the molecular composition of the gas as well. The value I_s decreases with the increase of molecular mass from helium ($M = 4$) to air ($M = 29$) and to argon ($M = 38$). This is associated first of all with the decrease of the vapour diffusion coefficient with increasing molecular mass M of the vapour-gaseous medium (see Fig. 1.9e).

The process which is the reverse of ice sublimation is termed *ablimation* (desublimation), i.e. crystallization of water vapour on soil surfaces accompanied by heat release and ice crystal formation in the form of frost. Ablimation takes place in the event that the soil surface temperature is lower than that of the surrounding vapour-gaseous medium and its intensity grows with increasing value of the temperature difference. In the early stage the process of vapour crystallization has a well-defined selective pattern. Formation of the first crystals of ablimation ice proceeds on edges, vertices and chips of crystals and on other surface faults with excess of surface energy. Then the crystals grow, coalesce and form a continuous cover of frost on the soil surface creating an additional resistance to heat transfer and causing in this connection a decrease in the temperature differences and, consequently, the ablimation intensity I_a decreases with time. The density of the growing ablimation layer varies with its thickness. It decreases with distance from the surface and changes with time because of the metamorphic process (recrystallization). With time the density of the ablimation layer increases, thus increasing its thermal conductivity. Concurrently with the increase in thickness of the layer under consideration from the surface of the water-unsaturated soil, the inward-directed moisture diffusion can pro-

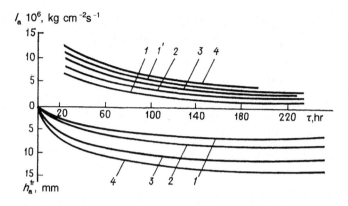

Fig. 1.10. Change in the ablimation intensity I_a and in deposited layer thickness h_a^{fr} with time q: 1 and 1' – ice-unsaturated and ice-saturated sand respectively; 2 – clay (polymineral, bentonite and kaolinite) 3 – ice; 4 – metal.

ceed. The moisture transfer proceeds in this case in vapour as well as in the liquid phase. At the same time I_a turns out to be dependent on the parameters responsible for the diffusion intensity (i.e. on free soil porosity, snow cover, temperature, etc). This is well followed in Fig. 1.10, reflecting the dependences $I_a = f(\tau)$ and $h_a^{fr} = \phi(\tau)$ obtained experimentally for sand, clay, ice and metal samples. The similarity of values of ice ablimation intensity for the moisture-saturated clays of different mineral composition depends upon the fact that their thermal conductivity does not vary much. The moisture-unsaturated sand, having a lower value of thermal conductivity, shows lower intensity of ablimation as compared with the moisture-saturated sand. In the case of moisture-unsaturated (for example, sandy) soil samples the moisture ablimation proceeds not only on the surface but also inside the ground. In clay soils in parallel with the processes of vapour transfer and ablimation the transfer of unfrozen water under the effect of grad t takes place, as was determined experimentally from the change in the initial distribution of the moisture content of the fine-grained soil samples. The process of water vapour ablimation in rudaceous rocks has its special features. Experimental investigations of metamorphic and igneous rock debris (detritus, 5–15 mm, and rock debris, 40–50 mm) have shown that the process of ablimation begins not throughout the whole ground surface but on the surfaces of the boulders which is then followed by a flux which propagates inward. A denser ablimation hoar is formed in this way on rock debris. The ablimation intensity is greater for larger debris components and this is likely to be associated with their higher heat conduction.

1.4   Freezing and thawing of ground

In Section 1.1 we considered temperature fields without regard for water freezing or thawing in the soil as the temperature went through 0°C. In other words, we analysed the processes of ground heating or cooling rather than freezing or thawing proper. Such cases are typical of soils which in effect contain no water. The zero isotherm does not differ fundamentally from any other isotherm of the temperature field in this case.

In actual conditions rather important thermal-physical and physico-chemical processes are responsible for soil properties, structure and composition as well as for the particular features of the soil temperature field in the moist soil system near or at the zero isotherm. The most important of these processes is the transition of water contained in the soil from liquid to solid state and vice versa, because release and absorption of a great amount of heat is associated with these phase transitions. As a consequence, additional heat sources or sinks appear in the region where the phase transitions proceed which affect fundamentally the ground temperature field and its dynamics. Thus, for example, heat of the order of 70 J is released in the course of freezing of the moisture contained in 1 g of the soil with 20% moisture content, when it is cooled below 0°C. Such an amount of heat is sufficient to heat this 1 g of rock by nearly 40°C. The amount of heat Q_{ph} being released or absorbed in the course of complete freezing or thawing of a unit soil mass with the moisture (ice content) by weight W_w and a skeleton (dry) density γ_{sk} is determined from the expression:

$$Q_{ph} = 334 \, W_w \gamma_{sk} \tag{1.17}$$

where $334 \, \mathrm{J\,g^{-1}}$ is the heat of phase transition of water into ice and vice versa.

It is obvious that the temperature fall (below 0°C) of moist soils in the course of their freezing must proceed slower than their simple cooling. The layer of freezing soil must in this case protect the lower layers against intense cooling, because the zero isotherm cannot propagate deeper until freezing of all the free water occurs and the appropriate amount of crystallization heat has been released at this boundary. The zero isotherm serves in a given case as the interface between thawed and frozen portions of freezing soils and is termed the *freezing boundary or front*. And vice versa, the thawing of frozen soils slows down their heating, because the greater part of the heat received by the soil is expended in the phase transition of ice into water in the neighbourhood of the zero isotherm, which is in a given case the thawing boundary or front. This effect, of the moisture phase transitions proceeding at the zero isotherm, on the temperature field of freezing and thawing soils

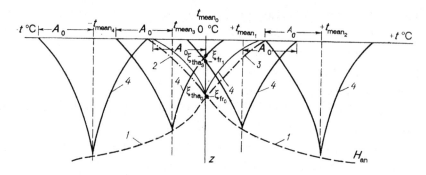

Fig. 1.11. Dependence of values H_{an}, ξ_{fr} and ξ_{tha} on the relationship between the mean annual temperature t_{mean}, and the amplitude of temperature fluctuations on the surface A_0: 1 – the depth of zero annual amplitude H_{an} at $t_{mean_4} < t_{mean_3} < t_{mean_0} < t_{mean_1} < t_{mean_2}$; 2 – the depth of seasonal ground thawing; 3 – the depth of seasonal freezing; 4 – the envelopes of the temperature curves: ξ_{fr_1} at $+t_{mean_1}$; $\xi_{fr_0} = \xi_{th_0}$ at t_{mean_0}; ξ_{th_3} at $-t_{mean_3}$; seasonal freezing and thawing are absent at t_{mean_2} and t_{mean_4}.

was termed the '*zero curtain*' by M.I. Sumgin.

In line with this phenomenon the depth of penetration of the harmonic (sine-wave) temperature fluctuations into the ground in which the moisture phase transitions proceed turns out to be essentially less than that in dry soils, where the processes of moisture freezing and thawing are absent. In the case when the amplitude of the temperature fluctuations on the surface proves to be smaller than the mean annual ground temperature $(A_0 \leq |t_{mean}|)$ the moisture phase transitions are absent and the depth of annual temperature fluctuations penetration turns out to be greatest (Fig. 1.11). Otherwise $(A_0 > |t_{mean}|)$, and within the upper part of the section either a layer of seasonal thawing $(t_{mean} < 0)$ underlain by the perennially frozen ground or a layer of seasonal freezing $(t_{mean} > 0)$ underlain by unfrozen soils is developed. Presence of the seasonal freezing (or thawing) layer causes a reduction of thickness of the layer with annual temperature fluctuations (see Fig. 1.11). This is associated with the fact that the essential part of annual heat storage which could go into heating or cooling of ground occurring below the layer of seasonal freezing or thawing is expended within this layer for phase transitions of water into ice and vice versa. Therefore the temperature wave penetrates to a smaller depth. At the same time the greater the thickness of layer ξ_{tha} or ξ_{fr}, the less is the thickness of the layer with annual temperature fluctuations H_{an}. And finally, at $t_{mean} = 0$ there is the case where the layer with annual temperature fluctuations is the same in thickness as the layer of seasonal freezing or thawing $(\xi = H_{an})$ (see Fig.

1.11), i.e. all the heat storage proceeds in the layer of freezing or thawing only, while the layers below are not subject to cooling or heating. We can observe the process of seasonal freezing as well as of seasonal thawing in this case depending on the state (perennially frozen or not frozen) of the underlying ground. At the same time it should be stressed that the thickness of the layer of seasonal freezing (thawing) at $t_{mean} = 0\,°C$ is proved to be maximal decreasing with increasing absolute values of t_{mean}.

Thus, given the process of seasonal freezing or thawing, the depth of penetration of annual fluctuations of temperature will be equal to:

$$H_{an} = \xi_{fr(tha)} + h_1 \tag{1.18}$$

Within the layer h_1 occurring between the base of the seasonal freezing (or thawing) layer and the depth at which the annual temperature fluctuations are damped out (see Fig. 1.11), moisture phase transitions are practically absent in the ground and thus Fourier's laws discussed above (see Section 1.1) turn out to be true for this layer. With the formula for calculating the depth of penetration of temperature fluctuations (1.8) we can calculate the thickness of layer h_1.

Given a layer of seasonal freezing or seasonal thawing, the change of mean annual ground temperature with depth differs from that when the phase transitions are absent. By and large, when the basic factors affecting the formation of the ground temperature field under a periodically steady-state temperature regime are considered, the mean annual temperature change throughout the depth ($t_{mean}(z)$) can be represented by a broken, linear relationship consisting of a number of straight portions. Let us consider this curve separately for the cases of seasonal freezing and seasonal thawing (Fig. 1.12), beginning with the mean annual air temperature t_{mean}^{air}, which usually is measured by the weather stations at the height of 2 m.

Within the layer of air adjacent to the Earth's surface (of 0 to 2 m in height) a fall of mean annual temperature with distance away from the surface is observed, i.e. $t_{mean}^{surf} > t_{mean}^{air}$ where t_{mean}^{surf} is the mean annual surface temperature of the cover. This is associated with the fact that it is the ground (day, or active) surface absorbing the solar radiation, which heats the air (and not the reverse). In the case when there is no cover on the ground surface, the mean annual air temperature turns out to be higher than the mean annual ground temperature t_{mean}^{0} by the value of a so-called radiative correction (Δt_R) which does not exceeding 1 °C usually. Given the various kinds of covers on the soil surface, the temperature change from t_{mean}^{air} to the mean annual ground temperature t_{mean}^{0} turns out to be more complex because the covers can exert warming and cooling effects.

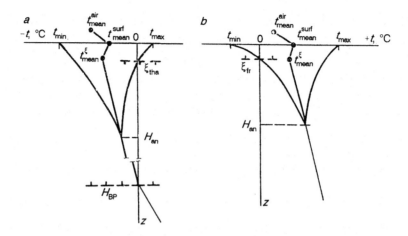

Fig. 1.12. The mean annual temperature change with depth for seasonal ground thawing (*a*) and freezing (*b*): t^{air}_{mean}, t^{surf}_{mean}, t^{ξ}_{mean} are the mean annual temperatures of air, at the surface and at the base of the seasonal thawing (freezing) layer; ξ_{tha} and ξ_{fr} are the depths of seasonal thawing, freezing, respectively; H_{an}, is the depth of annual temperature fluctuations; H_{BP} is the permafrost thickness.

The mean annual temperature is not constant in the layer of seasonal freezing (thawing) either. It decreases with depth from the first inflection of the temperature curve on the ground surface at $z = 0$ (see Fig. 1.12) The main reason for such a decrease of t_{mean} in the layer is the difference between the values of heat conductivity of this layer ξ when unfrozen λ_{unf} and when frozen λ_{fr}, this being in its turn the reason for formulating the so-called *temperature offset* Δt_{λ} in the layer, defined as the difference between the mean annual temperature at the base of the seasonal freezing (thawing) layer t^{ξ}_{mean}, and the mean annual temperature on the ground surface in conditions of a periodically steady-state temperature regime:

$$\Delta t_{\lambda} = t^{\xi}_{mean} - t^{0}_{mean} \qquad (1.19)$$

Of great importance in this case is the fact that there exists a heat balance at any depth of the annual temperature fluctuations, i.e. the ground receives as much heat for the half-period of heating as it loses for the half-period of cooling, otherwise the periodically steady-state regime could not exist. For example, if the ground received more heat during the period of heating than that being lost by it during the period of cooling, heating of the ground would proceed and the mean ground temperature would rise with time. The temperature offset is absent and $t^{\xi}_{mean} \approx t^{0}_{mean}$, when there is a balance of 'incoming' and 'outgoing' heat in ground in which there are uniform thermal physical characteristics (of heat conductivity $\lambda_{unf} = \lambda_{fr}$ especially) in the unfrozen and in the frozen state. If $\lambda_{th} \neq \lambda_{fr}$ such an offset exists. At the same time numerous calculations and simulation results have shown that the other thermal physical characteristics such as thermal capacity C_{vol} and heat of phase transition Q_{ph} do not effect the value of this offset. According to calculations and experimental data, Δt_λ varies usually in the range $0.5 - 1.5°C$.

The ground temperature below the layer of seasonal freezing (thawing) increases with depth on account of the geothermal gradient, resulting in the second inflection of the temperature curve at the base of the seasonal freezing or thawing layer t_{mean}^ξ. At the same time if $t_{mean}^\xi < 0$, there is perennially frozen ground in the profile, with the base of which one more inflection (the third) of the given temperature curve is associated (see Fig. 1.12a). At this point, unfrozen and frozen materials with different thermal physical properties (as a rule $\lambda_{fr} > \lambda_{unf}$) are in contact on the lower boundary of the perennially frozen layer. However if this boundary does not move the thermal fluxes must be equal ($g_{fr} = g_{unf}$) in frozen and unfrozen zones, i.e. λ_{fr} grad $t_{fr} = \lambda_{unf}$ grad t_{unf}. If $\lambda_{fr} > \lambda_{unf}$, grad t_{unf} > grad t_{fr}, i.e. the temperature increases with depth more rapidly in the thawed zone than in the frozen zone. As a result the temperature curve has a point of inflection at the interface. If $t_{mean}^\xi > 0$, i.e. there is no perennially frozen ground in the profile, the temperature curve is straight below the layer of seasonal freezing (see Fig. 1.12b).

Thus though the solution of *the heat conduction problem without considering the phase transitions* formulated in Section 1.1 in the form of Fourier's laws (1.6) – (1.8), is of great importance for understanding the processes of thermal conduction in the ground generally, it is suitable only for a very restricted class of natural objects when rigorous quantitative description of these processes is required. Among such objects are rock masses, for which the processes of phase transitions of water can be neglected. Among them are the masses of monolithic or drained, faulted hard rocks, of completely drained loose rocks (which occur rarely), and unfrozen or frozen masses in which phase transitions are absent for the time of observation. It is impossible in the majority of cases to define quantitatively and adequately, temperature fields of rock masses and their change with time without regard for phase transitions.

The processes of water migration in soil are associated in a particular way with phase transitions over a range of negative temperatures, which, on the one hand, affect the character of the moisture field (ice content) formed in the ground, and on the other hand, exert the reverse effect on the process of

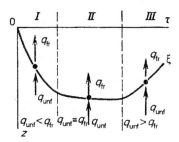

Fig.1.13. Dynamics of the phase boundary movement under the changing temperature conditions on the surface: *I–III* – stages of the temperature change – in time (*I* – lowering; *II* – stabilization; *III* – rising); ξ – position of the phase boundary; q_{fr} and q_{unf} are the intensity of the heat flux from ξ to the surface and from the unfrozen or thawed zone to ξ respectively.

phase transitions and consequently on the temperature field. And finally, the increase of water volume in the course of freezing as well as its migration into the freezing zone can cause changes in the stress field in soils, associated inversely with the process of freezing (thawing). The fact that the depth of the phase transition zone is time-dependent is a vital particular feature of this zone in which the processes discussed above proceed. This boundary (zone) is moving and we can observe the sharp changes in the values of heat fluxes arriving at this surface and leaving it (Fig. 1.13). It is these abrupt changes which are responsible for the *conditions of phase boundary movement in the ground.*

Let us consider such a case (see Fig. 1.13). Assume that the temperature at the surface of the ground mass to the base of which the constant heat flux q arrived from the Earth's interior, had passed through 0°C at the initial moment of time $q = 0$ and the process of freezing had begun. The temperature on the surface fell further for some time and then became constant (negative). And finally, after a further period it began to rise, but remained negative. The process of freezing (thawing) of this mass has three clearly discernible stages (see Fig. 1.13). At the first stage the value of heat removal q_{fr} from the phase boundary turns out to be greater than that of heat arrival from the Earth's interior q_{unf}, i.e. $q_{fr} > q_{unf}$. Consequently freezing at the front occurs and the boundary moves downward, i.e. $d\xi/d\tau = \xi' > 0$. At the second stage a balance of ingoing and outgoing heat is established at the phase boundary m, i.e. $q_{fr} = q_{unf}$. The freezing boundary turns out to be fixed in this case, while $\xi' = 0$. And finally, in the third stage, warming on the surface results in the value of heat being removed from the boundary ξ being less than that arriving from the Earth's interior, i.e. $q_{fr} < q_{unf}$. In this case the frozen body thaws from the bottom upwards. The thawing boundary moves upward while $\xi' < 0$. It is precisely in this manner that the process of perennial freezing and thawing of strata proceeds, in many cases.

It should be stressed that the inequality between q_{fr} and q_{unf}, i.e. $q_{fr} - q_{unf} = \Delta q$, with $\Delta q \neq 0$ is a necessary condition for the process of phase transition to proceed in freezing or thawing ground. At the same time for any period of time $\Delta\tau$, such a ground layer $\Delta\xi(\Delta\xi = \xi'\Delta\tau)$ freezes or thaws in which all the difference $(\Delta q\Delta\tau)$ between the amounts of heat incoming to and outgoing from the front will be spent on the phase transition of water $(Q_{ph}\Delta\xi)$. In other words, the heat condition at the mobile boundary of materials which are freezing or thawing can be written as: $\Delta\tau(q_{fr} - q_{unf}) = Q_{ph}\Delta\xi$. The rate of ground freezing or thawing can be found from this expression, i.e. the velocity of movement of the boundary of moisture phase transition is as follows:

$$\xi' \approx \frac{q_{fr} - q_{unf}}{Q_{ph}} \tag{1.20}$$

It follows from this that heating of ground below the phase front by phase transition heat being released in the course of freezing is impossible, as this heat can be released only when it can be removed, with the amount of released heat being equal to that of the heat which can be removed into outer space through the overlying layer of freezing ground.

Usually the previously obtained expression is written as $q_{fr} - q_{unf} = Q_{ph}\xi$, or if writing the expression for heat fluxes:

$$\lambda_{fr}\frac{\partial t_{fr}}{\partial z}\bigg|_{z=\xi-0} - \lambda_{unf}\frac{\partial t_{unf}}{\partial z}\bigg|_{z=\xi+0} = Q_{ph}\frac{\partial\xi}{\partial\tau} \tag{1.21}$$

where t_{fr} and t_{unf} are the temperature of the frozen and unfrozen zones. This expression is termed the *Stefan condition at the moving phase boundary*. This expression enables us to establish the law of phase boundary propagation. Thus in the case of constant negative temperature at the surface of the originally unfrozen ground, the freezing boundary $\xi(\tau)$ propagates through the mass in accordance with the law $\xi(\tau) = \alpha\sqrt{\tau}$, where α is a constant coefficient depending on the thermal-physical earth material properties, the initial temperature of the ground and the temperature at its surface.

It is possible to approach the problem of ground freezing (thawing) in several ways depending on the degree of consideration of the character of physical processes occurring at the phase transition boundary and in the freezing and thawing zones. Selection of one or another way for approaching a problem for practical calculations depends on the particular features of structure and properties of the ground unit under consideration. For the sake of simplicity we shall consider one-dimensional temperature fields in homogeneous ground masses.

The statement of the problem on ground freezing (thawing) with development of phase boundary (Stefan problem)

The statement of the problem of freezing (thawing) with the formation of one phase boundary, i.e. a boundary at which the temperature of phase transition is maintained for the given situation, is the simplest (i.e. all water-phase transitions proceed on this boundary (Fig. 1.14)). Then the thermal physical properties of rocks on this boundary will undergo an abrupt change: $\lambda_{unf} \neq \lambda_{fr}$, $a_{unf} \neq a_{fr}$ and $C_{unf} \neq C_{fr}$ where λ_{unf} and λ_{fr} are heat conductivities of unfrozen and frozen ground, respectively, a_{unf} and a_{fr} are their temperature conductivities [thermal diffusivity coefficients], C_{unf} and C_{fr} are the heat capacity coefficients. This statement describes very well the processes of freezing (thawing) in rudaceous materials and sands as well as in faulted (water-saturated) hard rocks. It is actually typical of these rocks that practically all the water phase transitions proceed at a freezing temperature t_{fre} equal or close to $0°C$. On further lowering of the temperature, phase transition is not observed and the thermal physical properties of the material remain constant.

Mathematical formulation of the problem includes two equations of heat conduction in this case (for unfrozen and frozen zones) and the Stefan condition for the moving phase boundary is written as: (1.22)

$$\left.\begin{aligned} &\frac{\partial t_{fr}(z,\tau)}{\partial \tau} = a_{fr}\frac{\partial^2 t_{fr}(z,\tau)}{\partial z^2}, \quad 0 < z < \xi(\tau); \ \tau > 0; \\[2mm] &\frac{\partial t_{unf}(z,\tau)}{\partial \tau} = a_{unf}\frac{\partial^2 t_{unf}(z,\tau)}{\partial z^2}, \quad \xi(\tau) < z < l; \ \tau > 0; \\[2mm] &t_{fr}(\xi(\tau),\tau) = t_{unf}(\xi(\tau),\tau) = t_{fre}; \\[2mm] &\lambda_{fr}\frac{\partial t_{fr}}{\partial z}\bigg|_{z=\xi-0} - \lambda_{unf}\frac{\partial t_{unf}}{\partial z}\bigg|_{z=\xi+0} = Q_{ph}\frac{d\xi(\tau)}{d\tau} \end{aligned}\right\} \qquad (1.22)$$

where l is the position of the lower boundary of the region under consideration. The last two equations are the conditions for conjugate solutions to the equations of heat conduction in frozen and unfrozen zones with a moving phase boundary (Stefan condition). They couple the heat conduc-

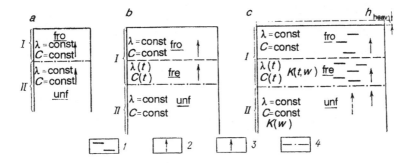

Fig. 1.14. Diagram of the ground freezing (thawing) process: a – with formation of the phase interface ξ; b – with the formation of the freezing zone; c – with allowance made for water migration into the freezing zone; I – frozen (fro) and freezing (fre) soil; II – unfrozen (unf) soil; 1 – segregated ice streaks; 2 – moisture flux I_w from unfrozen to freezing soil; 3 – the heat flux q in unfrozen, freezing and frozen soil; 4 – the phase interface; h_{heav} – the heave of the surface; k – the moisture conductivity of unfrozen and freezing soil.

tion equations and allow a solution. In order to obtain the conclusive solution to the set problem, it is necessary to supplement this statement with extreme (initial and boundary) conditions:

$$\left. \begin{array}{l} t_{fr}(z, 0) = \phi_1(z), \quad 0 < z < \xi(0); \\ t_{unf}(z, 0) = \phi_2(z), \quad \xi(0) < z < l \end{array} \right\} \text{initial conditions}$$

$$\left. \begin{array}{l} t_{fr}(0, \tau) = \Phi_1(\tau); \\ t_{unf}(1, \tau) = \Phi_2(\tau) \end{array} \right\} \text{boundary conditions}$$

This statement holds good for the process of soil freezing as well as thawing while the direction of the process (freezing or thawing) is determined in the course of the problem solution.

In this mathematical formulation of the problem the boundary conditions are of type I, taken as the upper ($z = 0$) and lower ($z = l$) boundaries, i.e. the change of temperature with time at these boundaries is given. Other types of specification of boundary conditions are also possible:

– of type II, when the value of heat flux or temperature gradient is specified on the boundary:

$$\left. \frac{\partial t(z_1\tau)}{\partial z} \right|_{z=0} = f(\tau);$$

– of type III, when the conditions on the boundary are specified as a combination of temperature and temperature gradient values on this surface, with the values of both the temperature and the gradient being unknown individually:

$$\alpha t(0, \tau) + \beta \frac{\partial t(z, \tau)}{\partial z}\bigg|_{z=0} = \Psi(\tau)$$

where α and β are the constant coefficients depending on the character of heat exchange on the surface.

It should be stressed that $\phi_1(z), \phi_2(z)$, $\Phi_1(\tau)$, $\Phi_2(\tau), f(\tau)$, $\psi(\tau)$, are the known functions of their arguments and are specified either in the form of specific formulae with the help of which we can calculate their values at any point or at any moment of time, or in the form of tables for a particular digital set of arguments. The boundary conditions in the statements of the Stefan problem most commonly encountered in geocryology are mixed: the conditions of type I are specified on the upper boundary as a rule, while the conditions of type II are specified on the lower one, i.e. thermal flux from the Earth's interior is given.

It is rather difficult to solve the Stefan problem even in the most simple form. The exact analytical solution, i.e. the solution in the form of formulae for calculating the position of the freezing (thawing) front and the temperature field for the particular moment of time, has been obtained up to now in individual cases only. In general the problem is solved numerically using rather powerful computers.

In the case when the temperature field does not vary with time, i.e. when the field is permanent, $\partial t / \partial \tau = 0$, the heat conduction equation 'regenerates' into the Laplace equation:

$$\frac{\partial^2 t}{\partial z^2} = 0$$

or in a multidimensional case:

$$\frac{\partial^2 t}{\partial x^2} + \frac{\partial^2 t}{\partial y^2} + \frac{\partial^2 t}{\partial z^2} = 0$$

If the conditions of type I are specified on the boundary of the region under investigation, it is termed a *Dirichlet problem*. If the conditions of type II are specified, such a problem is termed a *Neumann problem*.

Statement of the problem of soil freezing (thawing) over a range of temperatures (with the development of a freezing zone)

Investigations of the freezing (thawing) processes in soils (dusty sands, sandy silts, silty-clays, clays, etc.) show that the above mathematical model turns out to be too rough. Actually, a great amount of bound water is typical of such soils. As was shown above only free water freezes at temperatures close to $0°C$, while the bound water transforms into ice over a range of negative temperatures and these phase transitions turn out to be fundamental, practically down to temperatures of -5 to $-7°C$. Thus freezing soils can be divided by convention not into two zones (unfrozen and frozen) with a clear dividing line as in the previous case, but into three zones: unfrozen, freezing and frozen (see Fig. 1.14b). The interface between the unfrozen and freezing zones is the zero isotherm (strictly speaking, the temperature of the beginning of free water freezing t_{fre}, which turns out to be slightly below zero). All the free water freezes on this boundary. The freezing/frozen zones interface is less distinctive and corresponds to the temperature at which less-pronounced phase transitions come to a close. This temperature for various soils is in the range from -3 to $-7°C$. Within the freezing zone the phase transitions proceed with an intensity which falls with decreasing negative temperature and which can be determined from the curve of the unfrozen water content (see Fig. 1.8). This fact can be regarded as the presence of uninterruptedly distributed heat sources (or sinks):

$$q(z, \tau) = 334\frac{\partial W_{\mathrm{unf}}(t)}{\partial \tau} = 334\frac{dW_{\mathrm{unf}}(t)}{dt} \cdot \frac{\partial t}{\partial \tau} \qquad (1.23)$$

Phase transition of free water proceeds at the phase interface (between unfrozen and freezing zones) where the temperature is constant and equal to that of the beginning of phase transition. Therefore in this case the specific heat of phase transitions on this boundary should be understood as the value $Q_{\mathrm{ph}} \equiv 334[W_{\mathrm{vol}} - W_{\mathrm{unf}}(t_{\mathrm{fre}})]$ where W_{vol} is the initial (volumetric) soil moisture content; t_{fre} is the temperature of the beginning of freezing (or the temperature of the beginning of unfrozen water crystallization t_{cr}).

The essential dependence of thermal physical characteristics of soils in the freezing zone on the temperature, $\lambda_{\mathrm{fr}} = f(t)$, $C_{\mathrm{fr}} = \phi(t)$, is of great importance when approaching a problem, because the relationship between ice and unfrozen water in this zone is not constant and depends on temperature. This situation complicates the problem because the parameters which are responsible for the temperature field themselves depend on the required temperatures. So, mathematical formulation of the problem for the case of soil freezing in the range of negative temperatures takes the form:

$$C_{fr}(t_{fr})\frac{\partial t_{fr}(z, \tau)}{\partial \tau} =$$

$$\frac{\partial}{\partial z}\left[\lambda_{fr}(t_{fr})\frac{\partial t_{fr}(z, \tau)}{\partial z}\right] + 334\frac{dW_{unf}(t_{fr})}{dt}\cdot\frac{\partial t_{fr}}{\partial \tau},$$

$$0 < z < \xi(\tau);$$

$$\frac{\partial t_{unf}(z, \tau)}{\partial \tau} = a_{unf}\frac{\partial^2 t_{unf}(z, \tau)}{\partial z^2}, \quad \xi(\tau) < z < l$$

(1.24)

The next conditions are fulfilled on the moving boundary between the unfrozen soil and the freezing zone:

$$t_{fr}(\xi(\tau), \tau) = t_{unf}(\xi(\tau), \tau) = t_{fre},$$

$$\lambda_{fr}(t_{fre})\frac{\partial t_{fr}(z, \tau)}{\partial z}\bigg|_{z=\xi} - \lambda_{unf}\frac{\partial t_{unf}(z, \tau)}{\partial z}\bigg|_{z=\xi} = Q_{ph}\xi'$$

(1.25)

The boundary conditions are similar to those considered in the previous statement.

Formulation of the problem can be written in a more compact form, if the concept of effective heat capacity: $C_{ef} = C_{fr}(t_{fr}) + 334\,(dW_{unf}(t_{fr}))/dt$ is introduced. Then the equation for the frozen zone will take the more 'usual' form:

$$C_{ef}(t_{fr})\frac{\partial t_{fr}(z, \tau)}{\partial \tau} = \frac{\partial}{\partial z}\left[\lambda_{fr}(t_{fr})\frac{\partial t_{fr}(z, \tau)}{\partial z}\right], \quad 0 < z < \xi(\tau)$$

(1.26)

The solution to this problem is more complex than that in the first statement and is carried out by numerical methods using fast computers.

Both the statements considered take into account the conductive heat transfer only, with exclusion of the possibility of other modes of heat transfer (notably the convective one). In addition it is assumed in these models that all the water (both free and bound) freezes at the point of its occurrence, i.e. the possibility of ground water transfer by migration or filtration is excluded from consideration. Therefore there is no way, using these statements, to calculate the process of ice lens formation, i.e. to study the principles of the freezing soils cryogenic texture and structure formation. At the same time the solution of this problem is of great importance from theoretical and practical points of view. Suffice it to say that cryogenic texture is in many ways responsible for the strength properties of frozen soils used as foundations. Ground heaving in the course of freezing and thawing etc. is also associated directly with the processes of migration. Therefore

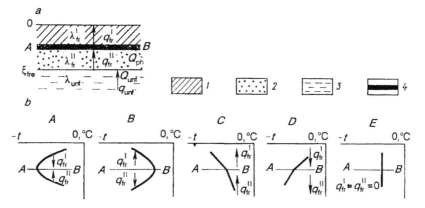

Fig. 1.15. The thermal-physical conditions for the formation and change of position of a segregated ice interlayer in the soil: a – diagram of small ice layer, AB, in the freezing soil; b – the direction of heat fluxes in the vicinity of the small ice layer AB and their dependence on the trend of the temperature curve; q_{unf}, q_{fr}^{I}, q_{fr}^{II} are the heat fluxes in unfrozen, frozen and freezing zones, respectively; 1–3 – soil (1 – frozen, 2 – freezing; 3 – unfrozen); 4 – the segregated ice layer.

investigation of the soil freezing (thawing) process in the context of the conductive problem turns out to be inadequate for the complete description of this phenomenon.

Statement of the problem of soil freezing with regard to moisture migration into the freezing zone

The process of moisture migration from the unfrozen portion of the profile into the freezing zone is often observed in soils in natural conditions. Penetrating into this zone, the bound water freezes out forming ice inclusions including those in the form of ice layers. When this process proceeds actively enough, ground heaving (surface deformation) is observed, i.e. the upper boundary does not remain stationary but moves upward. The proper allowance must be made for this in the mathematical formulation of the problem. Thus it is essential that the heat-moisture problem should be solved with regard to the possible ground heaving when considering the process of slow freezing of sandy-silty and silty-clay soils when moisture migration into the freezing zone is a possibility. This problem is very complex and even the physical model of this process has not been conclusively formulated to date, therefore each of the mathematical statements of this problem put forward by various authors has its essential disadvantages.

As a rigorous solution to the problem of soil freezing (thawing) with regard to moisture migration and the accompanying set of physico-chemical and mechanical processes is still under development, it is appropriate to consider the principles of the thermal physical aspect of the soil freezing process with regard to moisture migration and ice layer formation itself.

Water freezing in soils is possible only with the removal of the heat of crystallization i.e. with the availability of the appropriate temperature gradients in the freezing portion of soil (Fig. 1.15a, layer $0 - \xi_{fr}$). When high temperature gradients occur the freezing boundary (ξ_{fr}) propagates downwards rapidly while the moisture contained in the underlying soil layer has no time to migrate into the freezing layer. As a consequence, the pore moisture in the soil is fixed by phase transition into ice *in situ* and massive cryogenic structure is formed. The temperature gradients in the freezing portion decrease with increasing freezing depth allowing a sufficient amount of migrational water to arrive from the unfrozen portion of the soil into the freezing zone. Freezing of this water is distributed uniformly within the freezing portion of the soil or locally in the form of segregated ice interlayers (horizon AB), and it proceeds with liberation of a great amount of crystallization heat Q_{ph}. If the heat flux q_{fr}^{I} becomes equal to the sum of the heat flux q_{fr}^{II} being removed from the freezing soil and the heat being liberated in the course of phase transition, then the condition for ice lens formation is fulfilled, the movement of the isotherm AB will stop and the growth of a horizontal ice layer will be observed. In other words, the following thermal-physical condition will be fulfilled:

$$q_{fr}^{I} - q_{fr}^{II} = Q_{ph}\xi_{st}' \tag{1.27}$$

where ξ_{st}' is the rate of the ice lens (layer) growth. If a further lowering of negative temperature at the surface and the corresponding increase of the removed heat q_{fr}^{I} takes place, the left-hand side of the equation will be greater than the right-hand side and so the isotherm AB will begin to propagate downward resulting in the ice lens growth slowing down and ceasing. And only if the condition (1.27) is fulfilled at some new level (situated below the boundary AB), will the formation and growth of a second segregated ice layer be a possibility. It was thought previously that the condition (1.27) can and must be fulfilled at the freezing boundary only, because the unfrozen water migration is practically absent in the freezing portion of the soil (above the freezing front). However it has now been proved by way of experimentation that the unfrozen water migration can be great and sufficient for the thermal physical condition (1.27) to be fulfilled, and consequently, for segregated ice layers to be formed at negative temperatures inside the freezing layer. At the same time the lower the temperature gradient in the frozen portion of the soil the closer to the freezing boundary is the isotherm of ice layer formation. Both the formation of a new segregated ice layer at quasi-stationary positions of isotherms (fulfilment of the condition (1.27)) and the transformation or even disappearance of ice layers formed before are a possibility with changes in temperature distribution within the freezing portion of the soil. As this takes place, the moisture can migrate from one layer to another providing a way for one layer to grow at the sacrifice of another.

According to B.N. Dostovalov's studies five cases of thermal conditions on the boundaries of segregated ice layers and, respectively, of dynamics of their growth and change in their position in the soil are a possibility in accordance with the temperature gradients and form of the temperature curve (see Fig. 1.15b). Thus in case A the temperature minimum falls on the layer AB. Under this condition the thermal fluxes q_{fr}^{I} and q_{fr}^{II} and the migration moisture fluxes are toward the layer which can increase in thickness from the top as well as from the bottom. In case B the layer AB is situated at the temperature curve maximum. The heat fluxes q_{fr}^{I} and q_{fr}^{II} as well as the moisture migration fluxes are away from the layer upward and downward resulting in the segregated ice layer decreasing in thickness. In cases C and D the fluxes q_{fr}^{I} and q_{fr}^{II} have the same direction, with the moisture migrating in the same direction too. The segregated ice layer will respectively decrease from the top and grow from the bottom (case C) or grow from the top and decrease from the bottom (case D). In case E the temperature curve has no gradient, resulting in eliminating the possibility of heat transfer as well as of migration water transfer.

1.5 Methods for solving soil freezing (thawing) problems and approximate formulae for freezing and thawing depth calculations

The first attempts to solve the heat conduction problem taking into consideration the liberation of heat by phase transition at the moving phase boundary were undertaken in 1831 by the physicists Zh. Lyame and P. Clapeyron, members of the Russian Academy of Sciences. The problem of calculation of the depth of soil freezing had been solved for the simplest case by L.Zaal'shyuts as early as 1862. He obtained the simple formula for calculations, which has become well known as the 'Stefan formula', because the Austrian mathematician I. Stefan contributed significantly to the solution of this problem.

The exact solution of the Stefan problem was obtained for the first time by L.I. Rubenshteyn in 1947; however, this solution was not widely used because of the complexity of realization. In succeeding years V.G. Melomed's work (1957–1974) played an important role in setting up and solving the problem of soil freezing. Concurrent with this, approximate methods for solution of the Stefan problem were developed. L.S. Leybenzon, M.M. Krylov, V.S. Kovner, V.S. Luk'yanov, G.V. Porkhayev, I.A. Zolotar', G.M. Fel'dman made essential contributions in this direction. V.A. Kudryavtsev has made a great contribution to the solution of these problems. He put forward the approximate formulae for calculating the depth of seasonal and perennial freezing and heat storage in conditions of periodic temperature fluctuations at the ground surface which are especially valuable for the solution of many theoretical and practical geocryological problems. Let us consider some of the outlined problem solutions.

Stefan formula for determination of the depth of seasonal and perennial freezing (thawing)

The so-called Stefan formula is used for approximate calculations of the depth of ground freezing, given the thermal physical data and temperature conditions at the surface. It has been obtained using the following assumptions:

1) that consideration is being given to a homogeneous semilimited medium, the temperature of which is uniform with depth and equal to that of the phase transition, at the initial instant of time: $t(z,0) = \text{const} = t_{\text{fre}}$ at $z > 0$,

2) a constant temperature is specified for an initial instant of time on the surface and is maintained, $t(0,\tau) = \text{const} = t_0$, for the case of freezing $t_0 < 0$, and the medium is in the unfrozen state at the initial instant of time,

3) all the phase transitions proceed at temperature t_{fre}, i.e. the case of freezing with formation of a phase boundary occurs, while ignoring the phase transition in a range of negative temperatures as well as the processes of moisture migration in the course of freezing,

4) that a far greater amount of heat is released in the course of freezing than that being released on account of soil heat capacity C_{fr} with decreasing temperature. This is the most important condition in this statement simplifying solution of the problem as much as possible. Formally, this assumption is written as $Q_{\text{ph}} \gg C_{\text{fro}}|t_0|$. As a consequence of this assumption the heat capacity in the frozen zone C_{fro} can be neglected. The dynamics of the temperature field in the frozen zone are simplified sharply in this case; the freezing boundary $\xi(\tau)$ moves slow enough for the temperature field in the frozen zone to tend to come into the stationary state which is characterized by the linear pattern of temperature change with depth (Fig. 1.16a):

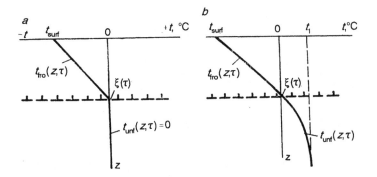

Fig. 1.16. Diagram of the conditions for the Stefan (*a*) and Leybenzon (*b*) formulae.

$$t_{\text{fro}}(z,\tau) = t_0 - \frac{t_0}{\xi(\tau)}z \tag{1.28}$$

It is easy to obtain the pattern of propagation of the freezing front m(q) in this case from the Stefan condition for the moving boundary:

$$\lambda_{\text{fr}}\frac{\partial t_{\text{fro}}(z,\tau)}{\partial z}\bigg|_{z=\xi(\tau)} - \lambda_{\text{unf}}\frac{\partial t_{\text{unf}}(z,\tau)}{\partial z}\bigg|_{z=\xi(\tau)} = Q_\varphi\frac{d\xi}{d\tau} \tag{1.29}$$

where Q_{ph} the specific heat of phase transition. Since

$$\frac{\partial t_{\text{unf}}(z\ \tau)}{\partial z} = 0,$$

because $t_{\text{unf}}(z,q) = 0$ while

$$\frac{\partial t_{\text{fro}}(z,\tau)}{\partial z} = -\frac{t_0}{\xi(\tau)}$$

so substituting the calculated gradients into the Stefan condition (1.21) we obtain the standard first differential equation, $Q_{\text{ph}}\xi' = \lambda_{\text{fro}}t_0/\xi$, the solution of which is the following function:

$$\xi(\tau) = \sqrt{\frac{2\lambda_{\text{fro}}|t_0|\tau}{Q_{\text{ph}}}} \tag{1.30}$$

The formula for calculation of the thawing depth is derived in a similar manner; however, λ_{unf} should be substituted for λ_{fr} in this formula. As mentioned above this formula can be used for an estimate of the freezing (thawing) depth. It gives too large a value as a rule. At the same time, the closer the freezing (thawing) conditions are to the model used as a basis for deriving this formula, the more accurately it describes the process.

The value $|t_0| \tau = \Omega(\tau)$ is termed *the sum of freezing degree-hours* (in the case of soil freezing) or of *heating degree-hours* (in the case of thawing). The Stefan formula has been derived from the assumption of the temperature constancy ($t_0 =$ constant) at the ground surface; however, it can be used when the values t_0 are variable also. To do this it is essential to determine the value of freezing degree-hours or heating degree-hours (Ω_{fr}, Ω_{th}) in a time interval during which freezing or thawing proceeds from the seasonal temperature change on the surface. Then the expression (1.30) will take the following form:

$$\xi = \sqrt{\frac{2\lambda_{fro}\Omega_{fro}(\tau)}{\varrho_{ph}}} \tag{1.31}$$

Given the depth of freezing ξ_1, caused by the known sum of freezing degree-hours on the ground surface Ω, the calculation from the Stefan formula can be improved. Then for the same soil the depth of freezing ξ_2 as a result of the effect of another sum of freezing degree-hours Ω_2 will be:

$$\xi_2 = \xi_1 \sqrt{\frac{\Omega_2}{\Omega_1}} \tag{1.32}$$

L.S. Leybenzon's formula for determining the depths of seasonal and perennial ground freezing (thawing)

As noted above, the essential error of the Stefan formula is associated among other things with neglect of the heat flux flowing from the unfrozen zone in the upward direction because of the difference of the temperature of this zone from 0°C. The model accepted in Leybenzon's method takes account of this by entering the initial conditions in the form $t(z,0) = t_1$ where t_1 is the constant ground temperature (across the section) before the beginning of the freezing (thawing) process. In the case of freezing $t_1 > 0$. In addition it is suggested that in the course of freezing (thawing) the temperature changes with depth in the unfrozen zone in accordance with the law $\mathrm{erf}\left(\dfrac{z - \xi}{2a_{unf}\sqrt{\tau}}\right)$ where $\mathrm{erf}\, x = \dfrac{2}{\sqrt{\pi}}\displaystyle\int_0^x e^{-s^2}\, ds$ is the error integral, with $\mathrm{erf}(0) = 0$ and $x \to +\infty\ \mathrm{erf}(x) \to 1$. Otherwise the statement of the problem is in line with that in the course of Stefan formula derivation (see Fig. 1.16 b).

Thus, the temperature distribution in both zones is taken as:

$$
\left.\begin{array}{l}
t_{\mathrm{fr}}(z, \tau) = t_0 - \dfrac{t_0}{\xi(\tau)} z, \quad 0 < z < \xi(\tau) \\[4mm]
t_{\mathrm{unf}}(z, \tau) = t_1 \operatorname{erf}\left(\dfrac{z - \xi(\tau)}{2 a_{\mathrm{unf}} \sqrt{\tau}}\right), \quad z > \xi(\tau)
\end{array}\right\}
\tag{1.33}
$$

where $a_{\mathrm{unf}} = \lambda_{\mathrm{unf}}/C_{\mathrm{unf}}$ is the diffusivity of unfrozen soil. The problem is solved in much the same way as in the previous case by substituting the expressions for heat fluxes in unfrozen and frozen zones into the Stefan condition at the moving freezing boundary and integrating the standard differential equation obtained.

The solution for evaluating the freezing depth m(q) is of the form:

$$
\xi(\tau) = \sqrt{\frac{2 \lambda_{\mathrm{fr}}|t_0|\tau}{Q_{\mathrm{ph}}} + \frac{t_0^2 \lambda_{\mathrm{unf}} C_{\mathrm{unf}} \tau}{\pi Q_{\mathrm{ph}}^2}} - \frac{|t_1|}{Q_{\mathrm{ph}}} \sqrt{\frac{\lambda_{\mathrm{unf}} C_{\mathrm{unf}} \tau}{\pi}}
\tag{1.34}
$$

In the case of thawing λ_{unf} and λ_{fr} change places and C_{fr} is substituted for C_{unf} in the formula (1.34).

When this formula is compared with the Stefan formula, it is apparent that the values of freezing (thawing) depth calculated from the Leybenzon formula will be lower than those calculated from the Stefan formula for the same period of time. This is to be expected because the Leybenzon formula takes into consideration the non-zero temperature distribution in the unfrozen (frozen) zone before its freezing (thawing). This slows down the front propagation as some portion of heat energy is spent on the temperature lowering (raising) in this zone up to the value at the beginning of phase transition. As one would expect, the Leybenzon formula transforms to the Stefan formula, if the initial temperature in the unfrozen (frozen) zone is $0\,°C$.

Among the formulae for approximate calculations, that put forward by V.S. Luk'yanov and M.D. Golovko in 1957 taking into consideration the heat capacity of freezing (unfrozen) ground as well as the presence of insulating covers on the surface is also worth noting. These researchers had derived a transcendental equation with reference to ξ, the solution of which presents calculation difficulties. Therefore Golovko developed a nomogram allowing the solution to be obtained easily to an accuracy sufficient for practical purposes. In spite of the fact that Luk'yanov and Golovko's formula is very useful for engineering calculations, the essential weakness of this formula is the uncertainty in assigning some parameters.

By and large there exist many (more than a hundred) approximate, often semi-empirical formulae put forward by various researchers for calculating the freezing (thawing) depth. However all of them have been obtained assuming the constancy of temperature on the ground surface with time and, what is more important, they do not make allowance for the clear-cut association of the thermal-physical aspects of ground freezing (thawing) and the geologic-geographical nature of this phenomenon. The approximate formulae put forward by V.A. Koudryavtsev and published in the *General Geocryology* textbooks in 1967, 1978 and 1981 do not have these disadvantages.

V.A. Kudryavtsev's approximate formula for determining the heat turnover and seasonal ground freezing (thawing) depth

The attributes (used for classification) of the freezing (thawing) processes put forward by P.I. Koloskov and developed by V.A. Kudryavtsev are the required parameters for these approximate formulae. There are four input parameters: annual temperature amplitude at the ground surface A_0, the mean annual temperature at the base of the seasonal freezing (thawing) layer t^{ξ}_{mean}, ground composition and moisture content defining all thermal-physical characteristics such as λ_{fr} and λ_{unf}, C_{vol}, Q_{ph}. Kudryavtsev's formulae have been derived for the periodic steady-state temperature regime at the ground surface and represent generalized Fourier's laws. They allow consideration of temperature waves to be extended to media with phase transitions.

The formulae for determining the freezing (thawing) depth, m, and heat storage through the surface for the layer of seasonal freezing (thawing), Q_{surf}, and the heat storage in the ground underlying this layer, Q_m, are written as:

$$\xi = \frac{2\left(A_0 - |t^{\xi}_{\text{mean}}|\dfrac{\sqrt{\lambda TC}}{\pi}\right) + \dfrac{(2A_{\text{mean}}C\xi_{2C} + \xi Q_{\text{ph}})Q_{\text{ph}}\dfrac{\sqrt{\lambda T}}{\pi}}{2A_{\text{mean}}C\xi_{2C} + Q_{\text{ph}}\xi + \dfrac{\sqrt{\lambda T}}{\pi C}(2A_{\text{mean}}C + Q_{\text{ph}})}}{2A_{\text{mean}}C + Q_{\text{ph}}}$$

$$(1.35)$$

$$Q_{\text{surf}} = \sqrt{2}((A_0 - |t^{\xi}_{\text{mean}}|)\sqrt{\frac{\lambda TC}{\pi}} + \frac{(2A_{\text{mean}}C\xi_{2C} + Q_{\text{ph}}\xi)Q_{\text{ph}}\sqrt{\dfrac{\lambda T}{\pi C}}}{2A_{\text{mean}}C\xi_{2C} + Q_{\text{ph}}\xi + \sqrt{\dfrac{\lambda T}{\pi C}}(2A_{\text{mean}}C + Q_{\text{ph}})}$$

$$(1.36)$$

where

$$Q_\xi = \sqrt{2} |t_{mean}^\xi| \sqrt{\frac{\lambda TC}{\pi}} \tag{1.37}$$

$$A_{mean} = \frac{\dfrac{A_0 - |t_{mean}^\xi|}{A_0 + \dfrac{Q_{ph}}{2C}} - \dfrac{Q_{ph}}{2C}}{|t_{mean}^\xi| + \dfrac{Q_{ph}}{2C}} \tag{1.38}$$

$$\xi_2 C = \frac{2(A_0 - |t_{mean}^\xi|)\dfrac{\sqrt{\lambda TC}}{\pi}}{2A_{mean}C + Q_{ph}} \tag{1.39}$$

where C is the soil volumetric thermal capacity; T is a period, equal to 1 year for the seasonal processes of freezing (thawing); ξ is the unknown freezing (thawing) depth; t_{mean}^ξ is the mean annual ground temperature at the depth ξ; A_0 is the amplitude of the temperature sine-wave fluctuation at the ground surface. The above expressions hold good for determining the depths of seasonal freezing as well as of seasonal thawing of ground in the case when the thermal-physical characteristics of soils in the frozen and unfrozen state are equal. Kudryavtsev has also obtained the solutions for the case of various thermal-physical characteristics; however, this adds complexity to the calculations. Therefore they are averaged out and the reduced characteristics are used in practice. Thus,

$$\lambda_r = [\lambda_{unf}(A_0 + t_{mean}^\xi) + \lambda_{fr}(A_0 - t_{mean}^\xi 0]/2A_0 \tag{1.40}$$

while

$$C_r = (C_{unf}\tau_{unf} + C_{fr}\tau_{fr})/T \tag{1.41}$$

where τ_{unf} and τ_{fr} are the duration of the thawed and frozen states, respectively, in the ground.

The reduced formula for determining the freezing (thawing) depth represents the quadratic equation in ξ. The roots of this equation are real and opposite in sign, if the condition $A_0 > |t_{mean}^\xi|$ required for the seasonal freezing (thawing) process to proceed is fulfilled. From physical considerations the negative root is neglected. At the same time the advantage of

Kudryavtsev's formulae over all the others discussed above for the case when Q_{ph} is small should be noted here. As already noted, all the previous formulae were derived on the assumption that Q_{ph} is large ($Q_{ph} \gg C|t^{\xi}_{mean}|$), and lose their meaning at $Q_{ph} \to 0$, as can be easily verified. Koudryavtsev's formula works very well in this case, because at $Q_{ph} = 0$:

$$\xi = \sqrt{\frac{\lambda T}{\pi C}} \ln \frac{A_0}{|t^{\xi}_{mean}|} \qquad (1.42)$$

—exactly in line with the Fourier solution for a medium without phase transition. The value ξ represents the depth of penetration of the zero isotherm or the depth at which $A_{\xi} = |t^{\xi}_{mean}|$. Thus Koudryavtsev's formula transforms to Fourier's formula in the limiting case of $Q_{ph} \to 0$. The fact that this transforms to the Stefan formula in the other limiting case (at great Q_{ph} and the zero initial conditions, i.e. with the assumptions for which the Stefan formula has been derived), as shown by V.Ye. Romanovskiy, is a further remarkable property of this formula. Thus, Kudryavtsev's formula not only is in line with the data which are obtained in the course of permafrost surveys as far as its basic parameters are concerned, but also is applicable at practically all the values of Q_{ph}. This is a unique property for the whole family of approximate formulae for calculating m.

Although Kudryavtsev's formula is the quadratic equation in ξ, the coefficients in this equation are very complex and it is rather difficult to carry out any calculations using this formula without special computers (minicomputers). Therefore nomograms, presented widely in literature, allowing determination of ξ from the predetermined $t^{\xi}_{mean}, A_0, \lambda_{fr}, \lambda_{unf}, C_{fr}, C_{unf}, Q$ were calculated on the basis of this formula. However such a form for presenting the calculation results has proved too cumbersome (three sets of nomograms have been constructed, with each of them including a set of seven nomograms) and involves the difficulty of interpolating the values C_{fr}, C_{unf} and Q_{ph}. Therefore Romanovskiy contructed a new nomogram through identical manipulation of Kudryavtsev's formula as a starting point. It turned out that this formula gave the more simple expression:

$$\xi = \sqrt{\frac{\lambda T}{\pi C}} \xi^* \qquad (1.43)$$

where a dimensionless value ξ^* depends only on two dimensionless variables:

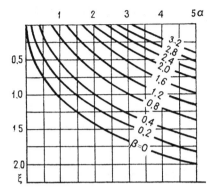

Fig. 1.17. V.Ye. Romanovskiy's nomogram for determining the depth of the seasonal ground freezing (or thawing) ξ from Kudryavtsev's formula.

$$\alpha = \frac{A_0}{Q_{ph}/2C} \quad \text{and} \quad \beta = \frac{|t^{\xi}_{mean}|}{Q_{ph}/2C}.$$

Such manipulation allows calculations with Kudryavtsev's formula that take 25–30 s using a microcalculator. The range of α and β changes in the new nomogram (Fig. 1.17) which includes all the possible combinations of values A_0, t^{ξ}_{mean}, Q_{ph}, λ_{fr} and λ_{unf}, C_{fr} and C_{unf}, for which the previous (Koudryavtsev's) nomograms were constructed. Given the mean perennial temperature t_{per} on the surface and the amplitude of the perennial cycle of temperature fluctuations A_{per}, Koudryavtsev's formula can be used for calculating the depth ξ_{per} of the perennial ground freezing also.

The problem is complicated by the fact that the mean perennial temperature at the depth $\xi_{per}(t_{\xi per})$, the value of which enters into Kudryavtsev's formula for calculating the value ξ_{per}, is in its turn linearly dependent on ξ_{per}. The value of the geothermal gradient g serves as a proportional constant in this case:

$$t_{\xi per} = t_{per} + g\xi_{per} \tag{1.44}$$

With some simplifications Kudryavtsev's formula can be presented in this case as:

$$\xi_{per} = \xi_{2C} + \frac{\left(2A_{mean}\xi_{2C} + \dfrac{Q_{ph}}{2C}\xi_{per}\right)\dfrac{Q_{ph}}{2C}\sqrt{\dfrac{\lambda T}{\pi C}}}{\left[A_{mean}\xi_{2C} + \dfrac{Q_{ph}}{2C} + \sqrt{\dfrac{\lambda T}{\pi C}}\left(A_{mean} + \dfrac{Q_{ph}}{2C}\right)\right]\left(A_{mean} + \dfrac{Q_{ph}}{2C}\right)} \tag{1.45}$$

where

$$A_{\text{mean}} = \tfrac{1}{2}(A_{\text{per}} + |t_{\text{per}} + g\xi_{\text{per}}|)\frac{\sqrt{\lambda TC}}{\pi} \tag{1.46}$$

$$\xi_{2C} = \frac{2(A_{\text{per}} - |t_0 + g\xi_{\text{per}}|)\sqrt{\dfrac{\lambda TC}{\pi}}}{2A_{\text{mean}}C + Q_{\text{ph}}} \tag{1.47}$$

This equation represents a fourth order equation in ξ_{per}. Its solution can be obtained by numerical methods.

2

Water transfer and ice formation in soils

2.1   Nature and mechanism of moisture migration in soils

Water migration in unsaturated soils is due to a complex mass transfer mechanism and a variety of water exchange driving forces. For geocryological problems the most interesting is the migration of bound and capillary water and vapour. Seepage (movement of free or gravitational water) in fine-grained materials is of minor significance and will not be discussed further. Water migration and vapour transfer in soils are related to the solution of many problems in earth science, engineering geology, pedology and geocryology (absorption and evaporation of water from the soil surface, its resorption by surrounding soil layers, capillary replenishment of soil, water migration towards the front of freezing and cooling, etc.). Thermodynamically, *water and vapour migration* in soil follow from the disequilibrium of the soil-water system caused by change in time and space of thermodynamic parameters (temperature, pressure, ion concentration, humidity, electrical, magnetic and gravitational potentials, etc.). It is usually impossible to measure directly the driving force of each mechanism separately. This is the reason to find a uniform (generalized) force comprising more or less fully all component forces. All this has resulted in an energy (thermodynamic) approach to mass transfer in colloidal and capillary porous bodies including the soil system.

All the water in soils, with the exception of free (gravitational) water, is held due to the free surface energy of the mineral soil skeleton E_s. Under the interaction of minerals with water, or more precisely, with a water solution, a part, E_w, of this energy is attributable to the bonding of ions of double electrical layers to water molecules. The difference $E_s - E_w = E_u$ is the part of the surface free energy of the soil system not spent on the interaction with water solution. Evidently the basic driving force of water transfer in the liquid phase (i.e. water migration in soil) is the gradient of this energy, E_u. This is a value of specific Gibbs free energy often called the *absolute chemical* or *isobaric-isothermal potential* of bound water μ'_ω. Because the absolute values of many thermodynamic functions, μ'_ω included, are not measurable, what is sought is not the absolute, but the relative thermodynamic potential of bound water: $\mu_\omega = \mu'_\omega - \mu_o$, where μ_o is the absolute chemical potential for free water. Since $\mu'_\omega < \mu_o$, μ_ω is a negative value.

The *relative thermodynamic potential* of soil water represents comprehensively the free energy reduction of free water when it interacts with a solid body. It is the work done in reversibly and isothermally converting 1 g of free water into bound water. Moisture potential, characterizing the energy of bound water in soil, is measured in units of work with respect to a unit of water mass (e.g. $J\,kg^{-1}$, $J\,mol^{-1}$ etc.).

The thermodynamic potential of soil moisture is a sum of particular potentials

$$\mu_\omega = \psi_\omega + \psi_o + \psi_z + \psi_p + \psi_e + \psi_m + \ldots \tag{2.1}$$

where ψ_ω is the matrix (capillary-adsorption) potential, i.e. the work spent on conversion of unit water mass from a solution, identical with that of the soil, to bound water (this potential comprises sorption and meniscus phenomena); ψ_o is the osmotic potential, i.e. the work spent on transference of unit of water mass from a volume of pure water to a volume containing a solution identical in composition and structure with that of the soil; ψ_z is the gravitational potential, i.e. the work needed to transfer a solution, similar to that of soil, from one elevation to another; ψ_p is the hydrostatic (or external gas pressure) potential expressing the work done on the soil due to the action of external pressure; and ψ_e, ψ_m are the electrical and magnetic water potentials respectively.

Note that a rise in temperature results in a higher water potential. This is associated with the increasing translatory movement of water molecules and a reduction in the energy of bonding with the soil matrix. Because the potential is characterized by a negative value, its algebraic value rises while its absolute quantity diminishes.

The soil-water potential μ_ω depends on the soil-water content W_{vol} ($g\,cm^{-3}$). By analogy with volumetric heat capacity for a temperature field, A.V. Lykov has introduced the notion of volumetric isothermal mass capacity or *differential soil-water capacity* C_ω showing how much water is to be added to the soil to change the water potential by a unit amount. It is a variable which depends on the soil-water potential, and on the composition, structural and mechanical characteristics of the soil and it is calculated from the relation:

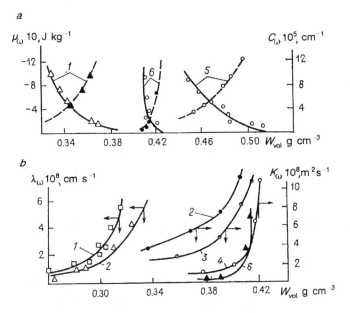

Fig. 2.1. Dependence of moisture potential μ_ω and volumetric differential moisture capacity C_ω (dashed line in *a*), and (in *b*) hydraulic conductivity coefficient of ground water λ_ω and diffusion coefficient K_ω, on water content of soils: 1–2 – clay-silty sand (1 – light; 2 – heavy); 3 – medium silty sand; 4 – clay; 5 – kaolinite clay; 6 – hydrous mica montmorillonite clay with $\rho_d = 1.56\,\mathrm{g\,cm^{-3}}$.

$$C_\omega(T) = \frac{\partial W_{\text{vol}}}{\partial \mu_\omega} \tag{2.2}$$

where T is the absolute temperature. Soil-water potential and differential soil-water capacity can vary by one or two orders of magnitude (Fig. 2.1a).

Relating to water potential is the notion, introduced by B.V. Deryagin, of wedging pressure of thin films of bound water P_{wed} which is the pressure difference observed in crossing the flat interface surface from the bound water, present in a thin layer, to the adjacent water, and equal to:

$$P_{\text{wed}} = \frac{\mu_\omega^w - \mu_\omega^{bw}}{V_m} \quad \text{or} \quad P_{\text{wed}} = -A\frac{1}{h^m} \tag{2.3}$$

where V_m is the molar volume of water; μ_ω^w and μ_ω^{bw} are the full thermodynamic potentials of the ground water and film-bound water respectively; A and m are coefficients representing characteristics of the mineral surface and water solution respectively; h is the thickness of the bound water films between two soil particles. The value of P_{wed} is negative.

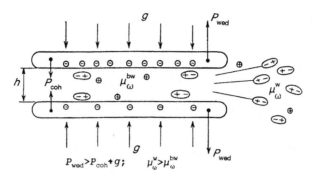

Fig. 2.2. Diagram of wedging action of thin films of bound water.

In the case when the absolute value of P_{wed} is greater than the interaction forces between two mineral particles P_{coh} and the external load on them, g, the particles will move away from each other due to the wedging effect of the film water accumulating under the action of the surface forces of the mineral particles (Fig. 2.2). In this case, the condition $\mu_\omega^w > \mu_\omega^{bw}$ with $|\mu_\omega^w| < |\mu_\omega^{bw}|$ is always met. This is what ensures water migration into the intersurface gap between particles until the potentials are equalized and a new equilibrium established in the system. In the case of $P_{wed} < P_{coh} + g$, even if $\mu_\omega^w > \mu_\omega^{bw}$ the particles fail to move apart. With the external load being greater than the wedging pressure P_{wed} and the repulsive force P_r between particles $(g > P_{wed} + P_r)$, the particles will draw closer and water will flow out of the intersurface gap. From this standpoint the pressure of swelling clayey soils should be regarded as resulting from the wedging effect of the thin films of bound water.

Water migration can occur in soils in the capillary form (capillary water proper) and in the film form (weakly bound water). Capillary water migration occurs mainly in the molar (volumetric) way due to meniscus (Laplace) forces. In frozen soils migration of bound water, occurring usually in the film form, is very important. In the case of the portion of the film transfer being comparable with that of the capillary water transfer in soil micropores, a mixed capillary-film transfer mechanism can be said to occur. The film mechanism of water migration becomes increasingly important with diminishing soil water content. Characteristically, soil water migrates in the films not as a compact mass (volume migration), but from particle to particle (molecular diffusion migration). Naturally, the film water migration velocity and, consequently, the water migration flow density, prove substantially lower than for the capillary water transfer.

Molecular-kinetic bound water migration in a dispersed soil is due, in the presence of an inhomogeneous force field, to the difference between the jumps of water molecules in forward and in backward directions along the force gradients. Migration by jumps is self-diffusion and is called translatory. For translatory jumps from node to node in the lattice to occur, it is necessary that the kinetic (vibrational-rotational) energy of the particle E_k be greater than that of its bonding energy (interaction) with other particles E_b, i.e. $E_k > E_b$. Note that heating increases the kinetic energy of particles and reduces the bond energy between them.

Calculated data have shown that each water molecule, at a temperature of 25 °C, for example, makes some 6×10^8 jumps per second. The jumps are separated from each other by time intervals of 1.7×10^{-9} s, i.e. the time of a molecule staying near the equilibrium centre. A water molecule makes in its equilibrium stay time (between jumps) some 1000 vibrations. Generally the frequency, j, of molecular jumps per second, is defined by the formula:

$$j = j_o e^{\frac{E_b}{RT}} \tag{2.4}$$

where j_o is some coefficient relating to the vibration frequency of particles near the equilibrium state; T is the absolute temperature in Kelvin; e is the natural logarithm base.

The mean time of a particle vibration near one equilibrium centre is the reciprocal of j, i.e. $\tau = 1/j$. In addition to the migration form of individual molecules in liquids and gases discussed above, there are also group molecule migrations pushed by surrounding particles. Such associations of molecules perform Brownian movement adding to the translatory migration of separate H_2O molecules.

The above concepts about the translatory movement of H_2O molecules and groups of molecules relate to chemically pure, bound water. They failed to take into account either the *ion hydration*, i.e. interaction of electrolyte ions with water molecules, or the effect of the binding of water molecules to active centres of mineral soil particle surfaces. Consequently, the bond energy of each water molecule should be increased by the energy of bonding of the H_2O molecule (ΔE_i) with an ion, on the one hand, and by the energy of binding a molecule with the mineral surface (ΔE_m), on the other. Thus, bound water, containing ions of the electric double layer, has the frequency of translatory H_2O molecule jumps reduced and this is expressed as:

$$j = j_o e^{\frac{E_b + E_i + E_m}{RT}} \tag{2.5}$$

Thus the migration of bound water in dispersed soils can be viewed as the difference of translatory H_2O molecule jumps in forward and backward directions. The rate of translatory molecule jumps, according to equation

(2.5), goes up with a rise in the temperature and a drop in the energy of bonding of water molecules to the mineral surface of particles or to the diffuse layer ions, i.e. with the increased film water mobility. Thus, with other conditions being equal, bound water will migrate from the thick films, where molecules are less bound, to the thin films, from the films of higher temperature (where molecule mobility is greater) to the films of lower temperature, from the films of lower concentration of dissolved salts to the films of higher ion concentration, etc. The intricacies of investigating (molecular-kinetically) the microscopic pattern of water migration and the impossibility of making quantitative estimates on this basis to forecast the water-transfer process in dispersed soils, leads us to apply generalized thermodynamic (phenomenological) laws. By analogy with such phenomena as the transfer of heat and electricity (Fourier's and Ohm's laws), vapour diffusion and water seepage (Fick's and Darcy's laws) and others, the intensity of the bound water migratory flow I_ω $(g(cm^2 s)^{-1})$ is taken to be, under the steady-state regime, directly proportional to the gradient of thermodynamic potential (full or partial):

$$I_\omega = -\lambda_\omega \operatorname{grad} \mu_\omega \tag{2.6}$$

where λ_ω is the coefficient of hydraulic conductivity $(cm\, s^{-1})$ equal to $C_\omega K_\omega$, while K_ω is the coefficient of soil water characterizing soil properties relative to the lag in (time for) the development of the potential field or of the distribution of moisture content $(cm^2\, s^{-1})$. The distinctive and substantial features of λ_ω and K_ω coefficients are, in addition to their dependence on dispersivity, chemical and mineral composition, and structural and textural characteristics, their variability with soil moisture variations and clearly non-linear dependence on water content (see Fig. 2.1b).

The above equation for the steady-state migratory water flow in soils is based on the thermodynamic water potential. It can be readily rewritten in terms of moisture content (with the simple relation between μ_ω and W assumed): $I_\omega = -K_\omega \operatorname{grad} W$.

In non-steady-state conditions for the water transfer regime in soils the following differential water transfer equations apply:

$$\frac{\partial W}{\partial \tau} = \frac{\partial}{\partial \chi}\left(\lambda_\omega \frac{\partial \mu_\omega}{\partial \chi}\right) \tag{2.7}$$

$$\frac{\partial W}{\partial \tau} = \frac{\partial}{\partial \chi}\left(K_\omega \frac{\partial W}{\partial \chi}\right) \tag{2.8}$$

2.2 Water transfer and ice formation in frozen soil

In terms of general thermodynamics, water transfer in frozen ground is due to the gradients of matrix-capillary, osmotic, temperature, electric and other potentials, which are components of the total thermodynamic potential of the unfrozen water and which result in the unfrozen water gradient (grad W_{unf}) in the soil system. An equation for the intensity of the steady-state unfrozen water flow in frozen ground is:

$$I_\omega^{fr} = -\lambda_\omega^{fr} \, \text{grad} \, \mu_\omega \tag{2.9}$$

where λ_ω^{fr} is the coefficient of conductivity of water in frozen soil. Where a simple relation is observed between the water transfer potential and the unfrozen water content of soil, the equation based on water content can be applied:

$$I_\omega^{fr} = -K_\omega^{fr} \, \text{grad} \, W_{unf} \tag{2.10}$$

where K_ω^{fr} is the diffusion coefficient of unfrozen water; W_{unf} is the unfrozen water content of frozen soil. Note however, the unfrozen water migration is due to the action of grad μ_{unf}, not of grad W_{unf}, which can only be used, as experiments have proved, in homogeneous soils.

Vapour transfer and the flow regime of vapour-gas mixtures in soil pores are mainly determined from the relation between the length of molecule free path ($l = 0.5 \times 10^{-5}$ cm) and sizes of pores – the soil capillaries of radius r. If $r < 10^{-5}$ cm, vapour transfer would normally obey Knudsen flow theory characterized by a molecular (effusion) transfer mechanism. The effuse vapour transfer mechanism is practically ineffective since the ultrapores and narrow sections of irregular capillaries, where vapour effusion can occur, are mostly filled with bound (unfrozen) water. With $r > 10^{-5}$ cm diffusion and molar vapour transfer occur, and with $r > 10^{-3}$ cm a viscous flow regime prevails, i.e. vapour transfer occurs in the molar (volume) way. The density of the vapour-gas mixture steady-state flow is determined by Poiseuille's law and proves directly proportional to the total pressure gradient and inversely proportional to gas viscosity. Molar vapour transfer is most important in large clastic rocks, in cavities and fissures of hard rocks and in karst holes. Vapour diffusion in micro- and macro-capillaries is due to molecular water migration resulting from the translatory movement of vapour molecules. Vapour diffusion (combining thermal diffusion, baric diffusion) is determined by Fick's law and caused by the gradient of saturated water vapour (grad d_{sat}) or the partial pressure gradient of vapour-air mixture (grad P). Such a vapour transfer mechanism is applicable mostly to water-unsaturated fine-grained soils. On the whole, the generalized equation for the steady-state vapour-flow intensity in frozen fine-grained soils can be given as:

$$I_v = - K_v^{fr} \operatorname{grad} d_{sat} = - \lambda_v^{fr} \operatorname{grad} P \qquad (2.11)$$

where λ_v^{fr} and K_v^{fr} are the coefficients of vapour conductivity and vapour diffusion respectively.

Water transfer and ice formation in frozen soil due to the temperature gradient effect

The temperature gradient, created and maintained in frozen soil results in the gradients of bound water potential (grad μ_ω), of water vapour potential (grad μ_v), and of saturated water vapour concentration (grad d_{sat}). These potentials result in turn in the migration of unfrozen water and vapour from regions of higher to those of lower water potentials, i.e. from regions of higher to those of lower freezing temperatures which have less unfrozen water and water vapour (18). The intensity of the total steady-state water flow in frozen soil is written in the moisture content form as follows:

$$I^{fr} = K_\omega^{fr} \operatorname{grad} W_{unf} + K_v^{fr} \operatorname{grad} d_{sat} = K_\omega^{fr}\delta_\omega^{fr} \operatorname{grad} t + K_v^{fr}\delta_v^{fr} \operatorname{grad} t \quad (2.12)$$

where δ_ω^{fr} and δ_v^{fr} are the thermal gradient coefficients of unfrozen water and saturated water vapour respectively with phase changes present. They are derived from curves $W_{unf} = f(t)$ as the relations $\Delta W_{unf}/\Delta t$ and $d_{sat}/\Delta t$. As follows from equation (2.12), the intensity of the water migration flow proves directly proportional to the gradient of freezing temperature and depends substantially on composition and structural and textural characteristics of soil and water as represented in the coefficients $K_\omega^{fr}, K_v^{fr}, \delta_\omega^{fr}, \delta_v^{fr}$.

As experiments show, the portion of vapour transfer becomes significant in the total water flow only when the soil pores are substantially empty of water and ice. For example, vapour transfer may be practically ignored in clay soil where [the coefficient of saturation] G is more than 0.5.

A gradient of temperature in freezing soil, results not only in water migration, but also in the formation of segregated ice and in complex physico-chemical and physico-mechanical processes. As water moves towards lower freezing temperatures the water flow in frozen ground diminishes, with the reduction of coefficients K_ω^{fr} and δ_ω^{fr}. This leads to the freezing out of the excess amount of unfrozen water in each subsequent frozen soil cross-section since its potential proves higher than that of the ice present. This freezing out of the excess amount of migratory unfrozen water is responsible for the additional ice formation and the gradual increase of ice

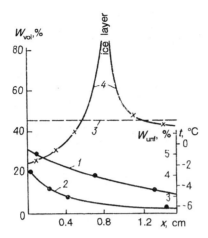

Fig. 2.3. The temperature (1), unfrozen water content (2), the total initial (3) and final (4) moisture contents along length x of frozen kaolinite clay sample without adjacent ice.

content of a frozen soil layer with a lower freezing temperature (Fig. 2.3). Evidently, the ice formation is most intensive at the section where there is a sharp decrease of density of the total water flow ($|\partial I/\partial x|$ = max), i.e. first of all in that range of freezing temperatures which has values of grad W_{unf} (or values K_ω^{fr} and δ_ω^{fr}) substantially reduced. The formation of ice, i, is calculated from the equation:

$$i = \frac{\partial I}{\partial x}\tau \tag{2.13}$$

where $\partial I/\partial x$ is the ice formation intensity.

Experiments showed that in the warmer part of the frozen soil a drop in the total initial moisture content occurred due to the steady outflow of unfrozen water. This was continually replenished in conformity with the rule of equilibrium W_{unf} content depending on t, by melting of pore ice. The marked loss of moisture from the warmer layer brings about intensive contraction resulting in a network of shrinkage fissures (Fig. 2.4a). Above this zone were some ice microstreaks which, gradually growing thicker and longer, and merging with each other, formed a continuous segregated ice layer (see Fig. 2.4b). The frozen sample expanded by the thickness of this ice layer, i.e. heaving deformation was recorded. Frozen samples of kaolinite clay (in comparison with samples of montmorillonite clay) showed, under similar values of grad t, a faster ice layer growth due to the presence in kaolinite clay of higher unfrozen water content gradients and higher coefficients of water diffusion which are responsible for higher density of flow.

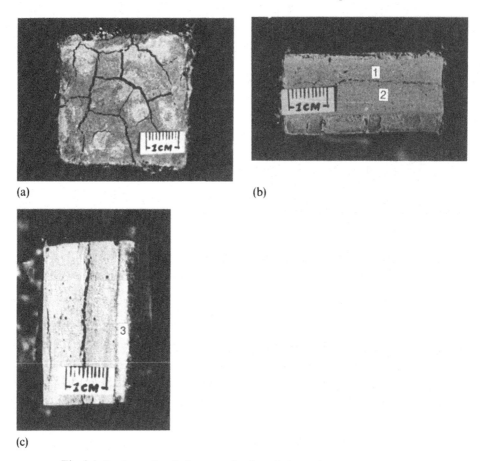

(a)   (b)

(c)

Fig. 2.4. Ice formation in frozen polymineral clay: *a–b* – without moisture inflow ($\tau = 8$ days, grad $t = 0.6\,°\mathrm{C}\,\mathrm{cm}^{-1}$) sections across and along the direction of heat and water flow; *c* – with moisture inflow from the ice plate ($\tau = 14$ days, grad $t = 0.6\,°\mathrm{C}\,\mathrm{cm}^{-1}$); 1 – ice layer; 2 – desiccated zone; 3 – ice plate.

On the whole, under 'closed' system conditions (with the sample not attached to a body of ice) water migration occurs only on account of the redistribution of moisture and this will decrease gradually. The density of water migration W_{unf} is substantially greater under the effect of grad t in 'open' system conditions (with ice adfrozen to a part of the frozen ground at a higher temperature), than in the 'closed' system, This results in the formation of a thicker segregated ice layer (see Fig. 2.4c). The flow of unfrozen water to this layer is not due to the loss of water from part of the frozen sample at a warmer temperature, but to the supply of water originating in the ice adfrozen with the sample (Fig. 2.5). As to the energy used, this process is obviously favoured since the quasi-liquid water film on ice crystal surfaces

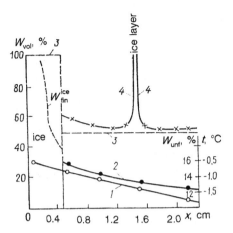

Fig. 2.5. Temperature (1), unfrozen water content (2), the total initial (3) and final (4) moisture contents along the length of frozen polymineral clay with adjacent ice (W_{fin}^{ice} is the final water content in the ice volume after the water outflow into freezing soil).

in the adfrozen ice will have a lower bond energy, and its thermodynamic potential value will be higher than that of the unfrozen water in the frozen soil sample. This variation of values of potential μ_ω has to result in unfrozen water migration from adfrozen ice to frozen soil. Consequently this ice will first of all replenish the unfrozen water storage in the warmer part of the sample, which is being expended in developing the segregated ice layer. The adfrozen ice loses mass and becomes loose, porous and light in colour (see Fig. 2.4c).

The above experimental results were obtained in less than a month. In field conditions where frozen ground is many thousands of years old and characterized by an intricate structure with alternating layers of segregated ice and minerals, it is natural to find as well some ice layers growing (with sufficient grad t available) at the expense of the others. In other words, ice layers formed under a lower freezing temperature will grow at the expense of higher-temperature ice layers due to the ice content redistribution. As a result, the original cryogenic structure of permafrost can change. Such a process is very likely in seasonally frozen soil layers, which have high temperature gradients. The segregated ice layers can change their thickness and new, less-thick ice layers may develop.

Water transfer in frozen ground and its interaction with air

With soils which are losing moisture under frost action, the external driving force of unfrozen water transfer is the difference of partial pressure of water vapour between the atmosphere P_m and the soil-air interface, P_{surf} (13).

As ice sublimates in frozen soil and water is transferred into the air, the soil gets a desiccated zone, recognized visually by its lighter colour (as compared with the lower soil layer not affected by sublimation). The front of sublimation (and the thickness of the desiccated zone) ξ_s ($\xi_s = \xi_d$) is most clear-cut in ice-saturated sands and not blurred as, for instance, in the case of clay soils. The ice sublimation intensity in soils ($I_s = f(\tau)$) is not constant in time; it naturally diminishes as the sublimation front goes deeper. Sublimation increases with an increasing proportion of clay and fine silt particles, of minerals of the montmorillonite group, of multivalent cations and with the degree of salinization.

This is associated with the increased W_{unf} towards the sublimation front, and consequently, with grad W_{unf} and the intensity of unfrozen water flow I_ω^{fr} (Figs. 2.6 and 2.7). The total water distribution in samples of soils is characteristically substantially different between sands and clays, which attests to the difference in the water transfer mechanism in coarse grained and fine grained soils (see Fig. 2.7).

In sands, which have practically no unfrozen water, moisture transfer occurs completely in the form of vapour transfer. In fine grained soils, which have a considerable amount of unfrozen water, evaporation is at work throughout all the volume of the zone of desiccation which can be sub-divided into layers with small or large gradients of total moisture content. Gradients of water content present in clay soils attest to the importance of an internal moisture transfer by unfrozen water migration towards the surface. The component of unfrozen water movement in clays can account for a considerable quantity, up to 50–70%, of the total moisture flow (vapour plus water). The curves of the distribution with depth of the total moisture in clay soil clearly display two distinctive water content points (see Fig. 2.7): near the surface of the sample (W_{cr}') which is some percent higher than the value of hygroscopic soil moisture at the temperature of the experiment; and on the sublimation boundary between soil layers with small and large gradients of the total moisture (W_{cr}''). These moisture values (W_{cr}' and W_{cr}'') remain practically constant when external conditions for the desiccation due to freezing of the clay soil remain unchanged. W_{unf} was found to be correlated with moisture content at the sublimation front (W_{cr}'') at given temperature, in soils of various composition, structure and proper-ties. This finding has proved to be the basis of a new method (the sublima-tion method) of estimating unfrozen water content of soils (9).

Ice sublimation intensity is limited under natural conditions not by exter-

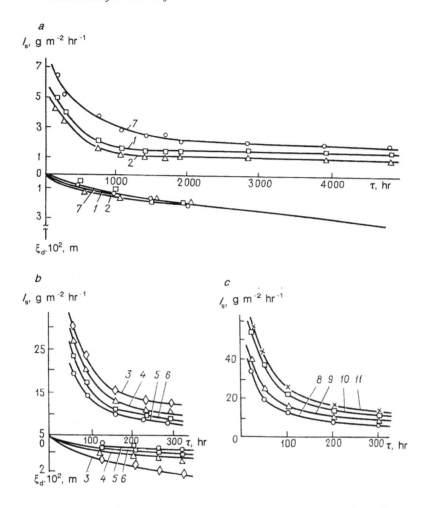

Fig. 2.6. Influence of granulometric (*a*) and chemical-mineral (*b,c*) soil composition on sublimation intensity I_s and the movement of the front of desiccation ξ_d: 1 – fine sand; 2 – medium clay-silt; 3–11 – clays of montmorillonite (3), hydromica (4), kaolinite (5), and polymineral (6–11) composition with differing concentration of $CaCl_2$ solution (6–8 – without $CaCl_2$; 9 – 1N sat., 10 – 2N sat., 11 – 3N sat.).

nal moisture exchange but by the internal. Thus one should treat the problem of moisture transfer independently and separately from the problem of heat transfer, in the desiccated soil layer. An equation for the intensity of the frost desiccation (ice sublimation) of soils is:

$$I_s = K_v \frac{P_{sat} - P_{ga}}{\xi_s} + K_w \gamma_o \frac{W_{unf} - W'_{cr}}{\xi_s} \qquad (2.14)$$

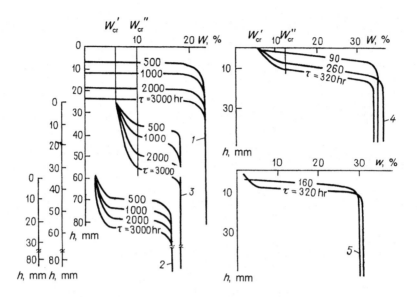

Fig. 2.7. Distribution with depth h of total moisture content as ice sublimation proceeds, in samples of soils of different granulometric and mineral composition: 1 – fine and clayey-silty sand; 2 – light sandy silty-clay mixed material; 3–5 – clay (3 – polymineral, 4 – hydrous mica, 5 – kaolinite).

where P_{sat} is the partial pressure of saturated vapour of the ice at the sublimation front; P_{ga} is the partial pressure of vapour at the 'ground – air' interface; $W_{unf} = W''_{cr}$ is the moisture content by weight, corresponding to the unfrozen water amount at a given temperature at the sublimation front; W'_{cr} is the equilibrium moisture content at the 'ground-air' interface: ξ_s is the thickness of the zone of desiccation. Equation (2.14) expresses the steady-state flow in the frost-desiccated soil on account of vapour transfer (first term) and liquid (second term) separately.

The velocity of the sublimation front in frozen soil, recorded visually from the changed hue of the desiccated soil and from the first inflection in the curve of the sample's total moisture content distribution with depth, is directly proportional to the value I_s and inversely proportional to the density of the soil matrix γ_0 and the initial total moisture content ($W_{in} = W_{nat}$) of soil. The thickness of the layer being desiccated ξ_s can be calculated from the equation:

$$\xi_s = \frac{\displaystyle\int_0^\tau I_s d\tau}{\gamma_0[W_{in} - 0.5(W'_{cr} + W_{unf})]} \tag{2.15}$$

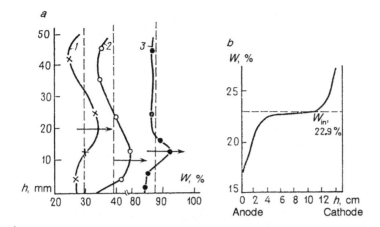

Fig. 2.8. Redistribution of moisture along the length of frozen samples (sample length h, mm, cm): a – as a result of shear in polymineral (1), kaolinite (2) and bentonite (3) clay (dashes show moisture prior to the experiment, arrows indicate the shear plane); b – in sandy-silty-clay after flow of electrical current.

Characteristics of moisture transfer and ice formation in frozen soils affected by gradients of mechanical stress, electrical fields and other external forces

Frozen soil masses have a *field of mechanical stress gradients* which originates and exists due to the natural and historical development of the soil and various engineering impacts. The difference of these stresses results in moisture migration from regions of higher to those of lower compressive pressures, from regions of lower to those of higher tensile or shear stresses. Widespread among the mechanical stresses arising in frozen ground are *shear stresses*. In the planes of shear, where the maximum stresses occur between soil particles with the minimum values of the thermodynamic potentials of moisture μ_ω, unfrozen water is under tensile stresses which result in its augmentation there under the effect of grad μ_ω (Fig. 2.8a). Clearly, this process occurs particularly under slow shear lasting for days or months.

The layers away from the shear zone lose water. The remaking of the structure and formation of ice micro- and macro-streaks occurs in the shear zone (Fig. 2.9).

The density of flow of the unfrozen water W_{unf}, can be written as follows:

$$I_\omega^{fr} = K_\omega^{fr} \operatorname{grad} W_{unf} = K_\omega^{fr} \frac{\partial W_{unf}}{\partial \sigma} \frac{\partial \sigma}{\partial x} \qquad (2.16)$$

Fig. 2.9. Cryogenic structure formed in the shear zone in kaolinite clay sample (initially of massive cryogenic structure).

where $\partial\sigma/\partial x$ is the stress gradient. Moisture transfer and ice formation in frozen soils affected by a shear stress gradient depend on soil composition. The thickness of the zone of ice formation and desiccation diminishes in transition from clays to silty-sands and from clays of kaolinite composition to polymineral and montmorillonite clays.

Moisture migration and ice formation in frozen ground affected by the gradient of electrical field

The moisture migration mechanism appears as follows. The electric field applied to the frozen soil sample disturbs the dynamic equilibrium of the liquid and solid phases of the water and results in the migration of hydrated cations of the electric double layer to the cathode. The bound water layers, which surround the cations, move in the same direction carrying along all other liquid.

These forces cause the migration first of the less bound water from anode to cathode. The equilibrium between water solid and liquid phases is disturbed but is reestablished by some ice melting and replenishing the water which has flowed from the anode region. The unfrozen water that enters the cathode region proves excessive, i.e. above the equilibrium value of water content W_{unf} at the given temperature. As a result it freezes out, enriching ice content and forming ice layers. An equation for the moisture flow in frozen soils due to the electrical field gradient is:

$$I_\omega = K_\omega^{fr} \frac{\partial W_{unf}}{\partial U} \frac{\partial U}{\partial x} \tag{2.17}$$

where $\partial U/\partial x$ is the electrical field gradient.

The moisture redistribution in frozen ground under the influence of the electric field is rather important (see Fig. 2.8b). In an experiment with a clay sample, after six days under a voltage of 2–3 V cm^{-1} and at a mean temperature of about $-2\,^\circ$C, heaving had occurred around the cathode and many thin ice layers (schlieren) had formed. These experiments have verified the linear dependence of the velocity of electrokinetic migration of unfrozen water on the electric field intensity and shown the existence of a threshold gradient below which moisture transfer is practically absent.

Moisture transfer and ice formation in frozen soil under the effect of osmotic forces (in the absence of temperature gradients)

This process is inseparably associated with salt diffusion in frozen strata and the migration of chemical elements. In fact, frozen soil interacting with salt solutions results in the simultaneous development of two interdependent processes: migration of salt ions and migration of unfrozen water. Both normal and reverse osmosis can take place. Under *normal osmosis* the moisture migration in the direction opposite that of ion flow results in osmosis involving the shrinkage of the soil sample. Yet this process occurs only in the interaction of frozen soil with highly concentrated solutions. More often than not frozen ground shows *reverse* osmosis of water, i.e. its migration in the same direction as the flow of the ions. This is due to the total thermodynamic potential of the solution being higher than that of the unfrozen water of the sample. The resulting driving force for the migration gives rise to the transfer of water molecules which are more mobile in the solution than in the film water bound by mineral particles. Water molecule migration from solution into frozen soil results in the formation of ice macro- and micro-segregations (Fig. 2.10). The sample of frozen soil is then somewhat strained, with its total moisture and salt content increased. As experiments performed at a temperature of $-4\,^\circ$C show, the contact of a saturated solution or of salt crystals with frozen samples of kaolinite clay results in osmotic transfer of unfrozen water from soil to the salt. Salt

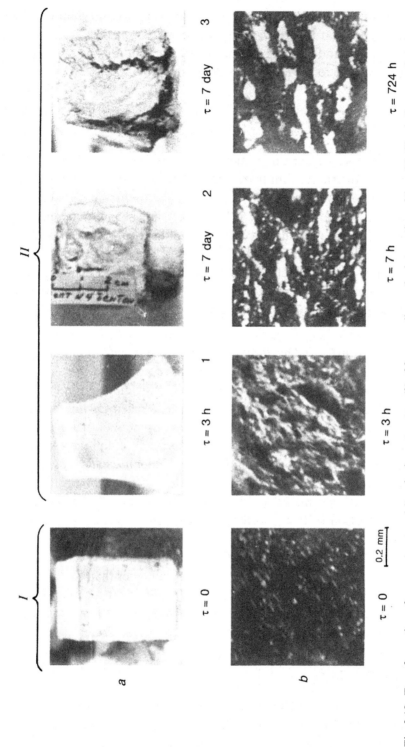

Fig. 2.10. Transformation of macro-structure (*a*) and micro-structure (*b*) of frozen soil samples interacting with an NaCl solution over time τ (according to Yu.P. Lebedenko and E.M. Chuvilin): *I* – soil samples prior to the experiment, $\tau = 0$; *II* – the samples after the experiment; 1 – kaolinite clay, deformed by mass exchange and ice segregation (NaCl solution 0.2N., $t = -1.5°C$); 2 – montmorillonite clay soaking and swelling (NaCl solution 0.4N., $t = -1.5°C$); 3 – polymineral sandy-clayey silt with the formation of rupture fractures (NaCl solution 0.4N., $t = -20°C$).

crystals thus dissolve, passing into solution, while the frozen soil near the contact loses moisture. As the concentration of the solution in contact with the frozen soil decreases, the normal osmotic flow diminishes as well and falls to zero when a critical concentration is reached. At this critical solution concentration the total thermodynamic water potential in soil is equal to the osmotic water potential in the solution and a thermodynamic equilibrium is attained. The critical concentration value for the solution C_{cr} will depend on the composition and structure of the particular soil, and also on external thermodynamic conditions. The value C_{cr} goes up with finer grain-size and lower temperature. For sands interacting with a NaCl solution, C_{cr} is under $0.1\,\mathrm{g\,mol\,l^{-1}}$ and for clays over $5\,\mathrm{g\,mol\,l^{-1}}$

2.3   Water transfer and ice formation in freezing and thawing soils

There are basically two different ways that soils freeze: with and without water migration. Soil freezing without water migration occurs either in the case of soils of low moisture content or a sufficiently rapid advance of the freezing boundary. For example, when a soil sample is frozen through quickly at temperatures of $-60\,^\circ\mathrm{C}$, $-70\,^\circ\mathrm{C}$, water freezes *in situ*, since the temperature field is by a factor of 10 more active than the moisture content field. Usually freezing in nature proceeds slowly enough for water to migrate. Then the frozen part of the freezing soil displays a certain pattern, as described, of moisture migration and ice formation brought about by the temperature gradient.

Soil freezing (or thawing) results in a sharp disturbance of any fully formed, thermodynamic equilibrium system and is seen as a dynamic co-existence of frozen, freezing and unfrozen zones, and with the development of a movable boundary dividing the phases, i.e. a front of freezing (thawing). Note that it is the frozen (not unfrozen) part of freezing and thawing soils that causes and determines the moisture migration. This lies in the fact that the existence and development of the gradient of freezing temperature in the frozen zone results inevitably and naturally in the development of the considerable gradient of thermodynamic moisture potential and the gradient of partial pressure of water vapour (grad μ_ω and grad P), and consequently, the gradients of unfrozen water and vapour content (grad W_{unf} and grad d_{sat}). The driving forces of moisture migration present in the frozen part of freezing (or thawing) soils cause the advance of liquid and vapour in the direction from the higher moisture potential (or moisture content) towards the lower, i.e. from regions of higher to those of lower freezing temperatures.

A moisture deficiency , arising thus in the high-temperature part of the frozen zone of a sample, will be replenished by migration out of the unfrozen part of a freezing or thawing soil. This proves more profitable energetically since the water here is less bound and more mobile than that in the frozen part. This in turn will result in the formation of gradients of thermodynamic potential of the moisture and of water content in the unfrozen part of the soil, which gradients in turn provide the frozen part with the necessary (for the frozen part) amount of liquid and vapour. The unfrozen part of the soil is thus a sort of 'reservoir' or moisture source for the frozen part. The temperature gradients in the unfrozen part do not lead to moisture migration driving forces. The explanation is that thermodiffusiion moisture transfer in unfrozen soils only becomes perceptible and is noted in tests where the temperature gradients are greater than $2\,°C\,cm^{-1}$ to $4\,°C\,cm^{-1}$. In natural conditions grad t in the unfrozen zone of freezing (thawing) soils proves to be smaller by an order of magnitude.

At the freezing (thawing) boundary, i.e. in the transition from the unfrozen part of the soil to the frozen part (according to N.A. Puzakov), the continuity principle of water flow I_ω and, thus, of bound water films is to be obeyed. Experiments verify this and show the continuous nature of the distribution with depth of the major moisture transfer parameters, i.e. $\mu_\omega(x)$, $K_\omega(x)$, $\lambda_\omega(x)$, $W(x)$ and $I_\omega(x)$. The values of temperature t_ξ and of water content W_ξ at the freezing-thawing interface are functions of the process and can be determined from combining equations for simultaneous heat and moisture transfer in the frozen and the unfrozen parts of soil. In the general case, with a decreasing rate of freezing of the soil v, value t_ξ goes up due to thermal inertia and, on the contrary, when the rate increases, the value goes down (Fig. 2.11). Soil moisture content, W_ξ, at the freezing front behaves in the opposite fashion.

Experiments prove that the origin and growth of ice layers occurs not on the very boundary of freezing (or thawing) but within the already frozen part of the soil and will be predetermined by both thermal-physical and physical-mechanical conditions of the soil system. Fig. 2.11 shows that the most intensive ice formation in the frozen part is noted on the sections of the sharpest inflexion of curves $\mu_\omega = f(x)$ and $W_{unf} = f(x)$, since this makes for a sharp change in the driving forces and thus in the intensity of moisture flow I_ω^{fr}. The curves of moisture distribution with depth demonstrate this all clearly enough (see Fig. 2.11b); visual observations of the freezing of soil samples of various composition and structure provide confirmation. Fig. 2.12 shows how a segregated ice layer starts growing and increasing in size during the freezing of kaolinite clay. Three distinct soil sections are conspicuous: *I*, the frozen section with an earlier formed schlieren cryogenic structure in which practically no formation of segregated ice occurs at

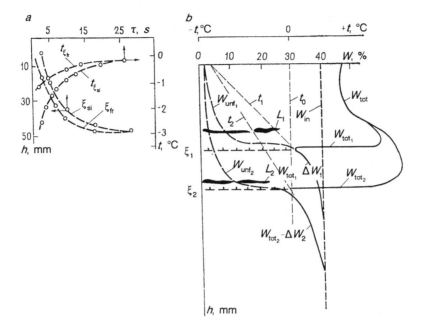

Fig. 2.11. Change with depth and time in sample of freezing kaolinite clay: a – of freezing front ξ_{fr}, of segregated ice formation ξ_{si} and of temperature at these boundaries ($t\xi_{fr}$ and $t\xi_{si}$); b – of the total moisture content W_{tot} and of unfrozen water content W_{unf}, at two times indicated by ξ_1 and ξ_2 with the formation of ice schlieren (L_1 and L_2 respectively) on account of moisture migration for curves $W_{tot1} - \Delta W_1$ and $W_{tot2} - \Delta W_2$ from the unfrozen zone.

present; *II*, the freezing section of intensive phase changes with massive or micro-schlieren structures where the initiation and development of ice schlieren occur; *III*, the unfrozen soil which is losing moisture. Sections *I* and *II* are light-coloured, while *III* is dark. This is due to the soil state, i.e. it is frozen in *I* and *II*, unfrozen in *III*. The presence of ice in section *II* is verified not only by temperature readings and movies and photographs, but also by microscopic investigations and special experiments with fluorescein.

Samples of montmorillonite clay, silty-clay and clayey, silt- rich sands produce similar results. They differ only in the intensity of ice formation in the frozen zone, the type of cryogenic structures formed and in some quantitative freezing indices. In all cases it was observed that with a reduction of the rate of freezing and with the freezing front subsequently stationary, the thickness of the actively freezing zone (zone of intensive phase changes) was reduced and the boundary of visible segregated ice formation approaches the freezing front and then merges with it (Fig. 2.11a).

With a linear pattern of temperature distribution in the frozen part of an

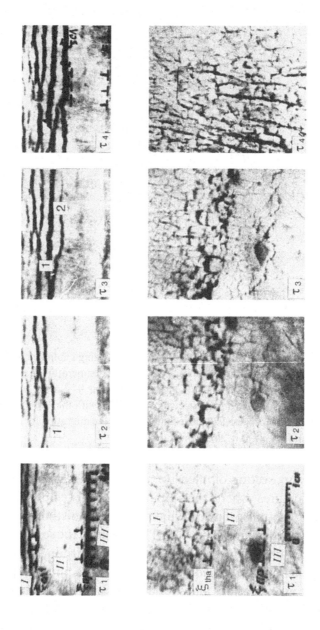

Fig. 2.12. Origin and growth of streaks of segregated ice (schlieren) in freezing clays, kaolinite (*a*) and montmorillonite (*b*), at various times (τ); *I, II, III* – frozen, freezing and unfrozen parts, respectively.

unsaturated freezing soil ($G < 1$), the intensity of moisture flow, as moisture migrates towards lower temperatures, decreases, which results in the freezing out of the excess amount of liquid and vapour to approach thermodynamic equilibrium. The intensity of free ice formation (with $G < 1$) at depth in the frozen part of unilaterally freezing soil $j = \Delta I_\omega / \Delta x$ will be different and is determined by the form of the curve of j and the freezing velocity v. Ice formation i in any cross-section of the frozen zone in time interval $\Delta\tau$ will be calculated from the expression $\Delta I_\omega / v$, i.e.

$$i = j\Delta\tau = \int_0^\tau \frac{\partial I_\omega}{\partial x} d\tau \tag{2.18}$$

Water transfer and ice formation in freezing soils are determined both by soil *composition and structure* and by the *conditions* of freezing. The composition of freezing soils is a basic factor responsible for differences in moisture transfer and ice formation in deposits. Thus there is practically no liquid moisture migration in gravel-pebble and sandy deposits where water transfer occurs chiefly due to vapour. When the deposits are fully saturated ($G = 1$) freezing usually gives rise to a volume increase due to a 9% increase in water volume on its conversion to ice, and often produces the so-called 'piston effect', i.e. a pushing of the excess water downwards. Segregated ice formation is noted only with mineral particles under 1 mm in size, when the adsorption-film moisture transfer mechanism commences. Ice formation here is directly correlated with the values of moisture-transfer coefficients and thermodynamic potential gradients in the freezing zone. Moisture transfer coefficients (K_ω^{fr} and λ_ω^{fr}) in frozen soils are reduced substantially over the range of 0 to $-1\,°C$, and from kaolinite clays to montmorillonite and frozen sands, while moisture potential gradients in the freezing zone increase as fineness increases, and from montmorillonite clays to kaolinite. Migration of moisture thus increases with an increase in fineness of soils and with a greater amount of kaolinite mineral. An increase in moisture migration results from higher silt content, and from an optimum combination of water conductivity properties and moisture transfer driving forces. Absorption capacity of cations increases also from kaolinite to montmorillonite. The cation-exchange influence on moisture migration depends on cation valency; thus moisture migration, ice segregation and heaving increase with the saturation of soil with multivalent cations and diminish with univalent.

Of great practical importance are the problems of moisture transfer and ice accumulation in soils freezing under different *thermodynamic* conditions. In a soil with freezing under 'open' system conditions the total migration of moisture is due to a water exchange comprising an internal part on account of the redistribution of the moisture in the soil itself and an external part on account of water migration from an external water-bearing layer. At the start of freezing the external migration flow is absent in soil, but then it appears near the water-bearing layer. As the boundary ξ_{fr} nears the water bearing layer, the proportion of the external migration water flow greatly increases relative to the internal one. Soil freezing under 'closed' system conditions results only in the internal moisture redistribution between the frozen and unfrozen parts of the soil. Therefore the intensity of migration of moisture depends here on the water storage in the unfrozen part. Ice accumulation in freezing soils depends on their freezing regime and increases with higher grad t in the frozen zone. However the increase of grad t in the frozen zone results in a higher freezing velocity which, on one hand, produces an increase in grad μ_ω and thus in the intensity of migration water flow v towards the freezing front, and on the other, a reduction in ice formation on account of the shortening of the water migration period t. Therefore freezing soil has an optimum relationship of the parameters v and grad t, under which the maximum ice formation will be observed. With a freezing velocity of over 8–10 cm day^{-1}, ice formation is either barely perceptible or absent altogether, since the freezing front advance is so fast that even with high values of grad t and grad μ_ω, ice formation in the frozen zone is not significant.

Also important in practice is the dependence of moisture transfer and ice accumulation in freezing soils on external load. The freezing of soil samples under 'open' system conditions, under pressure and previously consolidated in the unfrozen state, has shown that with a higher external pressure the density of moisture flow into the frozen part is decreased (Fig. 2.13). Depending on soil dispersion (amount of fine-grained material) there appear critical or limited values of external load P_{cr} under which moisture migration into the frozen zone and ice accumulation practically cease. For clayey silt-rich soils P_{cr} may be 0.5 MPa and for kaolinite clays 1.5 MPa. Hence moisture migration into the frozen zone of epigenetically freezing clay soils will not be significant at a depth of more than some 100 m where normally the load at the freezing front is over 1 MPa.

Moisture migration and segregated ice formation occur under temperature gradients in the frozen part of freezing and of thawing soils. Thawing of frozen soil is noted for simultaneous ice melting (with a part of the frozen soil passing into a thawed state) and ice formation in the frozen part of a sample near the thawing front. Moisture migration into the frozen part of thawing soil arises only if it has a temperature gradient. The thawed part of a

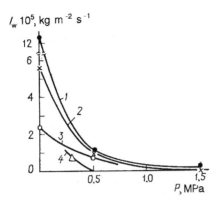

Fig. 2.13. Effect of ambient pressure P on migration of moisture I_ω into the frozen part of freezing soils (grad $t = 0.2 - 0.4\,°C\,cm^{-1}$, $u_{fr} = 0.1 - 0.4\,cm\,day^{-1}$). 1 – kaolinite clay; 2 – loam; 3 – montmorillonite clay; 4 – silty sand with some clay.

sample is then desaturated and consolidated and the total moisture content of the frozen part of the soil increases. When frozen soils are thawed rapidly the boundary of phase transitions is not always at 0°C. The temperature here is often above zero, but small ice layers (schlieren) are present for some time in that part of the soil where the temperature of mineral layers is already above zero. All this is due to the inertia of phase transition of ice inclusions. Ice thawing and water seepage sometimes turns ice layers into partly closed cavities.

The frozen part of a slowly thawing soil with initially massive structure (in the frozen state, prior to the experiment) can have an increase in its ice content and new ice layers under formation in a temperature range of 0 to −2°C. As the thawing front advances, the ice microlayers, formed under lower temperatures, find themselves in the region of higher negative temperatures. They grow much thicker, from fractions of a millimetre (under −2°C) to 2 cm, near the thawing front. The growth of ice layers, situated in the part of the sample with lower negative temperatures, occurs at the expense of other ice layers of higher temperature. Ice layers grow thicker most intensively near the thawing front where they are fed with water migrating from the thawed part of the soil. The thawing of frozen soils with original streaky (small ice layers) cryogenic structure results chiefly in an increase of thickness of the ice layers present in the frozen zone of the sample. Slow thawing results, as a rule, in soil 'collapsing' at the sites of the thawed-out ice layers without the forming of closed fissures in their place.

The influence of frozen soil *composition* and *thawing conditions* on water transfer and ice accumulation have not, so far, been studied sufficiently.

However on present evidence it may be said that their influence is similar to that on freezing. So the intensity of moisture migration and ice formation in the frozen zone increases with increasing soil dispersivity and from montmorillonite clays to those of kaolinite. The frozen zone of more compact and less wet soils usually has smaller ice accumulations. Decrease of grad t in the frozen part of the sample leads to reduced ice accumulation at the thawing front.

Let us consider in conclusion a water migration mechanism capable of acting in frozen and freezing soils and of forming injection and injection-segregation ice layers, lenses and sheets.

According to experiments the formation of a thick ice layer can occur only in cases where the hydrostatic pressure value of intruding water P_{in} is above the instantaneous rupture strength of frozen soil σ_{rup}^{ins} and normal (ambient) pressure P_n. It is necessary for the formation of an injected ice layer that $P_{in} > \sigma_{rup}^{ins} + P_n$. In an experiment with frozen ice-saturated silty sand (with some clay) with a recorded value of instantaneous strength $\sigma_{rup}^{ins} \approx 0.35 - 0.4\,\text{MPa}$ at a temperature of $-1\,°\text{C}$, no injection (using a metal needle) of water at a pressure of 0.1–0.3 MPa could be traced. It was only under a water pressure of 0.4 MPa that a *hydrorupture* of the sample occurred, 20 min. after the experiment started. Water began entering the rupture zone and froze forming a compact injected ice layer. The upper part of the sample was displaced upwards by the amount of the intruded water volume and by its expansion in going from water to ice.

In the case of the frozen soil being unable to expand (deform) laterally or vertically, i.e. with heaving due to hydrorupture and injection being impossible, when the condition $P_{in} < \sigma_{rup}^{lon} + P_n$ applies, the water coming under pressure seeps through frozen soil, fills ice-unsaturated cavities and freezes there. The necessary condition for water flow in frozen ice-saturated soil is here to overcome the initial gradient of unfrozen water seepage which will be determined by the ultimate shear strength σ_{sh} of loosely bound water.

When the conditions for hydrorupture with the development of an injected ice layer are not fulfilled in a frozen ice-saturated sample and if the experiment lasts long enough (several days or weeks) there is a small water injection into the soil, some increase in ice content and the formation of ice microlayers in the sample which, as a whole, sustains little heaving (Fig. 2.14). This is all possible when the hydrostatic pressure of intruding water exceeds the long-term rupture strength σ_{rup}^{lon} of frozen soil plus the value of normal load: $P_{in} > \sigma_{rup}^{lon} + P_n$. Thus experiments made on frozen ice-saturated clay and clayey silty sand (at $t \approx 1\,°\text{C}$ and $P_n = 0$) showed newly formed ice microlayers due to the freezing of water injected into the samples

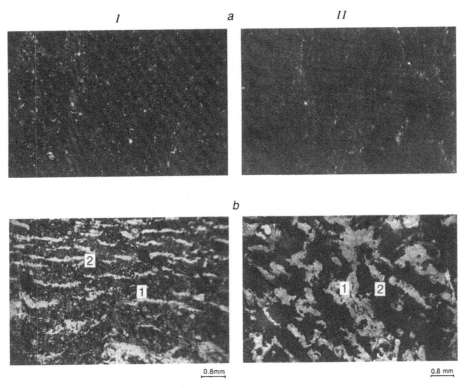

Fig. 2.14. Cryogenic structure of clayey silt-sand (*I*) and kaolinite clay (*II*) prior
to the experiment (*a*) and after water injection (*b*) into frozen samples: 1 – ice;
2 – mineral matrix.

under a pressure of some 0.2 MPa. The duration of tests was 3–7 days. The
long-time strength of these soils was 0.1–0.18 MPa. The increase in the total
moisture in the zone of injected streaky (fine lens) ice formation was 10–
20%. The injected water flow increased at first, but then as the ice content
grew and the coefficients of water transmission went down correspondingly,
it diminished to 0. Any subsequent rise in steps of the hydrostatic pressure of
water resulted in cyclic formation of seepage-injected ice which must in the
end bring about the hydrofracture of the sample.

All the above holds also for freezing soils in the case of the pressure-
actuated water flow into the unfrozen part of the sample (below the freezing
front). The formation and growth of injected ice layers are quite possible at
the freezing front if P_{in} is in excess of the rupture strength of soil at the
'unfrozen-frozen soil' interface. Otherwise ice-formation in the freezing part
of the soil will occur, due to the sum total of migration and injection flows.
The higher P_{in} the greater will be the portion from the injection flow in the
ice formation in the freezing part.

3

Physico-chemical and mechanical processes in freezing and thawing ground

3.1. Chemical reactions and processes in freezing and thawing soils

Essentially the same chemical reactions take place in soils during freezing and thawing and in the frozen state as in unfrozen materials. These reactions are solution, hydration, substitution, oxidation-reduction, ion exchange etc., but in the cold regions they have a number of specific features. For example, solution is less intense because under lower temperatures some salts dissolve at a much slower rate. Apparently, because of low temperatures the permafrost regions contain considerable amounts of the products of chemical interaction between the dissolved substances and water molecules, that is, hydrates and crystalline hydrates. The cation exchange reactions have, probably, a predominant importance for frozen soil, because unfrozen water is a rather concentrated solution, the ions of which actively interact with the ions of the mineral surfaces. Moreover the typical processes in frozen ground are coagulation of sols and formation of colloidal compounds. These processes are predetermined by the characteristic phase transitions of water in the ground (freezing or thawing) which cause dehydration in soils and, consequently, coagulation (on reaching the threshold of coagulation) of organic-mineral compounds. The geochemical processes occurring in the cold regions also have distinct specific features of geochemical processes that play certain specific roles. For example, free water affects seasonally frozen soil intensively only in the warm period of the year. A major role therefore belongs to bound (unfrozen) water which interacts with, and is in dynamic equilibrium with, the ice and soil.

Ground water normally has a higher carbon dioxide content because with the fall of temperature the solubility of gases (including CO_2) increases rapidly as does the organic matter content. For example, in the soils of the tundra on Bolshaya Zemlya the free H_2CO_3 content reaches $200\,mg\,l^{-1}$ and that of the ion HCO_3^- is $650\,mg\,l^{-1}$. Therefore, the concentration of hydrogen ions in ground water of the permafrost regions increases by several hundred times, which causes, perhaps, an acid reaction in the medium. The nature of many chemical reactions and the behaviour of soil components largely depend on the pH of the medium. An acid medium is more aggressive and chemically active; it decomposes silicates, and the hydrolysis reactions in it are stronger than those in neutral and alkaline media.

Geochemical processes in frozen ground are also influenced by the content of monatomic hydrogen (reducing) and oxygen (oxidizing). It is assumed that during phase transformations of water into ice the release of hydrogen may reach considerable values. For example, when $1\,m^3$ of water turns into ice, $120\,g\,mol^{-1}$ of monoatomic hydrogen are released. Since the soils in the permafrost regions are highly humid, oxygen cannot easily penetrate them. Therefore in the North, the oxygen surface, which shows the distribution of free oxygen by depth in the crust, rises and reaches the surface in bogs. As a result the frozen materials in the permafrost region are predominantly a reducing medium, thus having a higher content of divalent ferrous Fe^{2+} with formation of its mono-oxide compounds (siderite, pyrite, vivianite, etc.). Ferrous oxide in soils colours them in bluish-grey shades, and they are usually called *gley* soils. They are predominantly fine-grained, reducing and acid.

The processes of breakdown of *organic matter* are also different. Because of lower biological and biochemical reactions, the transformation of vegetal and zoogenic remains into organic matter is slowed, and decomposition of remains (formation of humus) terminates at a less mature stage. As a result of this process, light-coloured fulvic acids are formed rather than humic acids (the product of the final stage of decomposition). In the tundra soils, the content of fulvic acids may reach 70%, while humic soils contain only 10-15% of humus matter. Fulvic acids, like humic acids, compose a group of similar high-molecular-mass compounds, but they have less carbon and nitrogen and more oxygen and hydrogen than humic acids. Fulvic acids destroy minerals by their high acidity; they homogeneously saturate the soil and form a massive compact layer. More viscous and less mobile humic acids in the soil produce a lumpy, nutty structure typical, for example, of chernozem.

Chemical processes during a single freezing of soil

At the beginning of freezing, the water turns into ice, creating a new mineral. The gravitational, capillary and loosely bound non-saline water crystallizes at negative temperatures close to $0\,^\circ C$. Film water normally freezes within a wide range of negative temperature, determined from the curve of unfrozen water content. Salt waters, with mineralization of more than $30 \, \mathrm{g} \, \mathrm{l}^{-1}$, crystallize at temperatures of about -1.5 to $-2\,^{\circ}\mathrm{C}$ whereas brines may remain liquid at $-20\,^{\circ}\mathrm{C}$ and lower. The freezing of water usually causes a distinct differentiation of salts between the solid and the liquid phases. A part of the salts dissolved in water is enclosed in the ice, a part of the less soluble salts precipitates, and a part of the easily soluble salts is squeezed into lower water layers thus increasing their mineralization. The ice formed by freezing is several times less mineralized than the initial pore solution. Slow and gradual freezing produces the most 'pure' ice. During freezing, in accordance with the degree of solubility at negative temperatures, the most insoluble salts of $CaCO_3$ precipitate first (in the temperature range -1.5 to $-3.5\,^{\circ}\mathrm{C}$), and then Na_2SO_4, $CaSO_4$, etc. (at temperatures of -7 to $-15\,^{\circ}\mathrm{C}$), these salts forming the so-called *crystal hydrates*. Consequently, cryogenic layers are enriched with gypsum $CaSO_4.2H_2O$, mirabilite $Na_2SO_4.10H_2O$, and calcite $CaCO_3$; in other words, they are sulphatized and carbonatized.

Below the freezing boundary, the waters are highly mineralized due to easily soluble salts expelled from the frozen layer (chlorides of calcium, magnesium, sodium and hydrocarbonates of sodium). As a result of this cryogenic concentration, the rather highly mineralized subpermafrost water (up to $200 \, \mathrm{g} \, \mathrm{l}^{-1}$ and more) is formed (also occurring sometimes as interpermafrost water) and gives *cryopegs*. These waters remain liquid at negative temperatures. The thickness of layers with cryopegs below the base of the permafrost may reach several hundreds of meters.

Moreover, frozen ground also contains gases in the form of hydrates, the so-called *gas hydrates*, whose crystalline lattice is built up of water molecules. Molecules of hydrate-forming gas are distributed in cavities inside the lattice. The crystalline lattice of the water itself (without gas molecules) is thermodynamically unstable and cannot exist independently. In natural conditions the structure of the crystalline lattice is often filled with molecules of methane, ethane, hydrogen sulphide, and carbon dioxide. When gas penetrates into the water lattice, it becomes rigid and the water turns into a solid. Superficially the gas hydrate is very much like ice. Undamaged crystals of gas hydrate are transparent and homogeneous. The appearance in them of microcracks and different gas inclusions indicates commencement of decomposition (Fig. 3.1). The amount of heat generated by the phase transitions of gas hydrates is about $0.5 \, \mathrm{kJ} \, \mathrm{g}^{-1}$. Under the effects of warming and interaction with water, the gas hydrate decomposes with a hissing sound and intensively exudes gas bubbles. Negative and low positive temperatures in the permafrost regions are favourable for the formation of hydrate deposits at shallow depths.

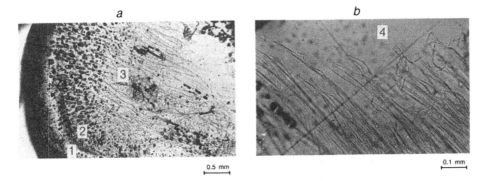

Fig. 3.1. Structural-textural features of a man-made agglomerate 'ice-methane hydrate': (*a*) general view of gas-hydrate thin section; (*b*) contact of pure ice with hydrate-bearing ice. 1 – zone of completely decomposed agglomerate with very few microchannels and microbubbles of gas; 2 – the zone of largely decomposed agglomerate with numerous gas bubbles; 3 – the zone of slightly decomposed agglomerate with typical microchannels; 4–5 – ice (4 – pure, 5 – hydrate-bearing with microchannels).

Chemical processes in frozen soils

Because in the frozen state (perennial or seasonal), the water phase in soils is not visible, for a long time it was believed that the soils were in a state of complete chemical inactivity. This misrepresentation underestimates the role of unfrozen water and would automatically lead to the application of Van't Hoff's law, which states that with a drop in temperature of $10\,^{\circ}$C the rate of chemical reactions is reduced by half. The fallacy of this has been demonstrated by many researchers who found considerable cation exchange reactions between soil minerals and bound water and who showed a much greater concentration of substances in film water compared to that in pore water. Though the absence of free water in frozen ground would seem to preclude the escape of chemical components from the unfrozen water, the mass-transport processes are sufficiently intense because of diffusion of ions in unfrozen water, to lead to adjustment of concentration of dissolved substances. In frozen ground, the process of flow is also active in unfrozen water films inducing convective transport of ions and soluble matter with migrating water. In the course of this process, phase transitions between pore ice and films of unfrozen water take place in accordance with the increase or decrease of concentration of ions. As a result there is a levelling out of the concentration of salts in unfrozen water with a continuously occurring decrease down to initial levels.

Chemical processes during repeated freeze-thaw of soil

The chemical processes developing during repeated (cyclic) freezing and thawing have been studied in great detail. Unlike perennially frozen ground, the chemical reactions in seasonally thawing soils are much more intense and are obviously periodic. The interaction between soil minerals and water (both free and bound) is a pulsating process, and the phase transitions of water into ice and vice versa should result in marked intensification of chemical weathering in seasonally frozen soils. These conclusions are confirmed by data obtained on 'cold' or 'cryogenic' soils and on weathering crusts in severe climates, which indicate that the process of geochemical reactions is qualitatively identical to that in warm humid areas. Moreover, an intense chemical transformation in seasonally frozen ground commences at the very first stage of weathering under the effects of hydrolysis, leaching, oxidation, hydration, and migration of colloids, and neogenetic clay and other minerals are formed. The studies of M.A. Glazovskaya in Antarctica have shown that in the 10-15 cm surface layer of soils, if the supply of oxygen is sufficient, oxidation takes place and MnO and Fe_2O_3 accumulate, colouring the ferric and manganese extractions on rock fragments into ochre-rusty or orange-red hues. Lower down in the layer, where there are manifestations of wash-out products and carbonatization, more mobile products of weathering accumulate, such as calcium carbonate and calcium, which do not effervesce in HCl. The study of surface weathering crusts under the microscope has established the stages of decomposition of the original minerals. At the initial stage, chlorite disappears, then hornblende and biotite, i.e. the banded and stratified silicates are the first to disintegrate. Feldspars become covered with yellowish-brown fine silty aggregates, i.e. secondary clay minerals.

The results given above are in agreement with the data on cold tundra and taiga soils, where non-gley cryogenic soils (brown soil, podzol, Al-Fe humus) dominate and poorly drained gley soils are in the minority. The chemical elements in non-gley soils are arranged by V.O. Targul'yan according to migration capacity as follows: $Si > Fe > Ti > Al$. The silicate forms, appearing as a result of hydrolysis, are fairly mobile in an acidic medium and are evacuated from the soil profile. Iron, titanium and aluminium in an acidic medium have low solubility and normally remain in the soil as oxides (Fe_2O_3, TiO_2, Al_2O_3) and hydroxides ($Al(OH)_3$, $Fe(OH)O$). In the course of humification in the permafrost regions fulvic acid appears as one of the most aggressive and mobile forms of humus. This acid moves downwards with the soil solution and destroys the hydroxides and silicate minerals by forming different kinds of organic-mineral compounds (oxalates, chelates, fulvates and adsorbed organic-mineral compounds).

Fulvates and oxalates, as the more mobile compounds, are removed from the soil profile, whereas chelates and adsorbed organic-mineral compounds soon lose their mobility and remain in the illuvial horizon. During this process the brown-coloured Al-Fe-humus coarse silty-clay horizons appear. At the same time, the real humus horizons and horizons of Al-Fe-humus and titanium compounds can be formed. The compounds of titanium, aluminium, iron (Ti-Al-Fe) and humus are accumulated in these alluvial horizons and this process, called, by Targul'yan 'tialferrisation', is typical of cryogenic soil formation. The leached horizon A_2, naturally, is depleted of Fe and Al hydroxides and oxides; therefore a relatively (not absolutely) higher SiO_2 content is observed as well as a higher shade of colouring due to decomposition and removal of dark-coloured compounds and minerals.

The chemical and physico-chemical processes are somewhat different in gley soils (poorly drained or boggy) typical of the north of the European part of the former USSR and of the Siberian coastal lowlands. Fine-grained (dust) matter is dominant in these soils with reducing conditions and acid reaction of the medium. The profiles of gley soils do not normally show distinct illuvial horizons but in gley and gley-podzol soils, on heavy sandy-silty-clay, for example, the content of Fe_2O_3 and Al_2O_3 is reduced concurrently with relative silica enrichment. The higher mobility of the iron is a result of its transition under reducing conditions into the monoxide form $Fe(OH)_2$ which does not precipitate from solution until the pH reaches about 5–6. The blue-grey monoxides of iron colour the profile of gley soil with the typical grey and blue-grey hues. This phenomenon is also promoted by the presence of fulvic acids which are the immature forms of humus and which are not brown (as humic acids), but light grey, thus making the colour of gley soils less intense than that of non-gley soils.

The chemical differentiation of the products of weathering, which is closely associated with the mobility of the chemical elements, is particularly important in geochemical processes in the permafrost regions and especially in soils with cyclic freezing and thawing. More mobile chemical elements are intensively removed by underground and surface runoff; other elements, on the contrary, are practically immobile and accumulate in watershed areas and on slopes increasing their relative concentration. For example, potassium, calcium, magnesium, sulphate and chlorine ions in the permafrost regions are very mobile and migrate in all waters in a truly dissolved state. The silicate form of silicon migrates mostly as mono- and polysilicic acids, which are removed in solutions by ground water. A certain amount of silicic acids (up to 40%) can be transported as gels and colloids in combination with organic matter. Non-silicate SiO_2 in the permafrost regions is practically immobile, as is illustrated by intensive formation of podzol soils. The low mobility of silica is attributed to the extremely low solubility of SiO_2 in the highly acidic medium typical of tundra and taiga soils. Up to 70–90% of aluminium migrates in the permafrost regions as colloids and as complex compounds with humic acids. Iron (Fe^{2+} and Fe^{3+}) has low mobility outside the permafrost regions. In cold humid conditions, 90–98% of the total iron content migrates as highly mobile colloids which are high-molecular organic-mineral complex compounds of the chelate type. Under the northern conditions some other microcomponents (Ti, Zn, Cu, Ni, etc.) also become more mobile and are transported not as simple ions but as colloids or as complex ions with a larger radius, which are formed with participation of high-molecular-mass organic matter.

3.2   Physico-chemical and mechanical processes in freezing and thawing ground

The freezing (thawing) of soils is accompanied by complicated physico-chemical processes, the nature and intensity of which are essentially different depending on whether migration of water occurs in the material or not.

Physico-chemical and mechanical processes in freezing soils

When ground freezes *without water migration*, then the free or semibound water in the pores increases its volume by about 9% as it turns into ice. If the pores are filled with water at $G = 1$, then the freezing ground expands volumetrically and local densifications or particle aggregates appear, caused by considerable crystallization pressure as water turns into ice. For example, if the volumetric expansion is obstructed, the freezing water may reach a pressure of 2200 Mpa. Between the large soil aggregates, experiencing the greatest densification (compression) are the largest pores where water freezes first (at negative temperatures close to 0 °C). Therefore liquid water is squeezed from the smaller pores inside the aggregates into the larger interaggregate pores and crystallizes there, destroying the interaggregate structure connections. As the negative temperatures continue to fall not only are the ground micro- and mesoaggregates compressed, but bound water in the intra-aggregate pores begins to freeze and turns into ice. If it has no free escape, it breaks up the structural intra-aggregate ties and disperses (fragments) the mineral aggregates of the soil. This destructive cryohydration process can take place simultaneously with the *breakdown* of the sandy and partly of the coarse silt fractions of freezing soils as a result of thermal weathering, which involves destruction due to different thermal expansion of the various minerals and elements composing the mineral component.

In soils the breakdown process can take place together with development of aggregation and stronger structural ties both between elementary particles of the ground and between small aggregates as a result of dehydration of aggregates (as part of the water freezes out) and of their approach to each other due to squeezing. For example, during freezing of moist soils, the concentration of ions increases in the unfrozen water. In this way, the threshold of coagulation can be reached and soil particles will coagulate (aggregate) with decrease of the general surface activity and dispersion. Therefore, a slightly more concentrated soil solution, which is normal during freezing, is sufficient to achieve deep mutual coagulation of soil particles and to form microaggregates of dust-size fraction.

The physico-chemical processes in soils that *freeze with migration of water* are much more varied, because the unfrozen layer, lying below the freezing front, is intensively dehydrated, so that the films of bound water are thinner and the particles and aggregates more closely packed. Dehydration causes considerable contraction and lowers the porosity of the unfrozen part of soils and forms larger aggregates and blocks of soil due to coagulation and aggregation. The soil particles, aggregates and blocks are more densely packed and are oriented in the direction of the water flow in accordance with the minimal hydrodynamic resistance. The pores inside the aggregates become smaller, whereas the volume and number of inter-aggregate pores grow and these acquire the shape of elongated slits.

Uneven shrinkage (both vertically and horizontally within individual aggregates and blocks of ground) in the unfrozen part of freezing soils causes the appearance of zones with various 'defects' of strength where contraction stresses are concentrated. Under the effect of the developing gradients of local stresses, the water migrates into these zones. If the zones of stress concentration receive insufficient amounts of water, then microfissuring may develop in the unfrozen dehydrated part of the freezing ground.

In the frozen part of the soils, as in the case of freezing without water transportation, intensive phase transition of water into ice takes place with an increase of water volume by 9% and the splitting up (disintegration) of large aggregates and blocks. Since ice appears first of all in large (interaggregate) pores, the size of aggregates is reduced and their density increases due to radial compression by the growing ice crystals. The size of skeletal interaggregate pores increases by several times compared to pores in the unfrozen dehydrated part of the soil.

Additional supply of water (due to migration) from the unfrozen part to the freezing part of the soil causes, on the one hand, a wedging pressure in films of unfrozen water, i.e. swelling of the whole water and ice saturated ground and, on the other hand, a progressive moving apart (disintegration) of macro- and meso-aggregates due to the increase of volume of the migrated water as it turns into ice. Consequently, in the frozen part of the freezing soil, the swelling-heaving is greatly intensified due to water migration, and macro- and mesoaggregates are fragmented and reorientated. Microblocks and mesoaggregates are rotated and the previous orientation (observed in the unfrozen part of the soil) is changed but inside them the orientation of particles remains the same.

As the negative temperatures continue to fall (far from the freezing front, in the zone of low-intensity of phase transitions of water) the unfrozen water freezes out in the thin intra-aggregate pores. This process causes dispersion of meso- and micro-aggregates (disintegration and peptization) and their orientation, observed in the unfrozen part of the soil almost completely disappears (see Section 4.3).

The differentiation of soil into a massive-frozen part (skeletal-mineral) and the visible fixed layers of migration-segregation ice is nothing less than the essential and unique physico-chemical and mechanical process of freezing of soils with water migration (compared with freezing without water migration). The ice layers appear in the zones of stress concentration, whose configuration in general corresponds to the future type of cryogenic structure. The bound water, supplied as a result of grad t and grad P activity, at first produces the wedging effect, which overcomes the local resistance of the soil to destruction and then freezes and increases its volume.

If the ice layers contain soil inclusions, then their displacement also occurs if grad t is active, but to the parts with higher negative temperatures, that is, the ice 'purifies' itself from admixtures and inclusions. This process is associated with uneven thickness of films of unfrozen water (there being a temperature gradient) on the opposite facets of soil inclusions. Migration of water to thinner films where it freezes causes squeezing out of soil inclusions towards higher temperatures. In general, the presence of solid, liquid and gaseous inclusions in ice largely depends on the rate of freezing of soils and of pore and migration ice. A lower rate of freezing reduces, and a higher rate increases, the amount of inclusions in ice.

The freezing of soils where there is a water supply to the frozen zone usually causes expansion of the zone with heaving of the surface of the ground. The unfrozen part of the soil is consolidated by shrinking due to dehydration and sometimes also by its compression under the effect of the heaving of the overlying frozen part (if this cannot be deformed upwards).

Physico-chemical and mechanical processes in thawing soils

The thawing of coarse clastic and sandy soils, both with low and high ice content, is usually associated with fairly simple physico-chemical processes, such as consolidation, dehydration, settlement and other processes connected with compression and re-orientation of fragments and sand particles, and with runoff down a tilted impermeable layer or infiltration of gravitational water into underlying horizons. The processes of thawing of fine-grained soils (silty sands, silty-sandy clays and clays) are different and more complicated. Two types of thawing are distinguished in the first approximation (as in freezing): with and without water migration from the thawing zone into the frozen part of the ground.

The *thawing* of fine-grained soils *without water supply* to the frozen zone of the sample normally occurs when either the thawing front moves quickly or the soils have low water (ice) content, which practically always results in deformation by consolidation. The water that is formed in soils under rapid thawing of pore or schlieren ice, either goes to hydration of soil particles (without water migration) or escapes from the soil under the force of gravity. It is obvious that during quick thawing of soils with large ice inclusions (for example, thawing of soils with large stratified-meshed cryogenic structure), cavities and empty cracks are formed in the thawed part of the soil, i.e. a specific post-cryogenic structure is formed. The melting of pore, contact and film ice as the negative temperature of frozen soils rises produces greater amounts of unfrozen water with its higher mobility, thus creating conditions for local migration of water within ground elements dehydrated during freezing (aggregates, blocks, particles) and for their *hydration and swelling*. The process of *osmotic swelling* along the partings in the soil dehydrated during freezing has a notable effect in the change of structural ties between the soil elements. This swelling is most clearly demonstrated at $0\,^{\circ}\mathrm{C}$ when the thawing of pore ice and ice inclusions terminates and the structural elements can move apart. In this process, the intra-aggregate ties become more relaxed than the inter-aggregate ones, and a transition takes place from close coagulation contacts to longer distance coagulation (aggregational) contacts. During thawing of soils with a high ice content, their moisture content may even exceed that at the liquid limit, which accounts for a wide distribution of *thixotropic* soils in the Far North.

Slow thawing of soils and water migration from the thawed part to the frozen part activates practically all the physico-chemical processes which occur in thawing soils without water migration. Several other physico-chemical processes, however, are also active. With penetration of migrating water into the frozen part of thawing soils, the ice content is increased and often migration-segregation layers of ice are initiated and grow. In the thawed part of thawing soils the soil aggregates are dehydrated, and shrinkage brings them closer: they become larger and denser.

In general, rapid or slow thawing of soils increases their fine-grained nature as a result of dispersion and peptization of soil aggregates and blocks and fragmentation of initial sand particles. The joint effect of *temperature and hydration mechanisms* of destruction is a determining factor in this transformation of structural elements.

During *cyclic freezing and thawing*, soils experience physico-chemical processes typical of freezing and of thawing ground. The specific feature of repeated freezing and thawing is accumulation in soils of particles of coarse silt fractions due to destruction of sand particles.

Cyclic thawing and freezing often results in differentiation of freezing soils by grain size, which is caused by displacement (heaving) of the larger fractions of the soil (fragments, large particles) as the temperature falls. The results of field observations and laboratory experiments indicate that after several hundreds of cycles of freezing and thawing, the soils become graded by size, i.e. the larger particles (more than 1–2 mm) move towards the source of cooling and accumulate in a surface layer, whereas the smaller particles remain in place and accumulate in the lower part of the layer. In this process, the facets of the moving particles are abraded and smoothed out.

Deformations and stresses of shrinkage and heaving in freezing soils

An analysis of deformations and consolidation (shrinkage) stresses in the unfrozen part of unilaterally freezing fine-grained soils has shown that these effects are most active near the freezing front in the region of maximum dehydration. This observation is confirmed by data on the shifts of deformation recorders in a sample with depth and time (Fig. 3.2a). For example, the deformation values increased with longer time periods of dehydration as the freezing front approached the recorder. Moreover, the total value of shrinkage deformation in the vertical direction is by an order of magnitude greater than that in the horizontal direction. This can be attributed to a considerable 'prevention' of horizontal shrinkage of the unfrozen part of the soil due to its adhesion to the frozen part which resists shrinkage. Therefore, as the water-saturated soil freezes with intensive migration of water and ice accumulation, a narrow 'neck' is formed in the frozen part of the zone of dehydration which extends into the unfrozen part of the sample (Fig. 3.2b).

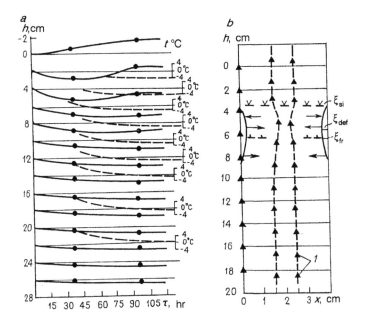

Fig. 3.2. Freezing of sample of kaolinite clay. (*a*) Change of temperature with time (dashed lines) and movement of deformation recorders with time in the vertical direction. (*b*) Vertical profile showing displacement of recorders (1) in the horizontal direction at the end of the experiment ($t_B \approx -5\,°C$, h = height of sample, ξ_{si} = segregation ice front, ξ_{fr} = freezing front)).

The 'prevented' shrinkage in freezing soils probably occurs not only as a result of interaction between unfrozen and frozen zones in the sample, but also because of the irregular distribution of moisture by depth, similar to the phenomenon in the unfrozen dehydrated soil. The irregular shrinkage deformations in the unfrozen part of the ground follow from volumetric-gradient all round stresses.

Experiments show that in freezing soils the shrinkage stresses grow with time to a certain maximum value determined at every moment of time by the relation of forces of structural adhesion of the soil with the forces trying to destroy it. The stress recorders indicate the beginning of the growth of stresses only when the frozen zone is already formed in the soil and water starts its migration to the front of ice segregation (Fig. 3.3 a). These stresses normally reach their maximum in the freezing zone (Fig. 3.3b) between the freezing front and the visual boundary of ice segregation, because here dehydration of soil is still active as the water slowly freezes.

Finer soil particles cause higher shrinkage stresses and higher stress gradients. Experiments show that the greatest values P_{shr} develop in clays

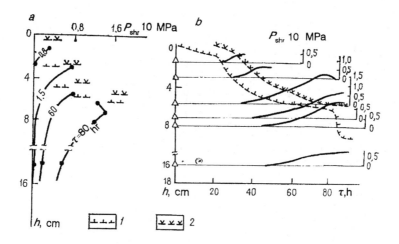

Fig. 3.3. (*a*) Distribution of volumetric stresses by depth for various durations of freezing of a sample of kaolinite clay; (*b*) development of the stresses and depth of freezing as a function of time τ. 1 – freezing front; 2 – visual boundary of segregated ice formation).

(0.08–0.2 MPa and more) and the smallest in sandy-silty materials (not more than 0.04 MPa).

Freezing conditions have the greatest effect on shrinkage in freezing ground. For example, a higher freezing rate produces lesser dehydration of the soil, with a thinner dehydration zone and, consequently, less shrinkage deformation. Moreover, not only is the general shrinkage small, but the 'prevented' shrinkage is correspondingly smaller, thus causing lesser P_{shr} values. In rapidly freezing ground, however, greater grad P_{shr} appears (Fig. 3.4a) as a result of greater grad W and a thinner dehydration zone. All these processes finally produce a larger number of ice layers (Fig. 3.4b). Three factors are active in the frozen part of the soil characterized by the gradual fall of negative temperature: freezing of part of the water, migration of unfrozen water, and intensive ice segregation. These three factors play an important part in the creation of stresses and heaving deformations: 1) crystallization of water on transition to ice and the increase of its volume by 9%, which cause stresses and heaving deformations, 2) the wedging effect of thin films of migrating unfrozen water causing stresses and swelling deformations, and 3) consolidation of the mineral part of frozen soils and dehydration of soil particles with lower negative temperature and freezing of part of the unfrozen water causing stresses and shrinkage deformations.

The shrinkage process in the frozen part of the soil is, apparently, most important only in the zone of intensive phase transitions, i.e. near the freezing front. This is probably the reason why maximum stresses and

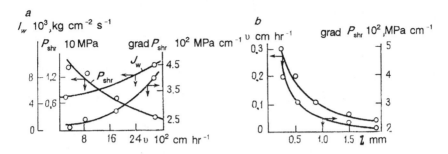

Fig. 3.4. (*a*) Relationships between volumetric stresses P_{shr}, gradient of stresses grad P_{shr}, rate of freezing and migration water flow I_w, for freezing kaolinite clay; (*b*) between stress gradients grad P_{shr}, rate of freezing and frequency of horizontal ice layers *l*).

shrinkage deformations are recorded not in the unfrozen dehydrating part of the soil sample, but in its frozen part between the freezing front and the visually observed front of ice segregation ξ_{si}, i.e. on the boundary where ξ_{def} deformations change their direction in the freezing soil (Fig. 3.2b). In actual fact, in the lower part of the freezing zone, in a sample of kaolinite clay (between ξ_{fr} and ξ_{def}), where intensive phase transitions and stresses (though weak) of swelling P_{sw} and of ice expansion P_{ho} develop as a result of water migration, the deformation recorders also show the development of a shrinkage process, that is, $P_{shr} - (P_{sw} + P_{ho}) > 0$. But in the layer between the boundary where deformations ξ_{def} change their direction, and the visual boundary of ice segregation or the boundary of transition of the freezing zone into the frozen, ξ_{si}, where considerable water (ice) accumulation is already active, the swelling-heaving stresses exceed the now weak shrinkage stresses, that is, $P_{shr} - (P_{sw} + P_{ho}) < 0$. As a result, in the segment of the freezing zone from ξ_{def} to ξ_{si} the swelling and heaving deformations develop, i.e. deformations with directions opposite to the directions of the shrinkage deformations (Fig. 3.2). It seems obvious that on the mobile and variable boundary, where the changes in the direction of deformations occur, the opposite forces should be equal, that is, $P_{shr} = P_{sw} + P_{ho}$, and this boundary will be the plane of stresses of equal value but opposite sign.

Special experiments have established that the freezing of water in a confined volume, which could be the pores of the soil, may produce significant stresses, if heaving deformation is not allowed. These stresses, if approximately estimated using the Clausius-Clapeyron equation, will be about 13.4 Mpa for every 1 °C of fall of negative temperature. Since natural soils cannot be 'closed' undeformable systems, the component of heaving stresses in the total stress and deformation value in freezing soils in most cases is not dominant, but occurs as a certain addition to swelling stresses, caused by the wedging effect of fine films of migrating water. The wedging activity of fine films of unfrozen water, at the same time also causing its migration, plays a large part in the formation of heaving stresses.

The amount of prevented deformation has a decisive effect on the heaving and settling stresses, i.e. the more the deformation is prevented, the higher are the stress values in the freezing soil. For example, results of a study of stresses using recorders with different rigidity values K_{def} have shown that, with the growth of K_{def} from 800 to 1700 MPa m^{-1}, the values of stresses in freezing soils almost double. The increase of rigidity value of the recorder in this case has reduced the allowed deformation of the sample.

The smaller size and compression of the unfrozen zone in a sample are significant in the development of the heaving stresses in soils. In fact at the beginning of freezing, as the unfrozen zone is dehydrated, it is easily deformed and becomes smaller. But later, the shrinkage deformations attenuate, whereas the general deformation of the ground system is reduced causing accumulation (growth) of unrelaxed stresses. This process is associated with the growth of heaving stresses with time (as the freezing front advances), which reach their largest values by the end of freezing (Fig. 3.5a).

The stresses and deformations in the freezing part of soils largely depend on their composition and structure. The greater the fine-particle content of soils and the more complete their water saturation, and the higher the initial water content, the more intensive are the stresses and the more active the migration water flows (Fig. 3.5).

The same relationship exists between the freezing conditions and the development of stresses and deformations in soils. For example, if there is an external supply of water to the freezing sample (an 'open' system), then it has higher values of heaving stresses. The same is observed during slow freezing of soils which is associated, as shown by experiment, with greater redistribution of water and ice accumulation in the freezing zone of the sample and, therefore, with a longer active deformation of the freezing soil due to wedging by the films of migrating water.

Heaving and shrinkage of freezing and thawing soils

The freezing and thawing of soils often causes deformation (rising or sinking), i.e. heaving or subsidence of the ground surface. During the economic development of the country, engineering constructions are particularly threatened not by the absolute values of these deformations but by their irregular occurrence over the area. The criterion of irregularity of, for example, areal heaving, is normally expressed as the excess of heaving at one

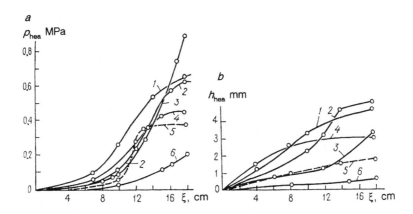

Fig. 3.5. (*a*) Development of heaving stresses P_{hea} and (*b*) heaving h_{hea} of the surface ($t_{surf} = -2\,^\circ C$) of soil samples, depending on depth of freezing ξ: 1–3 – clay (1 – kaolinite, 2 – polymineral, 3 – montmorillonite); 4 – peat, 5 – sandy silty material; 6 – sand).

point over the value at another point, as a percentage of the distance between them. In natural conditions, the values of this coefficient of irregularity of heaving may vary within the range 3–15%, while the change in absolute values of deformation over the area of heaving may change from fractions to tens of centimetres.

Heaving of freezing soils may occur in an 'open' system (with water supply from a water-bearing horizon) and 'closed' system (without external water supply) and when water is transported by injection to the front of ice segregation. In any one of these cases the mechanism of development of heaving deformations comprises several physico-chemical processes, the share of which in the total heaving value h_{heav} depends on the specific conditions of the freezing of the soil. The generalized value h_{heav} of the freezing layer may have the following expression:

$$h_{heav} = h_{ho} + h_I - h_{shr} \tag{3.1}$$

where h_{ho} is the heaving of the soil as a result of increase of water volume by 9% in the freezing part of the soil in the process of its transition into ice; h_I is the heaving due to water supplied to the frozen part of the soil by migration or injection; h_{shr} is the deformation value of shrinkage of the soil in its frozen and unfrozen parts.

Numerous studies have revealed that ice accumulation due to water migration into the frozen part is the principal process in the formation of the total heaving value of freezing soil (90–95%). Therefore, when the intensity of the migration water flow in the freezing soil I_ω increases and the rate of ground freezing falls, then ice accumulation and, consequently, the heaving value of soils, grows. This observation was used as the basis for practically all calculation schemes and methods of quantitative estimation of the heaving value of freezing soils (I.A. Zolotar, V.O. Orlov, N.A. Puzakov, and others). These methods allow us to determine the maximum possible heaving value, i.e. the actual value of total ice accumulation in the frozen zone of the ground.

In reality, the heaving value essentially depends not only on the total (summarized) ice accumulation value in soils but on the distribution pattern of ice content in it, on the type of the newly formed cryogenic structure, and on the shrinkage value h_{shr} here of the unfrozen zone of freezing soils. For example, if a massive cryostructure (pore ice only) is formed, then heaving may be small. There is almost no heaving in the freezing of soil of a typically cellular ('bentonitic') cryostructure owing to the specific kind of growth of vertical and horizontal ice layers with a decrease of volume of soil cells. The maximum heave usually occurs when an ice lens cryogenic structure is formed and heaving largely depends on the total thickness of ice schlieren.

The increasing intensity of water migration flows into the zone of intensive phase transitions in the series 'sandy-silt - silty-clay' causes larger heaving values due to migration ice accumulation in the freezing clayey soils. For example, the share of migration ice accumulation in the heaving of freezing kaolinite clay is 80–95%, whereas in the freezing of sandy coarse-silty samples it is rarely more than 50–60%. In clayey soils the massive heaving is normally not more than 20%, in sandy silty materials it may often reach 70–80% and more. Concurrently the settling deformations grow in the order 'sandy-silt – silty-clay', and consequently the heaving deformations are compensated to a greater extent. The shrinkage deformation h_{shr} depends on composition and structure of the soils and on freezing conditions and may reach a considerable value, sometimes in specific conditions even exceeding the value of h_f. Therefore, the lithological sequence of soils may change its heaving value, if the settling of the unfrozen part of the ground is taken into account. Though migration ice accumulation in clay has rather large values, the heaving value in it can be even less than in sandy silts as a result of large shrinkage deformations. For example, in montmorillonite clays shrinkage can reduce heaving by 80–90%.

Development of heaving deformations in freezing soils is largely dependent on freezing conditions, i.e. on the temperature gradient in the zone of intensive phase transition, on freezing rate and external water supply. For example, the higher temperature gradient usually coincides with the growth of intensity of the migration water flow I_w^{fr} and the total heave h_{heav}. A higher freezing rate is in all cases associated with lesser total ice accumulation and heaving. Freezing of soils in an 'open' system, that is, with a supply of water from the water-bearing horizon (migration or injection), is almost always associated with a sudden increase in heaving deformation compared to the 'closed' system. A load on freezing soil, i.e. freezing under pressure, causes less heaving because the density of migration water flow in the frozen part of the soil is less.

Subsidence of thawing soils has a more complicated mechanism and several peculiar features, which are different from the heaving of freezing soils. In the general aspect the subsidence value S_{sub} in natural conditions (without external load) may be written as:

$$S_{sub} = h_I^{fr} + h_{sw}^{th} - S_{sub}^i - S_{shr}^{th} \tag{3.2}$$

where S_{sub}^i is subsidence of the ground due to thawing of pore ice and ice layers; S_{shr}^{th} is shrinkage in the thawed dehydrated part of thawing soil; h_I^{fr} is heave deformation due to water migration and ice accumulation in the frozen part of the thawing soil; h_{sw}^{th} is swelling deformation in thawing soil in its transition from the frozen to the thawed state. It follows from equation (3.2) that, in special cases, the resulting deformation S_{sub} of soil may acquire positive values; for example, during slow thawing of frozen soils with massive cryogenic structure and high density of the mineral skeleton, when the components S_{sub}^i and S_{shr}^{th} are rather small, and the components h_I^{fr} and h_{sw}^{th} are considerable (due to sufficiently high and prolonged water migration into the frozen part of the sample and a large swelling value). In other words, in this case the ground surface does not subside (sink) but rises (heaves).

In natural conditions, the subsidence deformations (sinking of the ground surface) are dominant in thawing soils as a result of thawing out of pore ice with reduction of its volume by 9% (S_{por}), and of ice streaks or schlieren (S_{str}). In the course of quick thawing of ice schlieren, the cavities which appear in the ground do not always become fully closed and, therefore, the total S_{sub}^i value is often less (by several percent), than the total thickness of the thawing pore ice and ice schlieren ($S_{por} + S_{str}$). Only the slow thawing of frozen soils with ice schlieren can cause complete closing of cavities filled with segregation ice, that is, in this case, $S_{sub}^i = S_{str} + S_{por}$. On the whole, the total subsidence will be greater the greater is the ice content and the greater the content of ice schlieren, whose thawing makes up the greatest part of the S_{sub}^i component. Maximum subsidences are typical of frozen soils with laminated and meshy cryogenic structures.

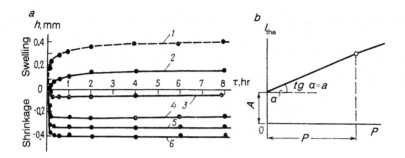

Fig. 3.6. (*a*) Effect of external load *P* on development of deformations with time during thawing of clay (after Yu.G. Fedoseyev) and (*b*) on the relative settlement: 1–6 – loads *P*, MPa (1 -0.01; 2 -0.05; 3 -0.1; 4 -0.15; 5 -0.5; 6 -0.4). Dotted lines: samples previously compacted and frozen under pressure of 0.6 MPa, solid lines: 0.8 MPa. (*b*) Relative compaction ($l_{tha} = e_{tha}$) of the thawing soils as a a a function of external load *P*).

For practical purposes $S_{sub} \approx S_{sub}^i$ can be adopted for sandy and coarse clastic frozen soils. In most cases, however, S_{sub} is seldom equal to S_{sub}^i especially in clayey soils. In fact, swelling is typical of the mineral part of clayey frozen soils even without water migration into the frozen part of the thawing soil (when $h_l^{fr} + S_{shr}^{th} \to 0$). In this case swelling deformations may reach considerable values (Fig. 3.6a), particularly in soils with a mobile crystalline lattice. In mining, cases are known when thawing of super-saturated frozen clayey sediments did not cause subsidence but a positive deformation of the surface due to swelling.

The cases of subsidence deformations of thawing soils described above belong to the so-called 'thermal' subsidence, when frozen soils were not subjected to any substantial loads (dwellings or engineering constructions). The notion of stabilizing subsidence, or S_{tha}, is used in engineering practice; it takes place under the effect of a continuous load *P* on thawing ground until the moment of complete thawing and stabilization of subsidence due to consolidation. Stabilized subsidence in this case is composed of the thawing subsidence S_{sub} or 'thermal' subsidence unaffected by any external pressure, and subsidence by consolidation S_{comp}, which is a direct function of normal pressure, i.e. $S_{sub}(P) = S_{sub} + S_{comp}$. A comparison of compression curves of unfrozen and thawing soils shows that the greatest deformations and changes in the porosity coefficient ε appear not in the process of consolidation of the thawed soil, but in the course of its thawing, i.e. in most cases $S_{sub} > S_{comp}$. The same dependence is shown by plots of changes in relative subsidence ($e_{tha} = (S_{sub} + S_{comp})/l = S_{sub}(P)/l$, where *l* is the thickness of the thawed ground) of frozen soils during thawing and their concurrent densification by the compressing-densifying load P (Fig. 3.6b). The equation of the straight line $e_{tha} = f(P)$ will be $e_{tha} = A + aP$ or

$$S_{tha} = S_{sub} + S_{comp} = Al + alP \qquad (3.3)$$

where A is the coefficient of thawing during thermal subsidence ($P = 0$); a is the coefficient of relative consolidation of the frozen ground on thawing.

The process of thawing in frozen ground greatly changes its texture and structure; adhesion becomes many times less in thawed soil ($C_{tha} \ll C_{fr}$), and permeability increases by several times compared even to unfrozen ground of the same composition. These effects determine the character and intensity of contraction of thawed soil, i.e. the consolidation subsidence value S_{comp}. The ice content i in thawing soils is an important index of estimation of their densification subsidence with time. The theory of filtration consolidation can be applied to ground with high ice content, and S_{comp} will be first of all determined by conditions of water filtration from the thawed zone of the ground. In the case of ground with low ice content, the theory of creep is primarily applied, and S_{comp} is determined by the structural transformations of soil as a result of visco-plastic displacement of particles and aggregates of the soil.

3.3 Physical-mechanical processes in frozen soils caused by changes in temperature

Temperature changes in soils cause fairly large volumetric deformations (contraction or expansion) and 'all-round' stresses (extension or compression); for example, as the temperature falls, the energy of the crystalline lattice grows owing to the lower intensity of thermal movement of its atoms and molecules and to corresponding reductions of the size of the lattice, which is manifested in a certain reduction of the coefficients of linear, α, or of volumetric, α_v, thermal expansion (Fig. 3.7a). Coefficients of thermal expansion are, apparently, higher in rocks and minerals with a lower energy of the crystalline lattice (Fig. 3.7b). Moreover, since quartz has a high value of coefficient α, then the greater the SiO_2 content in soils, the larger are their α values (Fig. 3.7c).

Temperature deformations of such a complex and multicomponent physico-chemical system as the ice-containing frozen soil are caused, on the one hand, by temperature deformations of the individual components of the soil (mineral skeleton, unfrozen water, ice) and, on the other hand, by volumetric deformations of frozen soils as a result of phase transitions of water in the range of negative temperatures and by development of such structure-forming processes as coagulation, aggregation and dispersion of

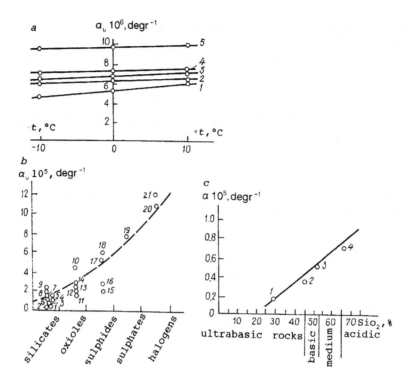

Fig. 3.7. (*a*) Dependence of coefficient of volumetric thermal expansion α_v on temperature, various rocks (1 – granite, 2 – diabase, 3 – limestone, 4 – tuffolava, 5 – sandstone). (*b*) Dependence on mineral composition (1 – orthoclase, 2 – zircon, 3 – sillimanite, 4 – bytownite, 5 – labradorite, 6 – topaz, 7 $\pm$ albite, 8 – augite, 9 – hornblende, 10 – quartz, 11 – corundum, 12 – rutile, 13 – hematite, 14 – magnetite, 15 – sphalerite, 16 – pyrite, 17 – chalcopyrite, 18 – halinite, 19 – gypsum, 20-sylvite, 21 – halite). (*c*) Dependence of coefficient of linear thermal expansion α on silica content in rocks (1 – dunite, 2 – gabbro, 3 – diorite, 4 – granite)).

soil particles, microfissuring, migration and redistribution in the volume of individual components of the ground system, changes in porosity, etc.

The resulting effect of the change in the frozen soil volume, caused by the changes in its temperature, does not, therefore, amount simply to the sum of temperature deformations of individual components, but varies greatly in soils with different chemical-mineral composition, dispersion, ice content-humidity, cryogenic structure and texture, because the soils have a strikingly varied development of structure-forming processes. For example, in the temperature interval from -1 to $-10°C$, the coefficient α in frozen clay will be $1 \times 10^{-2} - 1 \times 10^{-4}°C^{-1}$, in frozen sandy silts and silty clays $1 \times 10^{-3} - 1 \times 10^{-4}°C^{-1}$ and in sands $1 \times 10^{-4} - 1 \times 10^{-5}°C^{-1}$, whereas the coefficient of thermal expansion of the basic rock-forming minerals mainly keeps within $(1 \text{ to } 10) \times 10^{-6}\,{}^\circ\text{C}^{-1}$ and of ice within $(3 \text{ to } 6) \times 10^{-5}\,{}^\circ\text{C}^{-1}$. In other words, the mechanism of temperature deformation in frozen soils is much more complex than in continuous solid media. In sandy soils α values are additive and can be obtained as a sum of the deformations of the frozen soil components, except deformations appearing during phase transitions of water when its volume changes by 9%. In clayey frozen soils when negative temperatures change slowly and local temperature gradients appear in soil the unfrozen water is redistributed through the sample volume. For example, during all-round cooling the central part of a sample shrinks and dehydrates due to migration of unfrozen water into the peripheral parts. At the initial stage of water filling the pores ($G = 1$), the frozen soils heave, while at $G < 1$ their volume remains practically unchanged. The thermal deformations of soil components are secondary, while the values of thermal deformation of frozen soils can be several times as large as, and even exceed by several orders of magnitude, the values of the thermal expansion or contraction of its components. That is why the fall of temperature in non-ice saturated soils ($G < 1$) causes considerable reduction in the volume of frozen clay soils as a result of redistribution of water and of shrinkage. In ice saturated clay soils ($G = 1$), contraction deformations are essentially reduced, and the volume of soil can even grow due to the dominance of heaving. During this process, the stabilization of temperature deformations of frozen clay soils takes place rather slowly over several days (and even tens of days) after the samples have acquired an equilibrium temperature. This process is probably associated with the inertia of internal structural transformations of soil. This phenomenon was called the *consequential effect of temperature* by I.N. Votyakov and S.Ye. Grechishchev.

The basic factors determining the values and progress of temperature deformations of frozen fine-grained soils with concurrent contraction and expansion of their components are, therefore, the phase transitions of water, the presence or absence of free air porosity, and water exchange during temperature changes. The intensity of water exchange depends on the type of soil, its composition, structure and ice saturation as well as on external thermodynamic conditions (Fig. 3.8).

Development of thermal stresses in frozen soils depends on irregular changes of volume of soil elements. Therefore, these stresses should be grouped as *volumetric-gradient stresses* which are functions of restricted deformations in the volume of frozen soil. The authors of several works on that problem conclude that the value of these stresses may change within a wide range depending on temperature interval, composition and structure of frozen soils and conditions of cooling. It has been established that with

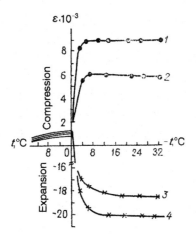

Fig. 3.8. Deformations of sandy silty material (1 and 4) and of polymineral clay (2 and 3) during cooling from $+20°C$ to $-30°C$ with water saturation G equal to 0.8 (1 and 2) and 1.0 (3 and 4).

the fall of negative temperature (from -2 to $-15°C$) in frozen soils the stresses grow, and their maximal values are reached at the lowest negative temperatures. With an increase in the content of fine material and of the content of montmorillonites in soil, the thermal stresses also increase (Fig. 3.9a). It is obvious that if the frozen soil samples (or, more strictly, the ground mass) are sufficiently large, the stresses, totalled along the length of the cooling mass, will always exceed the resistance of frozen soil to rupture, thus causing frost cracks and relaxation of stresses.

Temperature cracking of earth materials is particularly evident in frost shattering of rocks at low temperature in natural conditions under seasonal and diurnal variations of negative temperatures. The temperature changes in the ground may reach $100°C$, and annual temperature variations may penetrate tens of metres.

The diurnal and annual temperature fluctuations in the upper layer result in irregular *extension or compression deformations* which attenuate with depth. Due to its lesser deformation, the lower layer at a relatively lower absolute temperature holds back the complete development of deformations in the upper layer with large temperature deformations. In this way, in these layers 'unallowed' deformations appear, related to the temperature gradient with depth, which finally cause volumetric tensile stresses P_t^d or compression stresses P_t^c (Fig. 3.9b). When these stresses exceed the local resistance of ground to rupture ($P_t^d > \sigma_{rup}$), then temperature cracks, mostly vertical, appear and develop.

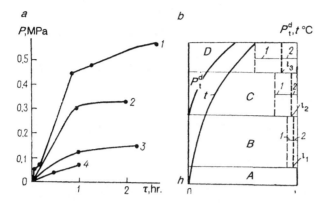

Fig. 3.9. Development of stresses and strains due to temperature in frozen soils. (*a*) Change in stress with time on reduction of *t* from $-2\,^{\circ}C$ to $-15\,^{\circ}C$: 1 – montmorillonite, 2 – kaolinite, 3 – sandy silty material; 4 – sand. (*b*) Change of temperature *t*, all-round stresses of distension P_t^d and linear dimension of frozen soil sample, *l*, against height, *h*. 1 – unallowable temperature deformations of soil, 2 – actual deformations, $l_0 - l_3$ – boundaries of soil sample; *A–D* – conventionally identified layers).

A striking manifestation of frost shattering in fine-grained soils is the formation of polygonal (in plan) nets of cracks and systems of intersecting wedgelike-vein underground ice. B.N. Dostovalov (17) has estimated the sizes of the polygons of cracking from data on temperature stresses developing in frozen soils. The distance between the vertical cracks (the size of polygons) can be determined using the equation:

$$x = \frac{2P_t^d}{\alpha G \operatorname{grad} t} \tag{3.4}$$

where P_t^d is the rupturing stress (volumetric-gradient tensile stresses equal to the instantaneous resistance of frozen soil to fracture); α is the coefficient of linear temperature deformation of the ground; G is the elasticity modulus during displacement; grad *t* is the temperature gradient with depth of the frozen soil. If the temperature gradients in homogeneous materials are small, then large rectangular blocks are formed. As these gradients grow, the rectangular cleavages are split consecutively by frost cracking of the second and higher generations, thus producing progressively smaller blocks. In heterogenous frozen soils, the process is more complicated, and frost clefts cannot be strictly parallel, but are mostly tetragonal and polygonal varieties.

The system of general temperature cracks, which break up frozen ground into rather large blocks and lumps, can be combined with smaller frost cracks (microcracks), which break up minerals and fracture rocks (up to the silt fraction). Temperature disintegration, or temperature destruction of fragments of rocks or minerals, is produced by different values of thermal expansion of the components of frozen rock, causing 'unallowed' deformations, and, consequently, volumetric-gradient stresses P_a on the boundaries of these components. Macrocracks occur when these stresses exceed the local strength of the rock or mineral fragment ($P_a > \sigma_{coh}$). Naturally, rocks will disintegrate not momentarily, but gradually as microcracks appear, develop, merge, and grow due to repeated heating-cooling cycles.

Disintegration and fragmentation of debris of polymineral rocks are determined by volumetric-gradient stresses related to the maximum difference between thermal expansion coefficients of the components and minerals constituting the rock ($\Delta\alpha = \alpha_{max} - \alpha_{min}$). The greater this difference, the greater, evidently, will be the 'unallowed' temperature deformations and, consequently, the volumetric-gradient stresses P_α in the rock, i.e.

$$p_\alpha = k_t - \Delta\alpha_{max} - \Delta t \tag{3.5}$$

where k_t is the coefficient of proportionality, or transition of 'unallowed' deformations into stresses, Pa °C.

3.4   Physical and chemical processes in frozen soils caused by an external load

In frozen soils under the effect of external loads, mechanical stresses developed by means of interrelated and successive processes are: elastic deformations (conventional-momentary and elastic aftereffect), plastic deformation (attenuating creep, viscous flow, unattenuating creep or progressive flow) and destruction (brittle, when the body loses its continuity, or plastic with loss of soil stability). In these processes the frozen soils show *rheological* properties, caused by their characteristic inner relations and by the presence of ice, which is an ideal fluid body. The specific composition and structure of frozen soils, sharply different from other solid bodies, essentially results, when under even small loads, in a process of deformation which may continue for a long period of time. The type of deformation depends on the load value. Under small permanent loads and with free lateral deformation (compression, extension or dislocation) the process develops at a declining rate, i.e. there is attenuating creep (Fig. 3.10a; curves $\sigma_7 - \sigma_8$). With larger loads, when the stress in the frozen soil is above a certain limit called *the limit of long-term strength* σ_{lon} or *creep limit*, the relative deformations develop at an increasing rate, i.e., there is unattenuating creep (Fig. 3.10a; curves $\sigma_1 - \sigma_6$).

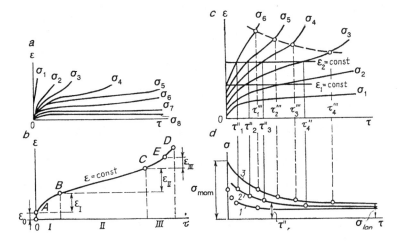

Fig. 3.10. Creep curves for frozen soil: (*a*) group of curves for various applied loads $\sigma_1 > \sigma_2 > \ldots > \sigma_7 > \sigma_8$; (*b*) ideal curve of creep (*I–III* – creep stages); (*c*), (*d*) compilation of relaxation curves from curves of creep (1 – at ε = constant; 2 – at ε_2 = constant; 3 – long-term strength).

A schematized curve of creep can be divided into several segments showing different stages of deformation (Fig. 3.10b). Segment $0A$ represents momentary deformation. Depending on the load value, this deformation can be elastic or elastic-plastic, and it entirely or partially disappears when the load is lifted. The segment AB shows deformation at a reducing rate, the unsettled creep stage (stage *I*). At this stage, the deformation of frozen soils after lifting of the load disappears only partially with time (plastic aftereffect), because the deformation includes structurally reversible and structurally irreversible and plastic deformations. The attenuating creep either continues until reaching a certain final value ε depending on stress σ, or develops continuously at a reducing rate. In this case the deformation does not stabilize, but grows without limit (the so-called secular creep). In both cases the deformation rate $\dot{\varepsilon}$ approaches zero. During unattenuating deformation, when $\dot{\varepsilon}$ = const (segment BC), the stage of stable or viscoplastic flow (stage *II*) occurs. This stage is entirely irreversible.

At high stresses, a progressive stage appears (segment CD), the stage of destruction (stage *III*). At this stage, the deformation rate grows and results in brittle or viscous failure of the frozen soil. Stage *III* should be divided into two sub-stages; during the first (segment CE) the plastic deformation still develops and failure is not reached, while the second substage (segment ED) leads to failure through intensive microfissuring and catastrophically rapid growth of deformations.

The periods and roles of individual stages of creep depend on the amount of load. The larger is the load, the shorter is the stage of settled flow and the sooner the progressive flow starts. The role and significance of different deformation stages also depend on soil properties. In frozen soils all three stages of creep develop, and the more plastic the ground and the higher its ice content, the greater is the significance of stages *II* and *III*. Unattenuating creep in ice appears practically at any stress; that is why the stages with constant and increasing rates are the most important. For example, the value of the critical displacing stress for ice, when its plastic flow starts, is not above 0.01 Mpa. The limit of the long-term strength of ice during displacement (at $-0.4\,^{\circ}$C) is not more than 0.02 MPa.

Relaxation of stresses is another, no less important manifestation of rheological processes in frozen soils. This means that to maintain a constant value (unchanging with time) of deformation, it is necessary to gradually (with time) reduce the load, i.e. to reduce stresses in the soil. In other words, the identical value of deformation of frozen soils ($\varepsilon = $ const) can be reached by applying different loads, but a longer time period is necessary, if the load is reduced. Consequently, the longer the influence of the load on the frozen soil sample (displacement, compression or distension), the less load is needed to reach failure.

The plotting of the stress relaxation curve $\sigma = f(\tau)$ is carried out with a group of creep curves $\varepsilon = f(\tau)$ (Fig.3.10c) at any fixed value of deformation, $\varepsilon = $ const (Fig. 3.10d; curves 1 and 2). By using the creep curves, a curve of long-term strength of frozen soil can be plotted (curve 3). For this purpose, the moments of time when progressive flow starts are marked on the curves of unattenuating creep (the beginning of stage *III* of deformation), and then curve $\sigma = f(\tau)$ is plotted from the obtained data. The initial ordinate of the curve corresponds to the momentary strength of frozen soil σ_{mom}. With a sufficiently long time period, when the change in curve $\sigma = f(\tau)$ can be practically disregarded, the stress value will correspond to the limited long-term resistance to failure of frozen soil σ_{lon}. This value is an important characteristic, because at $\sigma > \sigma_{lon}$ the creep will be unattenuating (which leads finally to destruction of the frozen soil), whereas at $\sigma < \sigma_{lon}$ the creep deformations will be attenuating (in time). Consequently, σ_{lon} is the highest stress at which progressive flow does not appear.

A study of behaviour of different bodies and materials under load shows that for elastic bodies this is sufficiently well described by the dependence between stress and deformation (Hooke's law). The deformation in viscous bodies, however, can grow with time under constant load, and the dependence 'stress-deformation' loses its significance. In this case, the 'stress-defor-

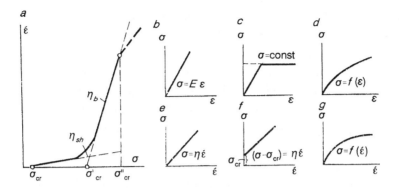

Fig. 3.11. (*a*) Rheological curve of fine-grained soil and (*b* – *c*) diagrams of deformation of bodies: (*b*) elastic (Hooke's law), (*c*) elastoplastic; (*d*) non-linear-elastic, (*e*) viscous (Newton's law), (*f*) viscoplastic (Bingham's law), (*g*) non-linear viscous).

mation rate' dependence is normally applied. As noted by S.S. Vyalov (4), this seems more reasonable, because there is an analogy between the diagrams of deformation of elastic and viscous bodies (Fig. 3.11), which allows application of viscosity theory and of plasticity theory for problems of viscous flow, i.e. linear in the former case and non-linear in the latter. These solutions, with appropriate assumptions, can be used by formal substitution of the deformation value ε by its rate $\dot{\varepsilon} = de/d\tau$.

The graphic dependence between the stress and the rate of steady flow is called the *rheological curve* and is described by Newton's or Bingham's equations (Fig. 3.11e,f). For thawed and frozen fine-grained soils however, the rheological curve is more complicated (Fig. 3.11a). The flow of such soils at constant rate starts only after the critical σ_{cr} is exceeded. Up to that limit the deformations usually attenuate. When the stress $\sigma_{cr} < \sigma < \sigma'_{cr}$, the rheological curve approaches a straight line, i.e. the flow in this segment of curve develops with constant maximal viscosity, called the Shvedov viscosity: $\eta_{sh} = (\sigma - \sigma_{cr})/\dot{\varepsilon} = \tan a$. When $\sigma'_{cr} < \sigma < \sigma''_{cr}$, there appears another linear segment on the rheological curve of frozen soils, where deformation (flow) occurs with constant minimal viscosity, called the Bingham viscosity: $\eta_b = (\sigma - \sigma'_{cr})/\dot{\varepsilon} = \tan \alpha'$. Therefore, the deformation (flow) of frozen soils at constant rate has two critical values of viscosity, i.e. the highest η_{sh} viscosity, which corresponds to a practically undisturbed structure, and the lowest η_b viscosity, corresponding to the ultimately destroyed soil structure.

Specific ice-cementing bonds and cryogenic textures and structures are of fundamental importance in the development of rheological processes in frozen soils. Deformation of frozen soils due to displacements of individual soil particles or microaggregates in relation to each other takes place in the films of bound water and in the ice inclusions dividing them; most of these films and inclusions are actually areas of lower strength, i.e. 'defects' in the body of frozen soils. Further development of the 'defects' and their transformation into microcracks or local dislocation zones, should play the principal role in frozen ground destruction and cause weakening of interparticle and interaggregate ties and lower local structural strength in some of its parts.

When ice content is high in frozen soils, then in the course of different types of deformations (shear, compression, extension) both momentary and long-term strength are reduced and reach the critical strength of ice. Consequently, during long-term deformations of frozen soils in the zones of displacement, microdislocation or extension, the local strength should decrease with time as a result of the supply of migrating unfrozen water and the increase of the ice content. The orientation of the forming ice microschlieren and the reorientation of soil microaggregates along the planes of displacement and microdislocation also cause reduction of resistance to dislocation, tension or compression in frozen soils locally. This process causes development of such deformations as unattenuating creep, flow or progressive creep, which terminate in destruction of the continuity and stability of the soil. In other words, a load which is constant in time causes greater deformations due to progressive reduction of frozen ground strength.

Concurrently with processes of structural transformation (weakening) during slow deformation of frozen soil described above, counter-processes and events develop in it. Among them are 'healing' of structural defects, closing of microcracks, reduction of interaggregate and aggregate porosity, rearrangement and denser packing of ground particles and aggregates, re-establishment of ties and greater molecular and cementing adhesion, etc. This combination of processes naturally results in strengthening of the frozen soil and its decreased deformation. According to S.S. Vyalov, N.A. Tsytovich and other researchers, the generation and development of rheological processes (creep and stress relaxation) in frozen soils are determined by two concurrent counter-processes of structural transformation: strengthening and weakening. If strengthening prevails, then creeping deformation attenuates. When weakening and strengthening deformations are mutually compensated, then the deformation develops continuously at a constant rate (viscoplastic flow). If weakening prevails over strengthening, then the deformation grows and unattenuating (progressive) creep develops.

The character and mechanism of slow deformation (creep) in frozen soils are different at different stages of creep (Fig. 3.10b). At the first unstable stage (attenuating creep) only small changes in the structure of frozen soil are observed, such as lesser inter-aggregate porosity, squeezing out of the air, a certain redistribution of ice content and moisture, healing of old and newly formed defects in structure and, as a result of displacement and denser packing of ground particles and aggregates, restoration of old and the appearance of new ties between particles. On the whole, the frozen ground is strengthened and deformations attenuate.

The second stage of deformation shows greater structural changes in frozen soil. The aggregates begin to disintegrate and fall into fragments, their basal planes start to reorient in the direction of the vector of dislocating tangential stresses or normal to the compressive stresses. This process reduces the resistance of the frozen soil to applied load. Destruction of structural bonds increases in the weakest areas of the frozen soil and structural defects increase. In this case the processes of formation of structures which strengthen the soils are subordinate. The principal role at this stage of deformation belongs to unfrozen water migration to the zones of concentration of dislocating or extending stresses. In the course of this process, the segregational ice schlieren, lying along the planes of dislocation (microdislocation) or tension, encourage sliding or fracture of the material and accelerate the flow process, which results in gradual and regular reduction of resistance of frozen ground to loads, i.e. the monotonous relaxation and deformation of ground at a constant rate.

Finally, a sufficiently long application of load creates areas in the ground with a maximum ice content, in fact an ice body with soil inclusions and the properties not of ground but of ice. With this kind of transformation of frozen soil structure its strengthening becomes practically impossible. Therefore, the deformation process passes to the stage of progressive flow which terminates with plastic loss of soil stability.

Not only the time period of load effect but also its value should be always taken into account when analyzing the mechanism of frozen soil deformation. For example, the creep curves (Fig. 3.10a) show that with the increase of applied load the area of constant deformation rate gradually disappears being transformed into the stage of progressive creep. Under the effect of high loads the unfrozen water migration and ice content redistribution in frozen soil are obviously insignificant.

If we disregard the cases of limited lateral deformations (triaxial test) or completely impossible lateral deformations (compression) of frozen soils under the effect of external loads, then the behaviour of ground in all other cases, i.e. during displacement, compression and extension (with free lateral deformation), will have a number of similar features. All kinds of deformations may in this case be classified as *momentary, long-term,* and *destructive*. Of momentary deformations in frozen soils the elastic deformations have the most practical significance, and of long-term deformations the attenuating creep and viscoplastic flow, degenerating under certain conditions into progressive flow, are the most important from the practical point of view. Among destructive deformations the brittle deformations are usually recognized as important, because they destroy the continuity of the frozen soil, and consequently inadmissible plastic changes of shape occur, resulting in the loss of the bearing capacity and stability of soils. These deformations are closely related to the ability of frozen soils to resist external load, i.e. resistance (momentary and long-term) to compression, extension, dislocation, etc.

The long-term deformations of frozen soils are of particular importance in both theoretical (geological) and economic aspects. These deformations develop under conditions of free deformation, and the longest of them and of the most practical significance is the settled or viscoplastic flow with constant deformation rate. The curves of creep during compression, dislocation and extension have on the whole a similar character (all three stages of creep and transitions from one stage to the next are clearly outlined), the difference is only in the quantitative property of the process. These conclusions were derived from the experimental data obtained by S.S. Vyalov (4) for frozen soils (Fig. 3.12), showing that, in the case of dislocation and compression, the transition from attenuating to unattenuating creep occurs during stresses which comprise 25–50% of the conventional-momentary strength of the frozen soil and, in the case of extension under loads, 8–10% of σ_{mom}. The stage of unstable flow can develop during several hundreds of hours and longer, the stage of the viscoplastic flow for several thousands of hours, and the stage of progressive flow (until the moment of destruction) from a few to hundreds of hours and more, depending on the amount of applied load.

For practical purposes, the effect of composition, cryogenic structure and properties (including temperature) of frozen ground on long-term deformations is of the greatest interest. In plastic frozen clayey soils with high contents of unfrozen water and ice, high negative temperatures and aggregational and ice-aggregational types of contacts, the stage of the steady flow can continue for a long time, and it passes into the stage of progressing or 'secular' creep at $\sigma \approx 0.1 - 0.5\ \sigma_{mom}$. In solid frozen soils with low ice and unfrozen water content, low negative temperatures, ice-crystalline, coagulation and ice-coagulation contacts, the stage of attenuating deformation

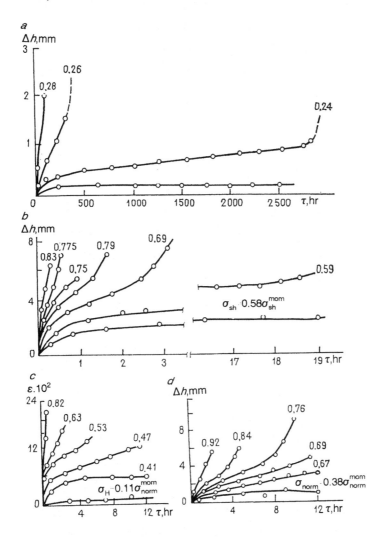

Fig. 3.12. Creep curves for frozen soils (4): (*a*) – silty clay (dislocation along a cylindrical rod frozen at $t = -0.4\,°C$); (*b*) – sandy silt (dislocation on shearing device at $t = -10\,°C$); (*c*) – sandy silt (uniaxial compression at $t = -20\,°C$); (*d*) – polymineral clay (uniaxial compression at $t = -20\,°C$). Δh = deformation, ε = strain).

usually dominates; it often transits immediately (omitting the stage of steady flow) into the stage of progressive flow. The progressive deformation develops at loads $\sigma \approx 0.5–0.7\ \sigma_{mom}$, continues for a relatively short time and results in mostly friable destruction.

The deformation of frozen soils during the stage of *visco-plastic flow* with constant rate can be well traced on rheological curves ($\dot{\varepsilon} = f(\sigma)$); it is determined by the coefficient of viscosity η. The presence of film unfrozen water in the soil is a determining factor in the development of creep and viscosity in frozen ground. According to N.A. Tsytovich (20), during experiments at $t = -0.8\,°C$, the viscosity coefficient in frozen sandy silty material at $W_{tot} \approx 19\%$ was $\eta \approx 1.9 \times 10^{11}\text{Pa s}$; and in frozen clay at $W_{tot} \approx 28\%$, $\eta \approx 0.9 \times 10^{11}\text{Pa s}$; this is almost an order of magnitude less than the viscosity coefficient of pure ice, which at $t \approx -0\,°C$ is $\eta \approx 1.2 \times 10^{11}\text{Pa s}$. These results can be, apparently, attributed to the fact that in the region of considerable phase transitions frozen clay contains much more unfrozen water, which causes greater rate of flow of clay compared to that of sandy silt and of ice. For this reason, as the negative temperature falls, the viscosity of the frozen soil grows. In this case, the viscosity coefficient is not a constant of the frozen soil but it is characteristic of the deformation process, essentially dependent both on the previous history of the stress-deformation of the soil and on the nature of the development of deformations, and on the kind of load (compression, dislocation, extension, torsion) and the method of load application (uniaxial, diaxial or triaxial tests) and conditions of loading.

The behaviour of frozen soils under external load in compression and triaxial tests essentially differs from their behaviour under conditions of free lateral deformation. For example, the creep deformation (without the possibility of lateral expansion of the frozen material) will always be attenuating under compression or under the effect of continuous load regularly distributed over the ground surface. As shown, however, by Brodskaya, Vyalov and Tsytovich, practically all frozen soils and particularly soils at high-temperature or with high ice content, experience considerable compressibility (densification) under load with time. This occurs owing to elastic deformation and closing of empty pores, cracks and other defects in frozen soil and also as a result of the lower porosity of the organic-mineral skeleton, as the unfrozen water and ice are evacuated from the frozen soil. Experimental research shows that compaction (consolidation) of frozen soils is inherently connected with the development of several extremely complicated physico-chemical processes. The most important are phase transitions of ice into bound water, migration of unfrozen water, redistribution of ice content in soil and transformation of microstructure, dislocation of ground particles and their aggregates, etc. On the whole, the pressure transmitted to ice-saturated ground is distributed between the mineral skeleton, ice and unfrozen water, which behave differently with time under load. Therefore, consolidation and compression of frozen soils largely depends on their composition, cryogenic structure, ice content, degree of filling of the pores with ice, and on the negative temperature and the active load (Fig. 3.13).

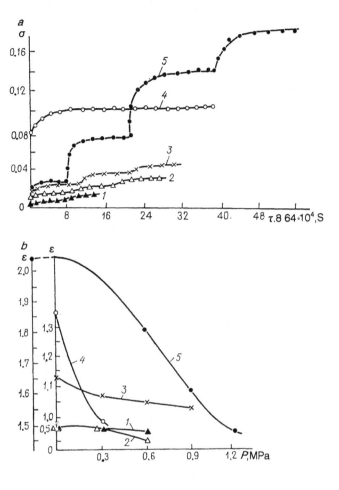

Fig. 3.13. (*a*) Curves of consolidation of frozen soils with time (*b*) compression of the same frozen soils for various loads (with load increments of 0.3 MPa): 1 and 2 – under temperatures of $-3\,°C$ and $-1.5\,°C$, respectively; 3–5 – polymineral clay: 3 – ice-saturated ($G = 1$); 4 – salt added ($Z = 1.5\%$), 5 – not saturated with ice ($G = 0.6$).

Transportation of unfrozen water and its evacuation from the ground system play an important role in frozen ground deformation under compression; as a result of this process the total water and ice contents in soils decrease. The deformation of frozen soils by compressional compaction due to seepage-migration in the frozen soil is often called *initial consolidation*, because it is assumed that deformation is the greatest in the initial period after application of load. In the course of time, its share in the total (stabilized) deformation of the consolidation of frozen soils becomes increasingly less. On the whole, seepage-migrational deformation is most

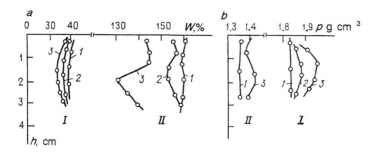

Fig. 3.14. (*a*) Distribution of total moisture content against depth of frozen samples, (*b*) density of same samples, of kaolinite (*I*) and montmorillonite (*II*) clays during compressional compaction by loads 0 MPa, (1), 0.25 MPa, (2), and 1.2 MPa, (3); $t = -1.5°C$).

obvious in water-saturated ($G = 1$) frozen clayey soils where it can be 40–80% of the total (stabilized) deformation.

The mechanism of unfrozen water movement in a frozen soil under compression is, apparently, mostly filtrational (under the application of an external pressure gradient), though the process of migration of film water to the periphery of the sample due to the gradient of the total thermodynamic water potential should be also taken into account. Seepage takes place when, at the boundary between a porous disc and frozen soil, the pressure in the unfrozen water is equal to atmospheric, whereas in the central part of the sample it is at a maximum. Concurrently, in the central part of the sample, due to higher pressure, the unfrozen water content grows as a result of ice melting and the water films get thicker compared to those in the peripheral parts. Therefore, seepage and migration induce the water to move from the centre of the sample to its periphery and beyond the limits of the frozen soil. The outflow of water from the soil system is accompanied by the redistribution of ice content and of the mineral skeleton through the height of the compressed sample, as is clearly shown by the curves of changes in water content and of density of the frozen soil (Fig. 3.14). Moreover, at first the initial consolidation involves the peripheral parts of the sample and later its central part as well. The cryogenic structure of frozen soil samples changed in the course of an experiment from massive (before compressive densification) to microlayered (the layers are oriented at right angles to the load acting) in the central part and micromesh in the peripheral part of the sample (after the compression load effect). A significant reduction of the total water content and first of all of ice as a result of seepage-migrational consolidation in the compressed frozen sample, can be clearly traced in Fig. 3.15.

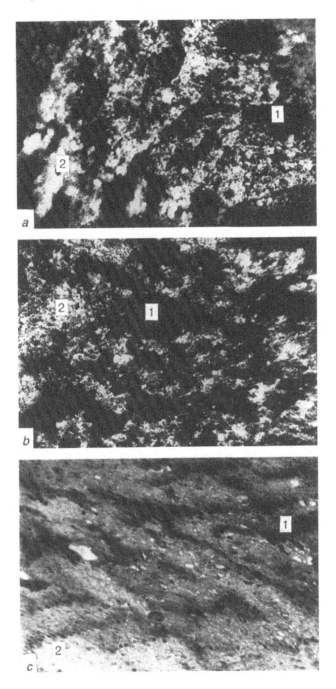

Fig. 3.15. Transformation of cryogenic microstructure and decrease in ice content in a sample of kaolinite clay during its compressional compaction: (*a*) – before the experiment (*b* – *c*) – under loads of 0.03 MPa and 1.2 MPa, respectively. 1 – mineral skeleton; 2 – ice.

Differentiation of the total compression compaction of frozen soils into seepage-migrational (primary consolidation) and attenuating creep deformation (secondary consolidation) is not easy because the two kinds of deformation are indivisible and continuous in time. Compaction of frozen soil as a result of secondary consolidation is usually associated with irreversible dislocations of ground particles, of their aggregates and of ice, which cause sharp changes in cryogenic microstructure. In the course of this process, the microaggregates of the ground approach each other and become larger, thus producing a denser packing. This process, however, does not exclude a possible breaking-up of individual larger aggregates. Concurrently with this dispersion and aggregation of the frozen soil skeleton in the course of attenuating creep, a reorientation of the basic planes of clayey particles and aggregates in the direction perpendicular to pressure, is observed. Densification of frozen soil, due to the decrease in porosity previously filled with ice and to more compact packing of the particles of the frozen soil, makes it much stronger owing to greater electromolecular bonds between the particles as they approach each other. It should be added that, when frozen soil is compressed, the structure of the schlieren and pore ice is also essentially reworked (into more fine-grained ice) with fusion of the sharp facets of ice crystals, and a larger number of viscoplastic flows of ice crystals and of their aggregates.

In general, we can state that the principal causes of compressional compaction of frozen soils under constant load and negative temperature are irreversible displacements of structural elements of the frozen materials which occur as a result of closing of empty pores and cracks and of viscoplastic flow of ice and because of volumetric (viscous) creep deformations of the mineral skeleton with concurrent squeezing out of unfrozen water from the frozen ground system.

4

Structure and texture of freezing and thawing soils

4.1 Thermal-physical and physical-mechanical conditions of development of migrational-segregated ice interlayers

The particular features of thermal-physical, physical-mechanical and physico-chemical processes in freezing ground create conditions for the formation of migrational-segregated ice layers with various orientation, frequency and thickness; in other words, they cause the appearance and development of *cryogenic structures*. These processes in freezing ground essentially transform the structure, density and strength of soils. For example, as a result of dehydration of the unfrozen part of the ground (migration of water to the frozen part), the processes of coagulation, aggregation and reorientation of soil particles develop intensively. These processes induce the formation of soil aggregates of different size and shape, thus creating various 'defective' zones which may later become zones of 'concentration' of significant stresses following from further development of shrinkage and swelling-heaving stresses in the freezing ground. They are generated in the still unfrozen desiccated part of the freezing soil. The bound water in the soil in the zones of critical stresses is under tension and, therefore, has a lower thermodynamic potential. This results in movement of moisture towards these zones (horizontal, vertical and tilted). As a result, the water migration in the unfrozen and freezing zones of the ground is determined not only by grad t, but by grad P_{shr} and grad P_{sw-ho} as well, i.e.

$$\mu_w = f(t, P_{shr}, P_{sw-ho})$$

The increase of water content in the zones of concentration of stress, which are located at the boundaries of the aggregates formed, continues only up to complete filling of the pores with water in the zone. Subsequent increase of water content is possible apparently only if ground cohesion is overcome by local microcracking, when a quick (jumplike) transition of water into ice takes place (the algebraic value of the potential increases sharply) in the former zones of 'concentration' of stresses. Owing to the increase of water volume by 9%, another force becomes active; the crystallization pressure of ice. The generation of a continuous ice layer is gradual (as is the formation of a gaping shrinkage crack in the unfrozen dehydrating soil). At first, in the freezing part of the ground near the freezing front, the discontinuous streaks of ice (seen only under a microscope) appear in jumps in the zones of concentration of stresses. At later stages, they become thicker schlieren, merge and form continuous linear layers of segregated ice, seen clearly in experiments – not near the freezing front, but in the already frozen part of the ground (Fig. 2.12).

Therefore, the process of shrinkage and swelling-heaving in freezing soils on the one hand causes changes in volumetric stresses, and, on the other hand, to a certain degree, determines the configuration, or the type of the future cryogenic structure.

It is obvious that knowledge of the thermal exchange and migration of water in freezing soil is essential for the study of the mechanism and kinetics of segregated ice formation. However, the role in the formation of cryogenic structure of physico-chemical and physical-mechanical processes should be accounted for in full measure. In this respect, the thermal-physical conditions (heat and mass exchange) are necessary but insufficient for the appearance of segregated ice layers. Needed for this purpose are those physical-mechanical conditions under which the local strength of soil is overcome (without gaping cracks) and the microlayers of ice appear.

At present, *the thermal-physical conditions* of formation of cryogenic structures in freezing grounds are studied in greater detail than the physical-mechanical ones. The previous chapters reviewed the heat and water transportation and ice accumulation in freezing soils. This chapter is devoted to the general analysis of thermal-physical conditions, which cause the growth of ice layers and which are controlled by the heat flow relations in the unfrozen and frozen parts of the ground. Since the formation and growth of segregated ice layers in freezing soils occurs not only near the freezing boundary, but also within the range of negative temperatures, the general expression of the thermal conditions for the formation of schlieren segregated ice will be:

$$Q_{fr} - Q_{unf} = \Delta Q = Q_{ph}^{is} + Q_{ph}^{im} \tag{4.1}$$

where Q_{fr} and Q_{unf} indicate the quantity of heat passing, respectively, through the cooling surface in the freezing zone of the ground and from the unfrozen zone into the frozen zone; Q_{ph}^{is} and Q_{ph}^{im} indicate the heat of phase transitions, that which escapes during formation of ice-cement from water fixed by the freezing process *in situ*, Q_{ph}^{is}, and that from formation of segregational ice layers or ice-cement from migrating water in the frozen part of the soil, Q_{ph}^{im}.

It is obvious, that at $\Delta Q > 0$ the ground will freeze and its cryogenic structure will be formed. If $\Delta Q = Q_{ph}^{is}$, freezing leads to a massive cryostructure. If $\Delta Q > Q_{ph}^{is}$, water migration and, possibly, formation of segregated ice layers or ice-cement will take place (when $Q_{ph}^{im} > 0$). The thermal conditions for the growth of segregated ice in the frozen zone of thawing ground can be similarly expressed, if a little more complicated.

From experimental data on the thickness of the forming ice layers, it is obvious that the layers, parallel $h_{|}$ and perpendicular $h_{\perp}$, to the freezing front, can be represented, in a general way, by the following:

$$h_{|} = f\left(\frac{I_{|}}{\upsilon fr}\Delta x_{|}\right), \qquad h_{\perp} = f\left(\frac{I_{\perp}}{\upsilon_{fr}}\Delta x_{\perp}\right) \tag{4.2}$$

where $I_{|}, I_{\perp}$ represent intensity of the water migration flow to the ice layers which are parallel and perpendicular to the freezing front; the intensity is a function of grad P and grad t in the frozen part of the soil; $\Delta x_{|}$ and $\Delta x_{\perp}$ are the intervals of the areas of growth of the ice layers which can be obtained (as shown below) only with allowance for the physical-mechanical conditions of segregated ice formation; υ_{fr} is the rate of freezing.

The generation of ice microlayers probably begins near the freezing front and reaches its maximum near the boundary where the direction of deformation changes (Fig. 3.3) and where intensive ice separation is observed (Fig. 2.12). This boundary is practically the boundary of equality of shrinkage and swelling-heaving stresses. It is not permanent with time, depends on the composition and structure of the soil and on conditions of freezing, and is usually located where temperatures are from -0.2 to $-4°C$.

Ice interlayers parallel to freezing front

These appear as a result of development of horizontal zones of concentration of stresses caused by dislocating (or shearing) stresses $P_{sh} = P_{shr} + P_{sw-ho}$ which appear during variously oriented deformations in the dehydrating and swelling layers of the ground. The horizontal shearing stresses naturally reach their maximal values on the boundary (ξ_{def}) of change in the direction (sign) of the soil deformations (Fig. 4.1). These stresses are directly observed in the whole freezing zone. In the frozen part and in the unfrozen dehydrated part of the soil this effect is indirectly active through the overlying and underlying layers of soil. The result of this

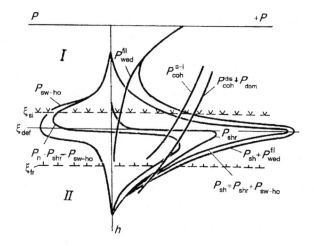

Fig. 4.1. Development in freezing soil of stresses of shrinkage P_{shr}, swelling P_{sw}, heave P_{ho}, horizontal shear P_{sh} and normal volumetric stresses P_n and thus the wedging pressure of film water P_{wed}^{fil}, local cohesion P_{coh} and cohesion at the boundary 'soil – ice layer' P_{coh}^{s-i}. Boundaries ξ_{fr}, ξ_{si}, ξ_{def} are those of freezing, visible segregated ice formation and change in deformation direction in the sample, respectively; *I*, *II* are frozen and thawed parts of the ground, respectively.

interaction is horizontal shearing (dislocation) stresses primarily along zones of defects parallel with the freezing front. Boundaries of aggregates and blocks of the ground become such zones. In this way, the horizontal zones of concentration of stresses in soil form and, when the shearing stresses and the wedging pressure of thin water films, P_{wed}^{fil} develop further, they can be transformed into zones of local destruction of the soil. This occurs as a result of overcoming of the local cohesion of the ground (resistance to dislocation P_{coh}^{dis}), when

$$P_{sh} + P_{wed}^{fil} = (P_{shr} + P_{sw-ho}) + P_{wed}^{fil} > P_{coh}^{dis} + P_{dom} \qquad (4.3)$$

where P_{dom} is the distribution of the external (including anthropogenic) pressure according to the depth of freezing ground. The satisfying of this condition means that water in the horizontal zones of concentration of stresses is no longer under stress, transforming by jumps into ice, generating horizontal ice microschlieren.

The area of the ground $\Delta x\|$, where the condition of horizontal segregated ice formation is satisfied (Fig. 4.2), is a potential zone of concurrent generation and growth of ice layers parallel with the freezing front. It is determined, according to the plot in Fig. 4.1, by intersection of curves: $P_{sh} + P_{wed}^{fil} = f(x)$ and $P_{coh}^{dis} + P_{dom} + f(x)$. In the upper part of the freezing

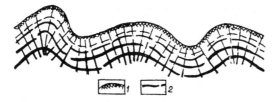

Fig. 4.2. Formation of superimposed cryogenic structures when the boundary of freezing soil is irregular: 1 – surface of the ground, 2 – ice streaks.

zone and of the frozen zone, generation of new horizontal ice schlieren of the second and subsequent generations may occur. This process takes place in any case however, mostly as a result of sharply increasing wedging activity of fine films of unfrozen water (P_{wed}^{fil}) because the shearing stresses diminish (Fig. 4.1).

The enlargement of already formed ice microlayers may occur also within the frozen zone of the freezing ground at sufficiently low negative temperatures. This is related not only to the sharply increasing wedging by thin water films, but also to the need to overcome at the ice schlieren boundary not the local ground resistance to dislocation (P_{coh}^{dis}), but the much smaller cohesion (P_{coh}^{s-i}) of the ice layer to the frozen soil ($P_{coh}^{s-i} < P_{coh}^{dis}$). When the condition $P_{sh} + P_{wed}^{fil} > P_{coh} + P_{dom}$ is not satisfied, then the appearance and growth of horizontal ice layers cease – for example, in the following cases, when:

1) the ground freezes at a very high rate, when water migration can barely take place and volumetric stresses do not develop in the ground;
2) a highly cemented rock is freezing the strength of which cannot be overcome by volumetric stresses;
3) freezing occurs under very high external pressure (for example, under high overburden pressure).

In the latter case, water migration occurs only until the pores of the ground are completely filled with ice, because after that the film water, due to lack of free space, will not move even if there are gradients grad μ_w or grad $P_{wed}^{fil} = f(\text{grad } t)$.

Vertical ice schlieren

These ice streaks in the freezing ground are associated with tensile (normal) stresses as a manifestation of irregular shrinkage varying with depth, in the unfrozen part of the freezing ground (to the level of the change in the sign of deformation m_{def}). The swelling-heaving part of the freezing soil which lies above, obstructs the shrinkage deformations of the dehydrating part. Consequently, in the case of freezing, the volumetric strain (normal) stresses P_n must be equal to the difference between the stresses of shrinkage and of swelling-heaving. In this aspect the boundary of the change of direction of deformations ξ_{def} must be the conditionally 'free' ground surface, where normal stresses are equal to 0 ($P_n = P_{shr} - P_{sw-ho} = 0$). Below this boundary lies the maximum of normal straining stresses (Fig. 4.1).

The stresses P_n, distending the ground horizontally, must first of all affect the vertical and tilted zones of defects which are the lateral boundaries of soil macroaggregates and blocks, formed during the dehydration of unfrozen and freezing parts of the soil. In this way the vertical zones of concentration of stresses appear. The water in these zones is in a state of tensile stress, which permits the water to be supplied there under the effect of grad P_n and provides the possibility of overcoming the local strength of the ground to the point of rupture, $-P_{coh}^{rup}$ (with further development of strain-inducing normal stresses). The vertical microlayers of ice appear as a result of this process. The area of the possible generation of their $\Delta x_\perp$ is determined by the intersection of curves $P_n = f(x)$ and $P_{coh}^{rup} = f(x)$, while a condition for generation of vertical segregated ice layers is:

$$P_n = P_{shr} - P_{sw-ho} > P_{coh}^{rup} \tag{4.4}$$

This condition (Fig. 4.4) does not exclude the possibility of generation of vertical microcracks and fissures in the unfrozen part of the freezing soil (during a particular kind of freezing process).

The wedging activity of thin films of migrating water does not participate in the generation and growth of the vertical ice schlieren (contrary to the horizontal ones). This is attributed to the movement of water to the vertical zones of concentration of stresses only under the effect of grad P_n and not under the effect of grad t, which would cause the appearance of grad P_{wed}^{fil}. This may relate to the frequent observation in nature of thinner vertical ice layers than horizontal ones.

A comparison between the generation of the horizontal and of the vertical ice layers shows that, in the first place, the probability of generation and growth of layers parallel with the freezing front is greater, because the total sum of P_{sh} stresses and of the wedging pressure of the film water P_{wed}^{fil} are involved. The formation of ice streaks perpendicular to the freezing front, on the contrary, is determined by the difference of stresses P_n. Secondly, the area of generation and vertical growth of horizontal layers in the freezing soil is larger and, therefore, the time of their formation is longer. Consequently, they must be thicker and more frequent than the vertical ice streaks. Thirdly, the vertical ice layers always appear at lower negative temperatures and grow in a wedge form pointing downwards, thus creating more favourable conditions for their rapid progression to the freezing boundary. Therefore, in experiments, as a rule, the growth of vertical ice layers and cracks is observed visually to overtake the growth of the horizontal layers and cracks in the unfrozen zone.

4.2   Basic types of cryogenic structures

To explain reliably the particular features of generation and growth of different types of superimposed and inherited cryogenic structures (Table 4.1) we should know the mechanism and pattern of formation of the vertical and horizontal ice layers in soil during freezing (thawing).

Superimposed cryogenic structures normally appear in soils which are relatively homogeneous (prior to freezing), as a result of the freezing process and not as a result of the primary (initial) structure of the unfrozen soil. Three basic types are usually distinguished among the superimposed cryogenic structures of freezing soils according to the presence, shape and location of ice layers in them. These structures are: the *massive*, where ice schlieren are not visually apparent; the *layered*, with elongated-oriented ice layers; the *mesh form*, when the ice schlieren in the section form a net or a grid.

The *massive* cryogenic structure is formed when the physical-mechanical conditions of segregated ice formation are not met or when the thermal-physical conditions for formation of cryogenic structure are in general not satisfied. The latter case is usually when there is either a high rate of freezing of water-saturated or unsaturated soils, or at any rate of freezing of low moisture content fine-grained soils or of coarse-grained soils – when water migration is practically absent, while the ground water in pores is fixed by freezing. In this case the contact, film, pore and basal types of ice cement formation are the most common.

The formation of massive cryogenic structure can also take place with water migrating in the freezing ground but only if the physical-mechanical conditions for segregated ice formation are not satisfied and the stresses in the ground do not overcome its local strength. The local strength can be responsible for controlling the general increase of ice content and for the absence of segregated ice formation in over-compacted or cemented soils and in soils whose initial moisture content is less than that of the shrinkage limit. In these cases, however, there is the possibility of formation of separate

Table 4.1. *Classification of cryogenic structures by conditions of their formation in freezing soils*

Lithological features of soils	Thermal physical conditions of formation of cryostructures $Q_{fr} - Q_{unf} = Q_{ph}^{is} + Q_{ph}^{im}$	Physical-mechanical conditions for the appearance of cryostructures		Type of cryostructure
		Condition of formation of ice schlieren ∥ to freezing front $(P_{sh} + P_{wed}^{fil}) > (P_{coh}^{dis} + P_{dom})$	Condition of formation of ice schlieren ⊥ to freezing front $P_n = P_{shr} - P_{sw-ho} > P_{coh}^{rup}$	
Soils with varied composition, structure, bedding and properties	Not satisfied	Not satisfied due to unsatisfied thermophysical conditions		Massive (contact, film)
	Satisfied	Not satisfied		Massive (pore, basal)
Fine grained soils homo-geneous in composition, structure, bedding and properties	Satisfied and provides migration of water into the frozen part of freezing soils	Sporadically satisfied		Porphyryitic
		The condition is only beginning to be satisifed providing for formation of only incompletely developed schlieren	Not satisfied	Discontinuous semi-layered
			Sporadically satisfied	Semi-layered phorphyritic
			The condition is only beginning to be satisfied providing for formation of only incompletely developed schlieren	Semi-meshy
				Layered with discontinuous vertical schlieren
		Satisfied	Satisfied	Meshy
			Not satisfied	Layered

ice schlieren and ice nests (porphyritic cryostructure) due to the instability (episodic nature) of the necessary physical-mechanical conditions and due to large pores, micro-cracks and other inhomogeneities in the soil.

The massive cryogenic structure sometimes appears when stresses in a soil could overcome the local cohesion of the soil but a sufficiently large external load (e.g., P_{dom}) prevents this.

The *layered* cryogenic structure is formed during freezing of soils, if the thermal-physical and physical-mechanical conditions are satisfied for formation of segregated ice layers, parallel with the freezing front, but the conditions are insufficient for formation of vertical ice schlieren. Generation and growth of the layered cryogenic structure usually take place in the interval of negative temperatures from -0.2 to $-3\,°C$. This type of structure occurs predominantly in fine soils (sandy-silty, silty and clay soils) and sometimes in dusty sands. The layered cryogenic structure, compared to other types of schlieren cryogenic structure, is more frequent in nature and is most evident at low freezing rates. Its formation begins with the appearance of individual thin lenses and ice microschlieren. Later their size increases, they merge and compose a single elongated ice schlieren parallel with the freezing front which can be traced visually. If the freezing rate v_{fr} decreases and the water flow towards the horizontal ice layer I_w increases, then thicker ice streaks are formed.

A large variety of layered cryogenic structure (layered, lens-like, lens-like-woven, etc.) is caused by the specific composition, structural and geological-genetic features of soils and the various conditions of freezing and intensity of thermal and mass exchange and the structure-forming processes. A number of varieties of layered structure is distinguished by the shape of schlieren, their length and orientation, by the relation between thickness of ice schlieren and mineral soil layers etc. (Chapter 7).

The *mesh* cryogenic structure is formed when the thermal-physical conditions are consistently satisfied and concurrent physical-mechanical conditions for formation of horizontal and vertical ice layers are also satisfied (conditions 4.3–4.4). Since horizontal shearing stresses are always greater than the normal (vertically oriented) stresses, the formation of purely laminated cryogenic structure is possible, whereas the formation of only vertical ice schlieren (without horizontal ones) is in principle impossible because of the migrational-segregational mechanism of their generation. In some soils, as the rate of freezing grows a gradual transition takes place from the layered structure to the mesh-like and massive cryogenic structures. Therefore, under certain freezing regimes, conditions are created for the formation more equally of vertical and horizontal zones of concentration of critical stresses.

Specific features of development and further growth of the mesh cryogenic structure, in contrast to the layered one, are determined by water migration within the block both horizontally and vertically. Moreover, the water moves to schlieren perpendicular to the freezing front only under effect of grad P_n, whereas it moves to schlieren parallel with the freezing front under the effect of the temperature gradient (or grad W_{unf}) and the stress gradient grad P_{sh}.

The mesh cryogenic structure varies both in shape (prisms, parallelepipeds, cubes, etc.) and in size of soil blocks as a result of different relations between normal and shearing stresses in freezing soils of different composition, structural-textural features and strength. The frequency of distribution or the distance between horizontal $l_|$ and between vertical $l_\perp$ ice layers are determined by the relations: $l_| = f(1/\text{grad } P_{sh})$ and $l_\perp = f(1/\text{grad } P_n)$.

It should be specially mentioned, however, that an increase in freezing rate (because of a low dehydration rate of the unfrozen part and small swelling of the freezing zone of the ground) leads to the end result of a decrease of values of shearing and normal stresses and a considerable reduction of the zone of generation and growth of vertical and horizontal ice layers. This process encourages (due to the short time of ice schlieren formation) the appearance of partially developed cryogenic structures (semi-mesh, semi-layered, angular-discontinuous, etc.).

With the increase of the rate of freezing, the conditions may arise when vertical schlieren can no longer be formed and an often fine but frequently discontinuous-layered cryogenic structure is formed (or discontinuous-lens-like, scaly, etc.).

The type of cryogenic structure is not determined only by conditions of freezing but depends substantially on the lithology of the freezing layer. The *cellular* cryostructure, which is a variety of block structure, proves this. It differs from the mesh and layered structures in the mode of formation and occurs mostly in bentonitic clays. For example, near the freezing front bentonite shows distinct cells of soil of irregular polygonal shape with a framework of ice. The cellular cryogenic structure of bentonite is entirely dependent on and predetermined by its specific microstructure and texture, which creates cellular or honeycomb structures in the unfrozen dehydrating part of the soil.

An analysis of a variety of superimposed cryogenic structures demonstrates that there is in principle the possibility of formation of independently developing inclined or even vertical ice layers, as observed by many re-

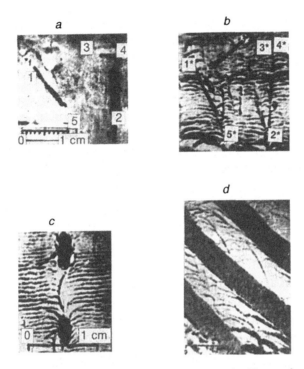

Fig. 4.3. Inherited cryogenic structures: *a – b –* Zones of structural defects, general view of soil with dislocated structure before and after freezing: 1, 2 open cracks before freezing, 1*, 2* – after freezing, 3, 4, 5 – closed (healed) cracks before freezing, 3*, 4*, 5* – after freezing; *c –* zones of influence of extraneous inclusions; *d –* zones of concentration of stresses (interbedding of kaolinite clay and quartz sand at 45° angle to direction of thermal and migration flows.

searchers, in Quaternary deposits. These ice layers are a result of the numerous cracks of various shapes in the unfrozen dehydrating soil (cracks due to bending, warping, flaking etc.). Moreover, in natural conditions, there is often an irregular boundary of freezing (which is tilted relative to the surface, undulating, angular, etc.). The fact that ice schlieren are generated and grow mostly parallel with or perpendicular to the front of freezing/thawing implies the existence in superimposed cryogenic structures, of a complicated system of ice schlieren (undulating, radial, ring, rhombic, etc.) caused by the irregular surface of the freezing or thawing boundary (Fig. 4.2).

Inherited cryogenic structures in freezing soils are quite common. They are caused by different kinds of lithological inhomogeneities and defects of loose deposits. The major genetic types of inherited cryogenic structures are associated with strength-defective and strained contact zones and with extraneous inclusions (Fig. 4.3). Their formation, as that of superimposed cryogenic structures, cannot take place without the thermal-physical and physical-mechanical conditions for segregated ice formation.

The inherited cryogenic structures in the zones of structural defects of soils (the strength-defective type) are confined to zones of mechanical crumpling, dislocation, sliding planes, closed cracks and other defects of structure and texture. The ice schlieren follow the shape of structural defects in the weakened zones of the freezing soil. The critical stresses for local destruction of the soil first of all are concentrated in these zones during deformation by shrinkage-swelling and the formation of structure in the freezing ground.

The inherited cryogenic structures in the zones of stress concentration at the contacts between different soils (contact-stress structures), in contrast to the strength-defect cryostructures, appear not in the weakened zones of the soil but in the zones of contact of soils which differ in thermal and mass exchange and in shrinkage-swelling characteristics. At the contacts, the differences in the amounts of deformation due to shrinkage and swelling give rise to critical stresses (stress concentration zones).

Consequently, the cohesion of the soil is overcome primarily in these places and ice schlieren are generated. For example, during freezing of stratified soils, as a rule elongated segregated ice streaks appear at the interfaces of layers (Fig. 4.4). They inherit completely the initial lamination (horizontal, vertical, tilted, or oblique) of the loose fine soil deposits.

In the freezing of soils, inherited cryogenic structures may be caused by extraneous inclusions and have the following characteristics. Firstly, the water resistant inclusions block migration flow to the freezing front, and downstream from them the ground is dehydrated and massive cryostructure is formed (or the ice schlieren become thinner), whereas behind them the ice content grows and the schlieren cryostructure appears. Secondly, these inclusions, which differ greatly from the soil in their deformation properties, concentrate stresses at their contacts with soil thus regularly causing the appearance of ice schlieren. Moreover, behind the extraneous inclusion (at a certain distance from it) the geometry of ice schlieren follows the contour of the inclusion, i.e. it repeats the character of the moving freezing boundary, which in its turn is predetermined by the presence of the inclusion with a thermal conductivity different from that of the ground.

In the frozen part of thawing soil which has a temperature gradient, the accumulation of migration-segregation ice and the formation of cryogenic structure are also possible (Fig. 4.5). Experiments show that the development of cryogenic structures in the frozen part of thawing soils has the same features as during freezing. During this process both the superimposed and

Fig. 4.4. Inherited cryogenic structures in marine lenticular clays (photo by G.I. Dubikov). Largest ice lens at bottom is *c.* 3 cm.

the inherited types of cryogenic structure can develop, in which the generation, growth and configuration of ice schlieren are predetermined by the initial composition and structure of the soil and by the conditions of its thawing. The formation of these structures of course requires the presence of thermo-physical and physical-mechanical conditions of schlieren ice formation during the process of thawing.

On the basis of this analysis, a genetic classification of cryogenic structures (based on appearance and development) was suggested grouping logically practically all known types and varieties of cryostructures (8). Table 4.1 illustrates this classification. Each class of structure (with and without schlieren) is divided into the types described earlier, and every type is subdivided into kinds of cryostructure according to the frequency of generation and thickness of ice schlieren. The number of kinds of structure (by frequency of occurrence of ice schlieren) within each type can vary and the different distances *l* between schlieren are determined in natural conditions both by the lithology of the freezing or thawing deposits and by external thermodynamic conditions, with which the stress gradients are connected.

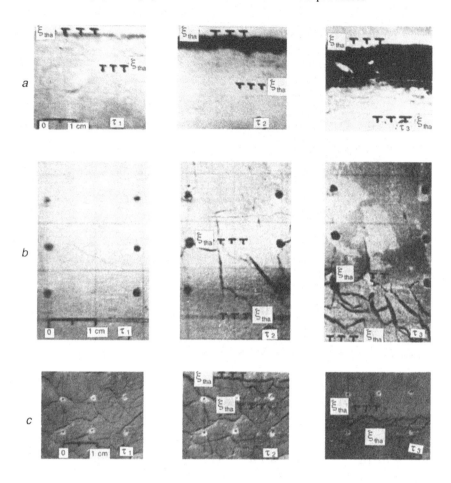

Fig. 4.5. Generation and growth of segregational ice schlieren in the frozen part of soils thawing from above (superimposed cryostructures): *a* – layered structure (slow thawing); *b* – block structure (rapid thawing) in kaolinite clay with initial massive cryostructure; *c* – thawing of clay-rich soil with initial schlieren cryostructure.

Division of the kinds of cryostructures into large-, middle- and small-schlieren (using numerical values) is not practical for all cases but only within any type of structure as applied to an actual type of ground. Other-wise this subdivision loses its purpose, because it will not relate to their formation. Therefore, in this classification of cryogenic structures, only the extreme variants are taken into account, in other words, the pattern of formation (direction of change) of horizontal and vertical ice schlieren depending on grad P_{sh} and grad P_n. Then for each of these variants, four extreme types of structure are recognized according to the thickness of ice schlieren, and between them there can be a great number of varieties. In fact, the transition of one type into another is smooth and gradual. The simple division of types into thick-, middle- and fine-schlieren structures (without due account of the type of structure, of the geology and genesis of the material and of the freezing-thawing conditions) is tentative, because it does not show, and even hides the patterns in the thickness of ice layers.

The cryogenic structures in freezing soils described above are of migration-segregation genesis. In natural conditions, however, the cryogenic structures may also appear as a result of other mechanisms. For example, in very wet weakly lithified soils the mechanism of cryogenic structure formation is associated with the freezing out of mostly free water and selective orthotropic ice formation. This mechanism is a result of the predominant matrix structure of silts and weakly lithified fine-grained porous soils, in which the larger part of water is free (unbound), while the soil particles cannot actively interact. Freezing of water produces an ice framework (block-framework cryogenic structure), in which the vertical and inclined ice layers appear earlier and much more rapidly than the horizontal crossmembers. The ice crystals push out the mineral particles in the direction of their growth thus concentrating them inside the blocks, i.e. the spaces between the walls of an ice framework are filled with the wet mineral part of the soil. As the ice walls of the framework grow (becoming thicker and longer), the mineral part of the freezing soil is dehydrated and densified by compression. This mechanism of formation of structure is called orthotropic-compressional (12). Moreover, with finer-grained material all other conditions of freezing being equal, the soil more often separates into ice and mineral layers and forms a cellular cryogenic structure. Block cryogenic structure with fewer ice layers is formed in coarser soils. The same type of pattern is observed as the rate of freezing decreases.

In natural conditions the injection mechanism of formation of ice layers is also active. The formation of injection and mixed (migration-pressure) cryogenic structures is possible only in the case when hydrostatic pressure of the water being injected exceeds the strength of the structural connections in soils. At high pressures, a hydroburst can occur even in frozen soil, with instantaneous injection of water and formation of larger layers of injected ice. Under hydrodynamic pressures above the critical long-term strength of the soil, the long-term injection of thin water films occurs with pressure-migrational segregated ice formation.

Cryogenic structures formed as a result of freezing change with time under the effects of negative temperature gradients, mechanical stresses, concentration of the pore solution, and of other, external fields. During this process, both in syngenetic and epigenetic layers of perennially frozen soils and in the layer of seasonal freezing-thawing, the already existing ice layers may grow and new ice schlieren can be generated and, in some cases, also disappear.

4.3   Formation of structure in freezing and thawing soils

The wide spectrum of physico-chemical and physical-mechanical processes which accompany freezing and thawing of soils causes considerable structural transformations of their organic-mineral skeleton manifested by change in the size, shape, relations and orientation of structural elements (primary particles, mineral and organic-mineral aggregates). The size of structural elements can increase or decrease during freezing. The decrease is a result of dispersion effects and the increase is caused by coagulation and aggregation.

When the ground freezes quickly at low negative temperatures (about -40 to $-60°C$) and there is no water migration, then concurrent generation of crystallization centres and the growth of ice crystals produces a process of disintegration of particles and aggregates. Chiefly, the large particles and aggregates of the mineral skeleton (sands and coarse dusty fractions) are intensively disintegrated, being more inhomogeneous and with a larger number of defects than the structural elements of smaller size (Fig. 4.6a). When the temperature of freezing is higher (above $-30°C$), enlargement of aggregates dominates over their destruction. For example, for polymineral sandy-clay material after freezing to $-30°C$, as with freezing to $-60°C$, the content of coarse silty and sandy aggregates grows but as a result of aggregation of smaller structural elements (Fig. 4.6b). Therefore, the intensity and duration of phase transitions of water are important for the direction of change in grain-size composition, i.e. lower intensity may cause migration of intra-aggregate water to centres of ice formation and can improve conditions for plastic rearrangement of mineral elements, thus contributing to their mutual approach and enlargement by coagulation and aggregation. Quick phase transitions cause only fragmentation along mineral boundaries and finer grain size.

In the case of water migration to the crystallization front, the leading role in structural transformation of freezing ground belongs to the processes of mass transport into the freezing zone, to differentiation and deformation of the soil mass during formation and growth of segregated ice layers and to dehydration and shrinkage of the unfrozen part of the ground. Reconstitution of structure in this process involves considerable compaction and strengthening of the mineral skeleton, which provokes greater strength of the soil in general, and formation of ice-cement adhesion, in particular. In

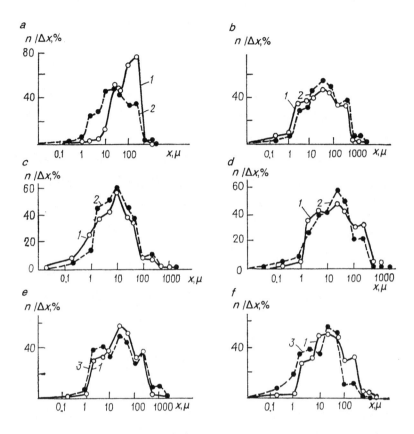

Fig. 4.6. Changes in grain size composition of clayey soils (from differential curves of small particle analysis): *a, b* – all-round freezing of heavy silty clay at *t* equal to $-60°C$ and $-30°C$, respectively; *c* – freezing polymineral clays with dominant Na^+ content; *d* – same, Ca^{2+} and Mg^{2+}; *e, f* – thawing polymineral clays consolidated under loads of 0.05 and 1 MPa, respectively. 1 – initial state; 2 – after freezing; 3 – after thawing. Weight content of particles by fractions, ordinate axis, and the size of particles, abscissa.

the unfrozen dehydrated zone of freezing soils concurrent dehydration of structural elements takes place, as they approach each other and form larger aggregates and blocks. Porosity is reduced and the mineral skeleton is densified; concurrently, the reorientation of particles and aggregates along the direction of migration flow and the formation of slit-like pores also take place (Fig. 4.7). The size of aggregates and blocks is controlled by the degree and intensity of dehydration and by the character and degree of development of deformations and shrinkage stresses. The shape of structural separations depends mainly on the mineral composition and crystallo-chemical features of the soil-forming clay minerals.

a   b

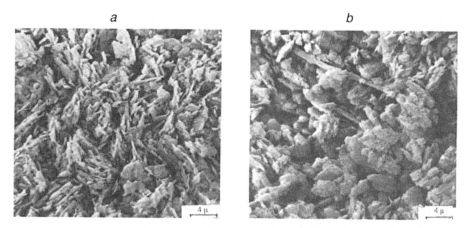

Fig. 4.7. Microstructure of the mineral skeleton of freezing kaolinite clay in the unfrozen dehydrated zone at the boundary of freezing (a) and in the heaved frozen zone (b) .

In the freezing part of clay-rich soils (i.e., in the region of significant phase transitions), it is not the initial unfrozen soil structure that is being reconstructed, but the structure already transformed by dehydration and shrinkage of the unfrozen part. The freezing out of water in large pores and the growth of ice crystals when the temperature falls causes heaving of the mineral skeleton and differentiation of the soil mass (when segregated ice micro- and macro-layers are formed). The structure of the mineral skeleton becomes loose, though the strength of soil is greatly increased by cementation by ice. The process of phase transition is often accompanied by partial fragmentation and reorientation of soil blocks and aggregates as a result of deformation by heaving but inside the aggregates the orientation of elementary (initial) particles usually remains intact. As a result of growth of ice inclusions, mostly in large pores and at structural boundaries, the pores retain their slit shape and enlarge (Fig. 4.8). In the process of further freezing (fall of temperature), the phase transitions of water occur in the smaller intra-aggregate and interparticle pores, thus causing disintegration of soil particles and their disorientation within aggregates and blocks of soil. As a result, the frozen soil has a disordered structure similar to that before freezing (in never-frozen ground).

Quantitative microstructural changes during freezing are, on the whole, the result of composition and initial structure. The mineral composition determines the shape of structural partings which appear during freezing. Since montmorillonitic soils are more easily deformed under the effects of physico-chemical processes concurrent with freezing, these show transformation of the initial, usually skeletal-matrix, microstructure into a turbulent

Fig. 4.8. Vertical orientation (parallel to migration flow of water) of mineral particles and aggregates in the unfrozen dehydrated zone of the freezing polymineral clay.

one even under conditions of modest dehydration and deformation of unfrozen parts of the soil far removed from the freezing front.

Crystallochemical features of the structure of kaolinite particles, making up the rigid structural elements, control the possibility of their fragmentation in the zone of ice formation and the reorientation of their basal surfaces along the boundary between the mineral and ice layers during growth of segregated ice. The initial (before freezing) grain size and the mineral composition determine the extent of structure-forming processes in freezing soil by influencing the intensity of water exchange, ice formation and dehydration of soil. Structure formation is more obvious in finer-grained (clay) soils and becomes less as the initial grain size increases. The chemical composition has an equally important effect on the formation of structure and on the direction of changes of grain size of freezing soils. In clay soils containing Na ions, under conditions of deep dehydration (caused either by low temperatures of freezing, or by intensive water exchange during slow freezing), the dominant processes are coagulation and aggregation of structural elements. Freezing of clays containing multivalent cations (Ca^{2+} and Mg^{2+}), however, is accompanied by dispersion along structural separations (Fig. 4.6c,d).

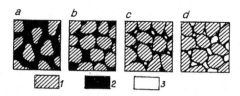

Fig. 4.9. Basic types of ice-cement in frozen soils: *a* – basal, *b* – pore; *c* – film; *d* – contact; 1 – soil particles, aggregates; 2 – ice cement; 3 – parts of pores free of ice and water.

The difference in formation of structure is caused by a higher content of clay-colloidal material in the Na-containing soil in the initial state (prior to freezing) resulting in a more active manifestation of physico-chemical processes, by development of a mostly local water exchange in the soil and consequently significant dehydration at discontinuities in the mineral assemblage.

If the water exchange in compacted soils is obstructed at low water contents, then disintegration of structural elements is observed. On the contrary, in relatively loose, water-saturated soils coagulation and aggregation dominate.

The formation of ice as a structural element in freezing soils cardinally changes the initial (unfrozen) structure. The resulting cryogenic microstructure is a result of both the initial composition and structure of unfrozen soil and the freezing conditions. The typical feature of microstructure of clastic and sandy soils is the presence of ice cement which binds the previously loose soil into a solid mass. Several types of ice cement can be formed depending on the initial water content in sandy soils; for example, the cuff (contact) type, the film (crustal) type and the porous and basal types of ice-cement (Fig. 4.9). The cryogenic microstructure of clay soils (clay-rich and fine-silty sands, clays) is similar to the macrostructure of frozen soils (massive, laminated, mesh and cellular). As the fineness increases, the thickness of microlayers of ice and their frequency increase (Fig. 4.10).

The specific features of formation of cryogenic microstructure in soils with different chemical composition of the pore solution and exchange complex is first of all evident in the change of aggregation of the mineral skeleton and in the phase composition of the water. The microstructure of salinized sandy soils shows a higher content of unfrozen water and a new structural element, the crystals of salts, which precipitate from the pore solution and cement the mineral skeleton, thus forming a new type of contact, i.e. the crystallization type. Structure-forming processes are much

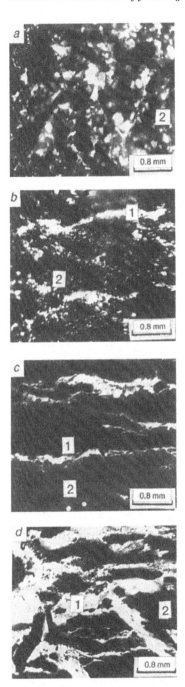

Fig. 4.10. Basic types of cryogenic microstructure in soils: a – massive;
b – lens-like; c – laminated; d – mesh; 1 – ice; 2 – soil particles.

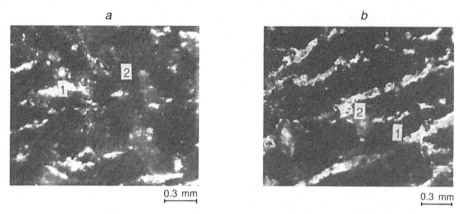

Fig. 4.11. Microstructure of frozen silty-clays, saturated with (*a*) Na^+ and (*b*) Ca^{2+}. 1 – ice; 2 – soil particles.

more complicated in clay soils where, moreover, ion exchange occurs; these processes, with allowance for the nature of the salts, cause either aggregation or dispersion of the mineral skeleton.

When clay soils with multivalent cations in the exchange complex freeze, the intensity of segregated micro-ice formation increases and laminated cryogenic microstructures appear. In soils saturated with univalent cations, discontinuous and thin ice micro-schlieren with a confused orientation are produced (Fig. 4.11).

The formation of cryogenic microstructure is influenced not only by the initial chemical-mineral composition, dispersivity, moisture content and density of soils but to a great extent by freezing conditions as well (rate of freezing, temperature gradients, the presence or absence of water supply, etc.). For example, in fine-grained soils as the freezing rate is increased the thickness of ice microlayers and the distance between them decrease, as do the average size of inclusions of ice cement and the size of its crystals. Inside the mineral blocks and aggregates, the inhomogeneity in microstructure increases from the centre towards the periphery of aggregates. During slow freezing, the aggregates acquire a more homogeneous microstructure, though a more distinct differentiation into purely mineral areas is observed (without ice inclusions), as are areas with higher ice content. The latter effect is, probably, caused by local redistribution of moisture/ice content (Fig. 4.12).

In the course of thawing, the structure of frozen soils is also transformed. On the basis of, as yet, a small amount of experimental data, we may assume that in most cases, and particularly during rapid thawing of soils, the general tendency is to disintegration of larger elements. Thawing is accompanied by weaker structural ties, lesser strength and greater permeability to water. This is shown in the lesser (to a tenth or less) coefficient of aggregation of the

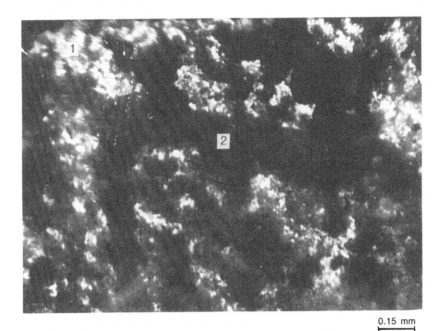

0.15 mm

Fig. 4.12. Differentiation of ice and mineral particles in freezing silty clay due to local redistribution of water: 1 – ice and icy areas; 2 – soil particles.

soil which is most conspicuous in dense soil with low ice content. For example, an analysis of changes in aggregation of thawing soils, consolidated before freezing by loads of 0.05 and 1 MPa, shows that during thawing of soils with relatively low moisture and high density, the microaggregate composition changes to a greater degree than in samples with lower density and greater moisture/ice content (see Fig. 4.6e,f). Transformation of structural elements of soils in this case mostly means disintegration during thawing along structural partings of large coarse (sand and dust) fractions and an increase in the smaller fractions. This process finally reduces the mean statistical size of structural separations of denser soils by 13.8 μm in polymineral clay and by 3.3 μm in heavy sandy clays. In the low density samples with higher moisture/ice content, the process of thawing caused a decrease of the mean size of aggregates by only 1 μm in clay and by 0.3 μm in sandy loam. This difference in the disintegration of structural elements in loose and dense thawing soils is attributed to more intensive wedging activity of fine water films in overdensified samples.

4.4 Structural associations and types of contact in frozen soils

During freezing of soils and in the frozen state, specific types of contacts and structural associations appear as a result of complicated processes of formation of structure. In hard rocks and in some cemented sedimentary rocks, the chemical relations dominate, whereas in fine-grained soils with high ice content the relations between the separate elements are more often molecular and ion-electrostatic, often called water-colloidal relations.

The formation of structural associations in such multiphase and multi-component dispersed systems as frozen soils always occurs not only between mineral surfaces, but between the surfaces of ice and mineral particles. Moreover, structural associations usually appear not on all the surfaces of elements composing the soil but only in places where they approach, i.e. at contacts. The number and character (nature) of various individual contacts have a certain influence on microstructure and properties of frozen soils. Every type and variety of contact is specified by its mechanism of formation and the nature of interrelating forces and by the geometry and size of contact.

At a first approximation, the whole variety of contacts in soils under negative temperature and with ice inclusions can be divided into point, area and volumetric and their primary feature is the area of contact interaction between organic/mineral particles and ice, and the other contact character-istics are aggregation (distant coagulative), coagulation (close coagulative) and dry (waterless) contacts, which differ in the energy of interaction and the distance between the contacting ice and ground particles (Fig. 4.13). The point contacts are the most common in clastic and sandy soils; the area contacts apply to clay-rich soils, whereas volumetric are typical of super-saturated soils with high ice content, when organic/mineral particles practi-cally 'swim' in the ice, i.e. are entirely surrounded by ice.

The sub-dividing of contacts according to the energy of their interaction largely depends on the distance between the interacting soil particles. For example, when two soil particles directly interact they constitute the most dry and strong (without water or phase) contacts. They have valent and ion-electrostatic interaction and directly bring together the elements of frozen soil; they are divided into types of contacts: dry mineral (mineral-mineral), dry mineral-cement (mineral-cementing matter), and dry mineral-ice.

The dry mineral contacts normally appear during diagenetic transform-ation of soils as a result of an increase of pressure with depth and loss of moisture of clay-rich sediments. 'Cold fusion' of minerals by chemical forces can be another possible way of forming these contacts between mineral particles when the soil freezes in a 'closed' system.

The mineral-cement contacts are formed in the process of recrystalliz-

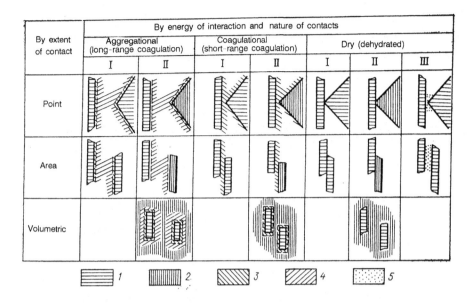

Fig. 4.13. Types of contact in frozen soils: 1 – organic-mineral particles; 2 – ice; 3 – bound water; 4 – semi-bound water; 5 – organic-mineral cementing matter; I – III – varieties of contact (I – mineral, II – mineral-ice, III – mineral-cement).

ation and extraction of a new phase from the pore solution. This process is associated with over-saturation of the pore solutions during diagenesis (including soil freezing, which increases their concentration) and the separating out from them of cementing matter, which forms strong 'bridges' between ground particles forming a rigid porous structure of the mineral component. Moreover, in soils strong cementation can appear not only as a result of crystallization of salts, but with polymerizing compounds as well, such as gels of silicic acids and hydrates of sesquioxides and humic or other organic compounds transient from sol to gel.

When the temperature is sufficiently low in soil (below -100 to $-150\,°C$), then the mineral-ice contacts may appear. This will occur if practically the entire bound water is frozen out, or migrates into other parts of the soil and there are formed 'mineral-ice' contacts similar to the dry ones. They may reverse during thawing and are the least stable of all dry contacts. Their strength is determined by cohesion forces in ice.

A typical feature of water contacts is the presence of films of unfrozen (bound) water between the interacting elements of frozen soil. Usually close coagulation (coagulation proper) and distant coagulation (aggregation) contacts can be distinguished. Aggregation contacts are formed mostly as a result of long-range molecular forces, in some cases due to magnetic and dipole (Coulomb) interactions. Coagulation contacts are determined mostly by the action of molecular and ion-electrostatic forces. The latter forces have a middle radius of action and appear at distances between particles of a few nanometres. Their effect greatly increases the strength of soils and is particularly obvious during dehydration (drying or freezing of bound water).

Aggregations, recognizing these points, should be most typical of mixed silty sandy materials and clays in the range of high negative temperatures (0 to $-5\,°C$). The strength of aggregation contacts is determined by the cohesion forces or 'glue forces' between molecules of semi-bound (unfrozen) water. This is caused by the circumstance that the cohesion forces (agglutination) between interacting components (mineral surface – bound water, ice – bound water) and the cohesion forces in ice and mineral are much greater than the cohesion forces in semi-bound water. With allowance for the area of the contact interaction, we may expect a greater strength from area aggregation contacts (typical of clay-rich frozen soil) compared with the point contacts (dominant in frozen sands). The maximum strength should belong to volumetric aggregation contacts, when an armour of ice (enveloping the mineral particles) is formed. This kind of contact is most frequent in soils with high ice content (water-saturated) with basal cryogenic texture.

An analysis of the character of structural relations and contacts in the ground provides an explanation for the sudden growth of strength (almost by an order of magnitude) of soils in passing from the unfrozen into the frozen state. This increase of strength is a result of freezing of a part of the soil water and thinning of bound water films when a new solid ice component and a new surface boundary: 'ice – bound (unfrozen) water' is formed. This process is, accordingly, associated with a considerable growth (due to smaller distances) of the energy of interaction of the contacting elements (mineral – mineral and mineral – ice). The distance between these elements in frozen soils even at high negative temperatures usually does not exceed hundreds of nanometres, whereas in unfrozen soils it may amount to thousands of nanometres.

Therefore, in soils the character, size and shape of ground and ice particles, the degree of water saturation, the negative temperatures, and the conditions of freezing between particles can serve as causes for the appearance of a great variety of types and kinds of contacts which, as a result of external effects, can transform one to another and which determine the structural relations between the elements of the frozen material.

5

Cryogenic geological processes and phenomena

5.1 Classification of processes and phenomena

Exogenous geological processes and phenomena in the permafrost regions are often referred to as cryogenic (frozen ground) geological processes and phenomena. They are due to the development of the thermal-physical, physico-chemical and mechanical processes that occur in freezing, frozen and thawing soils, which were considered in the previous chapters. The specific features of exogenous geological processes in the permafrost zone are determined by the cyclic recurrence of freezing and thawing, cooling and heating, by the properties of frozen, perennially frozen and thawing soils, by the variability with time of ground stresses etc. These lead to the different cryogenic phenomena (formations) that contribute to the geological structure of loose Quaternary deposits and manifest themselves in the topography and micro-terrain within the area of permafrost as well as in the area of deep seasonal freezing. With regard to the basic agencies of the environment which are the driving forces for the processes and phenomena, all the exogenous processes in the cryolithic (permafrost) zone may be subdivided into three major groups (Table 5.1).

The first group comprises *the cryogenic processes proper* (frozen ground geologic processes) caused by seasonal and long-term variations of heat and mass exchange over the ground surface and in the underlying materials. These are frost cracking and frost weathering caused by multiple recurrence of freezing and thawing, and cooling of frozen soils in the layer of annual temperature variations (the layer of seasonal freezing and thawing included) together with the dynamics of the state of stress in the variable temperature field; frost heaving of soils and icings conditioned by seasonal and long-term freezing of soils, ground and surface water (in icings), with the enlargement of soil volume during segregation of ice; thermokarst associated with seasonal and long-term thawing of soils with a high ice content and of

Table 5.1 *Classification of the principal exogenous geological processes in the permafrost regions (according to E.D. Yershov and L.S. Garagulia)*

Group of processes	Kind of process	Manifestations of the process	
		in the relief	in the deposits
I. Cryogenic processes proper	Frost cracking and formation of ice-wedge polygons	Polygonal fissured micro-relief	Open vertical fractures
		Polygonal and polygonal-dike relief	Systems of ice wedges, ice-ground, ground and sand veins in deposits of alluvial, slope, eolian and other origins
	Cryogenic weathering of rocks	Disintegrated polygon blocks, block fields	Fragmental-block formations of the upper layer of eluvial materials (golets eluvium)
	Frost heaving of rocks	Areas of heaving	Cavities and ice interlayers over soil bed
		Seasonal and perennial mounds, hydro-laccoliths	Ice inter-layers and ice cores in frost mounds
		Round spots and cryoturbations	Pot-holes of mixed ground (folded)
		Stone rings and stone wreaths	Soil-rock waste and rock waste-block veins
		Thufurs (small frost mounds)	Peaty deposits of mounds
		Mounds (formed with participation of frost cracking)	Mineral soil in mounds divided by open fractures, soil or peat veins
	Icing	Seasonal and perennial ice bodies	Alluvium with icings and ice sills
	Perennial ice and snow formation over rock surfaces	Glaciers	Glacial ice

Table 5.1 (*continued*)

Group of processes	Kind of process	Manifestations of the process	
		in the relief	in the deposits
		Eskers, kames, firn, snow-ice and snow-mudflows, snow avalanches	Glacial deposits
	Formation of under-ground massive ice beds		Ice beds in deposits of marine and glacial-marine genesis
	Thermokarst	Thermokarst subsidence	Ice-rich marshland deposits
		Thermokarst lakes and basins	Ice-rich lacustrine and marshland deposits
		Hummock and hollow topography	Outliers of deposits with earth veins
		Baidzherakhi	Outliers of deposits with ice veins
II. Fluvial, abrasion and water-related	Thermal abrasion	Thermal abrasion benches, thermo-cirques, coastal shallows	Lacustrine and coastal-marine deposits
	Thermal abrasion	Hollows, ravines,	Hollow alluvium
		Swampy lands,	Gleyey soils
	Paludification	Swamps	Peatlands
III. Gravitational	Solifluction, slumping	Solifluction flows, covers, terraces	Solifluction deposits
	Rock stream development	Rock streams, rock fields	Golets talus (according to Y.A. Bilibin)
	Collapse, scree	Block collapses, scree, scree cones	Colluvium
	Avalanching	Snow avalanches	
	Development of mud-flows	Mud and stone flows	

underground ice deposits, accompanied by the change in properties, consolidation and soaking of soils when thawing.

The second group comprises *fluvial, abrasion and water balance* processes arising from mechanical and thermal effects of water masses over frozen and thawing soils, annual variations of heat exchange over the soil surface and long-term variations of water balance over the surface. This group includes thermal abrasion, thermal erosion and waterlogging. For all these there are analogous processes outside the permafrost regions. However, these processes should be included in an independent group with regard to the specific nature of their occurrence in the permafrost. Development of all the above processes is always preceded by thawing of soils accompanied by disturbance of structural relations in the soils and changing of their physical-mechanical and thermal-physical properties. The specific nature of the processes is revealed in the particular forms of meso- and microrelief, in the transformation of the cover of loose deposits and in the formation of specific genetic types of new geological units (for example, in appearance of syncryogenic lacustrine and coastal-marine deposits, diluvial and talus deposits, in the accumulation of peat, etc.).

The third group of processes integrates gravitational processes, such as solifluction, sliding, slipping, talus, deserption, rock stream formation and the like. Some of the processes have analogies beyond the cold regions. However, all the gravitational processes without exception that occur within the cold regions are characterized by a specific nature of manifestation leading to the formation of specific shapes of meso- and microrelief such as rock streams, rock fields, solifluction terraces or covers, streams and terraces. Besides, these processes in the cryolithic zone are responsible for the formation of a specific composition and structure of slope wash (solifluction, talus, colluvial sliding and similar deposits); they are associated with burial of firn, ice bodies and glacier ice in these deposits, with the formation of cave and golets ice; and with cryogenic grading of the material and other special characteristics.

5.2   Frost heaving of soils

Frost heaving of soils arises due to the increase in volume of freezing moisture and the accumulation of ice (owing to water migration) at freezing. This process is widely developed both in the permafrost regions and in the regions of deep seasonal freezing of ground. The largest deformations due to heaving are observed in the freezing of an open system of highly permeable, usually sand-silty soils and saturated silty-sandy-clays at low rates of freezing and in the proximity of ground water (migration mechanism of heaving).

Often, heaving is also associated with freezing of coarse-grained soils in closed system conditions which arise in confined ground water-bearing horizons (intrusion mechanism of heaving). Usually, two kinds of heaving are distinguished, local and area. Area heaving is most often characterized by a high nonuniformity over the area as the magnitude of heaving within the limits of one landscape type can exceed by a factor of two the average value of heaving. Local heaving manifests itself more distinctly in the relief (mounds, hillocks, patches of heaving and other forms) which arises due to unsteady conditions of freezing, different water contents, soil composition and other factors of geological and geographical environments. In general, under natural conditions, heaving of soils may be associated with seasonal freezing of the seasonally freezing-thawing layer (seasonal heaving) and with perennial freezing of ground (long-term heaving).

Perennial heaving

Heaving of ground is the most widely developed cryogenic process, resulting in the formation of frost mounds differing in shape and size, on fine-grained soils and peat bogs. As a result of peat bog freezing, for example, in waterlogged depressions so-called 'inverse' relief is often formed rising for several metres over the surrounding surface. Local long-term heaving is always accompanied by formation of mounds reaching several metres in height and hundreds of metres in diameter. According to their formation local mounds are subdivided into segregation and intrusion forms, but a mixed intrusion-segregational formation is also possible.

The segregation-heaving type of mounds, referred to as palsas in the foreign literature, are formed as a result of soil moisture migration towards the freezing front under the influence of gradients of both temperature and moisture content. The local nature of their occurrence is confirmed by the fact that they are formed in particular places where, under the combined influence of a number of natural factors, more rapid and deep freezing of the soil occurs and where a lens of permafrost is beginning to form. It is here that migrating moisture begins to arrive from the surrounding unfrozen soil, causing heaving of the ground and elevation of the surface at the site. In winter snow is blown off the elevated surface; thus, the temperature of the ground is lowered, while the thickness of the lens of permafrost continues to increase relative to the surrounding sites, leading accordingly, to the enlargement of the mound. The growth rate of such mounds in the north of West Siberia during the initial stage reaches $10\text{-}30\,\text{cm}\,\text{yr}^{-1}$, then, with the bigger permafrost core and enlarged mound the growth rate is reduced to $1\text{-}2\,\text{cm}\,\text{yr}^{-1}$. The height of such mounds can reach $20\,\text{m}$ with a horizontal area ranging from several tens of metres to hundreds of metres dictated by their age and conditions of formation.

As a rule, long-term mounds of heaving of the *segregation* type accompany new formation of permafrost on unfrozen sites on flood-plains and limnotic basins (Fig.5.1a); often chains of mounds arise along the banks of lakes, over shallow water and on boggy areas.

It is often the case that mounds are confined to sites of peat deposits. This is associated with the presence in peat of a great amount of moisture which means that the thermal conductivity of the frozen peat is higher than that of unfrozen peat, and convex sites covered with peats are cooled more in winter than they are warmed in summer. Therefore, freezing is more intense in peats than in mineral soil which promotes development of local heaving and formation of peat frost mounds.

Long-term frost mounds of the *intrusion* type usually arise through injection of water (or fluidized soil) under the influence of the hydrostatic pressure that develops in closed systems in the course of freezing. They are mainly associated with freezing of sublacustrine (flooded) closed taliks surrounded on all sides by permafrost. In the North and in Central Yakutia they are called *bulgunnyakh*, or *pingos* in the foreign literature. Freezing of sublacustrine taliks is usually caused by shallowing or emptying of lakes. As a result the sublacustrine talik begins to freeze on each side, thus decreasing in size. The enclosed water is subject to hydrostatic pressure. Owing to this pressure the frozen roof warps in the thinnest place, forming a frost mound with a core of injection ice or of an ice and soil mixture (see Fig. 5.1b). Since freezing of a water-bearing talik is a long-term process, intrusion of water into the growing bulgunnyakh occurs many times. Along with this, the segregation type of ice formation can take place, leading to the complex structure of bulgunnyakhs.

The size of bulgunnyakhs is dependent on the amount of water in the closed taliks; they can be 30-60 m high, and 100-200 m horizontally. Often, another type of injection mound, the hydrolaccolith, arises when water is injected under the influence of the hydrodynamic pressure of subpermafrost and supra-permafrost water. Hydrolaccoliths are confined to sources (places of discharge) of ground water or to closed taliks, including those in flood-plains mainly in the southern zone of permafrost.

Seasonal heaving of ground

Such heaving accompanies seasonal freezing of nonpermafrost soils and freezing of the seasonally thawing layer over permafrost. The average heaving of a seasonally thawing layer is usually a factor of 1.5–2 less than

Fig. 5.1. Perennial frost mounds (photo G.I. Dubikov): *a* – migration-heaving peat mound, *b* – degrading heave mound (pingo).

that of the seasonally freezing layer. This is associated with the fact that the seasonally freezing layer is most often an open system and its freezing is accompanied by an active migration of moisture. Freezing of the seasonally thawed layer is more similar to that of a system of closed type because moisture migration out of the underlying frozen soil does not take place, with only vertical migrational redistribution within the layer and lateral to it – necessarily with nonuniform freezing. As shown by observations of the annual cycles of deformation, on freezing the ground surface rises in relation to an immobile bench mark, due to soil heaving as the thickness of the freezing layer increases, and reaching maximum elevation at the moment of maximum freezing. With the beginning of thawing the surface again sinks due to thawing of layers of segregation ice until it reaches the zero elevation of the bench mark. Such movement of the ground surface is also called *hydrothermal.*

Seasonal heaving of soils over an area is characterized by a high nonuniformity. An extreme case of expression of this nonuniformity is the formation of one-year (or seasonal) migrational frost mounds which exist only during the cold season of the year and disappear during thawing of the frozen layer. The width of these mounds can reach a few metres, and the uplift above the surrounding surface about 0.2–1.5 m. Intrusion seasonal frost mounds are usually associated with nonuniform freezing of soil of the seasonally thawing layer, resulting in development of hydrostatic pressure in the water moving above the permafrost. Such a phenomenon is most often encountered at the foot of slopes.

Manifestations of the seasonal heaving process are also observed in fine soil with inclusions of larger rock debris (rock waste, pebbles, blocks, boulders, etc.). Thermal conductivity of rock debris is higher than that of fine soil; accordingly, fine soil beneath the rock fragments is chilled more intensely and water begins its migration primarily towards the front of freezing there. When water is transformed into ice it uplifts and pushes out such rock debris. At seasonal thawing a rock fragment does not manage to return to its place as this space is already occupied, partially or completely, by water and fine soil, while the rock fragment itself is held by the surrounding ground. Multiple repetition of the cycle results in heaving (freezing out) of rock material with sorting (redistribution) of debris within the seasonally freezing layer: the upper part of the section is enriched with coarse material, on the ground surface. This is the way 'boulder fields' and 'boulder streams' are formed.

Development of this process in soil masses where freezing is accompanied by the formation of diagenetic fractures or frost cracks, with freezing along their walls, results in part of the coarse material being shifted with the formation of 'stone polygons' and 'stone rings' over the surface (Fig. 5.2a). On gently sloping hillsides in paragenesis with other slope processes 'stone polygons' are formed elongated downslope, and at gradients over 8–10 °C stone stripes appear (see Fig. 5.2b).

When piles, poles or pipelines are placed into the seasonally thawing or seasonally freezing layer, they are gradually heaved (frozen out). As a result of the annual repeating of this process poles are pushed out, become less stable, tilt and, in the long run, fall (Fig. 5.3). Heaving of poles leads to disturbance of communication lines, while piles pushed out of the foundations of different structures cause their deformation.

Fig. 5.2. Cryogenic sorting of fragments within the seasonally thawing layer: *a* – stone rings in Spitsbergen (photo A. Jahn), *b* – stone stripes on a gently sloping hillside in the Sopky range in the Urals (photo G.I. Dubikov).

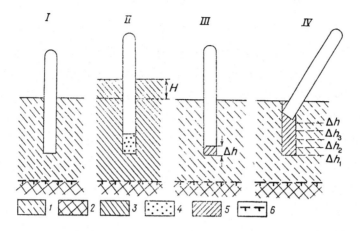

Fig. 5.3. Diagram of heaving (freezing out) of a pole extruded from a seasonally thawing layer composed of wet fine-grained deposits (from I.D. Belokrylov). *I–III* – stages of extrusion of the pole in an annual cycle, *I V* – falling of the pole out of the seasonally thawing layer after a series of years; 1 – thawed soil of the seasonally thawing layer, 2 – perennially frozen soil, 3 – frozen soil of the seasonally frozen layer, 4 – cavity, which is formed by withdrawal of the pole by the freezing to it of the soil of the seasonally thawing layer, and is filled with ice or ice-rich soil, 5 – cavity filled with water saturated soil at thawing, 6 – boundary of perennially frozen soil. Δh – pushing out of the pole; H – soil heaving in an annual cycle.

5.3   Frost cracking (fissuring) and polygonal formations, surface and underground

Frost cracking represents the process of temperature strain of frozen rocks in the variable temperature gradient fields. The volumetric tensions arising due to changes in temperature of ground within the annual range of temperature variations can lead to frost cracking of soil masses. Cracking takes place on the surface and penetrates the ground forming polygons in plan (Fig. 5.4). Multiple repetition of the frost cracking process gives rise to cryogenic formations and specific features of the terrain.

The phenomena associated with the frost cracking or fissuring process should be expressed in the size of polygons, width of fracture opening and depth of penetration, the nature of filling material and the interrelationship between the time of their formation and the time of sediment accumulation.

The stresses that determine the size of polygons are in proportion to the vertical temperature gradient (with depth), to the distance from a free vertical surface (break, another fracture), and to the modulus of elasticity during shear and coefficient of thermal expansion (compression) of ground. It has been established that the amplitude of the temperature variation on

Fig. 5.4. Polygons resulting from frost cracking of soils (photo F.N. Leshchikov).

the surface exerts the greatest influence on the size of the polygonal grid in the horizontal plane, and the mean annual temperature of the frozen ground serves as an indicator of severity for the depth of penetration of frost fissures into the frozen ground. The greater the amplitude of temperature variations the smaller is the distance between fractures and the smaller are the polygons. Therefore, under conditions of drastic and deep freezing of ground in winter, typical of extremely continental climate regions, there arise numerous frost fractures spaced at 0.5–2 m to 10–12 m. In conditions of less continental climate, frost polygons are formed, with cracks spaced at 20–40 m (Fig. 5.5) and even 50–80 m. The finest grids on the surface are formed at mean annual ground temperatures equalling or approaching 0 °C. In ground homogeneous with respect to composition (and mechanical properties), rectangular fissure networks are formed, whereas in heterogeneous ground the networks have a complex pattern, conjugating with each other at right angles. Polygons having four, sometimes five and more, sides can be formed in this way.

Frost fissures can penetrate rather deep into the frozen ground. With seasonal freezing of the ground their depth is limited by the thickness of the frozen layer; while in perennially frozen ground elementary cracks (of one winter season) can penetrate to a depth of 3–4 m and more. The lower the

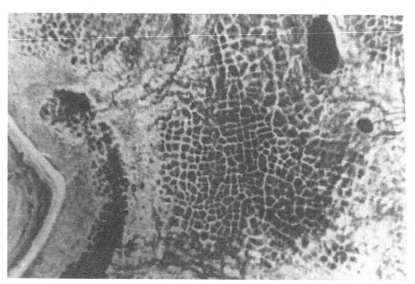

Fig. 5.5. Polygon-dike topography within the limits of high flood-plain: *a* – the Algan river of the Anadyr river basin (photo A.N. Kotov), *b* – the Gydan peninsula (photo G.I. Dubikov).

temperature of perennially frozen ground the deeper the elementary cracks penetrate. Frost fissures can reach 5–10 cm in width at the surface. As a rule, the frost fissures are filled with water, fine soil, sand, peat with the formation of polygonal-wedge structures which are subdivided into four types: recurring ice wedges, original ground wedges, primary sandy wedges and pseudomorphs replacing recurring ice wedges.

Recurring ice wedges

Frost cracks (fissures) are formed in winter. In the spring these cracks are filled with snow melt water which freezes, and elementary ice veins are formed in the frozen soil. In the layer of seasonal thawing the upper part of the vein thaws in summer, while in the underlying perennially frozen ground the ice vein is preserved. This is possible provided the depth of fissure penetration is greater than that of seasonal thawing. From year to year frost cracking is repeated at the same place; the cracks formed are again filled with water which freezes, each time. This is the way elementary ice veins are formed embedded one into another, promoting expansion of the ice wedge (Fig. 5.6). The structure of an ice wedge shows a vertical banding: vertically oriented air bubbles and inclusions of mineral particles such that the elementary ice veins can be distinguished. Their number corresponds to the number of years during which an ice wedge has been forming. Usually, it takes thousands of years. The deeper the frost fissures are and the longer the time of their growth, the larger is the width and vertical dimensions of the ice wedges. Further northwards, as the climate becomes more severe, the rate of growth of the ice wedges is quicker.

The growing ice wedges displace the ice-bearing soil. Around frost fissures there are ridges on the ground surface, while above the wedge itself (between the ridges) trench-like depressions are usually formed resulting from thawing out of ice wedges within the layer of seasonal thawing as well as due to erosion processes. The so-called polygonal-ridge pattern of micro-relief is formed.

The recurring ice wedges are found both in ground formed earlier under more severe freezing conditions, and in ground where the accumulation of deposits is proceeding. The former are named *epigenetic*, the latter, *syngenetic*. The latter are the deepest polygonal wedge formations, reaching 40–60 m and more in depth and 6–8 m in width. Epigenetic wedges, as a rule, do not exceed the thickness of the layer of annual temperature variations.

The recurring ice wedges form on sandy-silty, silty-clay and peaty sites of deposition terrain (river valleys, lacustrine basins) that are periodically covered with water, with mean annual temperatures below $-3\,°C$. The further north they are, the more widely they develop on gently sloping hillsides and on interfluves, and they are encountered even in the eluvium.

Frost cracking under conditions of insufficiently moist depositional plains can result in the formation of initial ground fissures and not ice

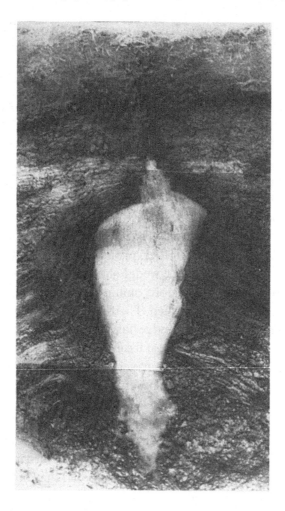

Fig. 5.6. Syngenetic ice wedge, up to 0.5 m wide, in the Pleistocene alluvial deposits of the Maya river of the Anadyr river basin (photo A.N. Kotove).

wedges, in the layer of seasonal thawing and freezing of the ground. In the spring the frost cracks are filled with liquified soil and not with meltwater. In the north these are rarely encountered, being found only where there is comparatively deep thawing. Southwards such sites are substantially more extensive. They are also encountered beyond the permafrost region.

Sand wedges

In conditions of severe and very dry climate with strong winds in winter, when frost cracks are formed sand fills them and small sand veins arise. Multiple repetition of the process results in the formation of sand wedges. Certain conditions can promote formation of *sand-ice* filled wedges.

Sand-ice wedges and sand wedges also form in less severe conditions than those of the Antarctic continent: in Central Yakutia, in the north of West Siberia, etc. At positive or negative mean annual temperatures near to 0°C, when frost fissures do not penetrate below the seasonally frozen or seasonally thawed layers, the sand wedges occur entirely within the depth of these layers.

Pseudomorphs of recurring ice wedges are secondary formations resulting from thawing of ice wedges and refilling with soil (Fig. 5.7). They are mainly encountered in such areas where recurring ice wedges thawed as a result of local or regional degradation of perennially frozen ground. They are observed to be widely developed beyond the limits of the regions of the recent permafrost: in the Ukraine, Poland, the former Czechoslovakia, Mongolia, China, etc. Within the limits of the permafrost regions they can be observed on sites where ground ice thaws as a result of thermokarst development or thermal erosion. They are frequently developed in the bottom of the 'dry' alases of Yakutia. Pseudomorphs of ice wedges are of great importance in the study of Quaternary deposits, elucidation of the history of permafrost strata and reconstruction of the palaeographic setting of the past.

Small polygonal forms of relief are also widely developed within the permafrost regions. They form under the influence of a variety of processes. First of all, this is associated with small polygon shapes of cracking and formation of closed (enclosed) systems of thawed ground in the layer of seasonal thawing when nonuniform freezing occurs from the top and laterally. In such systems hydrostatic pressure increases greatly, promoting transformation of thixotropic wet soil within blocks into a plastic-flowing state. Often, a surficial frozen crust is broken, and the liquified ground mass flows out onto the surface (Fig. 5.8a). Along with this an oriented heaving of stone material can occur (see Fig. 5.8b). These processes repeated many times result in the formation of 'spot medallions' and 'stone garlands'. Such shapes are widely developed in the north. Their dimensions vary between 0.3–0.4 to 1–3 m and more, they are mainly developed in the seasonally thawing layer in sandy-silty – silty-clay soils. On the depositional plains they take the form of 'spot medallions'. In the regions with a shallow cover of loose deposits enriched in stony material, 'stone garlands' are developed predominantly.

5.4 Thermokarst

When ice-rich frozen soils thaw, and the ground ice melts and drains, surface subsidence is observed. The process of thermal subsidence is widely developed in the permafrost regions. It occurs both on seasonal and perennial

(a)   (b)

Fig. 5.7. Pseudomorphs along ice veins in the Ob river downstream: *a* – structure of enclosure of subaqueous type in the Kazantsev deposits (photo V.V. Baulin and L.M. Shmelev); *b* – structure of infilling in the Sartan deposits (photo L.M. Shmelev).

thawing of the frozen soils. Such subsidence normally accompanies the thawing of ice-rich soils (which have undergone heaving) in the seasonally freezing layer as well. Nonuniform development of the process gives rise to a variety of landforms and types of microrelief (collapse swallow holes, flat bottom depressions, alases, frost mounds and the like). Among manifestations of thermal subsidence in the thawing frozen soils, the process of perennial local thermal subsidence is of the greatest interest, namely thermokarst, which is accompanied by land subsidence and formation of negative landforms (Fig. 5.9a). The mechanism of this process consists of the consolidation of the thawing ice-rich soils or soils that contain monomineralic ice deposits, under the pressure of the thawing layer, and the expulsion of the contained moisture to drain to the surface or by an aquifer. Consolidation of soils takes place when the soils are soaked with water above saturation or as a result of the displacement of water and subsidence of blocks and layers of soil into fractures and cavities earlier occupied by the ice.

(a)

(b)

Fig. 5.8. Spot medallions: *a* – in rick debris of alluvium of the Daldyno-Alakit interfluve (photo A.Y. Derevyagin); *b* – with polygonal heaving of stone material along the desiccation fissures (photo A.N. Kotov).

(a)

(b)

Fig. 5.9. Development of thermokarst: *a* – resulting from thawing of wedge ice after removal of sod cover on the I terrace above flood plain of the Tynda river (photo L.N. Maksimova); *b* – thermokarst lake in the foothills of the Cherskiy range – in the opening in the background an icing is seen (photo V.Ye. Afanasenko).

The prerequisite for the development of thermokarst is the availability of ground ice occurring in the form of monomineralic ice deposits or structure-generating ice in loose deposits. An adequate condition to give rise to thermokarst is such a change in heat exchange at the soil surface that either the depth of seasonal thawing begins to exceed the depth at which subsurface ice or ice-rich perennially frozen soils occur, or the mean annual temperature changes sign and perennial thawing of the frozen layers begins. General degradation of the frozen layers is associated not only with changes in heat exchange at the soil surface due to historical periods of climate warming, but also with changes in the constituents of the radiation-thermal balance of the surface, that is, with changes in the dynamics of vegetation development, modification of snow and water covers, with drying out of soils of the seasonally thawing layer and with other changes of elements of the geological and geographical setting. Man's activity is considered to be one of the main causes of the presently activated processes which are apparent in disturbance of the soil-vegetation cover leading to deeper penetration of the seasonal thawing (sometimes as much as 2–4 times).

The process of thermokarst development, as shown by V.A. Kudryavtsev, occurs in a different manner in the case of water withdrawal from a thermokarst depression and water inflow into a depression. In the case when water does not accumulate in the depression (draining type of thermokarst) the process slackens. Often, the thawed ice bodies are replaced by cavities that are not distinguishable in the terrain, i.e. they do not create thermokarst forms. When it gets colder such thermokarst usually ceases to develop or, if deposits of the seasonally thawed layer are eroded (are washed out by water), then thawing of the ground ice can resume and develop progressively. In this case thermokarst usually grades into the process of thermal erosion.

With an undrained thermokarst depression the process develops in another way. If water arrives in the depression, it accumulates solar heat thus leading to temperature increase of the water-body bottom materials; this, in its turn, usually promotes the deepening of the seasonally thawing layer. The latter is accompanied by further thawing of the monomineralic ice body, and the water body becomes deeper. In the long run, this can lead to complete thawing of the ground ice and the emergence beneath the water body of a closed sublacustrine talik (if the permafrost thickness is small the talik is open). Development of non-draining thermokarst is possible in any conditions, even with the most severe freezing.

Thermokarst landforms and types of microrelief are to a great extent dependent on the types of ice bodies and ice-rich soils subject to thawing, as

Fig. 5.10. Kettle lacustrine basin (khasyrei) and perennial frost mound. Southern Yamal (photo G.I. Dubikov).

well as on the peculiarities of ice development in the frozen soils, its forms and situation, etc. In West Siberia where thermokarst is mainly developed on sites of ice-rich marine, glacial-marine and glacial deposits, which comprise the strata with ground ice, the thermokarst basins are called *khasyrey* (Fig. 5.10). In Yakutia such basins which were formed after thawing of the 'ice complex' soils with syngenetic ice wedges are called alases. When syngenetic ice wedges are thawing (if there is no drainage from the subsidence depressions) thermokarst lakes are developed (see Fig. 5.9b) which are rather deep (up to 3–6 m) and vary in size (to several kilometres). With a flat bottom, the sites of active ice thawing are up to 10 m deep and more. When they dry out or are displaced (migrated) the alas basins form. Usually, alases and khasyrey develop on ancient depositional plains. Often, when thawing occurs in deposits with small polygonal-wedge ice, small lakes arise with depths of up to 1.5–2 m and having banks with rectangular outlines. When such lakes drain detrital polygonal mound-like landforms develop on their beds. If thawing out of ice wedges occurs under good drainage conditions and the ice wedge outlined blocks of soil are composed of rather hard material with low ice content, this gives rise to *baydzherakhi* – soil outliers (Fig. 5.11). If the ice wedges are replaced by soils then pseudomorphs of recurring ice wedges arise.

Fig. 5.11. Mounds (baydzherakhi) resulting from thawing of ice wedges in the Yuribey river valley, Gydan peninsula (photo G.I. Dubikov).

Thermokarst landforms are most widely developed in the subarctic belt of the northern coastal lowlands. Signs of the process become less distinct southwards and this is associated with less widespread ground ice. Beyond the limits of the permafrost regions only relict thermokarst landforms are encountered. On the depositional plain of Central Yakutia, characterized by a dry climate, the recent thermokarst landforms have a limited areal extent. Nevertheless, the great number of thermokarst lakes and alases in this region gives evidence of more active development of the process in the past.

5.5 Slope processes and phenomena

Slope processes and phenomena are caused by the action of gravity forces and lead to a variety of deformations – collapse, talus, landsliding, solifluction, stone streams, rapid flows and the like. The most distinct manifestation of gravitational forces occurs on steep slopes, with failure and displacement of rock waste and stone blocks resulting from the weathering processes. The rate of material displacement can vary, being high on certain sites.

The history of the sedimentary material within the permafrost regions is associated not only with cryohydric and temperature weathering of hard rocks, but also with the permafrost processes proper which lead to separation of the frozen soils. Such permafrost processes include frost cracking, thermal abrasion, thawing out of the texture generating ice and recurring ice wedges, etc. Among the most common depositional forms of the gravitational processes themselves are scree creep and accumulation of rudaceous materials at the foot of slopes. Thickness of these formations reaches several tens of metres. They are widely developed, for example, in the mountain regions of the North-East of Russia; they contain vast masses of congelation ice resulting from water infiltration and freezing. This type of slope process plays an insignificant role on flat and lowland territories of the permafrost regions.

Displacement of weathering products on landslide slopes is not of importance in the permafrost regions. This is explained by the presence of the perennially frozen ground and the comparatively shallow seasonally thawing layers of the ground. Landslides arising locally are mainly of small thickness and they are connected with the process of soil thawing or with sliding of ground masses along the tilting ice layers. Sliding of the deposits of the seasonally thawing layer is predetermined mainly by the emerging postcryogenic structure and texture. At the locations of thawed ice schlieren, failure-prone zones are formed, while soil aggregates are preserved as structural separations. As a result, hydraulic conductivity of the seasonally thawing clayey silts (for instance) increases by a hundred times. Moreover, during the warmest years, when depth of the seasonally thawing layer increases by 10–20%, the most ice-rich (lower) part of this layer becomes actively permeable. This gives rise to supra- permafrost water even in heavy clay rich soils and the possibility of flowing of whole blocks of ground.

Rock streams are slope formations widely developed in the area of permafrost, composed of scree-block-gravel materials of hard and semihard rocks. Their frequency among slope formations is rather great as is the role played by them in the displacement of weathering products along slopes. Rock streams are confined to slopes with steepness ranging between 3 and 40°. Their dimensions, shape and arrangement are heterogeneous. They arise on vast rocky slopes, can form as rock rivers and altiplanation terraces, fill narrow hollows, compose large-area block fields, etc. (Fig. 5.12a). The processes that lead to the formation of rudaceous materials, their displacement and accumulation within the limits of catchment areas are also heterogeneous. The main and permanent mechanisms of rock streams are thermogenic and cryogenic creep.

Thermogenic creep is conditioned by periodical (diurnal and seasonal) variations of temperature leading to the cyclic expansion and reduction of the sizes of rock waste and pulse-like displacement of rudaceous material downslope.

Cryogenic creep is associated with uplifting of rock waste in the direction

(a)

(b)

Fig. 5.12. Rock stream on Jurassic rocks (*a*) and golets ice in the base of rudaceous cover of rock stream (*b*) (photos A.I. Tyurin).

perpendicular to the slope (on account of ice lenses and layers arising in the body of the rock stream) and with the further sinking of debris together with infilling materials, resulting from the thawing of the ice. Multiple repetition of the cyclic process leads to a gradual movement of rock streams down-slope. Among the accompanying processes are: plastic-viscous flow of a fine grain matrix and creep of rock debris of the seasonally thawing layer along the overwet or ice-rich bottom. The above takes place when intense heaving of rock debris occurs in the seasonally thawing layer as well as a moving down and spreading out of fine soil. As a result, at the base of the layer there is more fine material, while in its upper portion there remains rudaceous material which is very permeable. In the spring, meltwater penetrates to the bottom of the coarse debris and freezes there, thus forming the so-called 'golets' ice (see Fig. 5.12b). During the warmest or rainy years when depth of seasonal thawing increases greatly, golets ice thaws, fine earth underlying the rudaceous rocks gets supersaturated and plastic-viscous flow begins, moving the whole overlying sequence of rudaceous material.

The rudaceous material in rock streams is conveyed at a rate of centimetres a year. Within the limits of a rock stream this rate varies both in time and in different parts. Often, rock streams are distinguished as a specific genetic type of slope formation of the mountain regions in Siberia – mountain or rock stream creep.

The processes of viscous and visco-plastic displacement of granular materials within the limits of the freezing regions occur both on slopes with a vegetation mat and over the predominantly smooth surfaces of accumulation. Among these the most common is the process of solifluction or viscoplastic (slow) flow of loose deposits on the slopes under the influence of their own weight, the component of which is directed down the slope thus causing plastic deformations of the ground.

Solifluction usually develops in fine silty clays and silty sands, often with a high content of rudaceous materials. The rate of solifluction is a function of slope steepness, depth of soil thawing, composition of deposits, durability of ground cover, terrain type, etc. Viscoplastic displacement of soils downslope manifests itself in the laminated deposits in which the layers of ice-rich silty clays and silty sands alternate with peat and humus interlayers (Fig. 5.13). The thickness of the deposits that form in a syngenetic way is greatest in the lower and least in the upper parts of slopes. Usually, two types of slow solifluction are distinguished, namely, covering and differential.

Cover solifluction is movement of ground that occurs more or less uniformly and rather slowly. In this case there is no appearance of flow on the slopes. This type of solifluction has typical rates of $2–10 \, \text{cm yr}^{-1}$ and manifests itself on slopes of up to $15°$ steepness. A specific feature of cover solifluction is that material movement occurs without substantial change of the inner structure of the ground. Moisture content of deposits does not exceed the liquid limit. In the upper part of the solifluction layer there is a layer that preserves its shape, usually making up 25–30% of total thickness

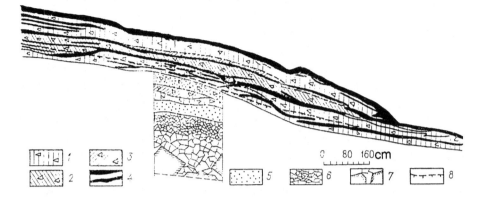

Fig. 5.13. Structure of solifluction terrace. North-eastern piedmont of Chukotka (according to L.A. Zhigarev): 1, 2 – light and heavy sandy silty materials, respectively, with fragments of porphyrite; 3 – heavy clay silty-sand with fragments of porphyrite; 4 – vegetation-peaty layer and buried peat interlayers; 5 – poorly-sorted sand; 6–7 – fine- and coarse-fragment eluvium of porphyrites; 8 – boundary of the perennially frozen soils.

of the ground flow, but at low velocities it can be as much as 90%. Slow solifluction is likely to be accompanied by sorting of material. The most apparent manifestations of this are alternating stripes of rudaceous material and fine earth.

Differential solifluction as distinct from the cover type shows itself on the surface in the typical landforms of micro- and mesorelief: solifluction lobes, terraces, stripes, etc. Such forms arise owing to differential rates of displacement both on the whole slope and within one solifluction flow. Different rates are largely conditioned by the ice content of soils varying in different parts of the slopes.

Within the limits of the freezing regions along with the slow solifluction considered above there is the so-called *rapid solifluction* or viscous flow of thawing soils on the slopes. This type of slope transfer, often called *mudflows*, develops on slopes of 15–25° steepness in highly supersaturated soils of the seasonally thawing layer. It is accompanied by disturbance of the internal soil structure as well as by breakage of the ground cover. The moisture content of deposits exceeds the yield point. Viscous flow develops during periods of intense melt or in cases of heavy precipitation. The layer of preserved shape is absent.

Mudflows occur frequently on cut slopes as perennially frozen ice-rich soils become exposed by slumping and thaw (Fig. 5.14). The rate of soil displacement in mudflows is rather great, reaching sometimes several metres per minute.

Fig. 5.14. Solifluction slumps on the slope towards the Yuribey river of the Kazantsev plain, the Gydan peninsula (photo G.I. Dubikov).

5.6   Processes and effects associated with the activity of water, glaciers and other geological agents

Of paramount importance in geological activity, is the role of flowing water, glaciers and other agents of the environment. Among these, the processes brought about by the action of intermittent and permanent watercourses prevail, along with those resulting from sheet wash by rainfall and meltwater, leading to conveyance of enormous quantities of sedimentary material.

Naturally, the impact of the flow of water is not merely the transport of material. Creeks and rivers contribute greatly to erosion and accumulation. Downcutting and lateral erosion play a key role in the formation of river valleys. Within the permafrost regions lateral erosion dominates over downcutting, both in rivers on plains and in mountains, thus leading to river meandering and development of meander lakes. The proportion of downcutting relative to lateral erosion is determined by a number of factors. The frozen state of the banks and the high content of ice of soils (up to 80%) that compose them contributes to the intensification of the lateral erosion process because of the thermal and mechanical impact of water flow and solar radiation on the frozen soils. Among manifestations of this impact is displacement of materials thawed from bank slopes into rivers as well as the formation of various niches resulting from thermal abrasion and thermal erosion due to undercutting of ice-rich soils by streams or waves. At the same time, downcutting (as compared with lateral erosion) is hampered by intense inflow in river catchments of talus deposits and solifluction material from slopes which leads to the overloading of river beds, thus causing stream migration sideways and bank undercutting. Small streamlets may thus become totally filled with the material.

Sheetwash of fine material in the permafrost regions has a number of specific features and occurs by transport of rain and meltwater from firn. The wash of fine particles along slopes is facilitated by the raindrops of summer precipitation and by slow thawing of snow cover. Downslope wash of silty-sandy-clay materials is intensified when the presence of higher ice content and schlieren cryogenic structures drastically reduce the structural strength of the thawing soils.

On slopes with rudaceous material a sheet wash of fine earth usually becomes dispersed (suffosion) into the slope (mainly on steep slopes) or grades into illuvial deposits (on gently sloping hillsides and in hollows). The processes of downslope wash and leaching may bring about formation of rather thick strata of deposits at the foot of slopes and on their gently sloping parts, deposits known as *detritus aprons*, *perluvium*, and the like.

A substantial role in regions of frozen ground is played by the action of intermittent watercourses, i.e. the action of stream flow from rainfall and melt water. This process of linear erosion named thermal erosion, along with sheet wash, greatly intensifies the formation of diluvial sediments and plays a significant role in conveyance of weathering products down slopes. The specific nature of thermal erosion by intermittent streams implies a combination of mechanical (cutting) action of water and thermal influence. The mechanism of denudation is associated with accelerated thawing of the frozen soils and their subsequent erosion.

Thus, in analysis of the process attention should be given to the resistance of soils to erosion, the interrelationship between erosion rate and rate of thawing of the underlying frozen soils as well as to the erosive energy of flowing water. It is evident that when there is a thawed layer of some thickness beneath a stream flowing in frozen soils, then with greater mechanical energy of flow the erosion front will catch up on that of thawing. In the longer term there will be the moment when the erosion front reaches the front of thawing. Further increase of the flow energy will not contribute to the intensification of erosion, since the process of erosion is completely controlled by the rate of thawing. Therefore, the capacity of frozen soils to erosion is determined, on the one hand, by the structural strength of thawing soils and on the other hand, by thermal-physical parameters of the frozen soils which determine the rate of their thawing at a constant water temperature. Accordingly, higher temperature of water flow leads to the intensification of erosion of the frozen soils.

The activity of intermittent streams in the permafrost regions is accompanied by the formation of dissected relief which is most typical of marginal parts of terraces lying above floodplains. Observations have shown that man-induced disturbance of soil and vegetative cover on sites having different elevations leads to erosional channels which sometimes grade into gullies. The latter land-forms are usually typical where vehicle tracks run down the sloping surface. In the beginning they are of canyon shape or V-shape when the rate of their growth lengthwise reaches 10–20 m yr^{-1}, and sometimes (in the bottom of ravines and hollows) it is as high as 100–150 m yr^{-1}. Rapid growth of gullies diminishes with time, which indicates their stabilization when the ever-increasing cryogenic washout from the slopes into bottoms of gullies prevents their further deepening. The bottoms of such gullies flatten while still retaining a steep slope, which is why their length seldom exceeds 1 km. Based on the available data, it is possible to determine the length of time for gulley stabilization. This takes 20–30 years after erosional landforms develop. Regions in which there is rapid growth of gullies include the northern part of West Siberia, the north of Yakutia, the region of the Baikal-Amur railway, etc.

Of specific interest is the process of destruction (erosion) in the permafrost regions of frozen coastal lands under the influence of the mechanical energy of waves and water temperature which cause cliffs to retreat due to washout and leaching. This process, which is widely developed in the permafrost regions along the banks of rivers, lakes and reservoirs as well as on the sea coast, is known as *thermal abrasion*. The process becomes active if ice-rich soils and ground ice are exposed. Owing to heat exchange with the water the frozen soils rapidly thaw, and the layer that has thawed gradually slides exposing the frozen ground. Thus direct contact is promoted between the frozen ground and water and there is rapid destruction. Such banks stabilize with time provided the ice content of soils that compose the banks is low (lower than a critical value). Stabilization of the banks enclosing water bodies and disturbed by thermal abrasion, is possible when lowering of water level results from the enlargement of the basin of the water body caused by the same thermal abrasion, or by erosional incision of outflowing streams.

When soils are uniformly saturated with ice, thermal abrasion activity is dependent on water temperature and, to a greater extent, wave action. The sea coast is subject to more intense wave action. As regards inland water bodies the wave action is dictated both by the force of winds and the size of these water bodies. In conditions of unidirectional prevailing summer winds the lakes are aligned in the same direction. These are usually called oriented lakes. The greatest intensity is observed on sea coasts and banks of large lakes in the extreme north-east of Russia and in the north of Yakutia where the widely developed ice-rich banks are subject to the most active wave cutting. Thus, lowland banks on the northern margin of the Yana river delta retreat at a rate of $16-20 \, \text{m yr}^{-1}$ and the coast of the Laptev Sea retreats at a rate of $4-6 \, \text{m yr}^{-1}$. The thawing rate of ice-rich banks of lakes in Central Yakutia amounts to $7-10 \, \text{m yr}^{-1}$, while thermokarst lakes in the Anadyr tundra migrate at a rate of $10 \, \text{m yr}^{-1}$. Intense erosion of the banks of the northern water bodies is associated with washing-out of the submarine bank slope composed of the frozen ice-rich soils, under the influence of both mechanical and thermal energy of the moving water and is characterized by three specific features. The first one is that the intensity of the washing-out of the frozen soils is a function of temperature; the second one is that the volume of sediments settling on the submarine slope due to washing-out of ice-rich soils is smaller than that of the frozen soils eroded. The third feature lies in the fact that settlement of the frozen soils at thawing leads to deepening of the water body thus contributing to the further development of thermal abrasion. The role of the thermal factor in thermal abrasion is conditioned by the ice content of the frozen soils. Thus, if banks are composed of soils containing no ice the thermal factor does not operate at all. The bank will retreat until a limiting equilibrium profile of the submarine bank slope is formed. If the banks are predominantly composed of ice, the thermal factor is practically 100% effective in the process of thermal abrasion. In this case the bank will retreat continuously.

Icings

A typical phenomenon of the permafrost regions zone are ice bodies which arise and grow only during the frost season of a year. They form from different sources of water: ground water, subsurface water, river water, lacustrine water (often icings have a mixed recharge). The flat-convex ice bodies – icings – result from multiple inputs of these waters onto the surface and their freezing in lamina. Icings have an impact on the re-distribution of surface runoff and terrain giving rise to specific deposits ('icing alluvium'), they are capable of exerting a detrimental impact on engineering structures. Often icings arise as a result of the altered freezing conditions associated with construction and operation of different structures and other features of land development. Severe and extreme continental climates with cold winters with low snowfall are most favourable for the development of icings.

Water flows onto the surface because of the increased hydrodynamic pressure as a result of seasonal freezing along watercourses; there is higher hydrostatic pressure with the freezing of lakes and sublacustrine taliks. That is, seasonal freezing reduces the cross-section for flow of the surface and subsurface waters of rivers and streams. There arises a hydrodynamic pressure. Under the influence of this pressure the frozen roof (river ice or frozen soil) is broken and water flows out onto the surface. Water flows in different directions and freezes, and flow is blocked until a further break occurs. With such cyclicity, laminated ice bodies of various size and thickness form on the soil surface. This depends on the number of cycles, volumes of flowing water, i.e. on the reserves of ground water and the soil-freezing situation. Similarly, icings can form because of hydrostatic pressure arising in the freezing of lakes and closed sublacustrine taliks. The rate of icing formation varies within an annual cycle. In autumn and early winter the rate typically is slow but steadily increasing (early stage of formation). During the winter period when the area and volume of the icing increase uniformly, the stage of maturity of the icing begins. The stage of maturity continues from mid-winter till early spring. After a persistent increase of mean diurnal temperatures above $0\,^{\circ}C$ the stage of erosion is initiated.

Icings are subdivided according to the sources of recharge, location in the terrain, time of existence, size, etc. There are three types of icings: hydrogenous (surface water), hydrogeogenic (subsurface water) and heterogeneous (mixed surface and subsurface water) These types are, in their turn, subdivided into kinds. Thus, for example, icings of hydrogenous type comprise the following kinds: water of rivers, creeks, lakes, snowmelt, glaciers and seas. In the majority of cases icings are confined to valleys of rivers and creeks (valley, floodplain or terrace), or to slopes, proluvial debris cones, glaciers and glacier-adjacent sites. There are annual icings which completely thaw by the end of the summer season, and perennial ones. When small thin, disconnected lenses of ice remain at the end of the summer season they are called *summer-surviving*. These icings are well distinguished in aerial and satellite images made after the complete disappearance of snow cover.

The size of such icings varies between very small (area under $10^3\,m^2$) and enormous (area over $10^7\,m^2$ and volume over $10^7\,m^3$). Enormous icings can be 10 m thick with a length reaching several tens of kilometres. If their thickness exceeds 5–6 m such icings are, as a rule, perennial (Fig. 5.15). The size of icings is dependent on the source of their recharge. Very small ice bodies form on account of the meltwater from the seasonally thawing layer and perched water, while big and enormous ice bodies ('tarynn') form on

Fig. 5.15. Icings in the river valley in Chukotka (photo V.Ye. Afanasenko).

account of the underground subpermafrost flow of water of mixed deep and subpermafrost flow origins.

Glaciers

These are masses of ice formed of precipitation as a result of snow accumulation perennially exceeding thaw at a negative basal ground surface temperature. There are areas of recharge and of ablation (thawing) in glaciers and they are divided by the recharge boundary. Glaciers form above the snow line, which is the level of earth surface above which snow accumulation prevails over thawing. The snow line is lower in cold and humid regions and goes up to high elevations in warm and dry regions. Thus, on the Arctic islands and Antarctic continent glaciers may extend below sea level, in the mountain systems of the equatorial zone they exist at altitudes of 4500 m and over.

The size and thickness of glaciers increases from the equator towards poles and from sea level towards high altitudes of mountains: size ranging between 0.1 km² and 1 million km² and thickness from several metres to several kilometres. A variety of processes and phenomena are associated with glaciers, which contribute to the formation of specific landforms and types of sediments through erosion and accumulation. The role of glaciers in formation of deposits is unique, because the vast volume of ice which is highly viscous moves along valleys or over the orographic elements thus eroding and conveying particles of practically any size without sorting. Glacial erosion occurs in the following way: 1) by means of removal of loose material resulting from weathering; 2) by means of corrasion – destruction and removal of bedrock as rock waste, either frozen into the bottom of glacier or dragged by the glacier along its bed; 3) by means of pulling out, when the glacier plucks the blocks of bedrock bounded by fractures and conveys them within the ice. Under the influence of glaciers the V-shaped young valleys are transformed into U-shaped ones having flat bottoms. Deposits conveyed by the glacier create different shapes, for example, ramparts stretching for many kilometres – *eskers* – and cone-like hills – *kames*.

Névé basins

These are stabilized accumulations of snow and ice (sometimes sliding slowly) that have remained after the thaw during the warm season of a year on sites of negative soil surface temperature. On tundra plains, snow patches remain in shaded hollows and beneath steep escarpments of valleys, whereas in mountains they survive on cornices and escarpments of slopes within the mountain-tundra belt. Snow patches reach 2–4 m in thickness, rarely more, while the thickness in perennial névé basins ranges between 5 and 10 m and over.

The impact of névé basins manifests itself in the formation of deposits and is associated with various processes known under the common name of *nivation*. One group of the processes is responsible for the preparation (making available) of soil material for conveying it downwards from the snow patch, i.e. the formation of weathering products. Such preparation is facilitated by negative temperatures existing beneath the snow patches, by total absence of vegetation and highly wetted soils resulting from thawing of snow patches. The second group of processes is associated with the conveyance and accumulation of the weathering products.

Avalanches

They are composed of loose snow formed mainly in the upper parts of erosional cuts. The initial stage of snow accumulation starts if equilibrium is maintained between the angle of slope of the snow surface and that of internal friction in snow. If a critical value of snow thickness is exceeded and the shear strength of the snow is small there arise zones of the snow cover in an unsteady state thus causing development and fall of snow avalanches down the slope. On hillsides of the Cherskiy and Alatau ranges, for example, avalanches on the average carry the same amount of rudaceous material per year as rock falls usually do.

Snow and ice mudflows

These are short-term rapid flows carrying water, ice, snow and muddy materials. Conditions that promote snow-ice mudflows are similar to those of ordinary mudflows: accumulation of large volumes of loose material, inflow of water and the slope of the hillside. The carrying capacity of mudflows is rather great. Rudaceous materials can be in a suspended state in the snow mudflow; moreover their density exceeds greatly that of the mudflow itself.

In conditions of polar and extreme continental climate, wind-born (*aeolian*) *transportation of fine earth* occurs widely. The process takes place both in winter and in summer, being most intensive in winter when fine-grain material is blown out by strong winds on sites with disturbed or no vegetation cover and where frozen soils are less cohesive (ice-cement cohesion) due to ice sublimation in them. Therefore, the amount of material transported in winter is dependent on wind velocity, intensity and on depth of sublimation. The conveying capacity of wind is one three-hundredth that of water, that is why wind can transport only sandy or more fine particles. In conditions of severe climate aeolian processes play an important role in the formation of a thick ice-rich sequence of deposits on coastal lowlands in the north-east of Russia which is called the 'yedoma' complex.

As a conclusion it should be noted that a general dependence of exogenous geological processes in the permafrost regions on heat exchange gives rise to features that are specific in their development and appearance. Of importance is the latitudinal and altitudinal zonation of the occurrence of the freezing-geological processes: heaving, icing formation, thermokarst, solifluction and the like which are modulated by the zonation of mean annual temperatures of the ground, depths of seasonal thawing and freezing as well as by the zonal distribution of thicknesses of the frozen ground, development of ice-rich frozen soils etc. Important, too, is the role of regional factors, especially with respect to the forms of the processes.

II

Composition, cryogenic structure and properties of frozen rocks

6

Formation of sedimentary materials in the permafrost regions (cryolithogenesis)

The most general and fundamental geological principles of the formation of sedimentary deposits in the permafrost regions are associated with the questions of grain-size distribution and chemical-mineral composition, structural-textural features and composition of the frozen sediments, the nature of which is elucidated in the course of study of chemical, physical-chemical and physico-mechanical processes that occur in the sediments of the permafrost regions in their stage-by-stage transformation. These questions are dealt with in works by B.I. Vtyurin, E.A. Vtyurina, Sh.Sh. Gasanov, I.D. Danilov, E.D. Yershov, Ye.M. Katanosov, V.N. Konishchev, Yu.A. Lavrushin, A.I. Popov, V.O. Targulyan, I.A. Tyutyunov, P.F. Shvetsov, I.A. Shilo, P.A. Shumskiy *et al.* In general, sedimentary formations reflect a number of factors and processes of lithogenesis, the development and manifestations of which occur specifically under various geological and geographical conditions. Certain combinations of factors, conditions and processes of lithogenesis predetermine the composition, structure and properties of sedimentary formations. The idea was to distinguish specific and strictly definable types of lithogenesis.

As early as 1957, N.M. Strakhov distinguished types of lithogenesis as specific forms of the lithogenetic process at the stage of sedimentation and diagenesis, giving rise to the quite varied sum total of deposits. This author distinguished four types of recent lithogenesis, namely, *ice, humid, arid and volcanogenic-sedimentary.* The ice type of lithogenesis was defined by Strakhov as the cases when there was an ice cover on the soil surface and negative mean annual temperature; the humid type was defined by the intensity of degradational processes of organic matter; and the arid one by the intensity of evaporation, i.e. according to climatic attributes (air temperature and humidity). The volcanogenic sedimentary type is not of a zonal nature; it develops on the most mobile sites of the lithosphere which are the most permeable for magma. At present the majority of researchers emphasize the necessity to define one more type of lithogenesis – cryogenic. This is associated with the fact that in the cold regions at each stage of formation of sedimentary materials (i.e. at the stage of weathering, transfer, accumulation of sediments and diagenesis) the factors and processes of lithogenesis are characterized by such wide differences in a qualitative respect that in the long run they give rise to specific deposits which are the products of a *specifically cryogenic* type of lithogenesis. They are cemented with ice at negative temperatures and are characterized by a high content of silty (aleurolite) and sandy material as well as ice breccia and ice conglomerates. They also have a specific and unique cryogenic structure, which is that of sedimentary rocks, and a specific typomorphic complex of minerals and rocks (ice, hydromicas, montmorillonite, substances and compounds with oxides of ferrous iron, manganese and other elements, a specific complex of mineral deposits and the like).

6.1   Sediment genesis in the permafrost regions

The formation of sedimentary material (sediment genesis) in the permafrost regions is dictated by the nature of processes occurring at the stages of weathering, transfer and accumulation of this material on catchments and in the final bodies of water of surface runoff. An essential stage is the formation of sedimentary material as a result of weathering in the general framework of the lithogenetic process. It is at this stage that sedimentary products are formed both qualitatively and quantitatively, arriving subsequently into the final water bodies, or remaining at the same place thus giving rise to a weathering crust. The following mechanisms of physical weathering of rocks have been thoroughly studied and are considered to be important: tectonic mechanical, temperature dependent, hydration and cryohydration.

The *tectonic mechanical* mechanism of physical weathering manifests itself in those magmatic, metamorphic and cemented sedimentary rocks in which, typically, a complex system of fractures is present long before they come within the area of influence of the weathering processes. The origin, outlines and length of fractures in rocks in the near-surface horizons of the Earth's crust may be associated with mountain pressure and tectonics and the relaxation and removal of such stresses. The latter effect leads to opening of previously existing ultra- and micro-fractures, their size and length increasing with the formation of a series of new fractures due to stress relaxation. The fractured near-surface zone of the Earth's crust is called the *zone of pre-degradation*. Its thickness ranges from several tens of metres in the platform areas to hundreds of metres in the tectonically active regions.

Thus, long before the interaction between rocks and agents takes place these rocks are in a way prepared for degradation and this predetermines the course and nature of their further supergene transformation. It is evident that this mechanism of physical weathering is characteristic of all types of lithogenesis including the cryogenic type of sedimentary rock formation.

The nearer the bedrock is to the ground surface the more active is the influence of the temperature factor or, to be pedantic, the more active is the role played by the *temperature mechanism*. The temperature variation on the rock surface can be as great as hundreds of degrees. Accordingly, the above is conditioned by the specific composition of the rocks as well as landscape and climatic features of the region.

Temperature differences bring about nonuniform deformations which die down with depth and gives rise to stresses of stretching and compression thus leading to initiation and growth of horizontal or vertical temperature fractures. There is a complex network of temperature-induced fractures in inhomogeneous rocks, caused by different values of thermal deformation of the components, along with the vertical and horizontal fractures: along, in other words, with a system of general temperature fractures that divide the massif into relatively big blocks and boulders so that there are smaller and shorter fractures (of a second generation) which prepare and promote breakdown of elements existing inside the blocks and boulders, into smaller fragments down to the aleurolite fraction, with degradation of individual minerals.

Degradation or breakdown of inhomogeneous rocks by temperature-induced weathering is likely to take place along weakened sites among which are the contacts between minerals in the rock and microdefects in minerals themselves, as well as along the boundaries of minerals in the rocks. The intensity of temperature-induced weathering of rocks is much dependent on the external environmental conditions (climate, orography, ground surface cover and the like) and is indicated by the amplitude of temperature variations on the rock surface and the number of transitions through 0°C to which rocks are subjected in the course of supergenesis.

Along with the temperature mechanism a substantial role is played in physical weathering by the *hydration mechanism* of rock degradation promoted by the impact of wedging out by thin water films. Indeed, all the ultra- and micro-fractures of rocks under natural conditions are usually filled with the finest films of bound water resulting from hydration or adsorption of vapour.

Gradually, the expanding and deepening micro-fracture is, in the long run, capable of storing a certain amount of free bulk water. In this case a qualitatively new mechanism of physical weathering is put into force, the one associated with phase transition of this water into ice – the so-called *cryohydration mechanism* of weathering. In the course of the transition of free water into ice the water volume increases by 9% causing an excess pressure of ice P^i_{wed} which has a wedging out effect. In a closed system (where volumetric expansion is not possible) the pressure caused by freezing water can be as high as several thousand megaPascals. Under natural conditions the value of such pressure is assumed to reach tens of megaPascals. Experimental studies by V.L. Sukhodrovskiy have shown that basalt and dolerite debris degrade much more rapidly in the presence of water (by 160 times) than without it, in the course of the cyclic impact of negative and positive temperatures. In particular, with 100 cycles of cooling and heating of basalt debris (with temperature varying from $-25°C$ to $+20°C$), submerged in water, about 6 mg of fine earth was obtained per 1 cm² of a sample. For dry samples this value was 0.035 mg. Outside the permafrost regions a 1 cm thick eluvium layer is formed over some hundred years, whereas within the area of intense influence of the cryogenic factor it is formed in several years. Thus, physical weathering in the regions of ground freezing is a complex process including a number of mechanisms of rock degradation, and is much intensified by cryohydration leading to increased rates and productivity of weathering, which is more effective compared with other types of lithogenesis.

Along with physical and physico-chemical transformations and degradation of rocks, the biological factor also plays an important role in weathering giving rise to a specific kind of *biogeochemical* weathering. B.B. Polynov was the first to note the essential role of microorganisms at the primary stage of weathering; much attention was attached by him to 'the fragmentary soil of the initial development stage' or 'the embryo stage of soil formation'. Later, it was discovered that the development and the course of biological weathering in the permafrost regions had a number of successive stages. For the first stage it is typical that mosses, algae, different bacteria and fungi in the course of their activity become established in the top layer. At the second stage they penetrate not only the surface layer, but also deep into the rock (for tens and hundreds of centimetres) splitting and breaking it down. On the surface the sod layer is discontinuous with its thickness reaching 10 cm. At the third stage an intense biochemical interaction with the bedrock takes place. The participation of biomass in the transformation of rock material manifests itself in oxygen generation, accumulation of chemical elements in live organisms, acceleration of chemical processes, formation of humus and organic acids and initial organogenic deposits.

The seasonally frozen ground differs from the permafrost rocks in a greater intensity of chemical reactions which have an evident periodicity. Between warm and cold half-periods the nature of the reactions varies both qualitatively and quantitatively. Such a pulselike nature of the interaction between soil and water (bound and free) and, most importantly, transitions of water into ice and vice versa, should promote a drastic intensification of chemical weathering of the seasonally frozen materials.

In general, it is assumed that chemical weathering reaches the depth of the standing water level, while physical weathering is controlled by the depth of penetration of annual temperature variations. P.F. Shvetsov followed by A.I. Popov, Sh.Sh. Gasanov and other cryologists, distinguish three or four horizons or stages in the profile of perennially frozen strata.

The top (from the surface) horizon comprises the layer of seasonal thawing and freezing: it is characterized by the occurrence of intense phase transitions of moisture and by a leaching regime in the summer period, with development of the processes of physical, biogeochemical and chemical weathering. Over two-thirds of the annual heat cycle and almost the whole soil-ground water cycle occur in this horizon.

The second horizon comprises the layer covering the boundary of seasonal thawing and extends to the total depth of penetration of annual temperature variations. Its thickness can vary from one to ten or more metres. It is identified with a rather small water cycle and about one-third of the annual heat exchange. In the upper portion of this layer of annual temperature variation, transitions of small intensity still take place: unfrozen water – ice.

The third horizon of the *weathering crust* in the permafrost regions is characterized by the apparent predominance of physical weathering which develops mainly on account of long-term (4 to 6- and 11-year) variations of temperature.

It is known that formation of weathering crusts and the composition of the materials is dependent on tectonic, climatic and hydrogeologic conditions, on composition of the bedrock and a number of other geological and geographical factors. Thus, for example, polymineral and coarse-grained rocks degrade more rapidly due to looser contacts between individual mineral grains. Ultra-basic and basic rocks which are more prone to weathering than acidic rocks usually have a thicker weathering crust. In such a case an important role is played by terrain, as on watersheds the thickness of weathering crusts is usually greater than on slopes, and among the latter maximum thickness is observed on the exposed south- and east-facing slopes owing to greater differences of temperatures and reduced moisture content of the ground. The climate of subpolar latitudes exerts a noticeable impact over the intensity and orientation of weathering. Physical weathering apparently prevails over chemical weathering.

The profiles of weathering crusts of the permafrost regions have relatively small thicknesses of the several horizons or zones (Fig. 6.1). Zone I is often represented by a poorly distinguishable mixture of rudaceous material (the so-called 'protecting layer'). The availability of rudaceous material near the surface is explained by its being heaved up from below and by the leaching of fine-grained material into the lower horizons of the freezing sequence. Such rock arrangements are known as *patterned ground*. In the mountain areas of the permafrost regions, golets ice serves as a texture-generating agent. On plains, with a finer composition (smaller units) of structured ground the cryogenic textures vary from massive to lens-like layers as dictated by temperature and humidity. Zone II is characterized by an unsorted mixture of fine-grained material with scree and rock debris. In the clay fraction there is montmorillonite and hydromica. In this zone concentration of potassium, magnesium, and monoxide iron compounds is observed, while at the permafrost table there is a higher content of easily soluble chlorine compounds ($CaCl_2$, $NaCl$, $MgCl_2$ and the like). Segregation ice here has the role of structure-generating ice. Zone III, represented by the horizon of poorly cemented blocks of low fine soil content, grades into the zone of initial degradation of bedrock (zone IV) and is well defined to the depth of zero annual variation of temperature. The two latter zones already occur beneath the layer of seasonal thawing and typically have vein ice along bedrock fractures. A common feature of the zones distinguished is the presence of ice as a new mineral specific for the permafrost regions. At the same time each of these zones of the weathering crust profile should be characterized by a typomorphic complex of supergene minerals.

Based on the above, Sh.Sh. Gasanov, following A.I. Popov *et al.*, distinguishes another specific cryogenic type of weathering as regards the profile of the crust – another such zone is included along with that of the rudaceous material – namely, the hydromica-montmorillonite-beidellite zone. It is evident that, depending on tectonic and hydrogeological regimes, continental climate and other specific natural conditions, the sections of the cryogenic weathering crust noted above may be altered (either getting more complicated, or, on the contrary, more simplified – by losing one or another zone).

The rate of physical weathering is usually estimated according to the thickness of eluvium or the degradation (retreat) of different bedrock outcrops, escarpments, benches, coastal lines, etc. For the permafrost regions

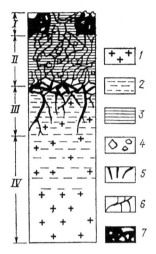

Fig. 6.1. Section of the cryogenic weathering crust (according to Sh.Sh. Gasanov):
1 – bedrock; 2 – fractures resulting from stress release; 3 – fine-grained material
with montmorillonite-hydromica clay minerals; 4 – debris of little altered
bedrock; 5 – vein ice in the bedrock fractures grading towards the top into
ice-cement; 6 – structure-forming segregation ice; 7 – golets ice or
structure-forming ice-cement.

these rates vary within a wide range. Thus, for example, according to data provided by V.L. Sukhodrovskiy, the maximum rate of weathering reaches: on the bench of the basalt escarpment of Franz-Josef Land 0.05 mm yr^{-1}, on the bench slopes of limestones and schists of Scandinavia 0.04–0.15 mm yr^{-1}, while on limestone benches of Spitsbergen island up to 5 mm yr^{-1}.

In general, owing to the predominance of mountain-folded regions in the permafrost regions, the coarse-grain share is over 50% of the eluvium and it is represented by gravel, rock debris and blocks. Sandy and aleurolite fractions can be 40%, on lowlands 70–80%. Pelite and argillaceous fractions rarely exceed 15% making up 3–5% on the average. The particular composition of eluvium formations suggests that the latter belong to a specific type of eluvium – *cryoeluvium*.

As is known, loose products of weathering and solutions formed in eluvium not only migrate within the limits of catchments to accumulate in the final water bodies, but also become concentrated in transit. As regards the stages of transport and accumulation of sedimentary material in areas of permafrost, wide differences are observed in quantitative indices of sediment transport and accumulation compared to noncryolithic regions. This is associated with the existence of quite distinct processes of transportation resulting from various kinds of freezing processes and in the formation of specific genetic types of continental sediments with respect to chemical-mineral composition, and cryogenic structure and the kind of freezing – whether syngenetic or epigenetic.

The matter resulting from weathering is conveyed for considerable distances to the final runoff water bodies, in the form of true and colloidal solutions, suspended solids of rudaceous material and as mudflows and similar forms of displacement. In the course of such displacement within the catchments weathering products are differentiated, i.e. the components of sedimentary material are separated one from another. At least three aspects of the matter of differentiation are to be considered: a) differentiation of weathering products into mechanical and chemical components; b) differentiation of weathering products by composition and content of chemical elements as they move towards the final runoff water bodies (chemical differentiation); c) differentiation of weathering products by grain size and mineral composition of granular deposits encountered over catchment areas (mechanical differentiation).

Initially, the separation of materials usually occurs in the foothills and lower slopes of catchment areas such as mountain uplands, hills, terraces etc. In the regions of perennially frozen ground such differentiation is more or less limited to the seasonally thawing layer. Note the extremely low rate of displacement of fine clastic material within the limits of watersheds as compared with the migration rate of pore solutions and runoff waters containing chemically dissolved and mechanically suspended substances. Owing to this, rudaceous material is separated from dissolved matter in the course of transport. The main volume of fragmented material remains on the slopes and lower slope regions, while dissolved and mechanically suspended matter is leached into rivers and water bodies. The mechanical composition of the deposits – talus, alluvium, prolluvium – is associated exactly with this process as is the formation of predominantly chemical-biogenic deposits in the final runoff water bodies.

Mechanical and chemical denudation is controlled by both climatic zonation and structural-tectonic and geomorphological features of a region. The ratio of mechanical to chemical denudation varies between mountain rivers and those flowing on plains: in mountain rivers mechanical denudation prevails, while on plains – chemical. The analysis of such types of denudation in rivers of the Euro-Asian continent, conducted by I.D. Danilov (6), showed the predominant significance of dissolved matter over sediment flow in allogenic rivers of Siberia and in rivers of the Arctic, sub-Arctic and northern temperate zones.

The predominance of dissolved matter over suspended solids in rivers flowing on plains is conditioned by low gradients of terrain, the gentle nature of the summer rainfall and the existence of dense and persistent sod cover. Accordingly, the mechanical denudation is slowed down in conditions of tundra landscapes. However, in general, the intensity of both mechanical and chemical denudation in the permafrost regions is lower by a factor of ten than in the warm climate regions.

At the stage of transportation weathering, products are differentiated by grain size and by composition of minerals in the granulometric ranges. Differentiation is a function of distance of material displacement and the geologic-geographic conditions for separation of materials in the course of transport.

The initial stage of differentiation takes place within the limits of watersheds giving rise to various types of slope deposits. Thus, the processes of gravitational denudation are intense on slopes (talus and rock falls); gradually, they are replaced by movement of boulders caused by freezing water which pushes them apart and outwards, subsequently leading to downslope movement as the cementing ice thaws out. This is the origin of the rock stream belt widely developed in the permafrost regions which is represented by rock debris flows and boulder fields. Fine debris cones are formed on the margin of this belt. Further down the slope a qualitatively new process starts to develop intensely, this process being one of the leading among those developing in the permafrost regions – namely, solifluction – giving rise to the zone of solifluction deposits. Since solifluction movement of soils develops at gradients of about 2–3°, it is evident that such displacement forms will be found at the foot of slopes accompanied by further separation of these deposits by grain size. Along with differentiation of weathering products by grain size and mineral composition in the course of their movement towards the final water bodies there is also chemical differentiation. Certain chemical elements become more mobile and are intensely leached by the surface and ground water; other elements, on the contrary remain practically immobile and, retained within the limits of watersheds and slopes, begin to become concentrated.

Thus, for an area of permafrost (the Aldan upland) the migration of a series of chemical elements was determined by I.B. Nikitina with respect to fissure-ground water, soil-ground water, surface water and rills. It was shown that in the permafrost regions in both oxidizing and reducing conditions the mobility of F, Fe, Ti, Cu, Ni, Zr, Ag, Mo and a number of other elements increased substantially as compared with their mobility in the noncryogenic regions.

Flowing water plays an important role in the transportation and accumulation of sedimentary material. This is contributed to by the activity of intermittent and continuous watercourses as well as by sheet erosion caused by meltwater and rainfall, which account for short-distance transportation of enormous amounts of sedimentary material. Continuous watercourses make possible concentration of the rudaceous material arriving from a vast area with long transportation distances from the place of formation. Therefore, this mode of transportation is considered to be the main one in the formation of sedimentary deposits.

A predominant role in the formation of river valleys is played by downcutting and lateral erosion. In the permafrost regions the processes of lateral erosion prevail over downcutting both in mountain and plain rivers, which predetermines meandering of rivers and formation of ox-bow lakes – a typical process for plains and flat territories of the North. The frozen state of the banks and high content of ice (up to 80%) in the soils that compose such banks, promote intensification of lateral erosion owing to the thermal-mechanical impact exerted by water flow and solar radiation on frozen soils. Among manifestations of such impact are collapse of thawed soils into rivers and formation of numerous thermal-abrasion and thermal-erosion niches in banks with the cutting of ice-rich soils by the flow or by waves. At the same time downcutting (as compared with lateral erosion) is hampered by the intense inflow of talus-solifluction material from slopes, which overloads river beds causing meandering and cutting of banks. Small rivers can be filled up and cease to exist temporarily, forming the so-called 'spoonful alluvium' consisting of poorly sorted and graded slope deposits. And, finally, downcutting in the permafrost regions can be effectively hampered by the processes of bottom ice formation in shallow watercourses.

Comparative study of alluvium in warm humid regions and in the permafrost regions identified a qualitative distinction of the northern-rivers alluvium, which is characterized by a wide range of grain size and heterogeneity, by the increased content of heavy mineral fraction (pyrite, arseno-pyrite, gold, etc), often by a two-member layer of alluvium, by salinized floodplain deposits and the like. Associated with these characteristics and with further cryogenic transformation, fluvial sediments acquire typomorphic features of composition and structure not typical of other zones. This was the reason for the defining of an independent geographic origin of the alluvium by E.M. Katasonov, Y.A. Lavrushin and other investigators.

Sheet erosion of fine soil in the permafrost regions involves a variety of specific features of rainfall and meltwater transportation. Downslope movement of fine particles is facilitated by drizzling summer rains and thawing of snow cover. Downhill wash of sandy-silty and silty-clay soils increases substantially with high ice content and with schlieren cryogenic structures, which greatly reduce the structural strength of thawing soils. On slopes with rudaceous deposits sheet erosion of fine earth usually grades into its suffosion. The processes of downhill creep and washout may lead to the formation of a rather thick layer of deposits at the foot of slopes and on flat sites, known as talus trains, perluvium, dells and the like.

Of importance is the activity of intermittent fluvial flows in the area of permafrost or the action of steady runoff of rainfall and snowmelt. This process of linear erosion, which is called thermal erosion, contributes greatly, along with sheet erosion, to the formation of talus sediments and plays a significant role in the transportation of weathering products over the slopes.

In polar and severe continental climatic conditions wind-borne transportation of fine earth also occurs widely, both in winter and in summer. Most very fine particles are blown out by strong winter winds on sites with disturbed or missing vegetation cover and where frozen soils become less cohesive due to ice sublimation. The volume transported in winter is a function of wind velocity, intensity and depth of sublimation. The transporting capacity of wind is one three-hundredth of that of water and, therefore, only sand or finer particles can be blown. However, the content of these moderately fine particles is rarely as much as 10%, while the clay content is generally insignificant. This is associated with the fact that ice sublimation in sandy soil leads to reduced cohesion between particles which can be easily carried away by wind. In frozen clayey soils, however, cohesion is not reduced after sublimation, but is increased.

Modifications of sedimentary material within the permafrost regions involve not only temperature and cryohydration weathering of hard rocks (magmatic, metamorphic and cemented) but also exogenous processes, freezing proper included, which cause breakup of the frozen soils into separate aggregates. Such processes include frost cracking, thermal abrasion, thawing of sheet deposits of structure-generating ice and recurring ice wedges, etc. At present the role and particular share of each process in transportation and accumulation is not evaluated, but it is evident that indices should be defined for this purpose. Such indices can include the following: the area of development in the permafrost regions or region of the given freezing-geological process, the volume of material and distance of transportation, rate of movement, intensity of occurrence of different accumulation forms, etc.

Thus, for example, in glaciated areas transportation by glaciers prevails over other types. Most widely developed in the permafrost regions is downslope transportation by solifluction, gravitational, rock stream and thermal erosion processes. As regards distance of transportation, the predominant role is shared by the fluvial and eolian types. The highest displacement rate is typically in aeolian movement, and also in mudflows, slumps and rivers. Such disturbance of the mass balance by the transportation of weathering products from catchment areas leads to more intense flattening of terrain and more rapid shaping of the smooth profiles of slopes. This is likely to bring about the formation of flatter and smooth outlines of macro- and meso-relief in the permafrost regions with less difference in relative elevations.

Sedimentary material arriving in the final water bodies includes the dissolved and colloidal substances and suspended solids mixed with rudaceous material. As a result of mixing with sea and lacustrine water the colloids and weakly soluble salts are precipitated, clastic material and suspended solids are disseminated and other processes and phenomena also take place. All these processes result in the formation of sediments which are characterized by a specific chemical-mineral and granulometric composition, structure, properties and regular concentrations of ores. The final product is determined by the peculiarities of the sedimentary differentiated material arriving from land and having a specific regime and conditions of sedimentation in water bodies. For example, the Arctic basin has, typically, low positive temperatures, sometimes even negative ones, and is characterized by the occurrence of ice cover on the surface throughout the year. Such a regime hampers wave propagation and is favourable for the precipitation of fine-grain material even at shallow locations. Apart from this, floating ice in icebergs in water bodies in the permafrost regions makes possible spreading of rudaceous material over long distances with sporadic deposition (even in deep water areas), thus giving nonuniformity of basin deposits with their typical nonsorted nature across the section.

The specific nature of hydrochemical and hydrobiological regimes of the final water bodies in the permafrost regions is revealed by the intense accumulation of carbon dioxide in the bottom layer, poor access of oxygen and other gases to the bottom sediments, in the predominantly neutral and low alkaline water environment, etc. The salinity of the water of the Arctic ocean reaches 35‰ diminishing from west to east down to 20‰ owing to fresh water inflow from the Siberian rivers. Lacustrine water is brackish or fresh. In basin sedimentation chemical precipitation usually prevails over the biogenic.

It is difficult to determine the rate of accumulation of sediments in recent water bodies as knowledge of the intervals during the process is lacking. The maximum rate can be as high as tens of millimetres per year. In Lake Onega (Onezhskoye Ozero) this rate is $2-5\,\text{mm yr}^{-1}$. The recent deep-set silts are being accumulated at a rate of 0.02–0.06 mm yr^{-1} on the average. For ancient geosynclines the average rates of sedimentary strata accumulation amount to a hundredth or tenth of a millimetre per year, while for ancient platforms – a hundredth or thousandth of a millimetre per year.

Sedimentary material arriving in the final water bodies has been transported by rivers, glaciers, icebergs, wind, and it also results from thermal abrasion of the banks, the action of sea currents, transfer of ash material and a number of other processes. In this way accumulation of morainic material similar to that of continental moraines occurs in the bottom of the Bering Sea, Barents Sea, Kara Sea, etc., Bukhta Provideniya, Zaliv Kresta and other northern water areas. This material is characterized by coarse-grained sediments, high content of boulders locally prevalent, and homogeneity of petrographic composition. Simultaneously, over the whole water area of the Arctic basin transportation of rudaceous material by icebergs promotes deposition of coarse-grained material. The concentration of icebergs and the pattern of their distribution over the water surface are dependent upon ice content, the rate of drift and thawing, bottom relief, bank structure and the like.

It was found that the concentration of rudaceous material drastically increased on sites with higher temperature seawater associated with intense ice thawing. In such conditions even continuous horizons of coarse sediments can be formed. Thus, for example, on the area from Greenland to Spitsbergen the layers of sediments contain much stone-sized material.

An important role in the delivery of sedimentary material from land to the bottom of water bodies is played by thermal abrasion of banks. The volume of sediments resulting from bank degradation constitutes about 20% of the whole volume conveyed by Arctic rivers. Virtually the whole of this material is deposited in the water surface of coastal seas giving rise to new accumulative forms of bank relief, causing the expansion of spits, shoals and islands. Only a fine-grained suspension is carried away into the ocean.

The rate of retreat of banks and sea coasts in the permafrost region is estimated in single figures and tens of metres per year. Thus, lowland banks on the northern margin of the Yana river delta retreat at a rate of 16–20 m yr^{-1}, on the Laptev Sea coast not usually more than 6 m yr^{-1}. The thawing rate of ice-rich banks in the Central Yakutia reaches 7–10 m yr^{-1}, while thermokarst lakes in the Anadyr tundra migrate at a rate of 10 m yr^{-1}. Predominant average values of coastal retreat of continents and big islands range between 2 and 6 m yr^{-1}. Banks of water bodies composed of non-frozen soils retreat at a rate of 3–10 m yr^{-1}.

Thus, annual values of rate of retreat of both frozen and unfrozen banks can be similar. This suggests that the thermal abrasion rate of the banks composed of ice-rich fine-grained soils, under comparable conditions, exceeds by 3–4 times the abrasion rate of banks composed of unfrozen soils of similar composition. Because the Arctic seas are for a greater part of the year, compared to the southern ones, covered with ice which hampers the development of thermal abrasion in this period, the total annual effect for coast retreat is similar in both cases.

Sedimentary fine-grained material arriving in the final water bodies is subject to considerable displacement and spreading owing to winds, tides, perennial oceanic and marine currents as well as by the movement of icebergs, coastal and pack ice, etc. Distribution of sediments by mechanical composition in the northern seas is in conformity with Strakhov's outline and is characterized by the replacement of terrigenous material (sand-aleurolite) in the coastal zone by chemical-biogenic-carbonate material and in deep water by the siliceous material.

Organic matter is mainly carried into the marine basins with continental run-off and, in addition, arrives as sediments resulting from the life cycle of plankton and benthos. In the rivers of the North organic matter content can be as high as 70% of the whole amount of the dissolved and colloidal matter. However, the major portion of this organic matter does not move as far as to the sea and is precipitated in river mouths and shelves.

For the recently formed bottom sediments of the Arctic seas, oxide-ferriferous-manganese concretions, for the formation of which bacteria are responsible, are very typical. Their concentrations are more or less confined to areas of occurrence of brown silts and to the warm Atlantic waters. Iron is brought into marine and oceanic basins of the North with continental runoff, dissolved or suspended in rudaceous material and organic compounds. Iron content in the sediments of the Arctic basin makes up 3% on the shelf to 10% in basins, while in the Bering Sea it is as high as 11%. Iron is present as colloidal $Fe(OH)_3$ and ferriferous-organic compounds.

The granulometric, chemical-mineral composition and structure of the bottom sediments of seas and water bodies of the permafrost regions are characterized by a variety of specific features dictated by the peculiarities of sedimentation and regime of the final water bodies of runoff. In seas and water bodies where there is long-term ice cover fine-grain sediments prevail while silt accumulation is already possible at a depth of 10–15 m (Laptev and East-Siberian Seas). If the ice regime is less severe, more coarse sediments accumulate at such depths (Chukchi and Bering Seas).

Grain-size distribution and structure of bottom sediments of fresh-water closed and semi-closed basins in the permafrost regions is influenced greatly by the duration of a persistent ice cover which diminishes wave action, disturbing the usual hydrodynamic sorting of the basin deposits and allowing the cyclic nature of the delivery of sedimentary material to be more apparent leading to the formation of banded-laminated sediments.

A very typical feature of shallow fresh-water basins of the cryolithozone (lakes and bogs of the tundra, the north of Tyumenskaya Oblast' [Tyumen region], the European part of Russia etc.) is also the accumulation of sediments rich in organic matter, mainly peaty sapropel and sapropel peaty silts.

6.2   Transformation of loose deposits of the permafrost regions into rock

Transformation of the basin and continental sediments into rock is considered to be the most specific stage of lithogenesis in the permafrost regions. Sedimentary formations are subject to single or multiple freezing-thawing and adapt to the changing thermodynamic and physico-chemical conditions of the environment (as they progress deeper). As a result of tectonic movements and denudation processes these formations may approach the Earth's surface many times (and even be found within the weathering zone) and then sink again, passing once more through the stages of diagenesis and epigenesis.

In the freezing and thawing of the sedimentary strata, processes associated with phase transition of water into ice are fundamental in the cryogenic transformation of composition, structure and properties of rocks. During freezing, the development of crystallization and structural relationships, the 'petrification' of soils takes place (according to P.F.Shvetsov). Simultaneously concentrations of dissolved matter become redistributed in the freezing strata (cryogenic desalination, and concentration, sulphatization and carbonization, formation of cryopegs, etc.); moisture migration towards the front of freezing and ice separation also takes place by segregation, injection and ablimation with shrinkage, swelling and heaving of the soils subject to freezing, etc. Of specific importance for the process of lithogenesis in the permafrost regions are volumetric gradient tensions that arise in the freezing and frozen soils causing substantial transformation of the initial structure and texture of the mineral portion of the unfrozen soil, its density and strength. The process of perennial freezing for a rather short-term period of time is sufficient to make such transformation of structure and texture of the soil's mineral portion as would take tens and hundreds of millenia with diagenesis of the normal type. Therefore, a specific type of diagenesis can be distinguished, namely, the cryogenic one that gives rise to a qualitatively new rock – cryogenic rock.

Transformation of the basin sediments of the permafrost regions by diagenesis and epigenesis occurs progressively, as in the warm humid areas, and is associated with a cycle of physical and physico-chemical transformations: dewatering and compaction of sediments, formation of typical fault and plastic deformations, alteration of chemical-mineral composition and the like. In this case there is no drastic transformation of clay minerals.

A quite different transformation of the basin deposits is typical of regression diagenesis and epigenesis in the permafrost regions, when deposits influenced by the conventional diagenetic physico-chemical processes move upwards from beneath the water level thus approaching the ground surface. They can freeze immediately after coming from beneath the water level (*synchronous-epicryogenic*) or exist in the unfrozen state for a long time, becoming frozen later if the climate becomes colder (*asynchronous-epicryogenic*).

Asynchronous-epicryogenic strata of the basin deposits

These usually develop on sedimentary formations that have undergone the stages of sedimentation, diagenesis and epigenesis outside the extent of permafrost development. Accordingly, it is typical of them to have oligo- or mesomictic composition and a high content of minerals of the kaolinite group, while their structural-textural features and properties are typical of the sedimentary formations in the warm humid or arid regions. Coming from beneath the sea, the weathering crust of these sediments is not yet influenced by the cryogenic factor. Then, owing to general or regional cooling these strata are subject to unilateral freezing from the top in the course of which grain size and chemical-mineral composition change. In the upper section of deposits which have not undergone deep lithification, high ice content will develop, characterized by a variety of types of superimposed and inherited segregation-migration cryogenic structures. Lower in the section with more dense and dewatered deposits, as well as lower freezing rate and lower temperature gradients, superimposed relatively thin horizontal layers will be formed, occasionally together with inclined ice schlieren, as well as injection and ablimation varieties of ice. In general, asynchronous-epicryogenic strata of the basin deposits are characterized by a three-member structure (Fig. 6.2).

Synchronous-epicryogenic strata

These can be formed over the basin sedimentary formations which have undergone the stages of sediment genesis and progressive diagenesis and epigenesis both within and beyond the area of development of perma-

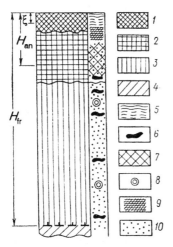

Fig. 6.2. Structure of asynchronous-epicryogenic strata of basin deposits: 1–3 – frozen materials (1–2 respectively, of recent and ancient crusts of weathering, 3 – of the zone of regression cryodiagenesis); 4 – unfrozen; 5–10 – cryogenic textures (5 – fine and frequent-banded, 6 – thick-and rare-banded, 7 – thick- and big netted, blocky, 8 – inherited, 9 – fine- and small-netted, 10 – massive); ξ, H_{an}, and H_{fr} – respectively, thicknesses of the seasonally thawing layer, annual temperature fluctuation layer, perennially frozen layer.

frost. The latter is of importance since it influences chemical-mineral composition and grain size of materials which will be further subjected to cryogenic transformation at a stage of regression diagenesis. Beyond the permafrost regions these rocks will be frozen only from the top, whereas in the permafrost regions (as they are coming from beneath the water level) they will be frozen both from the top and laterally, giving rise to specific obliquely laminated cryogenic structure. In general, such synchronous-epicryogenic strata are characterized by the following features.

Prior to continental (subaerial) freezing of the basin deposits on shallow and shelf sites they are subject to epigenetic submarine (subaquatic) freezing without having been much transformed diagenetically. Formation of cryopegs is possible in this case.

Above these epicryogenic (subaquatically frozen) materials a comparatively thin (a few metres) layer of syngenetic subaquatically frozen materials can be formed as shelf sites are drained and their freezing beneath the water takes place simultaneously with accumulation of sediments. Subaerial freezing from the top, accordingly, takes place through the layer of the subaquatically frozen materials uplifted from beneath the sea level. Their thickness is, as a rule, small (a few tens of metres) and is greatly dependent on the value of the negative temperature of the bottom water layer, geothermal gradient, composition and moisture content of the bottom sediments and other factors. On shallow parts of fresh-water bodies the formation of syn- and epigenetic subaquatically frozen sediments is also possible. In general, the composition and structure of synchronous-epicryogenic strata of the basin deposits differ greatly from those of asynchronous-epicryogenic strata. Four specific layers of the frozen materials are distinguished in the profile (Fig. 6.3).

Grading of continental deposits into rocks occurs differently in subaerial and subaqueous conditions.

Subaerial and subaqueous continental deposits

The two groups of deposits differ substantially in the pattern and intensity of sedimentation which brings about a wide range of chemical-mineral composition and structural-textural features. These differences exert a noticeable impact on the process of cryogenic transformation of sediments into rocks with formation of epi- or syncryogenic strata.

Subaqueous deposits (of flood-plains, mud flats, dead lakes, bogs and the like) are characterized by a stable regime of sediment accumulation, the rate of which is measured by fractions of or a few millimetres per year. In the majority of cases these deposits are silty, highly porous (owing to the presence of fine pores), supersaturated, well graded, with well-defined orientation, dark-grey colour and contain a great amount of poorly decomposed vegetation residues and organic matter. Usually, these sediments are poorly drained owing to supersaturation and the high content of fine particles and there are traces of gleying. The latter is evidence of a predominantly reducing environment as well as of the presence increasingly of proto-oxides of iron, hydrogen sulphide, methane (in marshlands) and high carbon dioxide content. Among new mineral formations are typically hydromicas, vivianite and siderite.

The formation of subaerial continental deposits occurs irregularly with time. Thus, for example, accumulation of solifluction deposits can occur either annually, or with 2–3-year intervals. However, despite such a nonstable nature of accumulation, the thickness of subaerial sediments can much exceed that of subaqueous sediments for a series of years, since the accumulation rate for a season (cycle) can reach centimetres and tens of centimetres (and even metres). These deposits are mainly represented by poorly graded mixtures of fine-grained and coarse material with poorly defined orientation and inclusions of buried peat. Their porosity is characterized by big and inter-aggregate pores. Subaerial deposits are, as a rule, little saturated and of

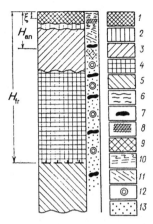

Fig. 6.3. Structure of synchronous-epicryogenic strata of the basin deposits:
1–4 – frozen materials (1 – epicryogenic of the recent weathering crust,
2–3 – respectively, syngenetically and epigenetically frozen under water,
4 – epigenetically frozen during regression cryodiagenesis); 5 – unfrozen;
6–13 – cryogenic textures (6 – fine- and frequent bands, 7 – thick- and sparse
bands, 8 – fine- and small-netted, 9 – thick- and big-netted, block,
10 – underdeveloped netted, 11 – inclined bedding and scaly, 12 – inherited,
13 – massive).

low humus content which gives rise to a predominantly oxidizing-neutral
medium. Typically, in connection with this is the presence of iron compounds as oxides, quartz fragments and new formation of montmorillonite
minerals.

Continental deposits can undergo diagenetic transformations either simultaneously with seasonal or perennial freezing, or prior to the action of the
cryogenic factor. Based on the above, three types of frozen strata can be
distinguished which reflect three different transformations of continental
deposits; 1) asynchronous-epicryogenic; 2) asynchronous-epicryogenic
palaeocryoeluvial; 3) syncryogenic.

The structure of asynchronous-epicryogenic strata of continental deposits is rather simple. Their section presents a recent and comparatively thin
weathering crust of cryogenic type that grades into the poorly defined
ancient (noncryogenic) weathering crust which has undergone a single
freezing. The lower, thickest part of the sequence under consideration is
represented by the epigenetic frozen continental deposits. The chemical-
mineral composition of these rocks reflects cryogenic transformations
caused by freezing and by changes in geochemical setting.

When accumulation and diagenesis of continental deposits occurs beyond the limits of permafrost but with a well-defined process of seasonal

freezing, these deposits undergo multiple (cyclic) seasonal freezings. Practically the whole sequence of such continental deposits at the stage of diagenesis should experience the multiple impact of the cryogenic factor on the background of general weathering. There is a correlation between the depth of the deposits and the length of time of the cryogenic eluviation of the unfrozen rocks. This gives the *palaeocryoeluvial* sequence of unfrozen continental deposits.

The process of cryoeluviation in the given case is characterized by a number of specific features differing from those typical of frozen ground. Thus, as the duration of the warm period is longer than that of the cold one, more heat and moisture is obtained by eluvial rocks within a year cycle; they are characterized by better drainage with the absence of underlying permafrost, and a predominantly oxidizing medium which leads to the formation of non-gley soil. The medium is usually neutral and even alkaline, with intense leaching of iron and aluminium elements. As a rule, these palaeocryoeluvial materials are slightly salinized with a high humus content (mainly, humic acids), contain great amounts of silty particles and carbonates and have light colours. Among the newly formed clay deposits minerals of the montmorillonite group and hydromicas prevail. The so-called asynchronous-epicryogenic palaeocryoeluvial units are formed under global cooling and freezing of palaeocryoeluvial unfrozen continental deposits.

Finally, accumulation of continental deposits in the permafrost areas promotes formation of syncryogenic strata. These are formed in cyclic layers, varying in thickness (mainly, during the warm season of a year) and with seasonal freezing (during the cold season). Accordingly, with a constant thickness of the seasonally thawing layer the upper surface of the permafrost rises annually, by an amount equal to the thickness of annually accumulated continental sediments (taking into consideration their heave while in the frozen state).

In the given case there is an apparent distinction in the formation of syn- and epicryogenic strata. The epicryogenic strata grow due to the deepening of the lower boundary of the permafrost while the syncryogenic strata increase due to the rising of the upper boundary of the permafrost. The problem and mechanism of syncryogenic formations were thoroughly studied by E.M. Katasonov, A.I. Popov and I.D. Danilov *et al.*

In the course of formation the syncryogenic strata are somewhat influenced by the usual processes of diagenesis. Thus, for example, as shown by Yu.A. Lavrushin, there is a specific type of diagenesis traceable in the alluvial deposits of the Arctic and sub-Arctic areas, namely permafrost diagenesis associated with multiple freezing and thawing and with accompanying physico-chemical reactions, phenomena and processes. The absence of diagenesis of the usual type is confirmed by the uniform colour of the sediment: there is no authigenic formation of minerals (excepting the mineral ice), and the processes of degradation of organic and vegetation residues are slow. The most drastic transformations of syncryogenic rocks are found in changes in structural-textural aspects. Consequently even before their transition into the permafrost state (at depths that do not exceed some metres) continental formations undergo the stage of specific frozen diagenesis and grade into materials of cryogenic type in respect to their mechanical and structural-textural properties. Accordingly, transformation of continental sediments and deposits into syncryogenic deposits takes place at the stage of cryogenic diagenesis, occurring generally within the horizon of intense phase transitions of moisture which corresponds roughly to the layer of seasonal thawing.

The intensity of the transformations of continental deposits of syncryogenic type differs substantially depending on whether these are subaqueous or subaerial sediments and is determined by tectonic regime, climatic conditions, the pattern of terrain, the rate of sedimentation, depth of seasonal thawing and the like. Thus, subaqueous sediments that are accumulating at a steady slow rate grade into the permafrost state undergoing a number of freezing cycles within a layer of seasonal thawing. The above leads to greater transformations of syncryogenic strata of continental-subaqueous origin by the processes of cryoeluviation as compared with continental-subaerial deposits.

In general, syncryogenic strata, unlike the epicryogenic ones, are characterized by a specific appearance, by composition, structure and properties. They are, as a rule, more silty and have a higher humus content which restricts the possibilities of absorption. Their composition mainly comprises monoxides of iron, hydrous micas and montmorillonite. Higher porosity, plasticity and thixotropy are typical of these deposits with uniform distribution of ice in the section, whereas epicryogenic strata are characterized by ice content diminishing with depth. Furthermore, the occurrence of syncryogenic strata in the profile indicates two important features: 1) syncryogenic strata, as a rule, cannot exist independently and are underlain by epicryogenic rocks, i.e. if syncryogenic rocks occur in the profile, the permafrost sequence is in general polygenetic; 2) in practice, it is unlikely that syncryogenic strata will extend deeper than the first few hundred metres. Usually, the thickness of syncryogenic materials of continental subaqueous origin does not exceed several tens of metres. Thus, if the maximum rate of accumulation of subaqueous-continental deposits is assumed to reach

1 mm yr^{-1}, then with the long-term conditions of severe climate necessary for the formation of syncryogenic sediments their thickness can reach about 100 m.

The features of the transformations on freezing of basin and continental deposits differ substantially between elevated sites and depressions. Thus, in the European North of Russia and in West Siberia the percentage of the frozen strata is much greater in the region of tectonic depression than in regions of uplift. The soils that underwent numerous cycles of freezing and thawing are loess-like in composition and texture and, as shown by experimental studies, it is possible to determine the relationship between loess formation and the number of cycles. The greater the number of cycles undergone by the initial sediments the more closely they resemble the typical loess-like, loess cover formations. As regards epicryogenic strata, the thickness of such a layer is limited by the layer of seasonal thawing. The formation of loess-like soils down the whole profile is possible in the syncryogenic and palaeocryoeluvial strata of fine-grained soils that have undergone multiple freezing-thawing. Their formation is assumed to take place in the course of syngenetic freezing of continental sedimentary deposits and, especially, subaqueous continental sediments.

At negative mean annual temperatures syncryogenic strata are formed with distinct attributes of loessification, while at positive temperatures palaeocryoeluvial strata of loess soils are formed. In both cases the highest degree of cryogenic transformation is observed in the range of mean annual temperatures near to 0°C given that the deepest layers of seasonal thawing and freezing then occur. The degree of loessification of soils resulting from their alternating freezing and thawing is in inverse proportion to the rate of accumulation of the sediments. At $V_{acc} \geq \xi$ the process of loessification comes to a full stop. The process is most intense at $V_{acc} \leq \xi$ when the newly deposited layer of sediments is undergoing some tens of thousand cycles of freezing-thawing. Therefore, there is a complicated and interrelated dependence of cryogenic transformation of sediments depending on the values of V_{acc} and ξ.

Formation of loess and loess-like soils of cryogenic origin is promoted by low rates of sediment accumulation with maximum depths of seasonal thawing or freezing of silty-sandy and silty-clay subaqueous-continental soils.

6.3   Formation of useful mineral deposits at different stages of cryogenesis

As known, all mineral formations are subdivided into endogenous (magmatic and metamorphic) and exogenous (or sedimentary). Among the latter the following groups can be distinguished in accordance with conditions of formation and stages of lithogenetic process: weathering, transfer, continental sediment accumulation, basin sediment accumulation and after-sedimentation transformation of deposits (10). Each of these groups can be further subdivided into classes of mineral deposits (according to the origin of the mineral inclusions or by the nature of the rock-forming process) and subclasses (by the mechanism of the formation of the deposits).

Mineral formations of the *weathering* group are directly connected with development of relatively thin weathering crusts of cryogenic type and provide mineral masses for a number of other types of mineral deposits (placer, sedimentary and the like). The mineral formations of this group are a result of physical, chemical and biochemical weathering of bedrock and washing by surface and ground water; or of redeposition of part of these products.

Eluvial sources of mineral deposits of this group comprise the residual and redeposited products of physical weathering, the intensity of which exceeds that of chemical weathering in conditions of cryolithic genesis. Physical weathering gives rise to the formation of thick rock waste-fine earth and block-fragmental deposits, block eluvium and sandy-aleurolite deposits. On one hand, these are deposits suitable as construction materials and widely developed in the permafrost regions (rock waste, sand, very silty sandy, and clay rich materials for wall, foundation and road construction, etc.) and glass-ceramic raw material (sand, clay); on the other hand, there are eluvial placer deposits of cryogenic type. Eluvial placers of gold and diamonds, cassiterite-columbite, platinum and the like which may be formed with the participation of the cryogenic factor, are found at present in the permafrost regions and are intensely transformed under the influence of cryogenic processes.

Presumably one can describe the deposits of iron, copper, uranium etc., of the permafrost regions, which are highly typical of humid conditions, as *infiltration* deposits. A key role is played by geochemical and mechanical barriers, i.e. situations that hamper the migration of chemical elements thus leading to their accumulation and concentration.

The specific nature of weathering in the permafrost regions manifests itself in the processes of near-surface change of mineral deposits. For instance, according to N.A. Shilo (21), transformations of sulphide deposits in the permafrost zone have various distinct features. Thus, with substantial changes of the profile of the weathering crust, the 'iron cap' has a smaller thickness than in similar conditions outside the permafrost regions. The presence of hydroxides and hydrosilicates with relics of primary sulphides is typical of the profile and this is evidence of the underdevelopment of the crust and stagnation of the weathering processes at an intermediate stage. Due to the restricted water exchange and slow leaching of sulphates they are present in greater quantity to depths of hundreds of metres in the north-east regions of Russia, Yakutia and the Kola Peninsula (Kol'skiy Polvostrov).

The sources of mineral deposits of the *transfer and continental accumulation* group of sediments are most strikingly represented by placer deposits, the formation of which occurs with accumulation of ore minerals in the coarse-grained deposits during displacement of the weathering crust material.

Talus and proluvial deposits form as a result of downslope movement of loose fragmental material under the influence of gravitational forces, thermal erosion, solifluction, mudflows, rock streams and the like. As with eluvial placers, frost heaving of fragmental material can take place in these deposits thus causing nonuniform distribution of valuable minerals across the section. However, commercial-scale concentrations of mineral deposits in the talus and proluvial placers of the permafrost regions have not been found.

As shown in the works of N.A. Shilo, the cryogenic type of lithogenesis is characterized by displacement of the concentration zone of placer deposits from slopes towards rivers and seas – which is the reason for the wide development of alluvial placers within the limits of the permafrost regions (placer deposits of Aldan, Yakutia, Chukotka, etc). The alluvial placers known at present are predominantly concentrated and confined to valleys of the northern rivers, while beyond the permafrost regions they are less distinct.

Of a certain importance among placer deposits are glacial placers, associated with morainic or fluvioglacial deposits. Some diamond-bearing and metalliferous glacial deposits are worked on a commercial basis. So are the gold-bearing moraines of Alaska, diamond-bearing moraines of Brazil, fluvioglacial placer of gold in New Zealand, platinum placers of Canada, etc. Among biogenic deposits of the permafrost regions, the most developed are peats, usually poorly degraded, with well-preserved vegetation residues and peaty or humic sandy silty to clay materials, and deposits of bogs and marshlands. These deposits, as a rule, are characterized by a higher content of hydrogen compounds (methane, hydrogen sulphide), vivianite and bog ores.

The group of *sedimentary* mineral deposits is subdivided into lacustrine, marine and oceanic. Their formation occurs in different ways: mechanically, physically, chemically and biochemically. Mechanically formed sedimentary deposits are mainly represented by gravel, sand and clay. A major role in the accumulation of sediments within the permafrost regions is played by terrigenous material and this serves as a source of construction material in different deposits. Of great practical importance are lacustrine and marine sands of monomineral and polymictic varieties. Marine sands are well graded, homogeneous and useful for construction purposes. Sandy sediments in lakes are limited, being represented by discontinuous bands and lenses of littoral sands. They are poorly graded and to a significant extent covered with silt.

Chemical-biological sedimentary deposits of lakes are rather widely developed in the permafrost regions, since, for instance, there are some 400 000 lakes over the territory of the European Russian North, characterized by a cryogenic type of lithogenesis. Among the clay minerals of lacustrine sediments (both recent and fossil lacustrine-glacial varved clays) the most developed are hydromicas, chlorite and mixed-laminated formations of the type mica-montmorillonite and chlorite-montmorillonite. The area of montmorillonite development is often the largest. Principal types of the recent formations in lakes are peaty and algal sapropels. Among iron forms in sediments ferrous iron prevails; its increased content is associated with simultaneous increase of residual organic matter. Lacustrine (Fe-Mn) and lacustrine-bog ores are widely developed and typical of the lacustrine deposits of the European Russian North.

Typical lacustrine ores (concretions of crustal, globular and irregular shape) are characterized by a significant manganese content (over 4%) and iron (up to 10%) and are mainly represented by iron hydroxides with inclusions of manganese hydroxides. The size of Fe-Mn nodules does not remain constant, varying both laterally and vertically. Usually, manganese and CO_2 content increase towards the top of the section and in the direction of shallow water, taking the form of manganese hydroxide and calcium rhodochrosite – manganous calcite. Iron and phosphorus in this case are concentrated in the lower parts of the section.

Typical lacustrine-bog ores, the formation of which is associated with acidic water of peats, have smaller concretions, massive structure, higher iron content (up to 50%) with insignificant content of phosphorus (up to 2%). Mineralogically, they are represented by iron hydroxides and ferriferous vermiculite-montmorillonites. Sometimes, concentrations of vivianite crystals are observed.

Chemical and biochemical sedimentary deposits of the northern seas and oceans are practically unexplored. The specific nature of their formation is likely to be associated with the low positive temperatures of the bottom sediments as well as higher solubility and washing out of part of the carbonates and authogenous silicates. The main minerals of clays of both recent and Pleistocene bottom sediments of the polar seas are hydromicas and montmorillonite.

In the course of diagenesis of marine sediments a substantial change of clay minerals is observed (beidellitization and montmorillonitization of hydromicas) as well as formation of a complex of authigenic minerals which are further redistributed and become concretions at certain local sites. According to the investigations presented by I.D. Danilov (7), in such cases iron sulphides, vivianite, iron-manganese compounds are formed with insignificant amounts of carbonates. Vivianite concretions, as a rule, have a ball shape, and a rough hummocky surface and diameter up to 5 mm. Their occurrence gives evidence of higher organic matter content of sediments and of the presence of a very reducing medium at the stage of diagenesis. Fe_2O_3 content in vivianite concretions is as high as 40%, while that of P_2O_5, 20%. Concretions of iron sulphides are typically oval, globular and ellipse-like shapes with diameters up to 10 mm. The core of concretions is usually pyrite surrounded by black amorphous matter consisting of colloidal ferric sulphide (hydrotroillite). The chemical composition of iron sulphide concretions of polar basins differs greatly from that of concretions occurring in warm-water sea ooze. In cores of all concretions SiO_2 content is high (60%) at the expense of quartz admixtures. Al_2O_3 content is on the average about 4% varying between 1 and 9%, that of iron sulphide varies between 9 and 25%, and sulphur as sulphide from 10 to 30%. The diffused organic matter admixture makes up 0.5 to 1% on the average. The character of these concretions is evidence of the oxidizing medium maintained by diffused organic matter. Iron oxide and iron-manganese concretions are encountered in the recent bottom ooze of polar seas as well as in coastal marine sediments (recent and Pleistocene) with sandy and sandy-gravelly composition, having small amounts of degrading organic matter. Usually, concretions have a globular shape, small size (2–5 mm) and are more or less uniformly distributed without substantial concentrations. The Fe_2O_3 content in iron-manganese concretions varies between 4 and 19%, and MnO between 2 and 18%.

The group of mineral deposits associated with *post-sedimentation cryogenic transformation* can be subdivided into deposits of epi- and syncryogenic classes which correspond to the basin and continental epicryogenic and continental syncryogenic and palaeocryoeluvial strata of sedimentary rocks.

The formation of cryogenic deposits proper, of cryolites, is noteworthy and among them the most widely represented are structure-forming ice, recurring ice wedges and injection and massive ice beds. An important feature of the freezing and formation of sedimentary deposits of epi-cryogenic strata is the occurrence of crystalline hydrates (calcite, mirabilite, gypsum, etc.) which precipitate at negative temperatures as poorly soluble compounds, as well as the formation of a variety of authigenic minerals and monoxide compounds (pyrite, siderite, marcasite, hydromica, montmoril-lonite, vivianite, carbon dioxide, methane, etc.). Cryopegs and gas hydrate deposits are also very typical of the permafrost regions.

Cryopegs are highly salinized underground waters having negative tem-peratures, which often lie below the base of the permafrost, although there are other types: intra- and above-permafrost cryopegs. Such subsurface water is formed as a result of downwards exclusion at freezing of easily soluble compounds with formation of thick zones containing sodium chlor-ide (more rarely calcium and magnesium chloride) brines. An intensified concentration occurs below the freezing front – cryogenic concentration. The thickness of the cryopeg layer varies from several tens to several hundreds of metres, and salinity of the water ranges from 30 to 300 g l^{-1}. Cryopegs are widely developed in the shelf area of the northern seas, the arctic islands, the Siberian platform, etc.

Gas hydrates or gases in the hydrate form belong to the solid crystalline hydrates and are represented by compounds of natural gases and water and have a complex structure. Their formation requires a certain range of temperatures and pressure, which creates conditions for the interaction between gas and water molecules such that there occurs a specific grid of crystalline hydrate with transition of the compound into the solid hydrate state. This process often leads to large accumulations of gas hydrate deposi-ts. Beyond the permafrost regions gas hydrate deposits occur at a depth often exceeding 1 km. In the permafrost regions they lie below the base of the permafrost and may occur within the permafrost sequence. A tentative estimate of possible depth of occurrence of natural gases in the hydrate state (*P-t-H* diagram) is presented in Fig. 6.4. As shown in the diagram, within the limits of the permafrost regions the range of depth for gas hydrates is dependent on the thickness of the frozen zone, since permafrost reduces the temperature of the underlying unfrozen rocks. Recently drilling of boreholes in West Siberia showed widespread shallow deposits of gas hydrates. Usually gas deposits arise in the presence of gas hydrates even without a lithologically impermeable cover. This is associated with the fact that gas hydrate concentrations usually occurring in the roof of gas deposits (nearer to the permafrost) serve as an impermeable barrier for stream-like leakage of

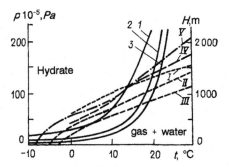

Fig. 6.4. Determination of the interval of probable deposition of natural gases in hydrate state (according to G.D. Ginzburg and Ya.V. Neizvestnov): 1–3 – equiweighted curves for formation of natural gas hydrates with different densities ($1 - 0.555 \, \text{g cm}^{-3}$, $2 - 0.6 \, \text{g cm}^{-3}$, $3 - 0.7 \, \text{g cm}^{-3}$) with coordinates of pressure and temperature; I–V – distribution of temperatures over depth H for the areas of the Yenisei-Khatanga trough with varying thickness of the frozen layer.

free gas from the underlying deposits. Therefore, the zone joining the permafrost regions to the ocean has favourable conditions for the formation of gas hydrate covers and often serves as a reservoir accumulating large volumes of natural gas.

The syncryogenic class of mineral deposits of sedimentary rocks is most clearly represented by loess and loessified formations, with a higher content of carbonates, a prevailing concentration of secondary silty particles and predominance of limonite iron among hydroxide compounds. Loess rocks and the covering layer of silty-clays similar to them have a practical interest as a raw material for glass and ceramics production. They are developed in the course of formation of paleocryoeluvial strata. The syncryogenic rocks themselves have a higher content of cryolites (in the form of thick strata of syngenetic recurring ice wedges, several metres wide at the top and several tens of metres high), structure-forming ice and ice-cement, and also having an intense accumulation of humic matter across the whole section of the sequence with predominant concentration of iron monoxides, carbon dioxide and different hydrogen compounds.

As a conclusion it should be emphasized that the particular features of formation of mineral deposits in the permafrost regions have not been explored adequately and require further long-term investigations.

7

Composition and structure of frozen earth materials

7.1 Characteristics of organic, mineral and chemical composition of frozen earth materials

The organic-mineral and chemical composition of frozen materials and their granulometry have not been well studied to the present time. Much attention was earlier given to the study of ice (as a mineral and a rock) as a component of big accumulations and deposits. However, seasonal and perennial freeze-thaw bring about not only transformations of ice, but also changes in mineral, chemical and organic composition of the mineral matrix.

Chemical processes that occur in the regions of development of frozen ground have a specific nature since the materials typical of these regions are mainly acidic-neutral and reducing, with increased content of carbon dioxide gas, dissolved carbon dioxide and fulvic acids. The processes of coagulation and peptization are widely developed giving rise to the formation of colloidal and silt-sized particles.

The mineral portion of a frozen soil usually comprises primary water-insoluble minerals, secondary water-insoluble minerals, secondary minerals soluble in water, organic and organo-mineral compounds. The specific feature of the frozen ground is the availability of a new structure-generating mineral, ice, the structure of which is determined by the conditions of formation of the frozen ground and its origin. Ice may fill the pores, form intercalations, lenses and cryogenic conglomerates, or substantial amounts of accumulated ice can form a monomineral rock as represented by ice wedges, injection ice and other types of ground ice.

Primary minerals and their aggregates which arise as a result of physical weathering of igneous and metamorphic rocks in the freezing regions differ greatly from the minerals formed outside these regions. This is demonstrated first of all by the increased content of minerals poorly resistant to weathering. Montmorillonite, hydromicas and beidellite are predominant, among secondary water-insoluble minerals which usually form in the freezing regions as a result of degradation and transformation of laminated and banded silicates and feldspars.

Secondary minerals soluble in water are represented in freezing ground by bicarbonates of calcium and magnesium, calcium and sodium sulphate as well as sodium chloride. Easily soluble salts (chlorides and sulphates) are found in solution, whereas poorly soluble ones (carbonates) are most often in the solid state. With lowering of negative temperature of frozen sediments, according to the degree of solubility, carbonates are the first to precipitate, then follow sulphates. The salts that include crystallization water form the crystalline hydrates which are solid components of the frozen sediments.

The availability of clathrate compounds is typical of the permafrost regions. These usually form at a depth of over 1000 m and constitute useful reserves of methane, ethane, hydrogen sulphide, carbon dioxide, etc. Among mineral deposits of the permafrost regions of importance are cryopegs – thick formations (over 1000 m) with highly saline water that does not freeze at negative temperatures and contains a number of easily soluble salts of calcium, magnesium, sodium, potassium, etc. A significant role in the national economy is played by the ice itself, in its different forms – sheets, veins, icings, etc.

Organic matter in the permafrost regions may take the form of poorly decomposed vegetation and wildlife residues and products of their degradation – humus. The products of degradation migrating in soil form the specific soil horizons of the northern regions. For soil with good drainage in the permafrost regions the development of soil is characterized by podzolization with 'tialferization' (Ti-Al-Fe) and formation of illuvial humus.

Gley soils formed in poorly draining ground are widely developed in tundra regions. Low permeability of clayey and mixed silty soils in the presence of a shallow table of perennially frozen soils leads to poor differentiation of soil horizons. A gley horizon underlies the peaty layer and gradually grades into the bed material. Sometimes beneath the peaty layer a discontinuous coarse horizon of humus accumulation is distinguished. At the boundary with the perennially frozen soils a higher Fe and humus content is observed.

In the freezing regions peaty soils are widely developed, resulting from degradation of marshy vegetation in conditions of excessive moisture, shortage of oxygen and low temperatures.

Peaty soil is usually subdivided into upper peats that cover elevated sites (watersheds, divides and slopes) and lowland peats that cover low places and depressions. The lowland peats form in more humid conditions with poor drainage and, therefore, their composition differs. The lowland peats contain greater amounts of humic substances (up to 30%) and vivianite, have a higher ash content and less acidic medium. Waterlogging, formation of peat and slow degradation of vegetation and wildlife residues promote formation of different hydrogen-containing compounds: methane CH_4, hydrogen sulphide H_2S, etc.

The grain size of the organic-mineral skeleton of the frozen soils and its mineral composition have a specific nature. In mountain regions the processes of physical-cryogenic weathering (cryoeluviation) occur most intensely, causing fracturing of hard rock giving big blocks, boulders and down to sandy and silty fractions. The lowland deposits, unlike the mountain ones, do not contain big blocks of fragmental material. However, in their fine-grain fractions, silty particles prevail. For the mountain regions it is typical that sedimentary material is poorly sorted and becomes even less homogeneous while grading from one genetic type into another (from eluvium into talus and alluvium). The main reason for such lack of differentiation is intense accumulation of big rock fragments in the upper parts of slopes with a rather intense transfer of fine-grained material to the lower parts of slopes. Another important reason is associated with prevalence of cryogenic degradation processes (physical weathering) in the sedimentary rocks in the wetter lower portions of slopes. Thus, for instance, eluvium of the region adjacent to Kolyma is represented by a narrow range of fractions (10–30 mm) with the absence of fine earth in the upper part of the section and occurrence of sand and coarse silty particles (Fig. 7.1). As regards talus the fractions occur in the range of 10–30 mm, but there are many fine-grained sandy and silty fractions. In the alluvial deposits the maximum is shifted towards fractions of 5–20 mm with significantly increased content of sand and silt, while in the fine-grained fraction silty particles prevail, in general, the materials become less homogeneous and less sorted, i.e. differentiation of sedimentary material takes place. On plains, where coarse fragmental material is absent, soils are more homogeneous and sorted.

Intense cryogenic breakdown of fragmental rocks conditioned by the processes of cyclic freezing and thawing with poor differentiation of weathering products leads to polydispersion and heterogeneous porosity typical of the northern regions. Differential curves of grain-size distribution of particles (both primary and secondary) usually have a multiple-mode pattern which reflects a high degree of dispersion of the mineral part of soils. The most common modes of fine-grained soils are in the range of colloidal, clayey, sandy, fine- and coarse silty fractions. Polydispersion is characteristic of

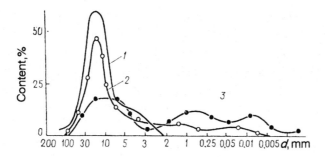

Fig. 7.1. Distribution of fractions of fragmental material of various origins (according to Yu.V. Shumilov): 1 – eluvium: 2 – talus; 3 – alluvium.

sands, sandy silts, silty clays and clays having, as a rule, a polymineral composition. Monodispersed soils characterized by one distinct mode are found amongst heavy clays and sands and are monomineral for the most part.

Cryogenic breakdown of sandy soils gives rise to the formation of loessified deposits most typical of soils with seasonal thawing and freezing. The granulometric composition of loess-like deposits is rather homogeneous: the coarse silt fraction (0.05–0.01 mm) being 60–90%. Loess-like soils are among the most developed deposits of plains and surrounding piedmonts. They are represented by surficial loessified fine material in Bol'shezemel'skaya Tundra; similar deposits are encountered in West Siberia, Northern and Central Yakutia, etc.

The organic-mineral part of the frozen ground is usually cemented with ice which results in the formation of ice-containing clayey and sandy soils, pebble gravel, ice breccia and ice conglomerates. The pores formed by the mineral skeleton of these soils are completely or partially filled with ice. Therefore, interstitial space and porosity are very important characteristics as is the degree of their filling with ice or unfrozen water. The void volume of the frozen soils is often small, differs in materials of different genesis and diminishes in going from eluvial to talus and alluvial deposits.

Grain-size distribution in the cryolithic zone is greatly influenced by the material's genesis. Thus, coarse fractions are typical of alluvial deposits, while in high mountain areas these deposits are mainly represented by fragmental material cemented with ice. In the cryoeluvial formations of plateaus, a rock waste fraction prevails, while the silty and clayey fractions are as much as 25%. On steep slopes rudaceous rocks are formed without fine-grained infilling owing to washing out of fine products of degradation, whereas on gently sloping hillsides fine material is continuously accumulated.

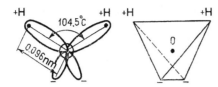

Fig. 7.2. Model of water molecule.

A distinct sorting is typical of the alluvial deposits of the permafrost regions: in the river bed – rudaceous and coarse-grain deposits; on the shallow places adjacent to the river bed – more fine-grained material; while in the floodplain silty and clayey deposits prevail.

The grain size of marine and lacustrine deposits in the permafrost regions is dictated by specific features of their genesis. Conveyance by river ice and by icebergs is of key importance in the distribution of rudaceous material in sediments of marine origin. The dispersion halo of this material is enormous. Therefore, deep-water microfractions of the bottom marine sediments of the permafrost regions often include gravel, pebbles and boulders.

Lacustrine deposits are represented by silty-clays and clays saturated with organic matter, having a specific, banded lamination. Intercalations of coastal sand and rudaceous inclusions, transported by river and lacustrine ice, are also encountered in these deposits.

7.2   Unfrozen water and ice in ground

The phase composition of moisture, i.e. the content of vapour, unfrozen water and ice in the frozen ground, predetermines its specific physical and mechanical properties and the pattern of occurrence of cryogenic-geological processes.

The liquid phase of water in the frozen sediments can be in different energetic and structural states ranging from the state of free water to the substantially modified structure in the immediate vicinity of the surface of mineral particles.

A triangle is formed by the atoms of the free molecule of H_2O with an angle H-O-H equalling about 104.5° (in ice 109°). The lengths O-H and H-H are equal to 0.096 and 0.154 nm, respectively (Fig. 7.2). Water, having a specific structure, is characterized by a number of anomalous properties. Thus, on melting of the solid phase the volume is reduced and not increased; the maximum density of water is at +4°C; the heat capacity on melting increases twofold and more – from 2.05 to 4.57 kJ kg^{-1}; the dielectric constant is rather high and varies from 88 at 0°C to 80 at +20°C, etc. The viscosity is about 1.8×10^{-2} Pa s at 0°C and diminishes with higher temperature to 0.28×10^{-2} Pa s at 100°C. Surface tension at the air boundary with pressure equalling 10^5 Pa is 75.6 and 58.9 mN m^{-1} at 0°C and 100°C, respectively. A sluggishness of structural transformations in water arising under the impact of external factors, is demonstrated by its properties. All the anomalous properties of water are associated with its structure, the distinctive feature of which is the availability of strong hydrogen bonds that form a quasicrystalline structure with triangular pyramid coordination of neighbouring molecules ('short-range order'). The energy of the hydrogen bond is 18.8 kJ mol^{-1} for the ice.

There are many models for the structures of water. The model elaborated by S.Ya. Samoylov is the one commonly accepted in Russia; it was confirmed by X-ray, spectroscopy and other methods of studying the structure and properties of water. According to Samoylov, in liquid water an ice-like lattice persists, somewhat disturbed by thermal movement, and its cavities are partially filled with monomer molecules. Transition of it into water occurs when heat is added at the rate of 0.334 kJ g^{-1} of ice. This takes place at breakage of 9–11% of bonds as determined by the ratio of melting to sublimation heat units (0.334/2.834 kJ).

The ice is an important soil-forming mineral and monomineral rock in the permafrost regions. Its presence in the frozen ground takes the form of ice cement, ice inclusions and masses of concentrated ice. In general, ground ice occupies 2% of the total volume of ice of the cryosphere (about 0.5 million km^3).

Ice is a specific mineral differing greatly by its composition and structure from other minerals and rocks. Some ten crystalline modifications of ice and amorphous ice are known. Under natural conditions a single ice form exists (ice 1), the crystals of which are arranged into a three-dimensional hexagonal grid and belong to the detriangular pyramid-like type of the triangular symmetry system. Such structure consists of six molecules of water forming a regular hexagonal cell with axis $b = 0.9$ nm (Fig. 7.3). Such a branched texture of the ice crystalline grid is dictated by the nature of hydrogen bonds that exist between its molecules. According to current hypothetical considerations tetrahedral molecules of water form the tetrahedronal aggregates of the ice structure. The coordination number of the ice structure is thus equal to four. The distance between the nearest centres of molecules is 0.267 nm. It was found that ice is built of discrete molecules of water connected by hydrogen bonds. The protons are arranged along bonds between the atoms of oxygen at a distance of 0.099 nm from the nucleus of one atom and 0.17 nm from that of another. Around each molecule six centres of voids are arranged at a distance of 0.347 nm, while voids them-

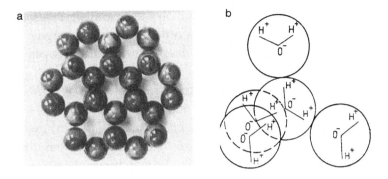

Fig. 7.3. Ice structure: *a* – model (top view); *b* – fragment of ice structure showing immediate surroundings of an H_2O molecule.

selves exceeding the sizes of molecules form the channels made by alternating hexagonal cells.

The molecules of water, and, consequently, of ice comprise stable isotopes ^2H, ^{17}O and ^{18}O, apart from isotopes of 'light' water $^1H_2{}^{16}O$. The content of stable isotopes of hydrogen and oxygen varies both spatially and in time, determined by different phase transitions (evaporation, condensation, thawing and freezing).

Ice in the frozen materials always contains admixtures in the form of solid, liquid and gaseous inclusions. According to origin, these are subdivided into primary (formed simultaneously with ice) and secondary, arising after the ice has been formed. The primary inclusions are authigenic (isolated from water or captured by ice during freezing) and xenogenic (formed of foreign admixtures in water). The secondary ones include hypergenic (penetrated through open fractures and pores connected with the surface) and hypogenic inclusions (that fill cracks and pores isolated as a result of regeneration).

Different composition and formation conditions of underground ice lead to a variety of structure and texture types. *Ice structure* is determined by the shape, size, type of surface, quantitative ratio and the nature of interrelationship of structural elements. For ground ice these are ice crystals, air- and organo-mineral inclusions. The structure of ice is characterized by the relation of crystallographic orientation to the external shape of the crystals and the relation of the same orientation to the elements of occurrence of the ice rock, i.e. by the degree of ordering of the structure.

The following structures of ice are distinguished as dependent upon the shape and crystallographic orientation of grains: 1) prismatic-granular, where the ice crystals have a regular and ordered crystallographic orientation proper to them (main axes of symmetry are parallel); 2) allotriomorphic-granular (irregular granular) with orderless crystallographic orientation; 3) hypidiomorphic-granular which is intermediate between the two aforementioned forms.

Ice texture is determined by the spatial arrangement of its components – crystals of different size and shape, air and mineral inclusions as well as by the degree of infilling. The most important attributes of the texture of ice are associated with the specific distribution of inclusions. In the absence of admixtures the ice texture is considered to be massive or glass-like, whereas with prevalence of gas in the ice volume it is called bubble-type; with intercalation of admixtures it is called laminated. Schistose texture is typical of ice composed of flat and prismatic crystals that form parallel layers. The structure and texture of ground ice reflect the conditions of crystal growth, availability of foreign admixtures in the form of insoluble inclusions and dissolved salts and gases as well as the thermodynamic conditions.

Ice crystals are characterized by anisotropy of mechanical, thermal-physical, optical, electrical and other properties, which are identified on measurement in different crystallographic orientations. These distinctions are associated with the crystallographic characteristics of ice in the spatial grid, in which the main role is played by epipolar planes that have a high reticular density of molecules.

Temperature regime changes cause thermal deformations of ice. Thermal deformations are measured by coefficients of linear (α) and volumetric (β) expansion. These coefficients are computed as a relative change of length or olume of a body on a temperature change of $1\,°C$. Coefficients α and β increase rapidly towards the point of melting and become extremely low at low temperatures (Fig. 7.4).

Distribution of heat in ice is determined by its thermal conductivity (λ). At an ice temperature of $0\,°C$, $\lambda \approx 2.22(\text{W m}^{-1}\,\text{K}^{-1})$. This value exceeds four-fold the thermal conductivity of pure water at $0\,°C$. With lower temperatures the thermal conductivity of ice increases (see Fig. 7.4). It can be estimated from simple empiric relationships, for example, for fresh (nonsaline) ice $\lambda = 2.22\,(1-0.0004t)$ where t is temperature ($°C$). With higher porosity and salinity, thermal conductivity diminishes. At a constant pressure the molecular heat capacity $C_{\text{m}} = 37.7\,\text{J mol}^{-1}\,\text{K}^{-1}$ at the melting point. The heat capacity diminishes with lower temperature (see Fig. 7.4). For nonsaline ice heat capacity $C = 2.12 + 0.0078t$. Heat capacity of ice is to a greater extent dependent on the amount of admixtures, especially at temperatures approaching the melting point.

The structural distinction existing between ice and water which is charac-

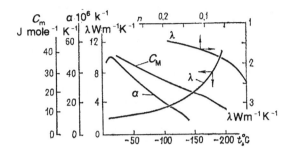

Fig. 7.4. Dependence of the temperature coefficient of linear expansion α, thermal conductivity λ, and molar heat capacity C_m of ice on temperature t, and of its thermal conductivity on porosity n (according to B.S. Rachevskiy *et al.*).

terized by the specific latent heat of fusion L_p diminishes with higher pressure. A certain quantity of heat is absorbed during ice sublimation; this value is the specific latent heat of ice sublimation. For nonsaline ice this value is $2834\,\mathrm{J\,g^{-1}}$, while for saline ice it varies between 2500 and $3000\,\mathrm{J\,g^{-1}}$.

Under the influence of external forces (loads) different properties – fragility, elasticity and plasticity of ice can manifest themselves. At certain conditions within the interval of temperatures from -3 to $-4\,^\circ\mathrm{C}$ ice behaves as an elastic body, according to Hooke's law. This takes place at compressive stresses up to $0.1\,\mathrm{MPa}$, with the time of load application less than $10\,\mathrm{s}$.

When the plastic limit of ice is exceeded its degradation begins. The maximum stress at which ice degradation begins is much dependent upon the rate of load application and conditions of deformation, because simultaneously with stress increase ice begins to creep. The strength of ice increases with lowering of temperature, nonlinearly with a diminishing rate of deformation.

When a load is acting on the ice causing plastic deformation there is no change in volume and no degradation. Ice plasticity depends on temperature, the nature of load and the rate of deformation. Ice resistance to flow is in direct proportion to the rate and is determined by ice viscosity. The data available show a wide spread in values of viscosity coefficient (from 10^3 to $10^8\,\mathrm{MPa\,s}$). This is likely to be associated with growth of stress in time and deviation of ice properties from those of a viscous (Newtonian) fluid. It is often assumed that ice possesses a yield strength of $0.1\,\mathrm{MPa}$. Actually, creep is observed in ice with stresses less than the yield strength.

The structure, composition and properties of *unfrozen* water in the frozen soils have not been explored thoroughly, having a complex nature. Therefore their consideration is to a great extent based on the knowledge of the nature and properties of bound water in unfrozen fine-grained soils as well as on the hydrophilic model systems.

It was found experimentally that the amount of unfrozen water in the frozen soil is a function of the soil composition and structure which, in turn, are determined by the origin and age of deposits. Phase equilibrium of moisture in the frozen soils is also influenced by the thermodynamic conditions (temperature and pressure) as well as various physical fields. A variety of characteristics of the composition of the frozen soil and structure that determine phase composition of the moisture can be reduced to a few physical-chemical factors such as specific active surface, structure of void space, concentrations and type of ions in the pore solution, as all these factors are in functional dependence.

Specific active surface S_{sp} reflects not only surface area, but also the energy factor of this surface onto which a conventional monolayer of water molecules is adsorbed. Accordingly, S_{sp} is a composite value and is a function of both the grain size and the water receptive capacity of its surface. Specific active surface increases in the series sand – kaolin – hydromica clay – bentonite Ca^{2+} – bentonite Na^+.

The unfrozen water in the frozen soils can fill the capillaries and be in the form of liquid films. At lower temperatures film water prevails, while at higher – capillary water does. Therefore, specific active surface serves as a key factor in the formation of unfrozen water only at temperatures below $-5°C$ and determines the total quantity of unfrozen water of island-like and multilayer adsorption and, partially, of the osmotic variety.

Assuming that thickness of the adsorbed layer is dependent on specific surface of soil and does not depend on its curvature, one may obtain a direct ratio $W_{unf} = W'_{unf} S_{sp}$ where W'_{unf} is the quantity of unfrozen water per unit of S_{sp}. The presence even of a small amount of clay and colloidal, especially, montmorillonite, particles causes an increase of S_{sp} and, accordingly, increase of the liquid phase component. Thus, for example, at $-6°C$ the content of W_{unf} in clays of different mineral composition increases in a series kaolin (3.5%) – hydromica-montmorillonite clay (9%) – bentonite (26%). Specific active surface in this case makes up 30, 109 and 560 $m^2\ g^{-1}$, respectively.

With higher temperature (above $-2°C$) the content of osmotic and capillary unfrozen moisture increases and a greater role in the phase composition is played by the structure of void space of the frozen soil, which it is convenient to express quantitatively through distribution of void volumes with radius. Fig. 7.5 shows typical curves of the $\Delta V/\Delta lgr$ relationship for soils of different composition. A distinct pattern is discerned: in the range of

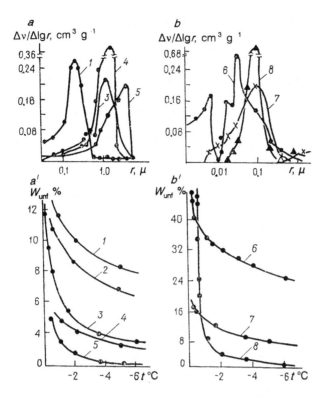

Fig. 7.5. Curves of differential distribution of void volumes by radius (*a, b*) and dependence of the unfrozen water content on temperature (*a', b'*) in soils of different granulometric (*a*) and mineral (*b*) composition: 1–2 – silty-clay material (1 – heavy, 2 – light); 3–5 – sandy clay-silt (3 – heavy fine, 4 – light, silty; 5 – light coarse, 6–8 – clay (6 – montmorillonite, 7 – hydromica, 8 – kaolinite).

high negative temperatures even for the same total porosity, the finer the pores, the larger is the liquid component.

The unfrozen water content in frozen soils usually follows the additivity rule which is confirmed, for example, by the linear dependence of W_{unf} on peat content. This principle applies because mechanical mixing of aggregates and particles of different composition does not, in effect, alter the main factors (S_{sp}, differential porosity, salinity) which are responsible for the contribution of each component of the frozen soil to phase composition of the moisture. Comparison of physico-chemical properties of frozen soils with different granulometry shows the expansion of the specific active surface of the mineral skeleton in going from sandy silty materials to clays, which is accompanied by an increase of the volume of fine pores and capillaries leading to a general increase in the unfrozen water content (see Fig. 7.5a).

Mineral composition is to a great extent a key factor determining the ratio of liquid and solid phases in the frozen soil. The manner of variation of composition within the range of negative temperatures (0 to $-6°C$) has a complex nature (see Fig. 7.5). Thus, at low temperatures (-1 to $-6°C$) W_{unf} increases in the order kaolin – hydromica clay – bentonite. At high temperatures (-0.3 to $-1°C$) the liquid phase increases in the order hydromica clay – bentonite – kaolin. In explaining this one must resort to results of investigations concerning void space and specific active surface.

Thus, for very dispersed montmorillonite clay with available hydrophilic internal basal surfaces in the fine pores, there are two peaks on the differential porosity curves, i.e. pore distribution by size has a bimodal nature (see Fig. 7.5b, curve 6). The first peak in the differential porosity curve corresponding to the finest pores of less than 8 nm, is determined by inter-crystalline porosity of minerals and inter-particle porosity. The second peak corresponding to pores of 30 nm is assumed to occur on account of inter-aggregate porosity.

Although there is no bimodal distribution of void space by radius of pores for hydromica and kaolinite clays, there are certain differences. Hydromica clay contains a greater amount of fine colloidal particles compared with kaolinite which predetermines its finer granular structure (curve 7), despite the fact that as regards average size of fine particles kaolin is more disperse. The homogeneous granulometric composition and regular shape of kaolinite clay particles leads to a higher monoporosity in kaolin among the soils compared (curve 8). Change in mineral composition brings about changes in the volume of ultra-capillary pores from, for montmorillonite clay, $0.3\,cm^3$ g^{-1}, to $0.07\,cm^3\,g^{-1}$ for hydromica clay and $0.02\,cm^3\,g^{-1}$ for kaolinite clay.

Specific surface and pore structure of clays being the function of their mineral composition determine the course of both quantitative and qualitative changes of the unfrozen water content at negative temperatures. Thus, at temperatures lower than $-1°C$ kaolinite clay contains a tenth of the unfrozen water content of montmorillonite and much less than hydromica clay (Fig. 7.5 b).

Within the temperature interval from $-1°C$ to the freezing point of clay the pattern of variation of unfrozen water content with mineral composition is quite different. In this case the greatest amount of the liquid phase is observed in kaolinite clay, the smallest in hydromica clay, in between being bentonite. This pattern is also explained by the structure of void space. Kaolinite clay contains a big volume of pores sized about 100 nm in which unfrozen water is found. A highly homogeneous structure of the kaolin void space gives a sharp variation of the unfrozen water content in going from

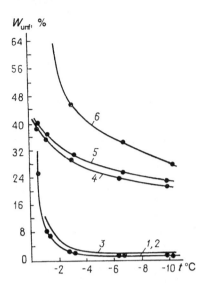

Fig. 7.6. Dependence of the unfrozen water content on temperature in soils of different chemical-mineral composition: 1–3 – respectively, Na-, Ca- and Fe-kaolin; 4–6 – respectively, Ca-, Fe-, and Na-bentonite.

$-2°C$ to higher temperatures, whereas the 'indistinct' pattern of pore distribution by size as seen on differential porosity curves for hydromica clay and, especially, for montmorillonite clays, is reflected in a smooth curve of $W_{unf}(t)$ typical of these clays' mineral composition.

The influence of *exchange cations* on the phase composition of moisture depends on soil composition and temperature. This influence manifests itself only in clays that include minerals of the montmorillonite group. As is shown by the experimental data presented in Fig. 7.6, the role of exchangeable cations in the formation of unfrozen water increases with saturation of bentonite with univalent cations. Within the temperature interval 0 to $-10°C$ the Na^+-bentonite contains the greatest amount of unfrozen water, while Fe^{3+} and Ca^{2+}-bentonite contains the smallest. This is explained by the fact that Na^+-bentonite is almost completely represented by microaggregates of colloid size. The fine granular structure of Na^+-bentonite brings about a small size of inter-aggregate pores approaching that between particles.

Salinity of the frozen soil, along with its chemical-mineral composition and dispersion, is a main characteristic that exerts a substantial influence on the content of unfrozen water and ice. The effect of salinity on the phase composition of the moisture in freezing soils depends on the concentration and type of salts.

Fig. 7.7 shows the dependence of the unfrozen water content on tempera-

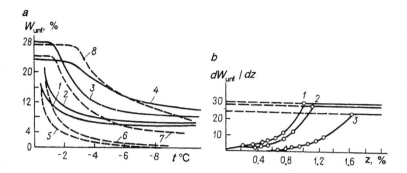

Fig. 7.7. (a) Dependence of unfrozen water content on temperature in heavy silty clay (solid lines) and kaolin (dotted lines) with different $CaCl_2$ concentration in pore solution (1 and 5 – 0.0N, 2 and 6 – 0.1N, 3 and 7 – 0.5N , 4 and 8 – 1.0N), and (b) intensity of unfrozen water content variation depending on salinization of medium-grained silty clay at $-2°C$ temperature (b) 1 – NaCl; 2 – $FeCl_3$; 3 – $Ca(NO_3)_2$.

ture and on $CaCl_2$ concentration in the frozen heavy silty clay and in kaolin. Salinization of kaolinite clay with $CaCl_2$ leads to a more drastic increase of the unfrozen water content than with polymineral heavy silty clay. The influence of salinity on the phase composition is dependent on the liquid phase content of the frozen soil; the smaller its initial content, the greater is the influence.

It should also be emphasized that the freezing point of a highly salinized soil is often similar to that of a free solution of the same concentration, since pore solution does not have in this case a physico-chemical bond with the soil skeleton. The experimental data (Fig. 7.7) show the significance of both concentration and type of salts. Thus, the addition of 1% NaCl into the frozen nonsaline silty clay is sufficient to change the unfrozen water content twofold, while addition of the same amount of $Ca(NO_3)_2$ has practically no effect on the content of unfrozen water.

Phase composition of moisture in frozen soils under natural conditions is defined by the existing thermodynamic conditions as well as by the petrographic characteristics of materials. Therefore, there is an intimate connection between W_{unf} and the geologic-genetic types of the frozen soils. Within the framework of each genetic complex, differences in phase composition of moisture are functions of the composition and structure of the soils. A large component of unfrozen water in soils of marine origin is associated with high salinity, fine-grained material, with availability of Na^+ within the ion-exchange complex and the fine-pored structure of these deposits (Fig. 7.8). There is a distinct difference in the phase composition of continental

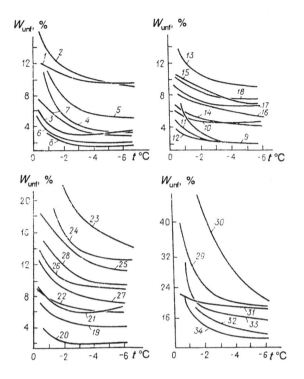

Fig. 7.8. Dependence of unfrozen water content on temperature in soils of alluvial (1–8), talus (9–12), eluvial (13–14), glacial-marine (15–20), glacial (21–22), marine (23–28), alluvial-marshy (29–30) and organic (31–34) origin. Soil composition: 3–14, 19–22 – sandy clayey-silt; 1, 2, 15–18, 25–27 – clay-rich silty sand; 23, 24, 28 – clays; 29, 30 – respectively, poorly and moderately degraded peat.

deposits of alluvial and eluvial origin which is likely to be associated with the presence of a large amount of hydrophilic organic matter and dissolved compounds of bivalent cations and active anions Cl^- in the alluvial sediments.

In unweathered water-saturated strongly consolidated rocks such as sandstones, aleurolites and argillites, phase transition of moisture into ice may occur at temperatures as low as $-10°C$, which is caused by the small size of pores and substantial density of the materials. Weathering leads to higher temperature of freezing owing to the relevant transformations of structure (Fig. 7.9).

7.3   Textural characteristics of the frozen material

Cryogenic texture of the frozen material is assumed to be that of the ice framework, consisting of inclusions, intercalations differing in shape and size, orientation and spatial arrangement, by which the structure of the

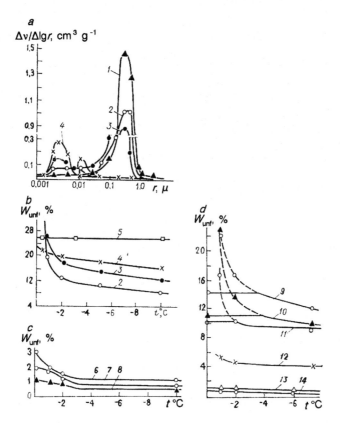

Fig. 7.9. Curves of differential distribution of void volumes by radius (*a*) and dependence of unfrozen water content on temperature in siliceous (*b*), sandy (*c*) and clay-rich (*d*) soils and rocks: 1 – diatomitic clay; 2 – diatomite; 3 – tripoli; 4 – opoka; 5 – silica gel; 6–8 – sandstones (6 – weathered medium-grained; 7 – unweathered fine-grained, 8 – unweathered coarse-grained); 9 – 11 – compact clays; 12–13 – argillites; 14 – aleurite.

mineral skeleton is subdivided into individual structural varieties. The concept of the frozen soil texture distinct from cryogenic texture implies study of textural features of both ice inclusions and the organo-mineral portion of soils.

The fine granular texture of soils forms in the course of their lithification. Specific features of cryogenic texture generation are dependent on chemical-mineral variability and variability of facies as determined by the different modes of freezing (syngenetic, characterized by cryogenic transformation of sediment simultaneous with its accumulation, and epigenetic, exemplified by freezing of lithified materials). The cryogenic texture of soil is much dependent on the soil's texture and structure prior to freezing. The availability of textural attributes in the initially unfrozen soils (lamination, fissuring, oriented inhomogeneity) leads to the formation of inherited cryogenic textures. In homogeneous materials (both in composition and structure) the superimposed cryogenic textures are formed in the process of freezing.

Heterogeneous cryogenic textures encountered in natural conditions are created by different mechanisms of cryogenic texture generation among which the main are migration-segregation, pressure-migration, injection and orthotropic compression. The different types of ice, segregation, injection, ablimation, etc., are dependent on the process of ice formation in cryogenic structures. The main mechanism of superimposed cryogenic texture formation is that of migration-segregation. This is associated with development of different processes including heat and mass exchange, physical-mechanical as well as physico-chemical ones (coagulation, aggregation, shrinkage, swelling, etc.) that substantially transform the primary structure of soils. These processes result in development of stresses, breakage of structural links and segregation of ice. Classification of migration-segregation cryogenic textures based on the study of their mechanism and conditions of formation was elaborated by E.D. Yershov (9) (see Table 4.1).

Cryogenic texture of rock is largely determined by its fissuring, saturation with water and freezing. There are, as a rule, inherited cryogenic textures. Size, shape, orientation and spatial arrangement of ice inclusions correspond to the geometry of fissures. In magmatic igneous rocks (granites, diorites, andesites, etc.) there are fissure and fissure-vein cryogenic textures. In fissures ice is in the form of films, crusts and nodules of crystals, with ice-cementing of loose infilling material which completely or partially fills up the cavities. Almost completely filled fissure cavities are typical of fissure-vein cryogenic structures looking like substantial veins.

In consolidated (cemented) sedimentary rocks the following types of cryogenic structures are distinguished: stratal-fissure, stratal-fissure-porous and stratal-fissure-karst (Table 7.1). The size of ice interlayers in the inherited structure is dictated by the opening of fissures and varies from fractions of a millimetre to tens of centimetres. With respect to ice thickness the following cryogenic structures are distinguished: thin, medium and thick fissure and fissure-wedge.

As regards spatial arrangement and orientation there are the following cryogenic structures : ordered-latticed (or regular latticed) and orderless-latticed (non-regular-latticed). The latter are, as a rule, confined to zones of tectonic crushing and intense weathering, while ordered-latticed cryostructures are remote from such zones. Differentiations in composition and structure of ice in the inherited cryogenic structures of rocks are primarily associated with different mechanisms of ice formation. The migration-

Table 7.1. *Classification of cryogenic structures of the perennially frozen rocks (after A.A. Kagan, N.F. Krivonogova)*

Texture name	Description	Main types of soil for which cryostructure is typical
Fissure	Ice occurs along fissures in the form of films, crusts along walls or cement and schlieren in the fissure infill	Magmatic – granites, diabases, andesites, basalts
Fissure veined	Ice fills the whole cross-section forming veinlets	Metamorphic – hornfels, quartzites, crystalline schists, gypsum
Stratal-fracture	Ice veins along fissures of bedding form the appearance of cryostructure; ice content of fractures of other systems has a subordinate nature	Sedimentary – dolomites, limestones, marls, aleurolites, argillites
Stratal-fissure porous	Ice veins along fissures of bedding and ice-cement in pores; ice content of other systems has a subordinate nature	Sedimentary – sandstones, clayey schists, marls
Stratal-fissure karst	Ice veins along fissures of bedding and cavities of leaching and dissolution form the appearance of cryostructure ice; content of other systems of fissures has a subordinate nature	Sedimentary – limestones, dolomites (karst), gypsum, rock salt

segregation type of ice in rock is encountered only in fine-grained water-saturated fissure infillings. However these are the main mechanisms of ice formation as distinguished by the authors A.A. Gagan and N.R. Krivonogova:

1) infiltration, dependent on the infiltration of surface and ground water; ice is polygranular with random orientation of crystals (allotriomorphogranular);

2) injection, associated with hydrodynamic intrusion of water into fissures and its freezing in them; injection ice formation usually leads to cryogenic expansion of fissures and development of highly fissured rocks; injection ice in big fractures is mostly transparent, while the textures themselves are called inherited-expanded;

3) ice cement, typical of fissured masses frozen below the level of

ground water; it is also typical of saturated rocks in the seasonally thawing layer;

4) ablimation (sublimation) ice which is formed due to freezing of vapour in bigger fractures reaching to the ground surface or other sources of recharge. Ablimation ice has a granular texture.

Cryogenic structures in the lithified materials of the *loose cover* differ in their composition and structure and conditions of freezing. At the present time many authors (B.I. Vtyurin, Y.M. Katasonov, E.D. Yershov, L.N. Maksimova *et al.*) have elaborated classifications of all known types and kinds of cryogenic structure encountered in the natural environment. Several principal types of cryogenic texture were distinguished in accordance with orientation and spatial arrangement; namely, massive, laminated, latticed and cellular. Different types of cryogenic texture were distinguished according to the size of ice layers and the spacing between them (Table 7.2).

One of the main processes determining the heterogeneity of types and kinds of migration-segregation structures is migration of moisture. The intensity of the migration flow of moisture and its duration determine the thickness of segregated layers while its intensity promotes development of shrinkage tensile stresses and deformations, their frequency and orientation. The intensity of moisture flows increases with higher dispersion of soils. Therefore, thick schlieren cryogenic texture is most often encountered in clayey soils. Changes in grain size of soils have a greater effect over the *appearance* of the cryogenic structure, i.e. the frequency of layers and their thickness, and a lesser effect over the *type* of cryogenic structure. The types of cryogenic structure are to a greater extent determined by mineral composition of materials. Thus, with higher content of minerals of the montmorillonite group in clays the type of cryogenic structure changes from horizontally laminated to cellular. With higher content of the montmorillonite group of minerals the moisture flow intensity and thickness of ice layers diminish, but, on the contrary, they increase with higher content of minerals of the kaolinite group.

At high rates of freezing a massive cryogenic texture is mainly formed; with less fine material, a lower rate is required for the formation of massive cryogenic texture. With a lower rate of freezing in fine-grained soils massive texture grades into streaky cryogenic structures. The lower the rate of freezing the longer the time required for the growth of ice layers and their thickness. At shallow depths of ground water under conditions of an open system, the formation of horizontally laminated thick schlieren cryogenic texture is most likely. In the process of cryogenic texture formation a

Table 7.2. *Classification of schlieren cryogenic structures (according to A. Vtyurina, B.I. Vtyurin)*

Type (according to schlieren arrangement and nature of ice cement)	Subtype (according to ice schlieren orientation)	Kind (by the interval between schlieren: rare laminated and big netted – over 100 mm; medium laminated and medium netted – 10–100 mm; frequently laminated and finely netted – 1–10 mm; microlaminated and micronetted – under 1 mm)	Variety (by the thickness of ice schlieren: thick schlieren over 10 mm; medium schlieren 5–10 mm; microschlieren less than 1 mm)
Laminated	Horizontally, obliquely, vertically laminated	Rare-, medium-, frequent, microlaminated	Thick-, medium-, fine-, microschlieren
Netted laminated		Medium-, frequent-, micronetted laminated	
Netted	Horizontally, obliquely, vertically netted, orderless netted	Big-, medium-, fine-, micronetted	
Cellular	—	Big-, medium-, micro-cellular	
Ataxitic	—	—	—
Pseudo-schlieren (predominately)	Horizontally, vertically laminated	Micro-, frequent laminated	Fine-, micro-schlieren

significant role is played by the initial structure and texture of soils. Cavities and fissures in fine-grained soils allow infiltration and other types of ice; they also exert substantial influence over spatial arrangement and orientation of layers of migration-segregation ice formed in the vicinity.

The influence of lamination and inhomogeneities, i.e. contacts between soils differing in thermal-physical and physical properties, lies in the fact that with different shrinkage and deformation characteristics the processes of heat and moisture transfer are modified. A specific setting is created in the zones of contact between layers and inclusions which experience to a great extent the difference in shrinkage deformation values. This leads to concentrated tensile stresses of shrinkage sufficient to overcome the strength of structural bonds of the soil and, consequently, results in the formation of ice layers that inherit the primary lithological inhomogeneity.

Ice formation in rudaceous uncemented materials is mainly conditioned by freezing of meltwater and rainfall that penetrate during infiltration and carry fine-grained material. Freezing of this water in rudaceous rocks with negative temperature often leads to heaving of blocks and formation of basal ice-cement (golets ice). With reduction of depth of seasonal thawing such ice may sometimes become buried in the permafrost sequence. Closed air-filled cavities in which ablimation ice is growing, may occur at contacts of soil with individual blocks.

Rudaceous material with fine-grained infilling often has a massive, crustal or composite lens-like laminated cryogenic texture in which ice interlayers are curved, not persistent in strike and often following the shape of the blocks. In the heterogeneous uncemented materials the inherited cryogenic textures are formed in discontinuities, fractures and cavities. The mechanism of ice formation is the same as that in rocks.

There are cryogenic textures of the seasonally thawing and seasonally freezing layers, existing only in the winter period and there are cryogenic textures of permafrost. The temperature regime during freezing and the chemical-mineral composition of soils have importance for their formation. The mechanism of structure generation in the layers of seasonal thawing and freezing is predominantly that of migration-segregation.

A dual (two-part) cryogenic structure is typical of the soils of the seasonally frozen layer following from variations of the freezing regime with depth. In soils of the seasonally thawing layer a second horizon of cryogenic textures is observed which arises due to freezing from beneath, because of the adjacent permafrost.

In the upper part of the permafrost the epigenetic cryogenic textures are formed at smaller gradients of temperatures and slower rates of freezing compared to the layer of seasonal thawing. Low rates of freezing and small temperature gradients lead to less intense migration flows of moisture, which however persist for a long time. Shrinkage processes and a thick zone of drying caused by long-term dewatering lead to smaller gradients of shrinkage tensions and, accordingly, to the formation of dispersed-laminated, big-latticed and block-like cryogenic textures (Fig. 7.10).

In the permafrost below the depth of zero annual amplitude generation of cryogenic texture took place at lower gradients of temperature and rates of freezing, under freezing of the epigenetic type. The pressure-migration and injection mechanisms of ice segregation are key factors in the formation of cryogenic structure determining its texture (size of segregational layers of ice, spacing between them). The horizontal-laminated cryogenic structure that forms in conditions of free inflow of moisture from below, grades into the block-type cryogenic structure with vertical ice schlieren oriented in the direction of moisture injection when hydrodynamic pressure is greater (Fig. 7.11).

Cryogenic textures of syngenetically freezing deposits form with (from the geological point of view) simultaneous freezing and diagenetic transformation of newly deposited sediments. The process of their transition into the permafrost state is associated with a cyclic (year to year) upward movement of the base of the seasonally thawing layer by the thickness of new sediments, as the freezing of the layer occurs from below. The depth of seasonal thawing is in this case constant. Thus, the cryogenic structure of permafrost soils frozen syngenetically is formed by the lower part of the seasonally thawing layer changing into the permafrost state. Cryogenic structures are formed in this layer only very rarely by freezing from the top (at high mean annual temperatures). In natural conditions there can arise different combinations of structure-generating mechanisms depending on the relation of the rates of accumulation of sediments to the rates of freezing and thawing. In one case soils may have a relatively homogeneous cryogenic structure following syngenetic freezing, while in another, cryogenic textures are repeated at a certain interval, i.e. cyclic cryogenic structure is typical (Fig. 7.12). The formation of cyclic horizons of cryogenic textures is associated with growth of thicker ice layers because of slower freezing from below or segregation ice formation on thawing from the top. Between the horizons of cryogenic textures formed by freezing from below, cryogenic structures may be found that have been formed in the course of seasonal thawing, i.e. in general, this type of cryogenic structure of the frozen soil has a mixed nature with respect to the mechanism and formation conditions. Cryogenic texture that has been formed in the frozen soil during thawing most often grades smoothly into the texture that has been formed on freezing from below. In this case ice is segregated in the permafrost in significant thickness (up to several tens of centimetres thicker than the accumulating sediment). Annually repeated seasonal cycles of freezing-thawing give rise to ice-rich horizons in the upper

Fig. 7.10. Cryogenic structure of epigenetically frozen marine Kazantsev clays. The Yamal, the Seyakha river valley. Block cryogenic texture; horizontal ice schlieren inherit the primary lamination of the clays (photo by G.I. Dubikov).

Fig. 7.11. Block cryogenic texture in silty clay. Vertical ice schlieren of injection origin can be seen. At the base there is tabular ice (photo G.I. Dubikov).

part of perennially frozen soils and ataxic cryogenic textures (Fig. 7.12). Such horizons of textures are typical of fine-grained soils in which the process of ice segregation is rather intense at thawing. On freezing from below the more homogeneous cryogenic structure is created by the relatively slow freezing. The thickness of ice layers in such structures diminishes from below upwards since the upper soils are more compacted.

The cryogenic structure of syngenetic soils is much dependent on the regime of freezing from below and thawing from the top. For soils having high mean annual temperatures (i.e. for the southern type of syngenetically frozen deposits) accumulation of small amounts of ice is typical under both winter freezing from below and summer thawing from the top. This is mainly associated with the low gradients of temperature in the upper layer of perennially frozen materials during both winter and summer. On the contrary, a high content of ice of the perennially frozen soils below the layer of seasonal thawing is typical for soils having low mean annual temperature (i.e. for the northern type of syngenetically frozen deposits).

7.4   Microstructure of frozen soils

The *microstructure* of frozen soils implies the totality of microstructural and microtextural features inherited from the unfrozen materials as well as those acquired during freezing and in the frozen state. The concept of *microstructure* includes the size, shape, pattern of surface and the quantitative proportions of elements that compose the frozen soil and the nature of their relationships. *Microstructure* is defined as the totality of attributes that characterize the relative arrangement and distribution of elements of the frozen soils in space.

The elements of the frozen soil microstructure include primary mineral grains, particles, their aggregates that compose the mineral skeleton, ice crystals, unfrozen water, air inclusions and foreign admixtures of a size, as a rule, of less than 1 mm.

Ice crystals in the frozen soil comprise ice-cement as well as various ice inclusions. The following kinds of ice-cement may be recognized in the filling of the pores with ice: cuff-type (contact) found in the adjacent cuffs and contacts between grains and aggregates of the skeleton; the film-like form that envelops the surface of grains and aggregates of the skeleton with part of the pores unfilled; pore type, that fills the pores completely; and the basal type, that accounts for the main mass of the material and divides the grains and aggregates of the mineral skeleton. There is also needle ice-cement of ablimation origin which is isolated in the form of ice spicules in pores and on the surface of grains and aggregates. Along with all these, there

Fig. 7.12. Lens-like laminated cryogenic texture of the syngenetically frozen silty clays (photo G.I. Dubikov).

are also massive, porphyry, ring-like, laminated, and latticed cryogenic microstructures. Their principal types are, in general, similar to the well-studied cryogenic macrotextures. However, unlike the latter, microstructures are to a greater extent curved, discontinuous and inconsistent.

The frozen soil microstructure is largely heterogeneous. Generally, the formation of cryogenic microstructure is controlled by the conditions of accumulation and freezing. In general, two principal genetic types of frozen soils are distinguished: epigenetic and syngenetic.

The microstructure of epigenetically frozen soils reflects their microstructural and microtextural peculiarities acquired before freezing. These are represented in the degree of aggregation, compactness of mineral skeleton and the inherited nature of ice segregation. Thus, for the silty clay ice-marine deposits of the Salekhard suite which have undergone a stage of lithification before freezing, a matrix-like microstructure of the mineral skeleton is typical, which is similar to the nonfrozen deposits. Distinctive features of the frozen soil are demonstrated by the morphology of ice-cement and ice inclusions, as well as by the microstructure of the adjacent zone (Fig. 7.13a).

The microstructure of syngenetically frozen soils is characterized by a

a   b

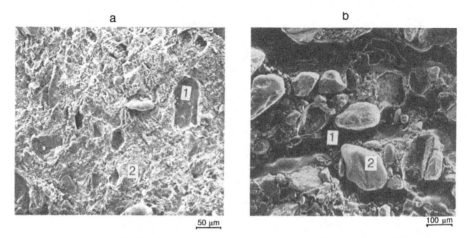

50 μm   100 μm

Fig. 7.13. Microstructure of epigenetically (*a*) and syngenetically (*b*) frozen clay
soils: 1 – ice; 2 – mineral particles.

specific nature with higher ice content, predominance of basal ice-cement
and the occurrence of a loose organo-mineral skeleton (Fig. 7.13b). Such a
microstructure, according to V.V. Rogov, is typical of the syngenetically
frozen deposits of the ice complex, which are characterized by the poorly
aggregated loose organo-mineral skeleton divided by the basal ice-cement
into aggregates sized 0.2–0.3 × 0.5–2 mm. Apart from ice-cement there are
ice lenses and bands 0.1–0.2 mm thick with crystal axes oriented almost
vertically.

The microstructure and microtexture of frozen soils are closely associated
with grain size (Fig. 7.14). Rudaceous and sandy materials which are non-
cohesive, have a massive cryogenic texture. Their mineral skeleton does not
experience any change at freezing. The contact between structural elements
is of a point type. Irrespective of the freezing regime, water is retained in
pores or is partially pushed out. The type of ice-cement varies from needles
and cuffs to basal.

Microstructure of frozen clay-rich soils differs from that of rudaceous
soils in having a composite nature caused by physico-chemical processes
which substantially transform the morphology and size of the mineral
skeleton, aggregates and void spaces, leading to the formation of porphyritic
inclusions of ice and segregation microschlieren. In the mineral intercalations with schlieren cryogenic texture, the ice is present in the form of
individual crystals arranged between mineral particles and their aggregates
as well as in the form of chains parallel to the longitudinal extension, which
serve as a transition between ice-cement and microschlieren (16). Ice crystals
in the microschlieren have a columnar, rarely tabular, shape with regular

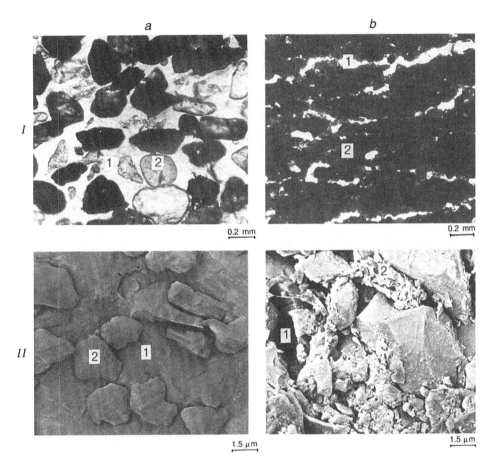

Fig. 7.14. Microstructure of frozen soils of different grain size according to data from optical (*I*) and electron microscopy (*II*): a – sand; *b* – silty-clay; 1 – ice; 2 – mineral particles.

straight edges, with their main optic axes being, as a rule, oriented perpendicular to the longitudinal extension of the schlieren.

Microstructure of frozen clayey soils is to a great degree dependent on their mineral composition. This is evident, firstly, when clay minerals of the montmorillonite, kaolinite and hydromica groups are present in the soil skeleton. Differences in their crystallo-chemical structure are reflected in the morphology of particles, aggregates and inclusions of ice-cement and in the quantitative ratio of ice-cement to unfrozen water, as well as in the intensity of heat and mass exchange with physico-chemical and physico-mechanical processes giving rise to a specific cryogenic microstructure (Fig. 7.15). A loose chaotic structure of tabular aggregates with isometric inclusions of ice-cement uniformly distributed in the inter-aggregate voids is typical of

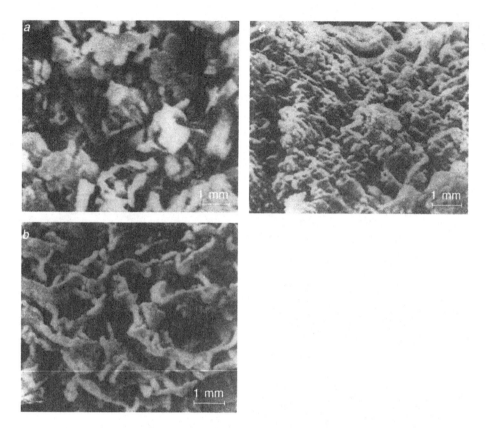

Fig. 7.15. Microstructure of frozen kaolinite (*a*), unfrozen montmorillonite (*b*) and frozen montmorillonite (*c*) clays.

kaolinite clays. Favourable conditions for moisture migration in the kaolinite clays promotes ice segregation with ice microschlieren varying in thickness from a hundredth of a millimetre upwards. A distinctive feature of the frozen montmorillonite clay microstructure is a cellular microtexture. The microstructure of the mineral blocks and the size of ice schlieren is determined by freezing conditions. In general, the blocks typically have a compact structure of aggregates and an insignificant amount of ice-cement, which is explained by the local contraction of the mineral skeleton resulting from moisture migration in all directions from the centre to the bordering ice framework. A characteristic feature of hydromica clay microstructure is slit-like pores of irregular shape conditioned by the morphology of particles and aggregates having a leaf-like form of skeleton with slightly curved surfaces and tortuous boundaries.

The microstructure of frozen fine-grained soils reflects the inherited nature of the microtexture existing before freezing. Among the inherited features are their structureless nature, laminated with availability of aggregates, etc. Thus, the undisturbed sample of a silty-clay from a seasonally thawing layer showed: compact soil aggregates; ice inclusions of porphyroid kind with part of the pores not filled with ice; and 'ice-covered sites' were practically absent. Microstructure of a disturbed silty-clay sample was characterized by a uniform distribution of the sandy-silty fraction, by the circular, isometric, less compact aggregates, by the absence of purely clay formations and by big pores as well as by the presence of both pore ice-cement and ice microschlieren.

Restructuring of the initial microstructure is much conditioned by freezing regime. As a rule, with higher rate of freezing, the more uniform is the distribution and correlation of structural elements; their average size is reduced; ice microschlieren become less thick and more frequent. Freezing of the soil in conditions of moisture inflow leads to a poorly aggregated mineral skeleton with small loose aggregates, a predominance of basal ice-cement and the presence of frequent ice microschlieren. In the absence of moisture seepage a compact skeleton forms having big aggregates with porous ice-cement and rare ice microschlieren. With a certain regime of cyclic freezing in the clay soil circular microstructures are formed, resulting from differentiation of the materials that comprise the soils (16).

8

Properties of frozen soils

8.1   Physical properties of the frozen materials

Moisture content, ice content, density and porosity are the main physical indices that characterize the engineering-geological aspects of frozen and thawed soils.

Moisture content of frozen soil is the water content, based on drying at a temperature of 100–105°C to obtain the constant mass of solid material. There are different indices – total moisture content, integral (natural) moisture content and volumetric moisture content. Total moisture content W_{tot} of the frozen soil is the ratio of water mass in solid and liquid state contained in the frozen soil to the mass of its skeleton, and in salinized materials, to the mass of the material's skeleton and of the salts present (as a percentage or fraction of a unit). The integral moisture content W_{nat}, is the ratio of all phases of water mass to the mass of the frozen ground: the volumetric moisture content W_{vol} is the ratio of water volume in solid and liquid phases to the volume of the frozen soil.

The total moisture content of frozen soils, unlike those unfrozen, can much exceed the value of total moisture capacity. The value varies within a wide range in frozen soils – from maximum molecular moisture capacity to values 3–4 times exceeding the upper limit of plasticity. In general total moisture content increases with finer grain size. The total moisture content with schlieren cryogenic textures is always higher than that of the soil with massive cryotexture.

Ice content (I) is an index that characterizes the ground ice content of the frozen soil (a percentage or fractions of a unit). For estimating ice content of soils contributed by texture and texturogenous ice use is made of the total ice content index. Quantitatively, it is the ratio of the whole ice mass to that of dry matter – the gravimetric ice content; the ratio of gravimetric ice content to total moisture is relative ice content; the ratio of the total volume of ice to that of the frozen soil is the volumetric ice content.

Depending on the value of the ice content, soils are subdivided into ice-rich, ice-bearing and those with low ice content. Ice-rich soils are those in which ice volume occupies more than a half of the frozen soil volume. After thawing such soils pass into a fluid or fluid-plastic state which accounts for their high subsidence. In the thawed state ice-rich soils are characterized by a low bearing capacity, low water resistance and high compressibility. Soils with a low ice content (lower than 25%) acquire a visco-plastic and semi-solid consistency after thawing, and are characterized by a high water resistance and low compressibility. Ice-bearing soils with ice content 25–50% have intermediate properties compared with the above categories.

Density ρ (volume weight, $g\,cm^{-3}$) of the frozen soil is the ratio of the frozen soil mass with texturogenous ice to the volume of the frozen soil with its structure undisturbed; density of the frozen soil skeleton is the ratio of soil skeleton mass to the volume of the undisturbed frozen soil with its cryogenic structures.

The greatest density is typical of the materials that contain iron oxides or pyrites, while low values are typical of the rocks in which montmorillonite and halloysite prevail. Higher content of organic matter leads to lower density. In approximate calculations the following average densities of the solid mineral component are usually assumed: for sandy materials, $2.65\,g\,cm^{-3}$; for silty-clay materials, $2.70–2.73\,g\,cm^{-3}$; for clays, $2.75\,g\,cm^{-3}$.

The density of the frozen soil skeleton varies generally from 2 to $0.62\,g\,cm^{-3}$. Since there are always pores filled with ice, frozen water or gas in the frozen material, the unit weight of its volume is always less than the density of the solid mineral component. The density of rock in the frozen state varies within the range of $1.0\,g\,cm^{-3}$ for ice-rich materials with ataxite cryogenic structure to $2.73\,g\,cm^{-3}$ and over for highly cemented argillites and sandstones with massive cryogenic structure.

One of the main physical indices of the structure of unfrozen soils is the *porosity* of the skeleton, the index which usually characterizes compressibility, i.e. total void volume per unit of soil volume irrespective of pore size and degree of saturation. In frozen soil where pores are filled with not only unfrozen water and gas inclusions, but also with ice-cement and ice that forms various cryogenic textures, one should distinguish between porosity (or void space) determined as a ratio of all the pore volume (voids) not filled with unfrozen water or ice to the volume of the frozen soil, and porosity of the frozen soil skeleton, which is the ratio of the whole void space volume not occupied by mineral skeleton to the frozen soil volume. Often, porosity is characterized as the ratio of volume of voids to the volume of the solid component of the soil and is called reduced porosity or *coefficient* of porosity. As regards unfrozen soil the porosity coefficient usually does not exceed 1.5–2, while in frozen and especially in highly ice saturated soils it often reaches 3–3.5, and with higher ice content increases several times. Usually, the porosity of unfrozen fine-grained soils is 20–40%, while the porosity of the mineral skeleton of the frozen soils can be as high as 60 and even 90%.

Washout capacity and soaking capacity belong to water resistance characteristics of the frozen soils that are indispensable in the evaluation of the thermal erosion hazards and of potential gully formation. These characteristics are to be used in estimating the rates of reservoir bank erosion and in calculations of stability of canal slopes and earth structures that interact with water flows.

Washout capacity of the frozen soils is a property that characterizes the capability to yield aggregates and elementary particles of the soils to the simultaneous thermal and mechanical effects of flowing water.

Washout of the frozen soils is dependent on a number of interrelated factors among which the main ones are the nature of structural bonds in the soil, its ice content and type of cryogenic structure (Table 8.1). In a general case, washout capacity of thawing soils increases when grains become smaller and density and cohesion lower, as well as with disturbance of natural structure. Washout capacity of syncryogenic perennially frozen soils is much higher than that of epicryogenic ones.

Soaking capacity of the frozen soils is such that soil loses cohesion and transforms into a loose mass while interacting with water. Soaking of the frozen ground is the result of ice melting and weakening of bonds existing between particles on swelling. Soaking is characterized by the rate and type of the process. It is accompanied by reduction of strength and determines the stability of the frozen soil with respect to degradation into mineral aggregates in still water (or water resistance). Unlike unfrozen soils, soaking of the frozen soils is dependent on not only lithological characteristics and type of natural cement, but also on ice content and its spatial distribution, i.e. cryogenic structure.

The frozen fine-grained ice-rich soils belong to colloidal systems that possess thixotropic properties on thawing. As shown by studies conducted in West Siberia, tundra areas of Bol'shezemel'skaya Tundra, Southern Yakutia and other regions, the fine-grained soils of the seasonally thawing layer are almost everywhere thixotropic in the absence of drainage on thawing. Under dynamic load their natural structure breaks down, they become liquified, losing their strength completely. After the load is removed a gradual recovery of the primary strength is observed. The main cause that

Table 8.1. *Classification scheme for frozen soils, according to washout capacity*

Washout category	Washout capacity	Allowable flow rates v, m s^{-1}	Shearing stress limit P 10^{-5} Pa	Ice content (I)%	Cryogenic texture Type	Kind
I	Poor	< 1.0	> 0.3	Low content 0–25	Massive	—
II	Medium	1.0–0.3	0.3–0.1	Medium concent 25–50	Laminated and netted	Rare and big
III	High	0.3–0.1	0.1–0.3	Ice rich > 50	Ditto	Medium
IV	Higher than usual	0.100–0.003	0.03–0.01	Ditto	Ditto	Small and fine
V	Catastrophic	< 0.003	> 0.01	Ditto	Ataxitic	—

predetermines the inevitable manifestation of thixotropic properties is a high content of silty and colloidal particles, enrichment of these with organic and organo-mineral compounds and high moisture (ice content) of mineral aggregates. The thixotropy of frozen soils on thawing promotes the washout and soaking processes.

Thermal expansion/contraction is a characteristic property of rock materials observed with changes of temperature and which is characterized by the coefficients of linear α and volume expansion β, representing, respectively, relative linear and cubic deformations arising due to a temperature change of 1 °C. Their relationship is reflected by the ratio $\beta = 3\alpha$. Thermal expansion/contraction in frozen materials is a key factor in the development of such processes as frost cracking – and formation of ice wedges, weathering and the like. Thermal deformations of frozen soils come about due to temperature deformations of soil components (minerals, rock fragments, water, air, ice), phase transitions water:ice and structural transformations of the material with temperature changes. The linear expansion coefficient of the majority of minerals that compose rock, is within $2–12 \times 10^{-6}$ °C^{-1}. Within this same range are the coefficients of linear expansion of magmatic, metamorphic and sedimentary-cemented rocks. Ice has higher values of α, of the order $30–60 \times 10^{-6}$ °C^{-1}, depending on ice structure, angle of the crystal optical axis, temperature interval and the like. The α value of unfrozen water estimated from the density of supercooled water, is $18–7.5 \times 10^{-6}$ °C^{-1} within the temperature range of 0 to -20°C.

Frost resistance of rocks is defined as their capability to sustain without collapse multiple freezing alternating with thawing. In practice, this characteristic is indispensable for the engineering-geological evaluation of magmatic, metamorphic and sedimentary-cemented rocks with rigid structural bonds. Frost resistance is estimated by the number of freeze-thaw cycles corresponding to reductions of strength. Usually, there are 25 cycles of tests, with specific studies requiring as many as 50 to 200. For construction purposes the number of freeze-thaw cycles (heating-cooling) resulting in 25% reduction of initial strength or 5% of mass is called the *frost resistance point*. Use is also made of the frost resistance coefficient K_{fre} – the ratio of the frozen rock compressive strength limit to that of dry samples.

Rock strength diminishes under the influence of negative temperatures with participation of the different factors of temperature, hydration and cryohydration. The most important factor of the frozen rock degradation under the influence of alternate cooling-heating is the cryohydration mechanism which is associated with phase transition of water into ice. Frost resistance of rocks is also dependent on thermal-physical properties and strength of rock-forming minerals, cohesion between individual grains, the type of rock wetting, the structural-textural peculiarities of the rock, degree of alteration etc.

Electric properties

Frozen rocks and soils are imperfect dielectrics, i.e. materials that possess simultaneously the properties of both dielectric and conductor. Under the influence of an electromagnetic field in these materials a directed translational motion of charge carriers (current of conductance) arises which provides their electric conductivity, as well as oscillations of bound charges (current of displacement) which determines their polarization. The chief parameters of electric properties of earth materials, the frozen included, are: specific electric resistance (SER) ρ or the inverse value – specific electric conductivity $\sigma = 1/\rho$, dielectric constant ε, coefficient of polarizability, η etc.

Specific electric resistance (Ohm m) is determined by the capacity of rocks and soils to conduct electric current, i.e. by their electric conductivity $\sigma = 1/\rho$ (Ohm^{-1} m^{-1}). The chief current-conducting component of frozen, unfrozen, and thawed soils is the pore solution. Its electric conductivity has an ionic origin. The gas phase, monocrystals of ice and rock-forming minerals belong to dielectrics and are characterized by a high specific electrical resistance.

The main cause of reduced electrical conductivity of soils in the frozen

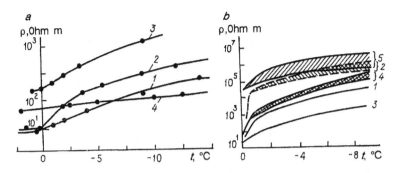

Fig. 8.1. Temperature dependence of specific electrical resistance, of frozen earth materials: (*a*) – rocks (according to M.S. King): 1–3 – sandstones of different moisture contents; 4 – clayey schists; (*b*) – soils (according to A.D. Frolov): 1 – sand with massive cryotexture; 2 – the same, with schlieren structure; 3 – sandy-silty materials and silty-clay materials with massive cryotexture; 4 – as 3, with schlieren; 5 – glacier ice.

state is the decreasing amount of the current-conducting component – the unfrozen water, with stretching and narrowing of conducting paths and their intermittent nature because of ice segregation in the material. Thus, the resistivity value of frozen ground is dependent on factors that determine the amount and pattern of unfrozen water distribution i.e., on composition, temperature, water (ice) content, salinity, cryogenic texture and the like (Fig. 8.1). With negative temperatures the resistivity may increase by several orders in a narrow temperature interval; the most drastic increase of values is observed in the interval of essential phase transition.

Dielectric constant (ε) of soils determines their capacity to polarize under the impact of an alternating electromagnetic field on account of the ordered orientation of bound electric charges available. Permittivity of the frozen materials is dependent on dielectric properties of their components. Relative dielectric permittivity of a gas component is equal to 1, as is that of a vacuum, while for the majority of rock-forming minerals it does not exceed 10. For free pure water ε is equal to about 80, i.e. an order higher than that of minerals, which explains the substantial influence of moisture content on the permittivity of earth materials. The dielectric constant of ice within the range of high frequencies (over 10^4–10^5 Hz) can be much lower than that of water.

With lower (negative) temperatures soil permittivity diminishes in general (Fig. 8.2) in conformity with the lower content of unfrozen water at lower temperature and with reduced ε values of the bound unfrozen water.

Frozen earth materials show imperfect elasticity. Under external

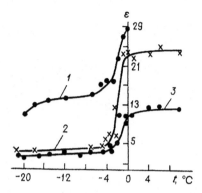

Fig. 8.2. Temperature dependence of dielectric constant of frozen soils of different composition and moisture content at $f = 10^6$ Hz (according to B.N. Dostovalov): 1 – clay $W_{tot} = 35.5\%$; 2–3 – sand with $W_{tot} = 9\%$ and 3%, respectively.

dynamic loads various elastic vibrations arise in them including longitudinal, transverse, surface-type and the like, differing by the pattern of displacement of surrounding particles. One of the dynamic methods to study elastic properties of frozen soils in natural situations and, mainly, in samples, is the acoustic (ultra-sound) method.

The most important acoustic parameters of rocks are the velocities of elastic wave propagation: V_p – longitudinal, V_s – transverse, V_R – surface. Velocities of longitudinal wave propagation have been studied most thoroughly.

As earth materials progress into the frozen state, velocities of propagation of elastic waves increase, being determined primarily by phase transitions at freezing and the development of the new component – ice. Ice is characterized by much greater velocities ($V_p = 3500$–4000 m s^{-1}) than water in a liquid phase ($V_p = 1450$ m s^{-1}). The velocity characteristics of frozen soils are also influenced by the type of ice segregation, although to a lesser degree than electric properties.

Taking into account the above velocities of elastic wave propagation in the frozen soils one may conclude that these are dependent on all the factors determining the amount and form of ice segregation and content of unfrozen water, namely, mineral composition and granulometry, porosity, water content, salinity of interstitial moisture, temperature, etc. (Fig. 8.3).

8.2   Thermal-physical properties of rocks

Thermal properties of rocks to a great extent control the occurrence of energy-mass-exchange properties, such as freezing and thawing,

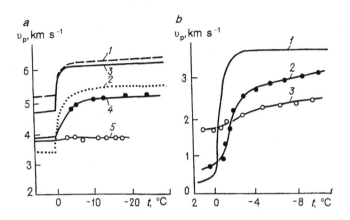

Fig. 8.3. Temperature dependence of longitudinal wave velocity v_p: (*a*) – rocks of different composition: 1 – monolithic basalt; 2 – the same, cavernous; 3 – granite, monolithic; 4 – saturated sandstone; 5 – the same, dry (according to 0.N. Voronkov, A. Timur); (*b*) – for soils of different composition: 1 – sand; 2 – silty-clay material; 3 – clay (according to F.F. Aptikayev).

cryogenic heaving, subsidence at thawing, thermokarst, thermal erosion, etc.

Thermal-physical properties of rocks with transfer of thermal energy by conduction are evaluated by three basic characteristics: heat capacity, heat conduction and thermal diffusivity.

Heat capacity is the amount of heat required for a unit of mass or volume to change temperature by 1°C. The following types of heat capacity are distinguished: specific heat C expressed in $J\,g^{-1}\,K^{-1}$, while heat capacity per unit volume $C_{vol} = CP$ in $J\,m^{-3}\,K^{-1}$. There is also a concept of *effective heat capacity* C_{ef} taking into account latent heat of phase transition. *Thermal conductivity* ($W\,m^{-1}\,K^{-1}$) characterizes the capability to convey thermal energy and is equal quantitatively to the heat flow passing through unit of area in unit time, when the temperature gradient is one unit. *Thermal diffusivity* is expressed by the coefficient (a, $m^2\,s^{-1}$), which serves as an index of temperature field inertia and is related to heat capacity and thermal conductivity by the ratio:

$$a = \lambda / C_{vol}$$

Thermal properties of earth materials are much dependent on composition, structure and state of the materials, i.e. on their genetic features and thermodynamic conditions of their particular setting. Table 8.2 gives data on thermal conductivity and heat capacity per unit volume of rocks, soils, water, ice, air and snow for general comparison and estimation. It should be

Table 8.2. *Thermal properties of earth materials*

Description	Thermal conductivity γ, $\mathrm{W\,m^{-1}\,K^{-1}}$	Thermal capacity C_{vol} $\mathrm{J\,m^{-3}\,K^{-1}}$
Water:		
+4.1°C	0.54	4180
+20°C	0.60	—
Ice	2.22–2.35	1930
Air:		
0°C	0.024	1–260
−23°C	0.022	—
Snow:		
loose	0.1	210
compact	0.3–0.4	420–630
Granites	2.3–4.1	1680–1810
Gabbro	1.74–2.91	2120–2240
Basalts	1.4–2.8	2270–2770
Peridotites	2.4–3.4	2240–2640
Quartzites	2.9–6.4	1780–1990
Slates	1.74–2.33	1850–1920
Sandstone	0.7–5.8	1130–2250
Limestone	0.8–4.1	1010–2010
Dolomite	1.1–5.2	1810–2840
Gypsum	0.8–1.3	1810–2020
Stone salt	7.2	1810
Coal	0.1–0.18	1560–1950
Rudaceous rocks:		
dry	0.23–0.35	1000–1900
water saturated thawed	1.1–2.1	2300–3200
the same frozen	1.4–3.1	1800–2300
Sands:		
dry 1200–1300	0.3–0.35	
water saturated thawed	0.7–2.6	1800–3200
the same frozen	1.5–3.0	1700–2200
Silty-clay loess like:		
dry	0.19–0.22	1200–1500
water saturated thawed	0.6–1.0	3000–3500
the same frozen	1.2–1.6	2000–2200
Clays:		
dry	0.8–1.0	1400–2200
water saturated	1.2–1.4	2800–3300
the same frozen	1.4–1.8	2000–2500
Peats:		
dry	0.012–0.14	100–150
water saturated	0.7–0.9	2400–3600
the same frozen	1.1–1.2	1600–2700

taken into account that the thermal conductivity of water, ice and air is linearly dependent on temperature. Thus, with a temperature drop of 1 K the coefficients λ of water and air diminish by 2×10^{-3} and 9×10^{-5}, respectively, while that of ice increases by 5×10^{-3} W m^{-1} K^{-1}.

Heat capacity values of rocks and minerals are rather steady and comparatively well studied experimentally. Specific heat capacities of soil components (mineral skeleton, ice, water-gas and peat) vary within a narrow range; $C_{sk} = 0.71$–0.88 kJ kg^{-1} K^{-1}; $C_i = 2.09$ kJ kg^{-1} K^{-1}; $C_w = 4.19$ kJ kg^{-1} K^{-1}; $C_g = 1.02$ kJ kg^{-1} K^{-1}; $C_{peat} = 0.8$–2.1 kJ kg^{-1} K^{-1}. Heat capacity of soils is an additive quantity and is the sum of the specific heat capacities of each component multiplied by the amount of their mass.

Thermal conductivity of frozen rocks and soils

Comparative analysis of experimental data shows that the thermal conductivity of intrusive rocks increases from 2 to 5 W m^{-1} K^{-1} in a series dunites – gabbro – syenite – diorites – granites, i.e. from basic to acid rocks. This is explained by the difference in SiO$_2$ content – the higher the content, the greater is the thermal conductivity. Thermal conductivity of extrusive rocks is also dependent on their chemical-mineral composition and degree of crystallization and varies, as shown by experimental data, within the range 2.0–3.6 W m^{-1} K^{-1}. With higher SiO$_2$ content their thermal conductivity increases in the series: porphyries – andesites – trachytes – basalts. As shown by analysis of thermal conductivity of metamorphic rocks, λ varies within a wide-range – from 0.8 to 7.4 W m^{-1} K^{-1}; it increases from slates to gneiss to quartzites which is explained by the gradual disappearance of schistosity in this series. Thermal conductivity of sedimentary cemented rocks differs substantially between three subgroups: 1) cemented rock waste; 2) cemented silty and clayey materials; 3) chemical and biochemical. The first one is represented by rudaceous and coarse rocks – conglomerates, gritstones and sandstones with λ from 1.5 to 4.5 W m^{-1} K^{-1}. The range is determined by the particular thermal conductivity of the fragments and cement.

Thermal conductivity of unfrozen silty and clay-rich cemented materials represented by aleurolites and argillites is, on the average, lower than that of rudaceous and fragmental rocks and varies in the range from 0.8 to 2.2 W m^{-1} K^{-1}. This is explained by their finer-grained structure for which a greater number of contact thermal resistances is typical.

The subgroups of chemical and biochemical rocks, for example, siliceous rocks of marine origin (tripoli, diatomite) are, in general, characterized by lower thermal conductivity as compared with all the above, 0.8–1.7 W m^{-1}

K^{-1}, which is explained by the high porosity in combination with low thermal conductivity of the skeleton of these rocks. Such monomineral rocks as dolomite and anhydrite are characterized by the highest thermal conductivity, respectively, 7.2–11.9 and 3.7–5.8 W m^{-1} K^{-1}, while lime-stones, 5.7 W m^{-1} K^{-1}, and marls 2.6 W m^{-1} K^{-1}, have a lower thermal conductivity.

Rudaceous rocks, being multi-component and multi-phase systems, have a wide range of thermal conductivities. The upper limit of thermal conductivity for rudaceous rocks is that of hard rock fragments (up to 3–9 W m^{-1} K^{-1}), while the lower limit (0.3–0.5 W m^{-1} K^{-1}) is set by the thermal properties of fine-grained material. With higher content of big fragments λ increases as the thermal conductivity of large fragments is greater than that of fine-grained rock waste. Rudaceous rocks with high moisture content have higher thermal conductivity in the frozen state compared with the thawed; this is associated with transition of ground water into ice and the four-fold increase of its thermal conductivity.

Variation of the thermal conductivity of rudaceous rocks is to a significant degree dictated by phase composition of moisture, dependent on temperature and type of infilling material. Thus, within the temperature interval from -10 to $-1\,°C$, λ of rudaceous materials with sandy infilling is practically constant. This is explained by the fact that the main phase transformations of moisture in this material take place in the range of 0 to $-1\,°C$. Below $-1\,°C$ the ice to unfrozen water ratio virtually does not change. In contrast, the thermal conductivity of materials with sandy silty and silty clay infilling diminishes with higher temperature, the rate of reduction depending on the liquid phase/ice ratio. A marked increase of λ, equalling 25–30% on the average occurs between 0 and $-5\,°C$. Within the -5 to $-10\,°C$ range thermal conductivity is assumed to be constant since phase composition of the moisture does not vary. Within the positive temperature range (0–25 °C) λ of rudaceous rocks increases linearly, but insignificantly, associated with the linear dependence on water temperature.

It should be noted that thermal-physical characteristics of rudaceous rocks with a different type of infilling become more uniform with higher content of coarse particles. This is explained by the fact that material characterized by different coefficients of thermal conductivity (sand, sandy silty, and silty clay-rich materials) are gradually replaced by rock waste having uniform (unchanging) values of thermal conductivity.

With other conditions being equal, thermal conductivity of soils diminishes with decrease of grain size in the following sequence: rudaceous – sandy – sandy-silty – loess – silty-clayey – clay – peat (Fig. 8.4). A smaller

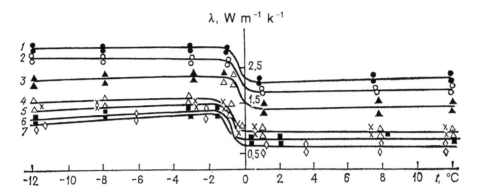

Fig. 8.4. Temperature dependence of thermal conductivity of materials of different grain size: 1 – rudaceous rock waste with sandy silty infilling material; 2 – fine sand; 3 – light fine sandy-clayey material; 4 – loess-like silty clay material; 5 – medium silty-clay; 6 – clay; 7 – well-degraded peat.

size of grains leads to a greater number of contact thermal resistances and is accompanied by the increase of hydrophilic nature and of ultraporosity which contribute to the higher content of liquid phase characterized by lower thermal conductivity compared to that of ice. This principle, as a rule, applies within the whole temperature range (from $+20$ to $-20\,°C$) including the range of intense phase transitions of moisture; it applies for soils with various moisture content.

The mineral composition of soils exerts a general influence over their thermal conductivity, although indirectly, via factors of rock structure. Thus, Fig. 8.5a shows an almost twofold difference in thermal conductivities of sandy silty materials of quartz and of vermiculite compositions in the temperature range -15 to $+15\,°C$. It is explained by the difference in thermal conductivities of quartz ($\sim 5\,W\,m^{-1}\,K^{-1}$) and vermiculite ($\sim 2\,W\,m^{-1}\,K^{-1}$). Apart from mineral composition the value is influenced by organic matter content, because organic matter is characterized by a comparatively low thermal conductivity: for peat $\lambda = 0.46\,W\,m^{-1}\,K^{-1}$. Therefore, the higher the peat content, the lower the thermal conductivity (see Fig. 8.5b).

One of the most important factors determining thermal properties of soils is their salinity, which influences their phase composition and structural transformations. Salinity leads to the increase of the liquid phase content thereby reducing thermal conductivity.

Since thermal conductivity of the soil mineral skeleton is, as a rule, higher than that of water and ice, compaction is accompanied by higher thermal conductivity. Wetting of a soil leads to substantial increase of λ since air

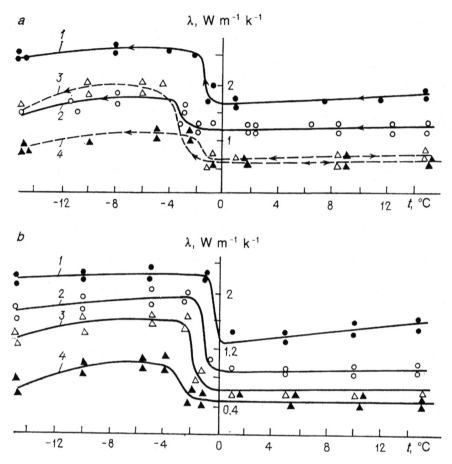

Fig. 8.5. Temperature dependence of thermal conductivity of soils of different composition: (*a*) – saturated soils of different mineral composition; 1–2 – heavy-sandy silty material (1 – quartz, 2 – vermiculite); 3–4 – clay (3 – montmorillonite, 4 – kaolinite); (*b*) – sand with various peat contents and peat: 1–3 – sand (1 – $W = 18.6\%$, $n = 0.94$, $\rho_d = 1.6\,\mathrm{g\,cm^{-3}}$, $q = 0$; 2 – $W = 66.5\%$; $n = 0.80$, $\rho_d = 0.81\,\mathrm{g\,cm^{-3}}$, $q = 0.25$; 3 – $W = 78.3\%$, $n = 0.80$, $\rho_d = 0.67\,\mathrm{g\,cm^{-3}}$, $q = 0.40$); 4 – peat, $W = 116.5\%$, $n = 0.80$, $o_d = 0.57\,\mathrm{g\,cm^{-3}}$.

having a low thermal conductivity ($0.023\,\mathrm{W\,m^{-1}\,K^{-1}}$) is replaced by water with a higher thermal conductivity ($0.57\,\mathrm{W\,m^{-1}\,K^{-1}}$) or ice ($2.29\,\mathrm{W\,m^{-1}\,K^{-1}}$).

Unlike heat capacity, thermal conductivity of soils is not an additive quantity which makes λ dependent on factors that determine soil structure, i.e. textural aspects. Of paramount importance is the manner of heat transfer in the soil – directly by particles, from particle to particle via contacts or from particle to particle through an intermediate medium.

Cryogenic structure also belongs to the factors responsible for soil texture and exercises substantial influence on soil thermal conductivity. As shown by experiments, at similar values of moisture content and density, clays having massive cryogenic structure are characterized by a higher thermal conductivity than those having schlieren structure. Account should be taken of the anisotropy of thermal properties for soils with schlieren cryogenic texture, which manifests itself in higher values of λ (usually by 20–30%) when heat flows along the ice lenses rather than across them. Peat has a noticeable anisotropy of thermal properties.

8.3   Moisture exchange properties of soils

Analysis of the state of moisture in the soil and its capacity to move is the first priority in the assessment of mass-exchange characteristics of hydraulic conductivity, mass capacity, coefficient of diffusion and moisture potential. It was shown by experiment that with higher moisture content the values of ground water potential and differential water capacity increased. Low absolute values of the chemical potential of the moisture of high moisture content materials indicate small amounts of bound capillary moisture. Being situated in the largest capillaries of the soil the water does not differ substantially from free water with respect to energy potential. Then, as moisture comes out of finer capillaries and pores, μ_ω falls greatly, i.e. moisture becomes more cohesive to the soil. The value C_ω (the volumetric differential moisture capacity) diminishes insignificantly. The transition from macro-capillary moisture to micro-capillary, for soils, is more distinctly revealed in inflections of the curves of the soil volumetric differential moisture capacity.

Grain-size variation (see Fig. 2.1) brings about changes in the energy of cohesion to the soil mineral fraction for a given moisture content – as much as one or two orders and more. Changes of granulometry giving smaller sizes of mineral particles enlarge the specific active surface of the soil and lead to ultra-porosity. As a result, the same amount of water is, energetically, more bound by the mineral surfaces of the soil, as it occurs in finer films and pores and is characterized by a lower value of moisture potential. Finer-grained soils contain more fine pores so that water-filled capillaries and ultrapores have menisci of greater curvature. The latter serve as an additional factor leading to the growth of cohesion energy of pore moisture in the case of small size of particles.

Chemical-mineral composition exerts influence over μ_ω and C_ω first of all through the differing surface energies of the mineral skeleton as well as through differences determined by mineral composition, in grain size, specific active surface, differential and total porosity of soils and other characteristics. Thus, hydromica-montmorillonite clay have much smaller values of moisture potential and differential water capacity than kaolinite clay (see Fig. 2.1).

Consolidation of unfrozen soil according to experimental data, leads to a substantial increase (at the same moisture content) of chemical potential of the moisture (see Fig. 2.1). Total porosity of soils is reduced in the course of consolidation at the expense of larger pores (those less strong), small pores being only slightly disturbed which leads to partial transfer of film moisture into capillary moisture, thus increasing the maximum radius of water-filled capillaries. With respect to frozen soils, lowering of temperature below freezing point is accompanied by gradual crystallization of water over a range of temperature with reduction of its amount and, accordingly, potential.

Coefficients of water transfer for soils are dependent on moisture content, density, granulometry and chemical-mineral composition as well as structure and temperature.

These coefficients increase when fine-grained unfrozen soils grade into coarse-grained (see Figs. 2.1 and 8.6). Thus, saturated sands which are characterized by the presence of poorly bound capillary moisture, have water-transfer coefficients exceeding those of clay rich soils by 100 times and more. While in clay-rich soils the K_ω values are lower than those of coarser (at the same porosity and moisture content) this is due to greater values of ultraporosity and larger specific active surface of mineral particles which is typical of fine-grained soils in general with more fine water-conveying pores. At low moisture content this leads to the situation where the same amount of film moisture in clay-rich soils (Fig. 8.7, curve 5, section III) is energetically more bound and, therefore, less mobile, characterized by less intense translational motion of molecules. At high moisture content when a capillary mechanism of water transfer prevails (section I) the energy of cohesion and mobility of pore moisture are in direct proportion to the square of the radius of the active water-filled capillary.

Water-conveying properties of a soil are much dependent on mineral composition. Thus, the diffusion coefficient in montmorillonite clay at $W_{vol} = 0.46 \, \text{g cm}^{-3}$ (see Fig. 8.6) was as low as one tenth that of hydromica-montmorillonite clay and one fifteenth that of kaolinite. The above is associated with the fact that moisture in pores of montmorillonite clay is energetically more bound and, accordingly, has a lower mobility thereby reducing the coefficient of moisture transfer. In general, with higher content of silty, clayey particles and peat in the soils as well as minerals of the

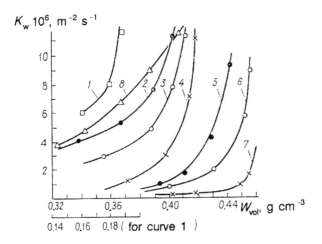

Fig. 8.6. Dependence of moisture diffusion coefficient K_w on granulometry and mineral composition: 1 – medium-grained sand with porosity $n = 33$–38%, 2 – sandy silty material, 3 – silty-clay material, 4 – clay with porosity $n = 42\%$; 5, 6, 7 – kaolinite, hydromica, montmorillonite clays respectively, with $n = 46\%$, 8 – peat – macellanicum, $\rho_d = 0.075\,\mathrm{g\,cm^{-3}}$ (degree of degradation 25%, mineralization 3%).

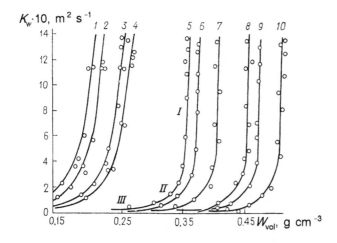

Fig. 8.7. Dependence of moisture diffusion coefficient on moisture content of soils of different granulometry and density: 1–4 – sandy silty material, $\rho_d = 1.98, 1.93, 1.82, 1.77\,\mathrm{g\,cm^{-3}}$ respectively; 5–10 – clay, $\rho_d = 1.69, 1.65, 1.55, 1.41, 1.35, 1.25\,\mathrm{g}$ $\mathrm{cm^{-3}}$ respectively.

montmorillonite group, the potentials of moisture and coefficients of moisture transfer diminish by one to two orders and more.

In considering the variation law of mass exchange properties one should bear in mind porosity of the material. Reduced porosity of the soil due to its compaction leads (at a constant moisture content) to higher moisture potential (see Fig. 2.1) and, consequently, higher coefficients of moisture transfer (see Fig. 8.6). As shown by experiment, the curves $K_\omega(W_{vol})$ are similar for soils having different densities and they are shifted on the abscissa in relation to each other in proportion to the change of density ρ. The effect of temperature on moisture mobility diminishes with greater dispersion of soils (13). Thus, with a temperature rise of $10\,^\circ C$ K_ω increases, in sandy silty materials by 15%, in more clay-rich materials by 10% and in clay by only 5%. This can be explained in that the same amount of pore moisture in clayey soils is subject to greater energetic influence of mineral surface than in more sandy materials and is therefore to a lesser degree subject to the effect of temperature.

The effect of temperature on the water-transmitting properties of frozen soils has much significance as the temperature determines the amount of unfrozen water in them, the thickness of liquid films and extent of water saturated capillaries and, correspondingly, mobility of the liquid phase. Fig. 8.8 shows the dependence of the moisture diffusion coefficient on volumetric moisture content, obtained experimentally for soils of various granulometry and mineral composition. The left part of these curves located within the range of highest moisture content refers to positive temperatures (10 to $20\,^\circ C$), i.e. to the unfrozen water-saturated materials. The right part was obtained for negative temperatures, that is, for the frozen soils with a degree of saturation, ice and unfrozen water, of one. The gently sloping pattern of the curve of K_ω against W_{vol} allows us to assume that K_ω is gradually falling on freezing. This is explained by a uniform film mechanism of moisture transfer in unfrozen and frozen soils. An extreme pattern of the K_ω versus W_{vol} curve for the frozen soils is associated with ice segregation in the frozen ground.

8.4   Mechanical properties of frozen ground

Mechanical (deformation and strength) properties of frozen soils are usually expressed by quantitative coefficients that determine a functional relationship between the value and type of mechanical action and response of the material. Deformation characteristics of frozen soils comprise moduli of total and elastic deformation, Poisson's ratio, indices of rheologic curves of flow and curves of creep, coefficients of viscosity and compressibility; strength characteristics are short-term and long-term values of shear strength (coefficients of friction and cohesion), compression, tension and equivalent cohesion.

The relationship between stress and deformation of frozen soils during

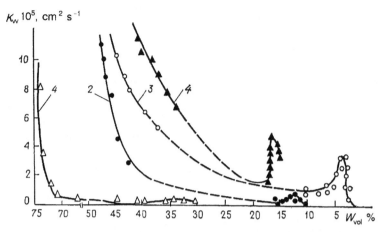

Fig. 8.8. Dependence of diffusion coefficient K_w on moisture content of unfrozen and frozen soils: 1 – silty clay material (see extreme right of graph); 2–4 – clay (2 – polymineral, 3 – kaolinite, 4 – montmorillonite). Freezing zone of the ground is shown by the dotted line.

creep at constant temperature is non-linear and can be described by a power function. With higher negative temperatures and with a progression from coarse-grained to finer-grained soils the same value of relative deformation is observed at smaller stress. Thus, when temperature is increased from -20 to $-5°C$ similar values can be observed at stresses 3 to 4 times less. The relationship between stress σ and deformation ε of the frozen material under uniaxial compression can be expressed by the modulus of total deformation as follows:

$$\sigma = E(\sigma, \tau)\,\varepsilon,$$

where E is the modulus of total deformation (Pa) on compression dependent on time τ and stress σ. Deformation ε is expressed in fractions of a unit. Modulus E represents the tangent of the angle to the deformation axis drawn through ordinate point σ_i corresponding to value τ_i of the isochronous curve.

Modulus of total deformation has a definite physical meaning directly reflecting resistance of the frozen soil to deformation. It diminishes with greater stress and longer time of load (Table 8.3). Higher dispersion (finer grain size) and higher temperature of the frozen soil also lead to reduction of total deformation modulus. Among factors that exert an influence on total deformation modulus are physical-mineral composition, moisture content, degree of saturation with ice etc.

The relationship between deformations and stresses in the elastic region is expressed by the modulus of normal (elongation) elasticity – Young's

Table 8.3. *Modulus of total deformation of frozen soils* ($t = -2°C$) *in uniaxial compression (after Ye. P. Shusherina)*

Rock (moisture content %)	Duration of loading	Load range MPa	E, Mpa
Sand (27)	1 min	0–0.5	213
		0.5–1.0	208
		1.0–2.0	39
	1 hour	0–0.5	102
		0.5–1.0	82
		1.0–1.75	6
	8 hours	0–0.5	78
		0.5–1.0	23
		1.0–1.25	3
Sandy-silty (26)	2 min	0–0.5	620
	24 min	0–0.5	450
Silty-clay (26)	2 min	0–0.5	280
	24 hours	0–0.5	188
Clay (26)	2 min	0–0.5	280
	24 hours	0–0.5	89
	240 hours	0–0.2	72
	480 hours	0–0.2	56
Ice	2 min	0–0.5	500
	24 hours	0–0.5	86
	240 hours	0–0.2	23
	480 hours	0–0.2	15

modulus (E_e, Pa) and the coefficient of transverse elasticity (Poisson's ratio μ) of the frozen soils.

Modulus of normal elasticity ($E_e = \sigma/\varepsilon$) of frozen soils is within the range 300 to 30 000 MPa, which is 10 to 100 times greater than this modulus in unfrozen soils and its value is dependent on a number of factors: composition, structure, temperature of the frozen soil and external stress. At an adequately low temperature the elasticity modulus of sandy-clayey frozen soils can exceed that of concrete.

The modulus of normal elasticity increases as the amount of the fine granular component of the frozen soils decreases. The highest values are typical of the frozen sand (from 820 to 22500 MPa, from -0.2 to $-10.2°C$), the lowest of the frozen clay (from 680 to 2780 MPa, from $t = -1.2$ to $t = 8.4°C$). The E_e values of silty-sandy-clay materials and clay-rich materials occupy the intermediate range. Temperature is the key factor influencing the value of modulus of elasticity of frozen soils (Fig. 8.9). The

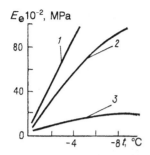

Fig. 8.9. Temperature dependence of the modulus of elasticity of frozen soils (according to N.A. Tsytovich): 1 – sand; 2 – silty sandy-clay material; 3 – clay at 0.2 MPa load.

modulus of elasticity of ice is less than that of ground with a rigid mineral skeleton (sand), but is much higher than that of frozen clays which have a greater amount of unfrozen water.

Upon axial deformation of a frozen soil specimen there is a change of cross-sectional area: at compression it increases, on stretching it narrows. Therefore, in order to describe the stress-deformational state of the frozen soil in the elastic region Hooke's law is not sufficient. One more relationship is necessary by which the ratio of transverse and longitudinal deformations can be expressed: according to Poisson's law, $\varepsilon_2 = \varepsilon_3 = -\mu\varepsilon_1$, where ε_1 is the relative longitudinal deformation, while ε_2 and ε_3 are transverse deformations; μ is the proportionality factor or Poisson's ratio. The latter serves as a second important characteristic of elastic material (Table 8.4).

Transverse deformation under the effect of longitudinal force arises owing to bonds existing between particles of the frozen soil, therefore the coefficient is also an index characterizing volumetric deformation of the frozen soil.

The data presented in Table 8.4 show the substantial influence of temperature on Poisson's ratio for frozen soils, which with higher temperature tends to a maximum value of 0.5 (as for ideally plastic bodies), while with lowering of temperature, to the values typical of a rigid body.

Among indices of deformability of frozen soils are the coefficients of viscosity and compressibility.

Viscosity represents one of the main rheologic properties of frozen soils. It is represented by the coefficient of effective viscosity η (hereafter viscosity coefficient) which is equal to the ratio of the stress value acting, σ, to the resulting rate of flow $\dot{\varepsilon}$, that is, $\eta = p/\varepsilon$. The viscosity coefficient or simply viscosity is measured in $N\,s\,m^{-2}$ (Pa s).

With respect to the two straight parts of the rheologic curve for the frozen

Table 8.4. *Poisson's ratio for frozen soils (after N.A. Tsytovich)*

Soils	W_{tot} %	t, °C	σ, MPa	μ
Frozen sand	19.0	−0.2	0.2	0.41
	19.0	−0.8	0.6	0.13
Frozen silty clay	28.0	−0.3	0.15	0.35
	28.0	−0.3	0.2	0.18
	25.3	−1.5	0.2	0.14
	28.7	−4.0	0.6	0.13
Frozen clay	50.1	−0.5	0.2	0.45
	53.4	−1.7	0.4	0.35
	54.8	−0.5	1.2	0.26

soils (see Fig. 3.11a) higher η_{sh} and lower η_b plastic viscosities can be distinguished. The larger viscosity (Shvedov) is determined from the equation $\eta_{sh} = (\sigma - \sigma_{cr})/\dot{\varepsilon}$, while the lower one (Bingham) from $\eta_b = (\sigma - \sigma'_{cr})/\dot{\varepsilon}$. Fig. 3.11a shows that plastic viscosity is equal to the cotangent of the angle of inclination of the corresponding straight part of the rheologic curve to the stress axis. Critical stresses σ_{cr} and σ'_{cr} are called, respectively, *conditional static* and *conditional dynamic yield points*. The ratio of the plastic viscosities is considered to be an important characteristic of the soil, reflecting the degree of its structural disturbance in the course of flow caused by variations of stresses. The greatest plastic viscosity is associated with the flow of soil having practically undisturbed structure, while the lowest one corresponds to the deformation of soil with effectively broken structure (4). The structure is assumed to be disturbed most intensively within the limits of transition (between Shvedov and Bingham) in the curved part of the rheologic curve. Wide variations of soil viscosity occur with lowering of temperature, especially in the range of the basic phase transformations of moisture. This is caused by general strengthening of soil as a result of the development of ice-cement bonds. Absolute values of viscosity accordingly increase by 100 to 1000 times and more. Simultaneously, there are variations of location and configuration of rheologic curves: they shift and stretch along the abscissa towards higher stress. Stretching of curves is associated with the widened range of stresses within the limits of which Shvedov and transitional sites are located. In many cases the interval of stresses for the Bingham part is reduced at the same time. These variations are induced by stronger ice-cement cohesion in the soil, diminished amount of unfrozen water and structural transformations of the mineral skeleton. The expansion of the stress interval for the Shvedov part of the curve, i.e. at low values of stress

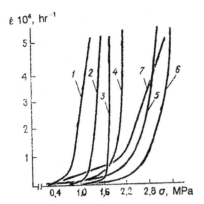

Fig. 8.10. Rheologic curves for flow deformation of soils of different grain size and mineral composition (uniaxial compression with constant velocities) (according to Y.V. Kuleshov): 1–3 – clays (1 – bentonite; 2 – *poly mineral*, 3 – kaolinite); 4 – silty clayey material; 5 – sandy silty material; 6 – sand; 7 – ice.

and rate of deformation, is caused by the slower rise of the yield point σ_{cr} with lower temperature compared with other bounding stresses. At the same time a diminished Bingham part is explained by higher fragility of the frozen soil which prevents visco-plastic deformations.

It has been confirmed by many studies that frozen soils of different granulometry and mineral composition have different resistance to the development of flow deformation (Fig. 8.10) and, accordingly, their effective viscosity coefficients differ. In a general case, other conditions being equal, the viscosity coefficient increases with coarser grain size and higher rigidity of mineral skeleton. First of all, it is induced by increase of their short-term strength. At the same time, according to the data furnished by N.K. Pekar-skaya, obtained after creep tests, at small stresses and fairly low negative temperatures to $-30\,°C$, long-term strength of clays usually exceeds that of sands. This is associated with the effect of primary structural cohesion of mineral particles, typical of fine-grained materials. At low rates of deformation the values of effective viscosity for clays and sands can become similar.

Data from investigations show that moisture content (ice content) affects viscosity of the frozen materials greatly and this influence has an extreme nature. In general, when pores are not filled completely ($G < 1$), soil viscosity increases with higher ice content, reaching a maximum value at $G = 0.8$ to 0.9. After that, viscosity diminishes and with still higher ice content tends to reach the value of the viscosity of ice.

Frozen fine-grained soils under the pressure of overburden or overlying engineering structures become consolidated as a result of complex physical-mechanical and physico-chemical processes. As shown in the experiments of S.S. Vyalov and N.A. Tsytovich and later A.G. Brodskaya, the ground has a substantial compressibility under load. According to the investigations, consolidation of the soil is conditioned by deformability and displacement of each component: gaseous, liquid (unfrozen water) and plastic-viscous (ice and solid mineral particles). In general, the compression curve (curve of consolidation) of the frozen soil has a pattern as shown in Fig. 8.11.

Three main parts can be distinguished in the compression curve of frozen soil $\alpha\alpha_1$, $\alpha_1\alpha_2$ and $\alpha_2\alpha_3$. The part $\alpha\alpha_1$ is the elastic and structurally recoverable deformation. The pressure value corresponding to point α_1 is near to the structural strength of the frozen soil and when this is exceeded compaction begins and the porosity decreases. The part $\alpha_1\alpha_2$ of the compression curve is the structurally irreversible main deformation and is up 70 to 90% of the total deformation. The subsequent part of the curve represents a strengthening at greater loads.

The total (stabilized) subsidence of frozen ground due to consolidation α_0^ε is determined in compression tests as the value of the relative compressibility coefficient of the frozen ground, using the equation $\alpha_0^\varepsilon = S_\infty / (h P)$, where S_∞ is the stabilized subsidence of the soil layer with the controlled negative temperature; h is thickness of the soil layer in the oedometer; P is the applied pressure.

Taking into consideration mineral and granulometric composition frozen soils can be arranged in the following series according to compressibility: montmorillonite clay > polymineral clay > kaolinite > sandy silty material > sand (Fig. 8.12, Table 8.5). Thus, at a pressure of 0.3 MPa and a temperature of $-1.5\,°C$ the compressibility coefficients of montmorillonite clay are 1.5 times that of kaolin, 1.2 times that of polymineral clay and almost twice that of sand. Coarser materials are less subject to deformation by compression. The high content of unfrozen water in montmorillonite clays contributes to more complete and longer development of creep processes in them. Of particular importance are irreversible displacements of mineral particles and aggregates associated with both inter- and intra-aggregate porosity. The microaggregates themselves undergo substantial transformations, losing their thin coating of fine particles, and becoming compact and dense.

Salinity of frozen soils drastically increases their compressibility owing to the development of seepage-migration stages of deformation. This is explained by the higher content of unfrozen water due to salinity and the reduced ice component which makes the properties of the frozen soils nearer to those of unfrozen ones.

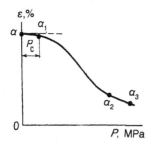

Fig. 8.11. Compression curve for ice-rich frozen soil (t = constant) (according to A.G. Brodskaya).

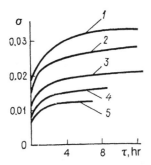

Fig. 8.12. Consolidation curves for soils of different granulometry and mineral composition at $t = -1.5°C$, $P = 0.3\,MPa$: 1–3 – clays (1 – montmorillonite, 2 – polymineral, 3 – kaolinite); 4 – sandy silty material; 5 – sand.

The compressibility of frozen soils of different lithologic type and salinity as well as length of time of consolidation, is reduced at lower negative temperatures. The reduced compressibility is explained by the reduced content of unfrozen water and thinning of the unfrozen water films, stronger bonds at contacts of structural elements and the higher strength of ice and frozen material in general.

The strength of frozen soils determines their capacity to resist disturbance. Disturbance occurs by development of microshears and microfissures in the frozen soil which accumulate to give rise to big fractures that completely break the frozen soil. Strength properties of frozen soils include the following: temporary and long-term resistance to compression (δ_t^{com}, δ_{lon}^{com}), tensile or rupture strength (δ_t^{rup}, δ_{lon}^{rup}), shear strength (S_t, S_{lon}) and equivalent cohesion determined by a cone penetrometer (C_t^{eq}, C_{lon}^{eq}).

Shear resistance of frozen soils is dependent on normal pressure P and, accordingly, is determined not only by forces of cohesion but also by

Table 8.5. *Values of summarized coefficients or relative compressibility for different frozen soils (after N.A. Tsytovich)*

Soil	W_{tot} %	W_{unf} %	ρ g cm^{-3}	$t°C$	$a_o^\varepsilon \times 10^5$ MPa^{-1} in the intervals of pressure steps, Mpa				
					0–0.1	0.1–0.2	0.2–0.4	0.4–0.6	0.6–0.8
Medium grained sand	21	0.2	1.99	−0.6	12	9	6	4	3
Ditto	27	0.0	1.87	−4.2	17	13	10	7	5
Ditto	27	0.2	1.86	−0.4	32	26	14	8	5
Heavy sandy silty, massive texture	25	5.2	1.90	−3.5	6	14	18	22	23
Ditto	27	8.0	1.88	−0.4	24	29	26	18	14
Medium silty-clay massive texture	35	12.8	1.83	−4.0	8	15	26	28	24
Ditto	32	17.7	1.84	−0.4	36	42	37	21	14
Medium silty-clay, netted texture	42	11.6	1.71	−3.8	5	10	18	42	32
Ditto	38	16.1	—	−0.4	56	59	39	24	16
Medium silty-clay laminated texture	104	11.6	1.36	−3.6	54	54	59	44	34
Ditto	92	16.1	1.43	−0.4	191	137	74	36	18
Banded clay	36	12.9	1.84	−3.6	15	22	26	23	19
Ditto	34	27.0	1.87	−0.4	32	30	25	20	16

internal friction and can be expressed by the equation $S = C = \text{tg } \phi \ P$, where C is cohesion; ϕ is the angle of internal friction.

In the majority of cases the total strength of the frozen soil increases with lower temperature. This is valid for practically all types of soils under all types of tests (Fig. 8.13). It is explained first of all by the smaller amount of unfrozen water due to the lower temperature and the simultaneous increase of close-contact ice, intensified ice-cement cohesion and strengthening of the ice itself.

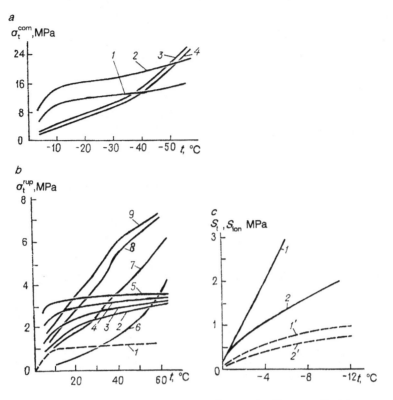

Fig. 8.13. Temperature dependence of strength of frozen soils: (*a*) – compression strength (1 – ice, 2 – sand, 3 and 4 – clay, with natural and with disturbed structure respectively; according to Ye.P. Shusherina, I.N. Ivashchenko, V.V. Vrachov); (*b*) – tensile strength (1 – ice, 2–5 – sand with moisture contents respectively 10, 12, 15, 18%; 6–9 – silty clay material with moisture contents respectively 12, 15, 18, and 20%; according to E.P. Shusherina); (c) – shear strength (1, 1′ – silty clay instantaneous strength and long-term resistance, respectively; 2, 2′ – clay, the same; according to N.A. Tsytovich).

The strength of frozen soils increases going from clay to sand. However, this applies only to a certain temperature point. Thus, strength of the frozen clay begins to exceed that of the frozen sand at a temperature of $-50°C$ and lower. The nature of the dependence $\sigma(t)$ for the frozen sand is to a greater extent dictated by ice content which is shown by similar curves (see Fig. 8.13a, curves 1 and 2, and 8.13b, curve 1, and curves 2 to 5). The higher strength of the frozen sand compared to ice is explained by more intensified adhesion bonds of unfrozen water, strengthening of the ice due to reduced defects and development of close-contact ice, as well as the hampering of the growth of microstresses in microfissure openings in the strong sand particles and viscous films of unfrozen water.

At a temperature below $-30°C$, when the role of phase transformations is reduced, a drastic increase of strength in clay-rich ground is observed which is due to a qualitative transition of coagulation-condensation bonds into crystallization ones as a result of cementation of contacts between particles of different chemical compounds. The ion-electrostatic interaction is likely to exert influence in the latter case. With higher total moisture content in the region of $W \ll W_o$ (where W_o is full saturation with water) resistance to compression for all frozen soils increases, whereas in the case of full saturation with ice and frost heave of the soil this value, as a rule, diminishes (20). The nature of the dependence of shear resistance on total moisture content at low temperatures (from -10 to $-55°C$), thoroughly studied by Ye.P. Shusherina, is essentially the same for all types of frozen soils, namely: shear resistance increases when saturation with water is not complete and structure is loose, which is explained by the development of ice-cement cohesion. In the case of full saturation with ice and where W_o is exceeded, the frozen soil resistance diminishes approaching the shear resistance of ice.

The salinity of frozen soils causes substantial reduction of their strength. This is explained by the greater amount of unfrozen water and reduced ice content of the frozen soils caused by the greater concentrations of dissolved salts (at a given temperature). In addition the structure and strength of the ice formed are dependent on the concentration of the pore solution.

Cryogenic structure has a fundamental effect on mechanical properties. At high negative temperatures, frozen clay, having a great amount of unfrozen water and massive cryotexture, is characterized by lower values of shear resistance, cohesion and friction compared to that with cellular cryotexture, for which it is typical that strength increases with greater thickness of ice schlieren if the shear plane is perpendicular to the lamination. Shear resistance of the soil with laminated cryogenic texture is greater, the larger the portion of the shear going through ice.

Frozen ground strength is to a significant extent dependent on time of loading, diminishing as it increases. Thus, at $t = -10°C$, S_t of a particular clay was 1.8 MPa, while $S_{lon} = 0.68$ MPa, i.e. one third. Shear resistance is reduced mainly because of reduction of cohesion forces of the frozen soil and, partially, by reduction of the internal angle of friction. Strength diminishes to a certain limit, the so-called limit of long-term strength.

As shown by experiments, the long-term resistance limit is much lower than, sometimes a fifth or a tenth of, the short-term resistance to compression. Thus, according to the data of S.Ye. Grechishchev, at a temperature $-3°C$, frozen sand ($W_{tot} = 19.8\%$), $\sigma_t^{com} = 75 \times 10^{5\,Pa}$ and

$\sigma_{\text{lon}}^{\text{com}} = 6.5 \times 10^5$ Pa, (i.e. one eleventh), frozen silty clay material ($W_{\text{tot}} = 31.8\%$), $\sigma_{\text{t}}^{\text{com}} = 35 \times 10^5$ Pa and $\sigma_{\text{lon}}^{\text{com}} = 3.6 \times 10^5$ Pa.

Experiments determining long-term resistance to compression of frozen ground under long-term load action showed that with greater load, deformation subsides more slowly: for example, at 0.25 MPa, deformation of silty-clay material ceased in 3 days; at 0.5 MPa, only in 10 days. Experiments by S.S. Vialov confirmed that frozen and permafrost ground possess long-term tensile strength. Thus, frozen silty sandy material at 31% moisture content and $t = -4.3\,°C$, having $\sigma_{\text{t}}^{\text{rup}} = 2.0$ MPa, was not broken during 6 years under a stretching stress of 0.18 MPa.

9

Characteristics of the basic genetic types of frozen ground

9.1 Features of the cryogenic types of frozen strata

Currently the strata of perennially frozen materials are subdivided into two basic types with respect to conditions of their freezing, namely, epicryogenic and syncryogenic. Under certain conditions (locally) a third type can be distinguished – diacryogenic perennially frozen materials.

In the majority of cases the frozen strata represent different combinations of these types and are then called polycryogenic (polygenetic) types.

Epicryogenic frozen strata are formed by freezing (usually from the top downwards) of lithified materials in which complex diagenetic physical-chemical processes have already taken place. With respect to their cryogenic structure they are subdivided into epicryogenic Pre-Quaternary bedrock (with rigid bonds, monolithic and others) and epicryogenic Neogene-Quaternary materials of the loose mantle.

Syncryogenic strata are formed by processes of sediment accumulation and freezing occurring simultaneously (synchronous in a geological sense). Therefore, frozen strata of this kind can be represented only by loose, Quaternary deposits. Their accumulation and freezing take place from below upwards. At the commencement of syngenetic freezing the base of the accumulating series would be composed of epigenetically frozen strata.

Diacryogenic strata are formed by the freezing from the top, from below and laterally of oversaturated non-lithified materials – newly deposited sediments and silts in which complex diagenetic physical-chemical processes have either just begun or are far from completion, being stopped by the processes of freezing.

The structure of epicryogenic strata

All the mountain territories of the permafrost zone are the areas of predominant development of epicryogenic strata of *hard and semi-hard rocks*. The cryogenic structure of epigenetically frozen bedrock is dictated by composition, structure, moisture content and fissuring at the commencement of freezing, mean surface temperature, variations of climatic conditions, the neotectonic and glacial setting, etc. Cryogenic textures, being inherited, are determined by the pattern of primary cavities in the rocks and peculiarities of their freezing.

The greatest depths of epicryogenic bedrock strata with negative temperatures that have been observed, exceed 1000–1500 m. Two or three zones can be distinguished from the top downwards depending on the conditions of formation of the cryogenic structure. The near-surface zone of highly fissured rocks is that of rather active cryogenic weathering, corresponding to the layer of annual fluctuations of negative temperatures where phase transitions of moisture, causing expansion or reduction of ice volume in fractures, are expressed. Below, the frozen rocks of the ancient weathering crust can occur, underlain by low ice content, often cryotic rocks which extend hundreds of metres vertically. Finally, at the bottom of the epicryogenic permafrost strata, in the fissured bedrock, strata can occur with saline water having negative temperature – cryopegs.

The cryogenic structure of epigenetically frozen *loose deposits* is to a significant extent determined by their lithogenetic type, moisture content before freezing, availability or absence of aquifers, degree of lithification and landscape-climatic setting, varying in accordance with the natural environment in the Cenozoic. Thick strata of basin deposits have been frozen epigenetically: marine, glacial-marine, lagoon and lacustrine deposits mainly having a fine-grained composition (clays, silty clays, sandy silty materials). The same type of freezing occurs in glacial accumulations of block-boulder/rock waste/fine grain composition as well as river-bed alluvial sands, pebble gravel, partially eolian deposits, peats, and ancient weathering crusts.

The strata of basin deposits have the most typical cryogenic structure. These freeze according to the 'closed' or 'open' type of system depending on the availability or absence of aquifers that provide the inflow of moisture on freezing.

The deposits of coarse composition – sands and pebbles – freeze under conditions of a 'closed' system with pushing of moisture ahead of the freezing front ('piston effect'). Therefore, the ice content of the frozen material is generally small (up to 10–20 %). Among the dominant types of cryogenic texture are the massive and the crustal. The availability of clayey layers serving as aquicludes gives rise to cryogenic pressure of ground water resulting in the formation of high ice content materials with basal cryogenic texture as well as layers and lense-like deposits of ice. The same effect is typical of coarse-grained deposits freezing in an 'open' system when there is inflow of pressurized or unpressurized ground water. In this case, series and horizons of ice-rich deposits often arise, with layered and lense-like deposits of ground ice usually containing particles of local rock material.

The series of basin deposits of homogeneous, predominantly fine-grained composition, can be 200–300 m thick, for instance in West Siberia or in the European North-East of Russia. These are composed of poorly sorted silty-clays with inclusions of gravel, pebbles, boulders or, rarely, relatively well-sorted clays and aleurites. The thickest strata are of glacial-marine origin and have the largest areal extent. A less thick series of lagoon, estuarine and lacustrine deposits typical of vast and deep water bodies have a smaller areal extent.

Typically, the strata of epicryogenic fine-grained deposits of homogeneous composition, having been frozen under a 'closed' system with upward migration of moisture and formation of segregated ice, have the structure shown in Fig. 9.1a. The horizon 5 to 10 m deep from the surface, rarely 15 m, has the highest ice content. In general, this horizon corresponds to the layer of annual temperature variation of the frozen stratum in which small transitions of unfrozen water into ice and vice versa occur. Grain size and chemical-mineral composition of this upper part of the epicryogenic strata are determined by the conditions of sediment genesis, diagenesis and weathering existing before freezing as well as by the conditions of freezing and the geochemical setting subsequently (10). Volumetric ice content here reaches 40–50 %, and the soils are often heaved. Cryogenic texture is predominantly fine-schlieren, often cellular and laminated-cellular, and in peaty deposits it is basal. The spacing between ice schlieren increases with depth while they become thicker. At a depth of 20–30 m from the surface total ice content of the frozen ground diminishes to about 20–30 %, with mostly big-cell and block-type cryogenic structures. The cross-section of the blocks is 0.5–0.7 m, sometimes 1–2 m, ice schlieren are 2–3 cm thick, sometimes 5–7 cm. At the depth of 30–40 m big-block cryotextures with open nets are developed, while farther down (sometimes about 100 m) only isolated broken schlieren are encountered, cryogenic texture being predominantly massive. Volumetric ice content diminishes from the top downwards from 20–10% within the interval considered above.

With a high ice content in the upper part of an epicryogenic series, structures may have either a homogeneous nature or several horizons with a high ice content. In the former case, according to V.A. Kudryavtsev, the higher ice contents of the upper third of the frozen epicryogenic strata are caused by long-term temperature variations of various periods and ampli-

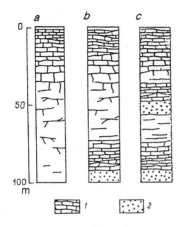

Fig. 9.1. Types of cryogenic structure of epigenetically frozen strata of fine-grained basin deposits: (*a* – of homogeneous composition; *b* – containing as a basal layer a previously aquiferous layer of sandy composition; *c* – containing two previously aquiferous layers of sandy composition): 1 – cryogenic textures of fine-grained deposits; 2 – previously aquiferous sands.

tude. Short-term temperature variations dictate the high ice content of the near-surface portion of the frozen strata to a depth of 5–10 m. Temperature variations of a longer period determine the higher ice content at a greater depth (from 15 to 30 m). Temperature variations of still longer period have an impact for still greater depths, etc. In summary, this leads to the situation where a third of the frozen series acquires a high ice content in fine-grained materials where cellular and laminated-cellular cryotextures begin to form. Since heat exchanges are drastically reduced with depth, the ice content cannot be high over the whole frozen series. Besides, fine-grained soils are usually more compact and dried with depth because of the normal diagenetic processes and migration of moisture towards the freezing front. Consequently, ice content diminishes downwards and networks of subhorizontal and subvertical ice schlieren become rare; still lower, the series is not rich in ice, and the cryogenic texture becomes massive.

In the second case several horizons are observed in the section of a series, having a higher ice content with cellular and laminated-cellular cryogenic textures. G.F. Gravis assumes this to be associated with development of cryogenic pressures, on freezing of homogeneous materials, in the lower (still unfrozen) portion of the series. As a result, repeated migration of moisture under pressure, towards the freezing front, often induces intense ice segregation at certain levels in the frozen series.

Epigenetic freezing of loose fine-grained deposits in the type of 'open' system when there are interlayers of aquiferous sands and pebbly gravel leads to a highly complicated picture of cryogenic structure (see Fig. 9.1b,c). Ice-rich layers characterized by fine-laminated and laminated-cellular cryostructure are created over the aquifers. Often, excess moisture is associated with the formation of ice strata and lenses of segregational origin. Nonuniformity of the freezing processes in different parts of the frozen series leads to cryogenic pressures in the aquifers and formation of intrusive ice lenses and ground ice which is often associated with fold dislocations. The zones of higher ice content with net and cellular cryostructures often immediately overlie lenses of intrusive ice and ground ice; creation of these cryotextures is assumed by G.I. Dubikov to be due to pressurized water injection into fine-grained materials.

Results of drilling in recent years show that in the north of the West Siberian Plate, within the limits of the Siberian Platform in the coastal lowlands of Siberia, below the epicryogenic frozen strata, cryopegs occur with highly saline water (up to $200\,g\,l^{-1}$ and over), the freezing point of which is much lower than $0\,^\circ C$. The thickness of the zone with highly saline water having negative temperature reaches several hundreds of metres. After the zone is formed, it is believed easily soluble salts are pushed out of pore solutions of the freezing strata (chlorides of sodium, calcium, magnesium, etc.) and they concentrate below the beds of ice-rich frozen and cryotic rocks. Cryopegs are also observed inside the frozen strata of loose deposits.

Formation of epigenetic ice wedges in the near-surface part occurs in epicryogenic strata. Vertically, they extend no more than 3–5 m, although 6–8 m high wedges are known, and theoretically, they can be as high as 12–15 m. Their width at the top usually does not exceed 1.5–2 m. The ice inside the wedges has a distinct vertical banding due to the arrangement of air bubbles and organic-mineral admixtures.

In the uppermost part of epicryogenic strata below the seasonally thawing layer the so-called transitional layer with higher ice content, 1–2 m thick, stands out in many cases. In this layer the degree of saturation with ice can reach 70–80%, cryotexture is mainly laminated, and often ribboned. The layer is formed either by transition of the lower part of the saturated seasonally thawed layer into permafrost, or by migration of moisture from that layer into the upper part of the permafrost.

In a number of cases the upper part of epicryogenic strata are represented by palaeocryoeluvial continental deposits formed outside the area of permafrost development but having been subject to multiple and deep seasonal freezing-thawing in the course of their accumulation (10). In the case of general climate cooling and one-time freezing from the top of these continental formations, epicryogenic palaeocryoeluvial continental deposits (see §6.2) are formed.

Syncryogenic deposits

These are always formed above a frozen substratum, i.e. above epicryogenic perennially frozen strata, by simultaneous, synchronous, accumulation and freezing of subaqueous and subaerial deposits. The extension upwards of the frozen series involves saturated soil; the bottom of the seasonally thawing layer freezes onto the top of the frozen stratum and ice schlieren are formed from the moisture from this layer. It is the manner of increase in thickness of the perennially frozen ground that distinguishes syn- and epicryogenic strata. The thickness of the latter increases by deepening of the permafrost while that of the former increases by the rise of the upper surface of the permafrost. Thus, with a constant thickness of the seasonally thawing layer, the upper surface of permafrost rises annually by the thickness of annually accumulated terrestrial deposits (with allowance for their heaving when transferring into the frozen state). It is also clear that sediments accumulating from the top grade into the permafrost state gradually, not suddenly, but undergoing a rather great number of freeze-thaw cycles until they come to lie below the layer of seasonal thawing. Tentative estimates show the maximum number of freeze-thaw cycles can be as high as 10 000 (10).

The process of syngenetic freezing is most apparent in the accumulating deposits of flood-plains, deltas, laida, alases, mainly of sandy-silty composition enriched in organic matter, fine-grained slope deposits – talus and solifluction as well as eolian. Much less developed are syncryogenic deposits characterized by a small thickness (the first metres), forming in shallow parts of fresh water bodies on drying shelf areas, the freezing of which both laterally and from below (when the frozen stratum is at $-3\,°C$ and lower) can occur with simultaneous accumulation of sediments.

Syncryogenic strata are most widely developed in such areas where a continental regime of sediment accumulation existed during the Quaternary period. Thus, the thickest (80 m and over) syncryogenic strata are developed in areas which were subject to gradual tectonic sinking during the Pleistocene compensated by the accumulation of continental deposits. Such areas are littoral lowlands and the Mesozoic-Cenozoic superimposed basins of the North-East of Russia, Novosibirskiye Ostrova [New Siberian Islands] and the plains of Central Yakutia. In addition, syncryogenic rocks are developed within river valleys and lacustrine-marshy basins in the continuous permafrost zone and at sites of low temperature, where their thickness is not more than some tens of metres.

Among syncryogenic strata there are 'southern' and 'northern' variants as regards their cryogenic structure. In the *southern* one, for each cycle, there are very small thicknesses of a frozen icy layer of soil, which have both icy soil frozen from below and from the side of the frozen strata and which overly a dewatered layer with low ice content. Uniform distribution of thin horizontal ice layers along the section is typical of such syncryogenic strata. This is conditioned by the predominant flow of cold from the top during autumn-winter freezing of the seasonally thawing layer, while the temperature of the frozen ground remains relatively high (not below $-3\,^{\circ}$C). As a result, the lower part of the seasonally thawing layer is drained and thin ice layers arise at its base. Sometimes, they can be absent altogether with formation of low-ice content massive cryogenic texture.

In the northern variant freezing of the seasonally thawing layer from below is considerable; there are syngenetically frozen strata for which a high ice content over the whole section is typical as well as the presence of bigger ice schlieren up to several centimetres thick on a background of thin laminated and lamellar-reticulate cryotexture, called by Ye.M. Katasonov 'belts'. Depending on the ground surface topography and, accordingly, on the outlines of the base of the seasonally thawing layer, these belts occur either horizontally or have a concave shape inheriting the shape of the thaw basin inside the dike polygons over the surface of flood-plains, laida, and bottoms of kettle lacustrine basins.

Formation of 'belts' of ice-rich cryogenic structure is, in the opinion of many investigators, associated with a continuous and rather uniform uplift of the surface of accumulating deposits, on the background of which there are cyclic variations of seasonal thawing depth induced by climate fluctuations. Climate cycles of 11, 40, 100 and 300 years, superimposed on each other, lead to the nonuniform occurrence of belts over the section.

Syncryogenic strata are also formed with accumulation of deluvial talus, solifluction, eolian deposits and their freezing from beneath. Syncryogenic fine-grained slope deposits which are formed in conditions of accumulation exceeding wash (with a constant thickness of the seasonally thawing layer) are mainly represented by peaty, sandy-silty and silty-clay materials and with inclusions of stone and sod material. Indistinct lamination, oriented in general along the slope, is characteristic of these. Layers are grouped together into bundles. Ice is irregularly distributed over the section of each bundle.

In the lower and upper parts of a bundle, laminated, reticulate and block or lens-like cryogenic structures are observed. The middle part is often characterized by massive cryotexture with infrequent ice schlieren. Syngenetic ice wedges are typical of the deposits; the depth they reach corresponds to the thickness of the stratum, and their distinctive feature is curvature along the slope gradient.

In general, syncryogenic strata have a number of attributes making it possible to distinguish them from epicryogenic strata. They predominantly have a sandy-silty composition, high content of ice, leaf-by-leaf enrichment in peaty and vegetation residues and diffused organic matter with different degrees of degradation, and the whole thickness is penetrated by vertical ice wedges. The shape of ice wedges is tongue-like with constrictions. Their lateral fringes are not smooth, typical is the presence of 'shoulders' to which are 'welded' thick schlieren – belts of ice-bearing strata. The latter are deformed irregularly near contacts with ice wedges. Elementary ice wedges inside the big wedges appear on their lateral contacts.

Diacryogenic frozen strata

These are formed when basin sediments (coastal-marine, tidal marsh, lacustrine, lagoon, estuarine) freeze at the bottom of a water body, at shallow places and on shelf areas as these basin formations emerge from beneath the water level. Sediments are poorly compacted and saturated with water being at the initial stage of diagenetic transformation. Diacryogenic materials are exemplified by the upper part of basin deposits of taliks, frozen for several metres (below the seasonal thawing layer) and underlain by unfrozen materials which further undergo epigenetic freezing. When the diacryogenic type of freezing takes place immediately after emerging from beneath water level, the sediments freeze as synchronous-epicryogenic, and if freezing occurs after a period in the thawed state – as asynchronous-epicryogenic (see §6.2). For materials frozen diacryogenically a high ice content is typical, with basal, reticulate and grating structure.

Polycryogenic strata are distinguished based on two approaches: 1) in a single frozen series the attributes of syn-, dia-, and epigenetic freezing are combined; 2) the frozen strata have a two-, three- and more member structure (stage-type): the upper part is syncryogenic and the lower one epicryogenic. There are many examples of different types of freezing of a single series of sediments at different stages of their formation. In particular, relatively deep-set lacustrine deposits freeze epigenetically and with shallowing of a water body and grading into a lacustrine-marshy regime the sediments can freeze and accumulate synchronously. Alluvial deposits can serve as a typical example of complicated cryogenic structure: their lower part (river bed facies) freezing epigenetically, the upper part (flood-plain), syngenetically, and crescent lakes (bottom), diagenetically. A composite polycryogenic structure is typical of coastal-marine deposits, tidal marshes, laida and beaches. In other words, the frozen strata may be genetically one, yet have the attributes of both syn- and epigenesis.

Among the most sophisticated problems of the so-called '*glacial complex*' or '*yedoma series*' is the genesis of deposits developed on plains of the North-East of Russia. The 'glacial or 'yedoma' complex of loess-like syn-cryogenic frozen sediments composes almost wholly (beyond the limits of river valleys) the Aldan-Olenek, Yana-Indigirka, Kolyma coastal lowlands, Novosibirskiye Ostrova [New Siberian Islands], the Mesozoic-Cenozoic superimposed basins of the North-East of Russia, and the plains of Central Yakutia. The ice-rich loess-like deposits are 80–100 m thick. The formation of the biggest syngenetic ice wedges is associated with these deposits (Fig. 9.2).

Loess-like ice-rich aleurites and ice wedges form an interrelated complex. 'Yedoma' is a smooth uplifted surface composed of high ice content sediments easily subject to wash-out. There are different points of view concerning their origin. Thus, there is the concept of their eolian genesis, i.e. accumulation of sediments under the influence of wind. This idea is mainly applicable to loess-like materials of similar type within the permafrost zone of North America and Alaska. It is applicable also to the thick loess series within the permafrost zone of Eurasia, both ancient and modern. Among the geocryologists in the former USSR, the view is that a thick series of ice-rich loess-like deposits are continental aqueous sediments that have undergone thawing and freezing in the course of their formation, mainly under conditions of flood-plains. The investigators supporting the flood-plain hypothesis describe the conditions of formation as follows: rivers issuing from mountains into the littoral plains are distinguished by slower flow, tending to meander or, on the contrary, remain steady, branching into channels and creeks ('forking') and forming vast deltas. In the opinion of Y.A. Lavrushin, the same deposits are alluvial facies of shallow bars in river beds and river laida. In the Lena river basin in Central Yakutia, a series of loess-like deposits resulted from carrying of silt from firn fields by glacial meltwater floods (A.A. Grigor'yev) with inundation of the lower reaches of the Lena river valley or from a combination of this with tectonic uplift at its mouth (G.F. Lungersgauzen) and purely tectonic factors and uplift of positive structures and sinking of depressions in which vast inundated basins arise. It is supposed that in the inundated basins fine elutriated sediments of predominantly aleurite composition are accumulating. The formation of a

Fig. 9.2. Syngenetic ice wedges in the series of aleurites of the yedoma in the North-East of Russia (photo by S.F. Khrutskiy).

loess-like aleurite series is considered to be associated with activity of glacial meltwater. G.F. Gravis and M.N. Alekseyev are of the opinion that their origin lies in talus-solifluction.

In the late 1940s and early 1950s the idea prevailed that the vast accumulations of ice in loess-like series in the North-East of Russia were buried remains of the former glacier cover supposed to have existed during the Pleistocene on the littoral lowlands of the Siberian North (E.V. Toll', K.A. Vollosovich, V.A. Obruchev, A.A. Grigor'yev *et al.*). Subsequently, a polygonal-wedge pattern of the principal monolithic ice bodies was found in strata of yedoma deposits (B.N. Dostovalov, P.A. Shumskiy, *et al.*); the answer was also obtained to the question of the big vertical dimension of ice wedges – they have grown upwards synchronously with accumulation of the flood-plain deposits (E.K. Leffingwell, A.I. Popov, Ye.M. Katasonov *et al.*). However, many questions still remain unanswered. Thus, alluvial strata where flood-plain facies have been accumulated to a height of 50–60 m and even 100 m above water level are not known in nature. Therefore, the concept of a flood-plain origin of the deposits that bear ice wedges is criticized by specialists in the structure of the alluvial deposits. The eolian hypothesis is developed by some investigators (S.V. Tomirdiaro *et al.*) as an alternative to the alluvial flood-plain hypothesis. It was found that in the late Pleistocene about 20 000 years ago, in the north-east of Eurasia vast areas of Arctic shelf emerged. America joined to Asia, represented a single continent and, with the growing land area, the climate became very severe and continental. This also led to desert conditions. Freezing winters with little snow and dry hot summers promoted intensified wind-borne transfer of fine sand and aleurite grains off the surface of the emerged Arctic shelf. By settling on various topographical elements mineral particles of fine sand and aleurite became mixed with peaty residues of the tundra grassy-sod-vegetation and these began to accumulate and freeze with a general uplift of the surface, thereby resulting in the possibility of parallel growth of ice wedges. The original concept of a lacustrine-thermokarst origin for the thick ice-rich aleurite strata with big ice wedges was suggested by N.A. Shilo (21).

Thus, at the present level of scientific progress, only one question has been solved concerning the polygonal ice wedge origin of big ice bodies in the beds of yedoma deposits. The question of what were the conditions of their accumulation remains without answer. The strata under consideration, ice-rich loess-like deposits with large vertical ice wedges, are characterized by homogeneity of granulometric composition with a continuously high content (to 60–80%) of coarse-aleurite fraction (0.01–0.05 mm) and high porosity which becomes evident in the thawing of ground and evaporation of ice in the walls of collapsing blocks and baydzherakh ('cemetery mounds'). However, what makes them distinct from typical loess is the low carbonate content, as a rule not exceeding 1.5–2% and only seldom reaching 4.5–5%, and the absence in the walls of prismatic cleavage breaks, columns, and sometimes a number of other attributes of loess.

A fine horizontal lamination is characteristic of the yedoma deposits. There are bands of relatively 'pure' aleurite composition and bands of highly peaty material containing much redeposited (allochthonous) peat alternating in the section, sometimes grading into 'leaves' of autochthonous peat, i.e. buried peats. In the beds of ice-rich aleurites there are horizons of fossil soil characterized by aggregation (granular texture), dark colour (humus horizon) and higher content of grass vegetation roots as well as the presence of shrub and underwood roots. Aleurite deposits of the yedoma complex are distinguished as those rich in ice, which makes up 70–75% and often

70–80% of their volume. The prevailing type of cryogenic texture is thinly laminated, horizontally oriented schlieren varying in thickness (1–15 mm), alternating along the section. 'Belt-type' cryotexture is characteristic. Between thickened interlayers there are bands of fine ice schlieren arranged either parallel to them, or forming a fine-lens network. Belts themselves can be arranged horizontally or inclined, often they are concave following the former shape of the bottom of the seasonally thawing layer in polygons between ice wedges.

9.2 Composition and cryogenic structure of the principal geologic-genetic types of sedimentary materials in the permafrost regions

Owing to the specific conditions and factors of lithogenetic processes in the permafrost regions there are a number of distinctive features in the composition, structure and properties of the frozen soils with development of essentially new genetic types of sedimentary formations, found only in the areas of permafrost development. A number of terms appeared in the literature, on the one hand specifying the genesis of sedimentary formations in the permafrost regions, such as 'cryogenic eluvium', 'cryogenic alluvium', 'cold and permafrost soil' and on the other hand, pertaining to formation of the frozen sedimentary layers proper, such as 'syngenetic' and 'epigenetic' strata, deposits of a 'yedoma' or 'glacial' complex, buried ice and others. These questions to a greater extent refer to the loose Cenozoic rocks than they do to Pre-Cenozoic bedrock as the cryogenic structure of the latter is determined by the availability of cavities and wetting before freezing. As a result, under a certain combination of conditions ice fills fractures, pores, caverns, small karst cavities, mainly in the near-surface (weathered) zone of bedrock masses.

Eluvial formations of the permafrost regions, representing accumulated residual products of hypergene bedrock transformation and found within the limits of recent or ancient crustal weathering, are characterized in general by modest thickness, controlled by the penetration depth of annual temperature fluctuations. Owing to predominance of physical weathering in the permafrost regions, the cryogenic crust of weathering on hard rocks is characteristically rudaceous accumulations (large blocks, block-rock waste, rock waste and debris-gravel) with fine earth admixture. At a mature stage of eluvium development, owing to the active removal of fine particles by water seepage and freezing out of coarser debris upwards, there is differentiation of weathered material in the vertical profile. The upper part of the weathering crust in high mountains and polar landscapes is practically devoid of fine particle infilling and acquires the appearance of a purely fragmental horizon, while blocky-rock waste material on the surface often forms stone rings and polygons. As a result, the lower part of the eluvium is enriched in fine material. Lower in the section sharp enrichment in rudaceous material is observed with gradual transformation into the horizon of 'disintegrated rock' – bedrock. Crustal, basal and branched-fracture cryostructures are typical of eluvial formations in the permafrost regions.

The cryogenic crust of weathering on soils is characterized by a high content of fine-grained material and, owing to this, by predominance of crustal, massive and lense-like cryostructures.

The cryogenic crust of weathering on loose sedimentary material is enriched in the silt fraction (0.01–0.05 mm), and eluvial formations have a distinctly silty composition. The overlying loess-like silty-clay materials here have a high homogeneity of granulometric composition, high porosity and little easily soluble salts. The basic rock-forming mineral of the sandy fraction is usually quartz, while those of the clay fraction are hydromicas and montmorillonite. It is because of this specific composition and structure of eluvial formations in the permafrost regions, compared with the eluvium of humid zones, that some investigators refer these deposits to a specific type of eluvium – 'cryoeluvium'.

Among deposits of the *slope series* along with slide, creep, collapse and talus formations, within the limits of the permafrost regions there are the frozen rock stream and solifluction deposits which are typical representatives of cryogenic formations proper.

Collapse and talus accumulations around the steep and precipitous slopes are mainly of rudaceous composition (blocks, rock waste, gravel) and enriched with fine-grained material in different amounts. The cryogenic textures of such rudaceous formations are predominantly crustal in poorly wetted soil and basal in highly moistened soil – as for talus formations, the crustal and contact cryotextures are typical of coarse-grained materials; in individual cases, in conditions of excess wetting before freezing (usually locally in the foothills of slopes) basal cryotextures arise. Rudaceous formations on relatively gently sloping or moderately steep slopes being mobile (cryogenic and thermogenic creep) are called the deserptium type (*rock stream*). Their thickness varies depending on geological-geographical conditions from several centimetres to 3–6 m. A typical attribute of rock streams is the near-surface rudaceous mantle without fine-grained infilling material, underlain by a horizon of fragmental material with silty-clay material and sandy silty clay, sand and gravel infilling, grading directly into fissured underlying bedrock. The cryogenic structure of the upper part of the rock streams is characterized by open cavities between blocks; in the lower part, with accumulation of infilling – textures become massive and crustal and still lower, there is an alternation of ice and soil layers and, as a rule, there is 'golets' ice filling the cavities between fragments and forming separate pockets, with bedding and lenses up to 1 m thick and more, sometimes several tens of metres long.

Solifluction deposits are characterized by consistent composition and structure, with typically a laminated pattern due to the buried turf layers broken and folded inside near the external margin of the solifluction formation. In the opinion of G.F. Gravis the cryogenic structure of these deposits is subdivided into two geographical variants, Arctic and subarctic. The Arctic variant is characterized by a high ice content and small inclusions of ice, and the subarctic by high ice content but with the presence of rather thick closely spaced indistinct ice schlieren and small lenses. Formation of cryogenic textures in such deposits can occur in freezing both from the top and from below.

Alluvial deposits of the permafrost regions are first of all distinguished by the type of their freezing. Their characteristic and distinctive features pertain to cryogenic structure and wide development of large accumulations of ice wedges. Thus, for the epigenetically frozen sands and pebble gravel of river-bed facies, crustal and massive cryogenic textures prevail. In the case of freezing of river-bed deposits with moisture inflow due to injection, the deposits are much enriched in ice with formation of lens-like and bedded ice deposits. Old lake deposits in the permafrost state are characterized by a high ice content and presence of broken lens-like bands of ice, which according to E.M. Katasonov, are usually inclined in the shore area and form oblique lens-like or cross-bedded cryogenic textures of the inherited type.

The flood-plain alluvium differs in the case of internal flood-plains compared to that adjacent to rivers, both in composition and cryogenic structure. The flood-plain adjacent to a river has a ridge-hollow topography, and is composed of sediments characterized by non-consistent, but in general coarse-grained composition (sands, sandy silty-clay materials) and sparse peat content. These formations when frozen syngenetically with low moisture availability show intermittent horizontal schlieren and lenses of ice 2–3 mm thick, and with abundant moisture – small reticulate, plexiform and fine net cryotextures and syngenetic ice wedges up to 1 m thick. The cryogenic structure of deposits of the internal flood-plain zone is closely connected with the polygon-ridge topography of its surface and is characterized by a high ice content owing to numerous fine lenses up to 1 mm thick, on the background of which thicker and persistent layers (belts) of ice

0.5–2.5 cm thick are observed. The most distinctive feature of the cryogenic structure of the flood-plain alluvial deposits with polygon-ridge micro-relief is the formation of the ice wedges (with width at the top of several metres and depth of over 10 m).

Composition and structure of *glacial* and *glacio-fluvial* deposits vary greatly. Thus, unsorted moraine is typical of glacial deposits – boulders, rock debris, pebble, gravel with sandy, silty and clay-rich materials, with different types of cryogenic structure and high ice content (30–50%). Fluvio-glacial deposits are basically more homogeneous, coarse-grained, and typically with lower ice content (10–20%) and massive,crustal and lense-like cryostructures. Fine- and medium-grained silty sands have fine schlieren, lens-like and reticulate-laminated cryotextures. Their ice content is greater, up to 30–40%.

Marine, lagoon and lacustrine deposits in the permafrost regions are also characterized by a number of specific features distinguishing them from the southern geographic variants of these formations. The main factors that determine the specific nature of the accumulation of sediments in the northern water bodies are the low temperatures of the aqueous medium and the surface ice. Thus, bottom water temperatures in polar seas are negative over most of the area ranging between -1.5 and -1.8 °C. Owing to this, biogenic carbonaceous and siliceous accumulations are practically absent in polar seas, and the deposits are terrigenous almost everywhere. The influence of surface marine ice on the sedimentation process is to enrich marine deposits with fragmental material brought in fast ice and icebergs from the coast. The cryogenic structure of the marine deposits is connected with epigenetic freezing. Prevailing cryogenic textures of clayey-mixed deposits are superimposed reticulate and net or block-type, and in the upper near-surface portions of the sections, laminated-reticulate and lens-like. In general, thick homogeneous marine strata of clays and silty-clays that have been frozen under a 'closed' system, i.e. without water inflow from outside, typically have an ice content decreasing with depth and wider spacing of the ice schlieren network. The upper zone of maximum saturation with ice is 10–15 m thick. Below lies the ice-poor zone of 'drying' with big net and block cryogenic textures; at depth they are massive, on the back-ground of which only fine oblique fractures filled with ice occur. The zone of segregation ice formation has, as a rule, 50–60 m thickness.

The strata of marine shelf deposits that had included aquifers before freezing, i.e. they were frozen under conditions of an 'open' system, typically show several icy horizons, mainly with laminated and bedded-reticulate cryostructures; sometimes they are highly saturated with ice, up to sheet ice formation. A feature of the cryogenic structure of marine shelf deposits is ice deposits having plate-like and lens-like shapes.

Continental basin fresh water (lacustrine) and mineralized (lagoon) deposits are mainly represented by silty-sandy and silty-clay materials with different amounts of organic matter. Specifically, there are banded-laminated clays and aleurites with alternating sandy or sandy-silty layers formed during spring-summer periods, and clayey layers, formed in winter while the lakes are covered with ice.

Thus, their formation is associated with cyclicity (winter-summer) in the accumulation of sedimentary material in the final water bodies under conditions of severe continental climate typical of the permafrost regions. In other words, the banded-laminated formations are typical deposits of the permafrost regions.

The cryogenic structure of lacustrine and lagoon deposits is determined by the type of their freezing and their specific lithologic composition. The deposits of large and deep water bodies with through taliks that have been frozen epigenetically (from the top) are similar to marine deposits in cryostructure. Near the banks of shallow lakes steeply inclined lens-like reticulate cryostructures arise, associated with lateral freezing. The ice-rich laminated-reticulate, reticulate, sometimes basal (ataxite) cryostructures arise far from the banks, when freezing is from below. In the uppermost part of lacustrine and lagoon deposits saturated with organic matter, there are mainly ice-rich, schlieren, laminated-reticulate, and reticulate (plexiform) cryotextures.

Among frozen deposits there is the specific genetic formation which is called the '*glacial*' or '*yedoma complex of the frozen loess materials*'. The latter compose the northern plains of littoral lowlands, deposits of the Mesozoic-Cenozoic superimposed depressions of the North-East of Russia, the plains of Central Yakutia and the like. The thickness of ice-rich loess-like deposits of predominantly aleurite composition can be as high as 80–100 m, and formation of the biggest syngenetic ice wedges is associated with them.

The strata of deposits of the ice complex are characterized by homogeneity of granulometric composition with a consistently high content (60 to 80%) of the coarse-aleurite fraction (0.01 to 0.05 mm), high porosity and often by the presence of fine horizontal lamination created by bands of relatively 'pure' aleurite and peaty layers. High ice content is typical of the glacial complex deposits (70–80% of their volume).

9.3 Natural ice as a monomineral rock

All kinds of natural ice that form independent bodies and accumulations can be regarded as a monomineral rock. They develop under conditions of negative temperature of land and sea surface, atmosphere and lithosphere. Distribution of ice on the Earth's surface and in the crust is extremely nonuniform. Major masses of ice are concentrated on the land surface, mainly as glaciers and ice sheets. With an areal extent of only 3.1% (in relation to the Earth's surface) they constitute over 97% of all the mass of natural ice. The bulk of glacier ice is found in the Arctic and Antarctic zones, where ice thickness can reach 4 km and over. In the lithosphere ice occurs in the uppermost horizons (first hundreds of metres) making up in total about 2% of the ice mass on the globe. Ice on water surfaces is mainly represented by sea ice, icebergs and snow cover. Their share of the total ice mass does not exceed 0.2%. Atmospheric ice formed in air or on various surfaces from the air comprises, according to the data of V.M. Kotlyakov, about 18% of the water vapour mass and 0.03% of the atmospheric mass. Their volume in the total mass of natural ice is minimal.

There are many classifications of natural ice. In the classification by P.A. Shumskiy, 28 kinds of ice are distinguished, and classified in three groups: ice resulting from freezing of water (congelation), sedimentary ice (snow cover) and metamorphic ice (glacier ice). Two large groups – land surface ice and ground ice – reflect the conditions of natural ice formation and occurrence of ice bodies in relation to solid earth materials.

Land ice can be subdivided, in the light of geocryological objectives, into newly deposited and re-deposited snow covers, metamorphosed snow covers and névé basins, ice of water bodies and streams, icings and glacier ice. Each group is distinguished by the mechanism of formation of the ice. Newly deposited and re-deposited snow covers consist of fine ice crystals (snow) and their fragments. Metamorphosed snow cover and snow patches comprise old snow transformed in the course of re-crystallization and having different size of grains. Ice of water bodies and streams forms under the influence of heat flow directed perpendicular to the freezing surface. Structural-textural features of this ice are dictated by the conditions of hampered growth of crystals, anisotropy in the rate of their growth, the presence of different admixtures in the water, as well as by the state of the water mass. Icings result from layer-by-layer freezing of water flowing out onto the surface of ground or ice, and they show a distinct lamination parallel to the surface of accumulation. Glacier ice formed of snow cover after its compaction and re-crystallization is characterized by a heterogeneity of structures and properties. As shown by observations in a glacier section, there are three stages of its formation: diagenesis of the snowpack, formation of firn and its transformation into ice.

Ground ice encountered in the form of large accumulations in the upper part of the lithosphere takes the form of ice bodies of thickness 0.3–0.5 m and more. Among these, different kinds of ice are distinguished: wedge, intrusive, migration, cave and buried ice. Wedge ice is formed in fractures arising in the frozen rock filled with moisture and has a heterogeneous composition with admixtures, and vertical and inclined banding. Intrusive ice forms in soils by crystallization of free ground water injected under pressure, and usually it comprises sheet-like, lens-like and stock forms, and often these have many air inclusions of various shapes and sizes. Migration ice is formed as a ground component if favourable conditions are created for freezing and migration of water towards the front of ice formation, this giving ice bodies several metres thick. Ice formations in caves have various origins. Among them are infiltration ice that forms in cavities, crust icing and ice stalactites and stalagmites; ablimation ice that forms ice strings and hoar frost in caves; snow ice forms snow patches, and in combination with infiltration often forms glaciers in caves; buried ice is the remnants of ice formed at the surface of the ground and having various origins (river, lacustrine, glacier, etc.), covered by a layer of sedimentary deposits that prevent its thawing.

Wedge ice comprises those types of ice which fill the fractures in weathered rocks (fracture-type) and form ice wedges in loose deposits. In the latter case they are the component of the Quaternary deposits called in the literature polygonal wedges or recurring ice wedges: they are called polygonal as they form a distinct polygonal grid; their section is shaped as a wedge or vein. They are also called recurring wedges owing to the multiple-repeated process of ice formation in vertical frost cracks that periodically develop in the same place (contraction hypothesis).

The types of ice under consideration (transparent in thin section) have various degrees of opacity, the colour is whitish, grey or brown, and they contain both mineral and organic admixtures, which according to the data of P.A. Shumskiy, form 3–5% of the total mass and 1–1.7% of the total volume of the ground. The ice has bubbles; the volume of gas-filled cavities makes up 4–6% of the total volume, and the bubbles have a globular or elongated shape (cylinder, pear-shaped, etc.). Among gaseous inclusions there are the authogenic (those separating from water during freezing) and xenogenic or foreign (air contained in interstices between ablimation crystals of ice). The latter type of gaseous inclusion is the most prevalent (2–4 % of total ice volume). Besides, xenoliths of surrounding materials are also encountered, being similar to fine mineral admixtures in composition. The arrangement of mineral admixtures and gaseous inclusions determines the vertical banding of the ice, but it is not always discernible. Vertical arrangement of mineral admixture bands is assumed to be the result of the expulsion of soil particles by ice crystals growing from the walls of the frost fissure inwards towards the axial plane where the water is freezing. The majority of wedge ice varieties are slightly saline with salt content ranging from 0.01 to 0.1 g l^{-1}, i.e. salinity is similar to that of ultra-fresh surface water of the permafrost regions and atmospheric precipitation. Wedge ice density is dependent on the amount of gaseous inclusions and mineral admixtures, ranging mainly between 0.85 and 0.90 g cm^{-3}. Porosity is usually 2–4%, in rare cases 8%.

The prevailing vertically banded structure of the wedge ice is determined by the pattern of accumulation and freezing of the elementary (annual) ice vein. The vein ice structure is xenomorphic-granular, plate-like and hypidiomorphic-granular. The orientation of the main optical axes of the ice crystals is often chaotic, and if it has an ordered pattern it is parallel to heat flow directed subhorizontally.

Therefore, water that fills a vertical frost crack is being frozen from both sides, from the walls towards the centre. The developing elementary ice vein consists of two vertical rows of crystals. Consequently, the maximum possible size of crystals is equal to half the width of the frost crack. There is a steady reduction of ice crystals from the top downwards, which corresponds to diminished frost cracking in the same direction. The size of the ice crystals is predominantly 1×1 cm, the maximum being 1.5×2.0 cm. Their size is also dependent on the temperature of wall cooling and diminishes with depth. There is also a dependence between ice crystal size and the age of wedge ice. Crystals enlarge with time owing to the process of 'ice metamorphism'. As shown by the observational data of V.V. Rogov, the size of crystals in recently formed wedges is 2–3 times less than that of the Pleistocene ones. Among the wedge ice types there are epi- and syngenetic.

Epigenetic ice wedges are formed in those sedimentary rocks which become frozen after having been accumulated and transformed (compacted) from above. The predominant size of the wedges does not exceed 3–5 m vertically, and 1.5–2 m wide at the top. The vertical dimension of epigenetic ice wedges is determined by the depth of the frost cracks' penetration into the frozen ground reaching 5–7 m, rarely 10 m; hypothetically, 12–15 m is possible. The ice in elementary frost fissures is formed of hoar frost crystals, snow pack in winter, and water that infiltrates in summer. An indispensable condition of ice-wedge formation is the penetration of the frost cracks below the maximum depth of the seasonally thawing layer. A cross-section of the typical epigenetic ice wedge looks like an inverted triangle with the base being smaller than the sides. In the lower part of the wedge, at the apex of the triangle, there are tongues, 'offshoots' of the wedge. Ice/soil contacts in the wedge are usually distinct, even sharp, often ferruginated and a fringe of pure transparent ice 1–2 cm thick is sometimes traced along them. The layers of the surrounding material in the vicinity of contact with the wedge, especially in the upper, widened parts, are often curved towards the top.

The following features are typical of the ice structure of epigenetic wedges. The colour is whitish, milk-like, sometimes brownish. The texture is distinct vertically banded; each layer begins from the more or less horizontal upper surface of the wedge. There is alternation of relatively pure ice 1-2 mm thick (up to 5 mm) and bands of ice enriched in mineral admixtures, vegetation residues and gaseous inclusions. The latter have an elongated shape being stretched, parallel to the axial suture of the elementary ice wedge. In the centre of these wedges the amount of gaseous inclusions is at a maximum reaching 4–5%, near the lateral contacts it is reduced to 1–3% of the volume. The diameter of the ice crystals does not exceed 1 cm. Often, subvertical bands are arranged fan-like; in the middle of the wedge the bands are vertical, while at the edges they are inclined, parallel to the lateral contacts with the surrounding soil.

Syngenetic ice wedges that grow synchronously with accumulation of sediment can have enormous size: 50–80 m (even more) vertically and 8–10 m horizontally. The shape of the wedges is usually composite, with expansions and narrowings and they often have a multi-stage form. At maximum growth, ice wedges become dominant constituents of the frozen material in general, soil is arranged in the form of vertical 'earth veins' or 'columns' narrowing towards the top between the ice network. In areal extent the ice occupies 60–70% of the ground at such sites.

Typically syngenetic ice wedges are in contact with ice-rich surrounding sediments. The layers of the latter, as a rule, are steeply curved upwards. This is assumed to be the result of the growing ice wedge squeezing the ice-rich soil near the contacts. In this case a direct correspondence should exist between the width of the ice wedge and the steepness of soil curvature near the contacts: the wider the wedge, the steeper is the curvature. However, this rule is not always valid, which is the reason why some investigators (Ye.M. Katasonov, A.I. Popov *et al.*) criticize the mechanism of deformation described.

The structure of ice of syngenetic wedges has a number of typical attributes making it possible to distinguish them from epigenetic ones (Fig. 9.3). There is always an admixture present in the ice of large quantities of soil particles and vegetation residues, especially in the lower and middle parts of the wedges. Bubble-structure is typical of the ice: gaseous inclusions are

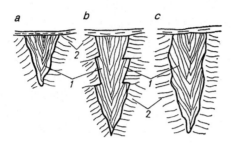

Fig. 9.3. Diagram of epi- and syngenetic ice wedge structure. Vertical cross-section (according to P.A. Shumskiy): *a* – epigenetic wedge; *b*, *c* – syngenetic wedges; 1 – annual layers of ice in the ice wedges; 2 – lamination of surrounding material.

shaped as spherical bubbles forming chains elongated vertically. At the same time vertical banding is poorly expressed and not always distinguishable visually, while in the epigenetic wedges it is revealed by accumulations of soil particles and vegetation residues. The ice texture of syngenetic wedges is similar to that of epigenetic wedges, but ice crystals in them are 2–3 times larger. The ice structure in the uppermost part of syngenetic ice wedges has a number of special features. Among these are vertical orientation of crystals, the small amount of soil and gaseous inclusions and the absence of vertical banding. This is explained by melting of the wedge ice near its upper surface forming migration ice layers which get frozen as if welded to its 'head'. N.N. Romanovskiy considers this process to provide 'frontal' growth of ice wedges upwards along with growth of vertical elementary wedges. A fringe of pure transparent ice about 10 cm thick is also traced near the lateral contacts. V.I. Solomatin and other investigators assume this to be associated with formation of segregated ice conditioned by the availability of horizontally oriented temperature gradients under the cooling influence of the open frost cracks.

The basic attributes of syngenetic ice wedges according to P.A. Shumskiy, B.I. Vtyurin, T.N. Kaplina, N.N. Romanovskiy *et al.*, are as follows: 1) a great vertical length that exceeds substantially the maximum possible fracturing, even in the most favourable conditions; 2) undulating lateral contacts and outcrops of the upper ends of elementary ice veins and 'welding' to lateral contacts of big schlieren (belts) of segregated ice in ice-rich ground. The belts look as if they rest on benches ('shoulders') on the side of the wedges, resulting from variations in the rate of wedge growth and sediment accumulation of the surrounding deposits. In addition, there are a number of distinctive attributes to syngenetic ice wedges to which different authors assign different significance. Apart from the characteristic attributes of the wedges themselves there are certain features in the structure of the surrounding deposits which give evidence of the simultaneous upward growth of wedges and accumulation of sediment: 1) change of facies composition of deposits between wedges horizontally from the middle of the inter-wedge space with maximum peat content towards lateral contacts where the soil admixture increases; 2) periodic development of peat lenses in the vertical section of 'earth columns' between wedges, with overlying mineral deposits; 3) in general similar conditions of accumulation of facies.

The southern boundary of occurrence of recent syngenetic ice wedges in alluvial, marshland and slope deposits is found much further to the north than that of epigenetic wedges. In the present natural setting, ice wedges grow syngenetically on flood-plains of rivers, periodically inundated laida coasts, bottoms of water-logged kettle lacustrine basins where peat is being accumulated and in the talus-solifluction trains at the foot of slopes. However, recent syngenetic ice wedges have much smaller size than the Pleistocene ones; their vertical length usually does not exceed 10–15 m. The majority of authors (Ye.M. Katasonov, A.I. Popov, B.I. Vtyurin, N.N. Romanovskiy *et al.*) are of the opinion that the biggest syngenetic wedges developed in the low-temperature ground of continuous permafrost were formed in the course of accumulation of high flood-plain alluvial deposits. Other authors (N.A. Shilo, Yu.A. Lavrushin, S.V. Tomirdiaro *et al.*) support the involvement of eolian, slope and other processes.

Intrusive ice represents intrusive accumulations resulting from the intrasoil freezing and crystallization of free ground water injected into the frozen or freezing ground under pressure. They form deposits of lens-like, layers and column shapes similar to intrusive bodies of magmatic rocks (batholiths, laccoliths, sills, dikes and the like; Fig. 9.4). Intrusive ice is most typically expressed in cores of frost mounds (hydrolaccoliths), the biggest among them are 40–50 m high with diameter ranging from several tens to several hundreds of metres. In Yakutia such mounds are called *bulgunniakh*, in North America, *pingos*.

Injection of ground water and soil/water masses can occur many times thus leading to the composite structure of intrusive ice formations. They have been found to be confined in the majority of cases to the contact of clay and sandy-silty-clay deposits with underlying coarse-grained sandy and sand-pebble formations; more rarely they are encountered within strata of fine-grained soils. Typically, intrusive ice appears pure and transparent, but in the base of the intrusion there are tongues and streamlets of fine mineral particles and near contacts there are xenoliths and isolated particles of soil,

Fig. 9.4. Sheet ice in marine Quaternary deposits. The Yamal peninsula, Ney-to lake (photo by G.I. Dubikov).

vegetation residues and peat. Sometimes, there are pebbles and boulders in the transparent ice.

The frost mound structure as distinguished by Sh.Sh. Gasanov has an external glass-like shell that contains a small amount of admixtures, and an internal core with a great amount of mineral admixtures and air inclusions. Sometimes, the layers of glass-like and bubble-containing ice alternate. Air inclusions encountered in the intrusive ice are often arranged layer-by-layer, parallel to the roof of the intrusion; they also form vertical, slanting and non-oriented accumulations. In the large ice intrusions of the laccolith type there is sometimes a radial arrangement which leads to a radiating-fibrous texture of ice resulting from bending of layers on heaving.

The structure of intrusive ice in the seasonal mounds differs from that in the large hydrolaccoliths. In the seasonal mounds the ice is pure, transparent and has a distinct vertical pan-automorphic-granular structure. In the large hydrolaccoliths the structure is allotriomorphic-granular with large grains of ice (1–16 cm cross-section) and chaotic crystallographic orientation.

Reformed intrusive ice is distinguished as a specific variety of intrusive ice. According to Sh.Sh. Gasanov, this ice forms intrusive layers 6–8 m thick with 300 m cross-section in the upper horizons of the frozen series. Initially, the ice is formed under the influence of hydrostatic pressure of ground water, then, as a result of pressure there is migration of water and liquified soil out of the closed system.

The basic attribute of the reformed intrusive ice structure is an alternation of layers of pure bubble-rich ice and ice-rich soil owing to which a laminated texture arises, which is often deformed. In this case the ice consists of folded parallel layers and those having different orientations.

Migration ice can also form large accumulations in the ground. Often they are called *segregation* (*i.e. separation*) *ice*, that is, ice forming schlieren. However, when they are large monolithic ice bodies, the term segregation seems to be inappropriate, as the ice is not separated into layers, but forms a single body.

This type of ice formation is considered at present to be relevant in the formation of large sheet, lens-like deposits of ground ice and even frost mounds (2) . However, this viewpoint is criticized by other investigators. A specific type of segregation was distinguished, characterized by the action of pressure migration. It can be stated for sure that migration ice is a predominant constituent of the frozen material within the limits of the so-called convex peat mounds.

The composition and structure of ice in the deposits associated with pressure-less migration of moisture are similar to those in the ordinary schlieren (segregational) ice. Its colour is, as a rule, white if it is transparent and pure. The ice contains bubbles of air, and soil and vegetation inclusions. The shape of the bubbles is irregular, pear-shaped and sometimes they form fibrous tubular accumulations. The soil in general shows layers of relatively pure ice alternating with ice containing admixtures.

The structure of the ice is hypidiomorphic-allotriomorphic and prismatic-granular. The prevailing orientation of the basic optical axes of the ice crystals is ordered and perpendicular to the freezing front with crystals mainly shaped as plates or columns. Their cross-section can be 3–10 cm. Ice density in the deposits of migration ice, like that of ordinary segregational-schlieren ice, ranges, according to P.A. Shumskiy, from 0.9140 to 0.9168 g cm^{-3}.

It is thought that in especially favourable conditions, when there is seepage of surface water from rather deep lakes (that do not freeze completely in winter), ice of similar type can form sheet deposits 20–30 m thick and several hundred metres long. Usually, they have no expression in the recent topography. The ice of the deposits is transparent and laminated by virtue of the nonuniform distribution of air bubbles and organic-mineral admixtures, and it is coarse-grained. Its structure is allotriomorphic-granular.

III

Principles of the formation and development of the frozen strata and layers of seasonal freezing and thawing

10

Thermodynamic and climatic conditions for formation of the frozen layers

10.1 Energy balance of the Earth

The thermodynamic (energy) approach to the study of the formation and development of frozen ground has an ever-increasing significance. However, up to the present time, the internal and external parameters that have an effect on the state of the frozen layers and the main thermodynamic functions (internal energy, enthalpy, entropy) have not been studied thoroughly. This is why the strategic efforts of investigators concerned with the study of the thermodynamic conditions of the development of frozen ground are directed towards determination of the thermal and water-thermal balances of the frozen layers.

Since frozen ground is a planetary phenomenon, its areal extent and development are dependent on the general state of the Earth and its variations determined by the planetary thermal (energy) balance. One should bear in mind that thermal balance of ground is dependent on zonal, regional and local distribution of the inflowing energy, in its turn conditioned by the geological-geographical setting. Therefore, the energy balance of ground should be studied in the planetary, zonal, regional and local aspects.

The thermal state of the Earth's surface is determined by the amount of heat energy arriving from external and internal sources to the surface. According to the current conceptions, the external energetic effect over the planet is exercised by the following:

1) radiation energy (electromagnetic field) of the Sun and stars intercepted by the Earth (10^{25} J year^{-1});
2) corpuscular, including neutrino escape, radiation of the Sun and stars (10^{18} J year^{-1});
3) energy of meteorites falling onto the Earth;
4) gravitational effects of the Moon, Sun and other celestial bodies (10^{20} J year^{-1}).

The inflow of energy from external sources increases the internal energy of the Earth. Internal sources of energy can be arbitrarily defined as those arising from:

1) nuclear reactions (10^{21} J year^{-1});
2) gravitational processes inside the Earth (10^{20} J year^{-1});
3) variations in rate of the Earth rotation (10^{20} J year^{-1});
4) exothermal chemical reactions (10^{19} J year^{-1}). In the long run, the internal energy of the Earth is reduced by radiation of energy from internal sources into external space.

The power of the energy sources varies greatly. At the same time the external energetic action is 100 000 times (greater by five orders of magnitude) that of internal sources. This leads to the conclusion that the basic source of energy for many processes that occur on the Earth's surface, including the formation of the temperature field of the upper part of the Earth's crust, is the inflow of radiation energy from the Sun. The ground surface serves as a 'heat kitchen' where the principal processes of heat exchange take place between the Earth and surrounding space. The temperature regime formed at the surface is the key factor that affects the dynamics of the temperature field of the upper lithosphere as well as the temperature of near-ground air. The dynamics can be determined by calculating balance ratios of the inflowing and outflowing energy.

Thus, the planet Earth is found in the flow of solar shortwave radiation (wave length about 0.5 μm). The parallel beam of solar rays delivers to the boundary of the atmosphere about 1.4 kJ m^{-2} s^{-1} of heat. The Earth's surface receives only a part of the solar radiation. The remaining part is reflected by the clouds into space, or diffused and absorbed by the atmosphere. The solar energy that reaches the Earth's surface consists of direct radiation, Q_{dir}, and that diffused in the atmosphere, q_{dif}. Owing to the globular shape of the Earth its surface receives different amounts of direct solar radiation per unit of area. Besides, owing to rotation of the Earth and its circulation around the Sun as well as the inclination of the rotation axis to the ecliptic plane, the amount of solar energy inflow to unit area of the Earth has a distinct daily and annual variation, and owing to many-year and many-century variations of the elements of the Earth orbit, has many-century variations, too.

Part of the radiation received by the Earth's surface ($Q_{dir} + q_{dif}$) is reflected from it, while the remaining part is absorbed. The ratio of the reflected part of the radiation to the whole inflowing radiation is represented by the albedo α of the surface. Total average albedo of the Earth as a planet is equal to 0.37–0.40. For natural surfaces of the Earth the value α varies within a wide range, from 0.05 for a water surface to 0.95 for new snow. The shortwave solar radiation absorbed by the Earth's surface can thus be presented in the following form: $(Q_{dir} + q_{dif})(1 - \alpha)$. However, that is not the whole of the inflow part of the energy balance. A large amount of heat in the form of long-wave radiation is received by the surface from the heated atmosphere (infra-red radiation with maximum energy in the band of $\lambda \approx 8$–$10\,\mu m$), since the surface of any heated body radiates energy in proportion to the fourth power of the absolute temperature of the surface $(I = f(T^4))$. The part of the energy radiated by the atmosphere and absorbed by the Earth's surface is represented by I_a. The inflow part of the balance should include the heat arriving at the surface from the Earth's interior, q. However, the value of this component is much less than others and, as a rule, is not taken into consideration.

The inflow part of the balance provides energy for the majority of processes that take place on the Earth's surface and in the ground below. First of all the inflow part of the balance maintains temperature on the Earth's surface, which differs substantially from 0 K. Therefore, expenditure items of the radiation-thermal balance of the Earth's surface consist, first of all, of the amount of long-wave radiative energy which the Earth, as a heated body, gives back to external space, I_r. The difference between I_r and I_a is often called in climatology the long-wave effective radiation of the Earth's surface $(I_{ef} = I_r - I_a)$. By using the indices discussed and assuming the radiation balance R of the underlying surface to be the difference between the absorbable shortwave radiation of the Earth's surface and the outgoing long-wave radiation the equation will take the following form:

$$R = (Q_{dir} + q_{dif})(1 - \alpha) - I_{ef} \qquad (10.1)$$

As regards other essential and better-studied processes, there are evaporation (condensation) at the surface (LE is the product of latent heat of evaporation-condensation and the amount of evaporating or condensing moisture), turbulent heat exchange at the surface with ambient air, p, and heat flows into the soil, B. During the summer period these processes lead to energy loss at the surface while in winter heat flows B in the soil are directed towards the surface and should belong to the inflow part of the balance. This is often valid for the turbulent heat exchange p. Thus, the form of the thermal or radiation-thermal balance of the Earth's surface is much dependent on the time of the particular balance.

In climatology the equations of the radiation-thermal balance are usually in a form that involves grouping of terms according to the manner of heat exchange, and not by their belonging to the inflow or expenditure items of the balance: one part comprises the components of radiative heat exchange, the other one – the components associated with convective and conductive mechanisms of heat transfer:

$$(Q_{\text{dir}} + q_{\text{dif}})(1 - \alpha) - I_{\text{ef}} = R = LE + p + B \tag{10.2}$$

where the left-hand term, indicated as R, is the radiation balance, while the right-hand one is the thermal balance.

Each member of the radiation-thermal balance is an integral characteristic. They represent the amount of energy that has arrived at the surface or left it during a certain time period (year, half-year, month, ten days, etc.). The variability of individual members and the whole form of the balance is largely determined by this time interval. Thus, for instance, if the balance is calculated for a year, then the value of B (with a regime near to steady with respect to period) is practically equal to zero and can be left out of calculations; the values LE and p, however, belong to the expenditure items of the balance. If the balance is calculated for a half-year, then B becomes comparable with other members of the balance. If calculated per month, all terms of the radiation-thermal balance have a distinct annual variation, i.e. they become functions of time in the yearly cycle. This is a result of annual variation of the amount of incoming solar radiation to the surface in accordance with the law similar to the sinusoidal one. The key role of the first member of the radiation-thermal balance arising from the influence of mainly astronomical factors thus manifests itself. The remaining terms of the incomings and outgoings of the balance are equal in their mutual influence and conditionality. Each of the balance components has its 'earth' factors (geological, geophysical, geographical), which determine their variability. But with the change of even one component of the balance the remaining ones respond and are modified until a new balance is established.

The mechanism of this relationship is based on the dependence of each component of the balance (with the exception, probably, of the first member), on the temperature of the Earth's surface for which the balance was calculated. Therefore, surface temperature t_{surf} is an objective index of heat exchange level on the surface, i.e. the greater the value of inflow or expenditure of the balance, the higher is surface temperature. Let us consider an example for the sake of illustration. Let us assume that over the Earth's surface there was equilibrium of inflow and expenditure parts of the balance at a certain temperature. At a given moment wetting conditions of the surface changed and the value E increased. Thus, the expenditure part of the balance increased too, while the inflow part remained unchanged. This will lead to reduced temperature of the surface, which, in its turn, will change many components of the radiation-thermal balance. But the most substantial change, a decrease, will take place with I_r. Thus, expenditure will be less, and a new equilibrium will be established, but at a lower level of heat exchange – at a lower t_{surf} value.

In the spring period the amount of absorbed radiation increases and the incoming part of the radiation-thermal balance equation exceeds its outgoing part. As a result of this the surface temperature increases and all the components of the balance change depending on the temperature. First of all, I_r increases, and expenditure tends to be equal to the incoming. Thus, following sinusoidal (yearly) variations of the absorbed radiation value there are sinusoidal variations of the Earth's surface temperature. It should be noted that thermodynamic equilibrium of radiative heat exchange is established rather quickly, but rapid fluctuations of surface temperature are hampered by the high inertia of the temperature field of the underlying ground. This inertia of the temperature field demonstrates itself in the cycles of heat B. Therefore, near-surface temperature fluctuations show a lag relative to variations of the absorbed solar radiation. The value of these variations and the value of the lag are determined by not only by the amplitude of the fluctuations of the incoming radiation, but also by the thermal-physical characteristics of the underlying ground (λ, C, Q_{ph}).

If the radiation-thermal balance is considered not only on a planetary and zonal scale, but also at regional and local levels, one should bear in mind that equations of the balance should represent a particular formulation of the law of conservation of energy. These equations can be compiled for a thin surface layer and for different volumes of lithosphere, hydrosphere or atmosphere where energy (heat) can arrive in different ways. In particular, heat transfer can occur by moving water which will require consideration of the relevant water balance along with the thermal. The water balance equation, for example, of the land surface, gives an expression of equality of incoming and outgoing moisture at the horizontal surface under consideration or into a certain volume of ground for a certain time period. For a land surface this equation can be represented as follows:

$$r = e + f + b \tag{10.3}$$

where r is precipitation; e is the difference between evaporation and condensation of moisture at the Earth's surface; f is runoff; b is variation of moisture content in the ground. It is important that lateral heat transfer by movements of air or water is taken into account in the consideration of the thermal balance of the hydrosphere or atmosphere regions. However, inadequate accuracy of measurements makes it necessary in practice to resort to estimation of thermal balance, ignoring many of the terms as these are poorly studied.

10.2   Thermodynamic conditions for development of seasonally and perennially frozen ground

As noted above, the particular value of surface temperature is formed in the process of interrelated and mutually conditioned variation of the individual components of the radiation-thermal balance in the 'ground-atmosphere' system. There are several methods of determining the functional relationship between the temperature regime of the Earth's surface and certain components of the radiation-thermal balance. One of them is associated, for example, with determination of the difference of mean annual temperature of surface and air Δt_p using the value of the turbulent component p of the radiation-thermal balance. It is assumed that the temperature field of the soil and air is a result not only of radiation-thermal exchange over the surface of the given site, but also a result of inflow and outflow of heat in the course of air circulation. With the known coefficient of convective heat transfer k from the surface the above relationship can be expressed by the equation:

$$p = k\Delta t_p = R - LE - B \tag{10.4}$$

then

$$\Delta t_p = 1/k(R - LE - B) \tag{10.5}$$

Another method of determining dependence of soil temperature on components of the radiation-thermal balance is based on solution of the balance equation as to I_{ef}:

$$I_{ef} = \sigma s T_{surf}^4 (0.4 - 0.06\sqrt{e})(1 - cn^2) = (Q_{dir} + q_{dif})(1 - \alpha) - LE - p - B \tag{10.6}$$

where σ is Stefan-Boltsmann's constant (2.08×10^{-7}); s is radiating power of the surface as compared with that of an absolutely black body (0.85–1.0); T_{surf} is the absolute temperature of the radiating surface; e is absolute air humidity; n is cloud cover in fractions of a unit; c is coefficient of variation of cloud cover by latitude.

By simple transformations and using the values of all parameters included in the equation (10.6) mean annual temperature T_{year} of the radiating surface can be obtained from the summer (index 'S') and winter (index 'w') periods (17):

$$T_{year} = \frac{1}{2^4\sqrt{\sigma s}}\left(\sqrt[4]{\frac{(Q_{dir} + q_{dif})(1-\alpha)_s - LE_s - P_s - .78B_s}{(1 - cn_s^2)(0.4 - 0.06\sqrt{e_s})}}\right.$$

$$\left.+ \frac{\sqrt[4]{(Q_{dir} + q_{dif})_w(1-\alpha)_w - LE_w - P_w + 0.7B_w}}{(1 - cn_w^2)(0.4 - 0.06\sqrt{e_w})}\right) \quad (10.7)$$

If the right-hand side of this equation is equal to or less than 273.1 K, then perennially frozen ground can be formed. Condition $T_{year} \leqslant 273.1$ K shows that mean annual temperature of the layer that underlies the ground surface is below 0°C, i.e. water will freeze and the ground will pass into the frozen state. In truth, that will be valid for a bare surface without snow cover and vegetation and in the absence of infiltrating water and of a temperature offset in the layer of seasonal freezing.

Finally, dependence of ground temperature on components of the radiation-thermal balance can be determined from the balance equation solved against annual heat cycles in the ground B. Indeed, since the heat cycles in soil and rocks depend on the temperature regime of the ground, and the thermal conductivity, heat capacity and phase transformations of moisture are known (17), then:

$$B = \xi(nA_{mean}C + Q_{ph}) + \sqrt{2}t_{surf}^{mean}\sqrt{\frac{\lambda TC}{\pi}} \quad (10.8)$$

and since $B = R - LE - p$, then:

$$t_{mean}^{surf} = R - LE - p - \xi(nA_{mean}C + Q_{ph})/\sqrt{\frac{2\lambda TC}{\pi}} \quad (10.9)$$

where ξ is depth of seasonal freezing or thawing of the ground; A_{mean} is the average amplitude of annual temperature fluctuations in layer ξ; C is volumetric heat capacity; λ is thermal conductivity coefficient of the soil; Q_{ph} is heat of phase transitions of moisture in the soil; t_{mean}^{surf} is mean annual temperature of bare surface of ground; T is a period equal to one year; n is a coefficient approximately equal to 2 at low values of Q_{ph} and $\sqrt{2}$ with greater Q_{ph} and λ.

Since the mean annual temperature of a bare surface t_{mean}^{surf} can differ greatly from that of soils at the base of the seasonally freezing layer t_{mean}^{ξ} owing to the warming effect of snow Δt_{mean}^{sn} and vegetation Δt_{mean}^{veg} cover and infiltrating rainfall Δt_{mean}^{inf}, on account of the temperature offset in the layer of seasonal freezing Δt_{mean}^{sh}, then the thermodynamic condition for the development of perennially frozen strata should be represented as follows:

$$t_{mean}^{\xi} = t_{mean}^{surf} + \Delta t_{mean}^{sn} \pm \Delta t_{mean}^{veg} + \Delta t_{mean}^{inf} - \Delta t_{mean}^{sh} \leqslant 0°C \qquad (10.10)$$

The equations above (10.8–10.10) show that the temperature regime of the ground is determined by the incoming amount of solar radiation and structure of the radiation-thermal balance (proportions of its components), but it is also dependent on composition and thermal-physical properties of the underlying earth materials, on heat cycles, temperature shift and phase transitions of moisture, as well as on surface cover, ground water regime, etc. In other words, the thermodynamic conditions of perennially frozen strata are determined by geographical, geological and hydrogeological conditions.

Of paramount importance is the determination of the thermodynamic conditions for the development of perennially frozen strata as well as of the thermodynamic conditions for the formation of seasonally freezing and seasonally thawing soil layers and thus of the conditions enabling or making impossible their co-existence. Account should be taken of the mean temperature of the bare surface t_{mean}^{surf} and the temperature deviation from the mean value during a year, i.e. the amplitude of temperature fluctuation at the surface of the ground A_0. For the sake of convenience let us assume that $t_{mean}^{surf} = t_{mean}^{\xi} = t_{mean}$. In such a case four essentially different situations can be analyzed concerning development of the seasonally freezing and perennially frozen soils (Fig. 10.1).

For condition $t_{mean} > 0$, and $A_0 < |t_{mean}|$, only unfrozen ground can exist (situation I), as for the whole year the soil surface temperature does not go through 0°C into negative temperatures. In the case $t_{mean} > 0$, but for a certain time (the cold period of the year) surface temperature and that of the ground beneath is negative due to $A_0 > t_{mean}$, then there is seasonal freezing of ground and formation of the seasonal freezing layer (situation II). In the case when $t_{mean} < 0$ the surface and underlying soil have a positive temperature during the warm period of a year (due to $A_0 > |t_{mean}|$). As a result, partial thawing from the surface of perennially frozen ground takes place with formation of the seasonal thawing layer (situation III). Finally, when during the whole year mean surface temperature does not exceed 0°C ($t_{mean} < 0$, $A_0 < |t_{mean}|$) there is perennially frozen ground without seasonal thawing from the surface (situation IV).

Thus, the indispensable condition for the existence of perennially frozen ground is $t_{mean} < 0$. The occurrence or absence of the processes of seasonal freezing and thawing and, accordingly, seasonally frozen or seasonally

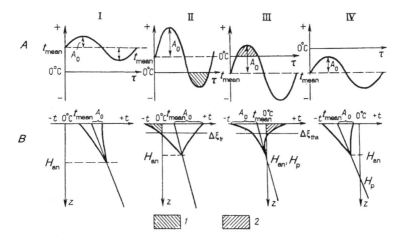

Fig. 10.1. Conditions for formation and existence of seasonally and perennially frozen ground: A – expressed in terms of amplitude A_0 of temperature and mean annual temperature t_{mean} at the ground surface; B – by following the envelope of temperature fluctuations in the layer of annual temperature variation H_{an}; ξ_{fr} and ξ_{th} – depth of seasonal freezing and thawing of soils, H_p – lower boundary of permafrost; z – depth, t – temperature, τ-time; 1–2 – layers of seasonal freezing and thawing, respectively; I–IV – for different values of A_0 and t_{mean}.

thawed layers are determined by the relation of mean annual temperature and amplitude of temperature fluctuations at the surface of the ground. They exist if $|t_{mean}| < A_0$ and they are absent if $|t_{mean}| > A_0$. All the above cases (situations) can be shown in a diagram of A_0 versus t_{mean} (Fig. 10.2). The bisectors correspond to the condition $|t_{mean}| = A_0$ and therefore serve as boundaries of the likely existence of seasonal freezing or thawing.

Of paramount importance in geocryology is the concept of potential thawing or freezing. Potential thawing is possible when $t_{mean}^{surf} > 0$ in the case of seasonal freezing of ground. The term refers to the depth of thawed ground developed during the summer period under the conditions $t_{mean} > 0$ and $|t_{mean}| < A_0$, when, at the commencement of thawing the whole ground mass is frozen. This can be explained by an example: in an area of unfrozen ground ($t_{mean} > 0$), there was excavation of material in winter to make an embankment of great thickness. The ground used was in the frozen state. Then in summer time this embankment thaws from the surface and, since it is rather thick, it will not thaw completely during the summer. Thickness of the layer thawed during the summer will correspond to potential seasonal thawing. A similar example can be given for the case of potential seasonal freezing ($t_{mean} < 0$).

With ground temperature near to 0 °C episodic increases of seasonal

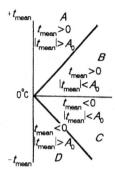

Fig. 10.2. Diagram showing conditions for development and occurrence of
seasonally and perennially frozen soils for various values of A_0 and t_{mean}:
A – perennially frozen soils and seasonal freezing are absent; B – perennially
frozen soils are absent, seasonal freezing occurs; C – perennially frozen soils and
seasonal thawing occurs; D – perennially frozen soils occur, seasonal thawing is
absent.

freezing that exceed the depth of potential thawing for the given ratio of
t_{mean} and A_0, can give rise to a thin frozen layer existing for more than one
year. Such frozen layers are called *pereletok* (short-term permafrost). It is
distinguished from the short-term existing perennially frozen ground by the
inconsistency of its freezing.

10.3   Frozen ground as a result of zonation of thermal- and mass-exchange processes on the Earth's surface and in the atmosphere

Zonal distribution of the radiation balance of the 'Earth-atmosphere' system is dictated by the key role of astronomical factors: altitude of
the Sun above the horizon and sunlight hours per year. As an example, Fig.
10.3 presents mean values by latitude of the thermal balance components of
this system. Analysis shows that the expenditure part of the thermal balance
mainly consists of heat consumption for evaporation at the surface and for
turbulent heat flow between the Earth and atmosphere. Shortage or excess
of incoming or outgoing heat are carried away by horizontal motions in the
atmosphere. An important item of the 'Earth-atmosphere' thermal balance
is absorption or yield of heat by oceans. In tropical or temperate latitudes
the absolute values exceed the value of residual radiation, i.e. the radiation
balance. Absorption or yield of heat by the lithosphere is only a small
portion of the system's thermal balance for a season.

Table 10.1 gives the values of the Earth's surface thermal balance obtained by different authors experimentally both in temperate and polar
latitudes. The data of Table 10.1 show that absolute values of each compo-

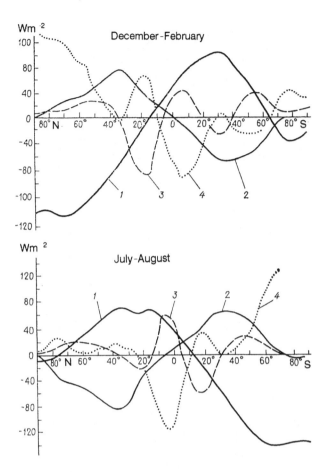

Fig. 10.3. Average latitudinal values of components of thermal balance of the 'Earth-atmosphere' system (according to G.N. Vitvitskiy): 1 – radiation balance of the 'Earth-atmosphere' system; 2 – variations of heat content (accumulation or loss of heat for the period under consideration) of hydrosphere; 3 – phase transitions of moisture; 4 – re-distribution of heat by horizontal motion of the atmosphere and oceans.

nent vary during a year. Maximum values of the radiation balance, heat consumption for evaporation and turbulent heat exchange are typical for the middle of the summer period, while minimum values are typical for the middle of the winter period. Maximum heat flow into the soil occurs in the early summer, while maximum yield by the soil occurs in the early winter. This displacement as to maximum inflow and outflow of heat at the surface is associated with a high heat consumption for thawing of snow cover $Q_{\mathrm{ph}}^{\mathrm{sn}}$.

On the territory of the former USSR the radiation balance of the surface varies from zero on the remotest Arctic islands to $200\,\mathrm{kJ\,cm^2}$ in the foothills of Central Asia. In mountains of the Russian North-East, its value is lower,

Table 10.1. *Seasonal sum and percentage relationship between arrival* ($+$) *and consumption* ($-$) *of heat balance of the Earth's surface on the open areas of plains (after M.K.Gavrilova)*

Components of heat balance	Summer 10^4 kJ cm^{-2}	%	Winter 10^4 kJ cm^{-2}	%	Landscape, soil composition
Zagorsk (Moscow region)					
R	120.2	$+100$	-8.8	-81	Various grass
LE	89.6	-75	2.0	-19	clover meadows,
P_B	23.5	-19	-7.9	$+73$	sandy silty
Q_{ph}^{sn}	7.1	-6	-2.9	$+27$	material
			0	0	
Vorkuta (North-East of European Russia)					
R	46.0	$+100$	-6.7	-76	Hummocky tundra,
LE	18.4	-40	0	0	heavy silty clay
P_B	20.9	-45	0.8	-10	
Q_{ph}^{sn}	6.7	-15	-8.8	$+100$	
			1.3	-14	
Solenyy (village) (north Western Siberia)					
R	58.6	$+100$	-19.2	-79	Moss and lichen
LE	28.1	-48	5.0	-21	hummocky tundra,
P_B	23.4	-40	-14.2	$+59$	silty clayey
Q_{ph}^{sn}	7.1	-12	-10.0	$+41$	material and
			0	0	sandy silty material
Yakutsk (Central Yakutiya, Lena valley)					
R	88.7	$+100$	-7.9	-76	Mixed-herbaceous
LE	25.5	-29	2.1	-20	grassy meadow,
P_B	57.8	-65	-0.8	$+8$	sandy silty
Q_{ph}^{sn}	5.8	-6	-9.6	$+92$	material
			0.4	-4	

less than 20 kJ cm^2; in mountains of Southern Siberia it is about 40 kJ cm^2. The annual sum of the radiation balance does not exceed 40% of the incoming radiation. The natural surface albedo of a bare site during the snowless season has a comparatively low value of 0.15–0.16. Effective radiation is 25–32% of the total radiation. Heat consumption for surface evaporation varies within a wide range compared to total radiation: 15–39% in summer and 11–33% in winter. Heat flow into the soil for a thawing season (and in the same way, for a freezing season) and correlations of the flow with radiation balance, vary from west eastwards, and from north southwards in the territory of the former USSR. Over the European Arctic coast (the settlement of Amderma) and north-east of European Russia it makes up about 15% of the value of R, in the north of West Siberia 12%, and in Central Yakutia 10%. Eastwards of Lake Baikal and in the south of Central Siberia the total heat flow into soil makes up 7% and in desert regions of Central Asia 2–4%, of the value of R.

Let us consider the characteristics of atmospheric circulation over the territory of the former USSR which have a substantial influence on the heat- and mass-exchange of ground with the atmosphere. It is known that since there is a difference in heating of the atmosphere at low and high latitudes there exists constantly a meridional gradient of pressure which is a driving force for the Earth's atmospheric circulation. Boundaries of zonal transfer do not coincide with boundaries of the radiation zones as the zonation of air flows is determined not only by radiation processes but also by such factors as blocking processes and Arctic intrusions.

Blocking is due to high anti-cyclones that disturb the western transfer of air masses typical of the temperate latitudes. Most often blocking processes develop above the European territory. In Eastern Europe cyclones come from the Mediterranean bringing warm and humid flows. Warm anti-cyclones block the usual paths of cyclones directed towards the east and cause them to deviate northwards. It is with these processes that the features of the radiation-thermal regime of European Russia are associated, the shifting northwards of the sum total of absorbed radiation, values of effective radiation and radiation balance, temperature zones, etc. Arctic intrusions are most often encountered in Eastern Siberia. These processes weaken the zonal circulation and are accompanied by the advance into temperate latitudes of high, cold troughs. The almost constant presence of high-altitude troughs is the main characteristic of the atmospheric circulation relevant to the formation of the radiation-thermal regime of the Asian part of the former USSR. As a result of winter intrusions of cold air, the mean temperature of the lower half of the troposphere, at the 40th parallel in the east of Asia is almost 4 °C lower than that in North America. The lower temperature is explained by the bigger size of the continent which, together with the orography of its surface creates favourable conditions for the radiation cooling of air, the increase of the latitudinal gradient of the total absorbed radiation, the effective radiation, the temperature of the ground layer of air, etc.

Heat and mass exchange at the Earth's surface with the atmosphere has an altitudinal-zonal or altitudinal-belt pattern in mountain regions, where, with higher altitude the smaller is the radiation balance owing to greater loss of radiative energy by reflection and radiation. However, of importance here is the role of the advection component in the 'Earth-atmosphere' system. A substantial role is played by the aspect of macro-orographic elements of the mountain regions – ranges, ridges and massifs. Usually, windward slopes in the path of air flows bringing heat and moisture (western slopes of the Urals, Putorana Plateau, Anabar massif and Verkhoyansk Range) receive more heat than leeward eastern slopes. In the areas of influence of monsoon air flows of the Pacific, on the contrary, the eastern slopes of the mountain structures receive more heat and moisture (Sikhote-Alin, Bureinskiy, Taykan, Al'skiy ranges and others). Such features of heat and mass exchange may lead to either shifting of altitudinal boundaries of belts or variation of their vertical extent on slopes of different aspects.

Thus, the radiation processes along with zonal components of the atmospheric circulation serve as prerequisites of the most characteristic latitudinal features of heat and mass exchange over the Earth's surface on the territory of the former USSR. Meridional components of the atmospheric circulation give rise to longitudinal and sectorial differences within the limits of the latitudinal zones. Circulation of the atmosphere associated with the formation and shift of cyclones and anti-cyclones both in zonal and meridional directions from the place of their origin, brings about fluctuations of temperature and air humidity. Simultaneously, there is transformation of air masses over the surface which is characterized by a variety of properties and conditions for incoming and outgoing radiation and thermal balance.

The role of the atmospheric circulation and of radiation-thermal processes in the development of weather conditions differs according to the season of the year. The radiation factor has the greatest influence in summer time, as is revealed in the intensification of the process of transformation of the thermal regime of the air in the south-eastern regions of European Russia, Kazakstan and Central Asia (more than a half of the radiation balance is accounted for in the heating of the atmosphere). However, the prevalence of the northern winds in summer over the European North of Russia, West Siberia and Kazakstan brings about a lowering of the general level of air temperature. The summer maximum of precipitation which constitutes 80–85% of the annual amount is associated with cyclonic activity in Eastern Siberia and the Russian Far East. In the Far East cyclones and thick cloud cover with sea winds lead to substantial loss (up to 60%) of solar radiation.

In winter the air temperature regime over the greater part of the former USSR forms largely under the influence of the atmospheric circulation as the incoming solar radiation is drastically reduced. The role of the radiation factor in winter is evident in the radiation cooling of the surface (on account of the effective radiation), which leads to higher temperatures in European Russia than in the east of West Siberia, Central and Eastern Siberia and the Russian Far East, at the same latitude. This difference may be as much as 10 °C. It is explained by the fact that cyclonic activity over the former USSR European territory and north-east of West Siberia brings about thick cloud cover that hampers the radiative cooling and reduces the depth of snow cover. Snow covers 70-80 cm thick are observed north of 60° N in the north-east of European Russia and West Siberia, and on windward slopes of the Central Siberian plateau (over 80 cm). On the contrary, in Eastern Siberia, during the cold season of a year, cold air masses prevail that are greatly cooled in the ground layer of air and become even colder than the Arctic masses (in November and March), showing a thermal inversion.

Thus the observed distribution of temperature, precipitation and other climatic elements is the result of a complicated interaction of all climate-forming factors. In accordance with the climatic zonation of B.P. Alisov, four climatic belts are distinguished by their radiation regimes: Arctic, sub-Arctic, temperate and subtropical. In general, the recent climatic conditions over the greater part of the former USSR territory reflect such a heat exchange between the soil surface and atmosphere as to give rise to the existence and formation of permafrost. In the north of European Russia, West and East Siberia, and in the Russian North-East these conditions in Arctic and sub-Arctic belts are associated with long winters and a predominance of Arctic air masses over the territory. In East Siberia and eastwards of Lake Baykal, within the limits of the temperate climatic belt the development of permafrost is associated with formation of a continental regime of atmospheric circulation with frequent Arctic intrusions of cold air masses, an anti-cyclone regime and radiative cooling of the ground surface in winter time. Within the subtropical belt, conditions of perennial freezing are observed in the mountain regions at elevations exceeding 3000 m and are related to intense radiative cooling of the surface and certain other factors.

Since latitudinal zonation and altitudinal belts together with their sectorial and longitudinal distinctions are the main factors in the differentiation of the radiation-thermal regime of the ground surface, then, the same patterns are found in the differentiation of the temperature regime of the ground and, accordingly, the differentiation of the areas of seasonally and perennially frozen ground.

11

Seasonal freezing and thawing of ground

11.1 Formation of the layer of seasonal freezing and thawing of soil

According to V.A. Kudryavtsev, seasonal thawing is thawing from the top, of frozen ground having a mean annual temperature below $0\,°C$. The layer of seasonal thawing is always underlain by permafrost and its thickness is determined by heat cycles in the layer that extend to positive temperatures of the ground. Seasonal freezing is the process of freezing from the top of unfrozen ground having a mean annual temperature above $0\,°C$. The layer of seasonal freezing is underlain by unfrozen ground and its thickness is determined by heat cycles that extend to negative temperatures of the ground.

The main features of the layer of *seasonal thawing* from the viewpoint of thermal physics are as follows. With the mean annual temperature being negative the layer is frozen for the greater part of a year. However, in summer the surface temperature becomes positive, for which an indispensable condition is $A_0 > |t_{mean}^{surf}|$. The beginning of the process of ground thawing is associated with this progression of negative temperature to positive values.

Seasonal thawing of ground presents a complicated process of thermal physics with phase transitions, migration and seepage of moisture in thawed and frozen zones of the layer, ground subsidence and the like. This process goes on, as long as sufficient heat arrives on account of positive heat cycles, at the front of thawing to continue the process of transition of the frozen soil moisture into the liquid state. The depth of seasonal thawing reaches its maximum by the end of summer. The direction of the heat cycles that go through the ground surface changes with transition from its heating to cooling. The soils of the seasonally thawed layer begin to cool, and the thawing process slackens and stops altogether (Fig. 11.1a). Subsequently the thawed layer starts to freeze, and depending on particular conditions freezing can begin either at its bottom boundary, or from the ground surface.

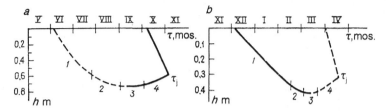

Fig. 11.1. Dynamics of the depth of seasonal thawing (*a*) and freezing (*b*) of the ground: 1–4 – stages: 1 – of rapid thawing (freezing); 2, 3 – respectively, of slower penetration and of relatively stable position of thawing (freezing) front; 4 – freezing (thawing) from below; τ_j – moment of joining.

It can be tentatively assumed (3) that freezing from below of the thawed layer starts later than from the top (at t_{mean} of ground from -0.5 to $-4\,°C$), or simultaneously from the top and from below (at t_{mean} of ground from -4 to $-5\,°C$) or it can occur earlier than that from the top (with t_{mean} of the permafrost below $-5\,°C$). With lower mean annual temperatures of the ground and higher temperature of freezing at the surface a greater portion of the seasonally thawed layer freezes from below. In natural conditions, at t_{mean} ranging from -7 to $-10\,°C$ usually 15–20% of the seasonally thawed layer freezes from below. Freezing from the top starts from the moment of stable negative temperature over the surface. Thus, freezing of the seasonally thawed layer is as a rule, characterized by development of two fronts or boundaries dividing 'frozen-unfrozen ground' resulting from the development of the freezing process both from the top and from below. Such a pattern of freezing leads to moisture migration to both fronts, schlieren ice formation in the freezing zones, the arising of tensions and deformations in the soil etc. Freezing of the thawed layer continues until the fronts join together (see Fig. 11.1a). From this moment (which is called the moment of joining, τ_j) and on to the new development in early summer of the process of seasonal thawing, the ground is found in the frozen state; although there can be processes of moisture migration, phase transitions and ice segregation, provided there are corresponding temperature gradients, which can bring about re-structuring of the cryogenic structure of this layer.

In *seasonal freezing* many aspects of the formation and development of the seasonally frozen layer are qualitatively distinct. Thus, freezing of this layer occurs only from the top, i.e. there is only one boundary dividing 'frozen-unfrozen ground'. The processes of moisture migration and phase transitions that occur in the seasonally frozen layer give migrational-segregational interlayers of ice; there is vertical heaving of the ground surface, deformations and tensions of shrinkage and swelling, etc. With deepening of the freezing front the rate slows to a complete stop, usually observed in the late winter (see Fig. 11.1b). Then, after the surface temperature has risen above 0 °C the process of thawing of the seasonally frozen layer both from the top and from below begins. Thawing from below can occur simultaneously, later or earlier than that from the top, determined by the value of heat flow from the underlying thawed (unfrozen) ground, i.e. by the value of the positive mean annual temperature of the ground. The higher the temperature, the greater the portion of the seasonally frozen layer that is thawed from below. In the southern and snow-abundant regions, in particular, the island of Sakhalin, the south of the Ukraine, Caucasus, Crimea and Central Asia, the seasonally frozen layer of the ground mainly thaws from below. Thus, two fronts of thawing are formed during thawing of the seasonally frozen layer, i.e. two 'thawed-frozen ground' boundaries. Complete thawing of this layer occurs when the two fronts join, which is usually observed in the late spring-early summer (see Fig. 11.1b).

The rate of movement of the fronts during freezing and thawing is dependent on a variety of factors. The rate of thawing ξ'_{tha} and of freezing ξ'_{fr} from the top is first of all dictated by the temperature regime of the ground surface. The more rapid the temperature increase (for ξ_{tha}) or drop (for ξ_{fr}) and the greater the amplitude of these fluctuations, the quicker and deeper the front moves from the top downwards. This rate is much dependent on the thermal-physical properties of the soils and most of all on the value of Q_{ph}. The greater is Q_{ph} (i.e. with greater moisture content of the freezing-thawing soils) the slower is the movement of the front ξ. The rate of advance of the freezing or thawing boundary from below is mainly determined by the mean annual temperature of ground at the bottom of the layer ξ.

The maximum depth of both seasonal freezing and thawing, considered over a long-term period, varies within a wide range. Annual deviations of the thawing depth from the mean long-term value at the same place increase southwards while those of freezing depth diminish southwards. Thus, in the north of the permafrost zone at mean annual temperatures of the ground below -5 °C, variations of the seasonally thawing layer depth at a point do not exceed 10 cm. In the Moscow-city region, for example, within the limits of the same site, for the period of 25 years seasonal freezing depth varied from several centimetres to 1 m, i.e. varied by an order of magnitude.

The dynamics of seasonal freezing and thawing depths is studied at weather stations and posts with the help of cryopedometers of various design. Under field conditions depth of freezing or thawing may be studied directly in bore holes, outcrops, etc., by geophysical methods (vertical electric sounding, electric prospecting, thermometry and the like). All the

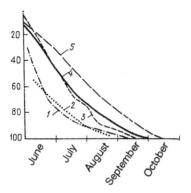

Fig. 11.2. Curves of thawing intensity for the ground in different regions of permafrost: 1 -Srednekolymsk; 2 – Yakutsk; 3 – Noril'sk; 4 – Anadyr; 5 – Salekhard.

data on depth of thawing ξ_τ over time obtained during field studies indicate depths of maximum seasonal thawing ξ_{tha} at the same points in accordance with the method of V.F. Tumel (for seasonal freezing – according to the method of L.D. Pikulevich). Relative seasonal thawing (freezing) versus time τ is plotted for this purpose ($I_{tha} = \xi_\tau/\xi_{tha} \times 100$). Despite the fact that there are annual variations of depth ξ, the intensity or rate of thawing (freezing) remains more or less constant. The availability of such empirical dependencies for different types of terrain makes possible tentative estimation of maximum depth of seasonal thawing (freezing) by a single measurement ξ_τ. As an example, Fig. 11.2 shows curves of soil thawing intensity obtained for the different regions of permafrost development.

Cyclic recurrence of seasonal freezing and thawing that occurs in a relatively thin stratum of the ground, the annual passage of positive and negative temperatures accompanied with phase transitions, to a great extent determines the specific composition and cryogenic structure of the seasonally thawing and seasonally freezing layers. The process of repeated freezing and thawing leads to the formation of primary silty particles. At the same time physical-mechanical processes that occur in the freezing of clay-rich soils cause coagulation of clay and colloid particles and formation of secondary silty micro-aggregates. As a result, the soils of seasonal thawing and freezing layers are characterized by a higher silt content.

The dynamics of the seasonal freezing and thawing processes in the permafrost zone leads to a differentiation in the composition of the profile of these layers. Thus, in the layer of seasonal thawing three horizons may be distinguished: a lower one that freezes either at contact with the permafrost (at low negative t_{mean} of the ground), or from the surface (at high negative

t_{mean}) which is subject to seasonal fluctuations of temperature; the middle one that freezes from above and is subject to both seasonal fluctuations and fluctuations over periods of days; and the upper one subject to seasonal fluctuations and those over periods of days, and diurnally. The deposits of the lower and, especially, upper horizons have, as a rule, increased fine-grain content. In the profile of the seasonally freezing layer, as regards grain size often only two horizons are discerned.

Despite the fact that chemical and microbiological processes in the seasonally thawing layer are much reduced compared with the layer of seasonal freezing, the soils of the permafrost zone have a rather large absorbing complex. They contain secondary minerals and there is a process of gleying that causes peptization of the previously formed micro-aggregates. Gleying is accompanied by the formation of considerable amounts of hydrophilic organic and mineral colloids that promote a thixotropic structure of the ground.

The features of the *cryogenic structure* of the seasonally frozen and seasonally thawed layers are determined by the composition, texture, thermal-physical and water properties of the soils, by the initial (pre-winter) moisture content and its distribution over the section, the depth to and regime of the ground water relative to the base of the seasonally frozen layer, and by the distribution of the permafrost and overlying water in the seasonally thawed layer and its regime, with the dynamics of winter freezing and the temperature regime of the freezing soils. Ice distribution through the section is different in the case of seasonal freezing and in the case of seasonal thawing.

In the case of seasonal freezing massive cryogenic texture is, as a rule, formed in the uppermost part of the layer, the freezing of which occurs under large gradients of temperature. Then, lower in the layer where the freezing rate diminishes there are more favourable conditions created for migrational ice segregation and here schlieren cryogenic structures are formed in which ice layers become somewhat thicker and more spaced with depth (see Fig. 8.3). In general the ice content and thickness of the ice layers are dependent on the depth to the ground water level. In conditions of an arid climate, where seasonally frozen soils occur, massive cryogenic structure prevails over the whole profile. Northwards, where the climate becomes less arid, pre-winter humidity is greater and mean annual temperatures of the soils are lower, there is intense re-distribution of moisture in the layers. In the fine-grained soils of the upper part of the seasonally frozen layer (below a thin horizon without schlieren) laminated, netted, cellular and other cryostructures are formed, while in the lower part there is frequently massive

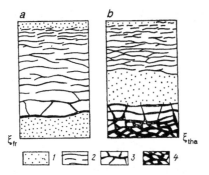

Fig. 11.3. Typical cryogenic structure of the soils of layers of seasonal freezing ξ_{fr} (*a*) and thawing ξ_{tha} (*b*): 1–4 – cryogenic textures (1 – massive, 2 – laminated, 3 – netted, 4 – ataxite).

structure. Only if ground water is nearby are schlieren textures and higher ice content found in the lower part (and rarely in the middle part).

Regarding the layer of seasonal thawing in fine-grained soils, ice distribution in the section is dependent on the ratio of freezing from the top to freezing from below, pre-winter moisture content and the possibility of infiltration of overlying water from the surface. Most often a three-part cryogenic structure arises here with maximum ice content typical of the lower and upper parts of the layer. The middle part appears relatively desiccated. For the upper part of the layer rather fine laminated and netted cryostructures are typical, while for the middle one, massive and rarefied fine lens-like cryostructures are usual (Fig. 11.3). For the lower part of the layer, thickened laminated, netted and ataxitic cryostructures are typical and often horizontal ice schlieren seem to follow the outlines of the base of the seasonally thawed layer. The ice content of this part of the layer is usually equal to or exceeds total moisture capacity. It is practically independent of pre-winter moisture content and is formed in the course of slow freezing of the layer from below, in the most favourable conditions of moisture migration towards the front of freezing because the migrational and gravitational directions of moisture movement coincide. As shown by V.A. Kudryavtsev (17), freezing of the seasonally thawed layer from below is conditioned by half-yearly heat cycles that go through its base:

$$Q = |t_{mean}|\sqrt{2\lambda TC/\pi} \tag{11.1}$$

Assuming that all these heat cycles supply only phase transitions of moisture, i.e. formation of ice layers, one may easily calculate the maximum possible total thickness of such ice layers in the bottom of the layer of seasonal thawing. With average values of λ and C for the frozen soils the value Q is equal to $|t_{mean}|6285\,kJ\,m^{-2}$. Each $1000\,kJ\,m^{-2}$ on freezing from below can give an ice layer 0.3 cm thick. Based on this, the maximum thickness of the ice layer at $t_{mean} = -10°C$ can hypothetically be as large as 15–20 cm, at $t_{mean} = -5°C$, 7.5–10 cm, while at $t_{mean} = -1°C$ it does not exceed 1–1.5 cm.

If in the case of freezing of the seasonally thawed layer there is inflow of suprapermafrost water, then ice is uniformly distributed over the section and there is no horizon of soil losing water. The maximum heaving of the soils of the seasonally thawed layer is observed in this case, with formation of frost mounds persistent for a year. In conditions of maritime climate, when freezing from the top occurs relatively slowly, the most favourable setting for ice segregation arises in the lower horizon of the seasonally thawed layer. In the case of sharp, continental climate, when the rate of freezing from the top is high, freezing from below is insignificant. Accordingly less favourable conditions are created for ice schlieren segregation in the seasonally thawed layer, and the ice content of the lower horizon is reduced.

11.2   Types of seasonal freezing and thawing of the ground

As early as the first years of the 1950s, V.A. Kudryavtsev pointed out that the processes of freezing and thawing are, on one hand, thermal-physical resulting from thermal interaction of the atmosphere and surrounding space with the lithosphere, and on the other geological-geographical, since the thermal processes take place in the particular geological medium and a complicated geographical setting. It was convincingly proved that determination of the features of seasonal freezing and thawing should be based on thorough study of each factor and condition of the given process, taking into consideration that the depth of freezing or thawing is an extremely variable and unsteady parameter. The characteristics of this depth and its variation can be determined on the basis of consideration of thermal-physical and geological-geographical aspects of the process of seasonal freezing and thawing. Consequently, the thermal-physical relations of depth of seasonal freezing and thawing should be represented by parameters which take a full and comprehensive account of the influence of geological, geographical conditions and their dynamics in the course of natural developments and man's activity. In other words, the types of seasonal freezing and thawing of soils determined by the particular combinations of contributing factors should be identified and mapped, and not only the depths that vary from year to year. Based on the above V.A. Kudryavtsev elaborated the *classification of the types of seasonal freezing and thawing* according to four dominating attributes: 1) mean annual temperature of ground t_{mean}; 2) annual amplitude of temperature fluctuation at ground surface A_0; 3) composition of materials of the seasonally frozen and seasonally thawed layers; 4) moisture content of the materials. A specific feature of Kudryavtsev's classification is distinguishing the types of seasonal freezing and thawing by the conditions of their formation, and not by the depth. Depths can be obtained by using the field observational data and calculations for mean perennial or any extreme conditions. The criterion of their validity is comparison with actual data for the period identical to the studied one.

On one hand, the two first attributes are geographical; on the other, they characterize the thermodynamic conditions for existence of seasonally and perennially frozen ground. These indices are easily studied in natural conditions and may be mapped both at large and small scales. The last two attributes are geological ones and reflect the results of seasonal and perennial freezing of the ground under the established thermodynamic conditions at the surface.

The boundary that divides the phenomenon of seasonal freezing from that of thawing is assumed in the classification to be a mean annual ground temperature of $0\,°C$. Further division of seasonal freezing and seasonal thawing in the classification of V.A. Kudryavtsev is based on stability of the process with regard to the transition from one into the other confirmed by the value of the mean annual temperature and its deviation from $0\,°C$. Due to short-term fluctuations of climate and accidental deviations of mean annual temperature from its mean long-term value, to the south of the permafrost zone there are periodic transitions of mean annual temperature around $0\,°C$ of $+1$ and $-1\,°C$ and episodic ones within $+2$ and $-2\,°C$. Therefore there are transitional and semitransitional types of seasonal freezing and thawing of rocks in the classification based on mean annual temperature (Fig. 11.4). Transition of mean annual temperature from the range of $+2$ to $+5\,°C$ into the negative or from the range of -2 to $-5\,°C$ into the positive is associated with drastic and deep changes in heat exchange over the Earth's surface. Consequently, within these intervals of mean annual temperatures persistent types of seasonal freezing and thawing are distinguished. In the temperature interval of $+5$ to $+10\,°C$ and -5 to $-10\,°C$ the transition of one process into another is associated with long periods of climate change, and, therefore, steady types are distinguished, of both seasonal freezing and seasonal thawing of the ground. In the case of temperature being higher than $+10\,°C$ and lower than $-10\,°C$ transition of one process into another is practically impossible. Therefore, the type of

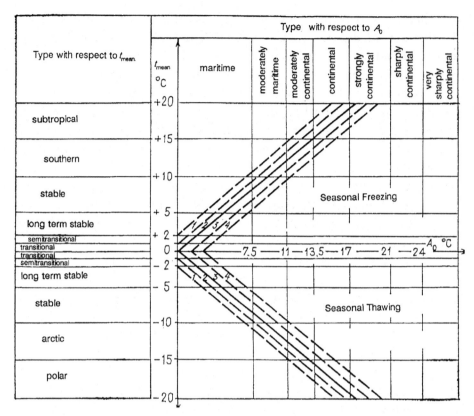

Fig. 11.4. Classification of types of seasonal freezing and thawing according to
mean annual temperatures and amplitudes of surface temperatures (according to
V.A. Kudryavtsev): 1–4 – types of seasonal freezing and thawing arising
episodically (1) and periodically (2); disappearing periodically (3) and episodically
(4).

seasonal freezing with t_{mean} from $+10$ to $+15°C$ is called southern, with
t_{mean} from $+15$ to $+20°C$ – subtropical, and over $+20°C$ – tropical,
whereas the type of seasonal thawing with t_{mean} from -10 to $-15°C$ is
called arctic and from -15 to $-20°C$ – polar.

The following types of seasonal freezing and thawing are distinguished by
the amplitude of ground surface temperature fluctuations the value of which
is dependent on continentality of climate: 1) maritime type with amplitude
less than $7.5°C$ typical of the sea coast in temperate latitudes; 2) moderate-
maritime with amplitude ranging from 7.5 to $11°C$ typical of the northern
seas coast; 3) continental with amplitudes ranging between 11 and $13.5°C$,
encountered in the European part of the former USSR; 4) continental, with
amplitude from 13.5 to $17°C$, typical, for example, of the West-Siberian
lowland; 5) somewhat more continental with amplitude from 17 to $21°C$,

typical, for example, of the Central Siberian tableland; 6) sharply continental with amplitude from 21 to 24°C (in the internal regions of the Middle Siberian tableland and in the north of Kazakhstan); 7) extremely sharply continental with amplitude over 24°C encountered in depressions of Russian North-East and eastwards of lake Baikal.

In total 85 geographical types of seasonal freezing and thawing of ground are distinguished with respect to t_{mean} and A_0, which are subordinated to latitudinal zonation of heat exchange at the ground surface. If the eight types of fluctuation are subdivided with respect to climate continentality, this number will become 133.

Regional characteristics of the processes of seasonal freezing and thawing are reflected in Kudryavtsev's classification for material composition and moisture content. The following main types of materials are distinguished by composition: 1) consolidated and semi-consolidated rocks; 2) gravel-pebble and rock waste; 3) sands; 4) silty-clay sandy material; 5) silty clay materials; 6) clays; 7) peat, as well as various combinations of these depending on regional conditions.

Four gradations are distinguished in the classification with respect to moisture content depending on the amount of moisture that participates in phase transitions in the course of freezing and thawing. The first gradation at $W_{nat} < W_{unf}$ is characterized by the absence of phase transitions of moisture on cooling of the material below 0°C (there will be no freezing or thawing). In the remaining three gradations phase transitions increase from 0 at $W_{nat} = W_{unf}$ to a maximum value at $W_{nat} = W_{com}$. The greater the moisture content of the material, the greater the amount of heat consumed for phase transitions of moisture, and, accordingly, the smaller is the depth of freezing or thawing. Therefore, depending on the value of moisture content different types are distinguished of a) deep, b) medium and c) shallow seasonal freezing (thawing):

a) $W_{unf} < W_{nat} < W_{unf} + 1/3\,(W_{com} - W_{unf})$;
b) $W_{unf} + 1/3\,(W_{com} - W_{unf}) < W_{nat} < W_{unf} + 2/3(W_{com} - W_{unf})$;
c) $W_{unf} + 2/3(W_{com} - W_{unf}) < W_{nat}$,

where W_{nat} is natural humidity of the ground at the moment of the beginning of freezing or thawing; W_{unf} is the amount of unfrozen water; and W_{com} is total moisture capacity.

Each type of seasonal freezing and thawing distinguished in Kudryavtsev's classification has the name that reflects its main classification features, for example: arctic, sharply-continental, sandy, deep; stable, continental, clayey-silt, fine; etc. The number of different combinations of the main parameters can be large. Therefore, there is a great number of different values of seasonal freezing and thawing depth. At one and the same time, the same depth in different regions is often produced by a different combination of conditions, i.e. with different conditions of formation, the values of seasonal freezing or thawing will be the same. The change of the same parameter or factor of the environment by a similar amount but in different regions will result in widely different depths of seasonal freezing (thawing).

Kudryavtsev's classification makes it possible to map the types of seasonal freezing and thawing of ground, i.e. the conditions determining depth, at a required scale. For each type shown in the map the particular depth values ξ can be given, obtained after processing of the field observation data and calculated according to equations based on using parameters A_0, t_{mean}, and taking account of soil composition and humidity. In its turn, analysis of such maps makes possible the identification of the general and regional conditions controlling ground temperature at depth ξ and the pattern of variation of the thickness of seasonally thawed or seasonally frozen layers in their dependence on geological-geographical conditions. In other words such maps can show the occurrence of particular characteristics of the seasonally thawed or frozen layers at given sites. Consequently, with knowledge of the variation of natural conditions and of the changes of the main (four) classification parameters, one can predict depth ξ, as well as exercise control over the depths of freezing and thawing of soils. This offers the opportunity of predicting the ground temperature fluctuations and conditions of formation of the layers of seasonal freezing and thawing under various field conditions caused either by the natural-historical development of landscapes or the economic development of the area.

The main principles in the variation of seasonal freezing and thawing depths in the most typical soil conditions are distinctly traced in plots of the dependence of ξ on mean annual temperature of the soils, amplitudes of surface temperature fluctuations and composition and humidity of materials (Fig. 11.5). The families of curves presented can be used for the tentative assessment of the variation of seasonal freezing and thawing depths. The qualitative effect of the classification indices (t_{mean} and A_0) on the values can be plotted in a convenient form (Fig. 11.6). It is shown in the diagram presented that: 1) higher positive or lower negative values of t_{mean} from t_1 to t_2 (at a constant value of A_0) lead to the reduced ξ value (from ξ_1 to ξ_2); 2) with higher amplitude of surface temperature fluctuations A_0, the depth of seasonal freezing or thawing ξ increases, with lower amplitude it diminishes. The diagram shows these dependences by the sequence of curves of variation of depth of seasonal freezing and thawing that correspond to the

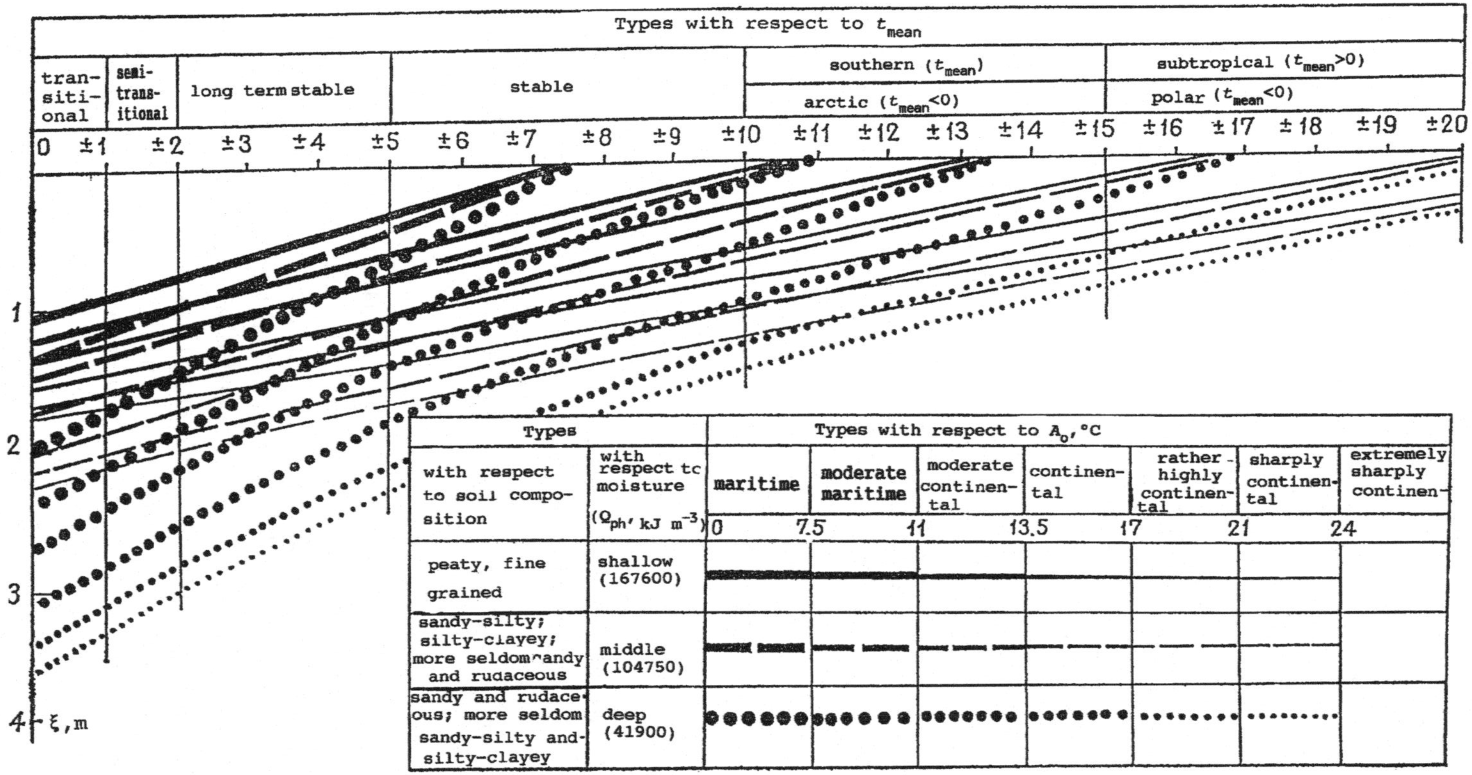

Types with respect to t_{mean}

transitional	semi-transitional	long term stable	stable	southern (t_{mean}) arctic ($t_{mean}<0$)	subtropical ($t_{mean}>0$) polar ($t_{mean}<0$)

0 ±1 ±2 ±3 ±4 ±5 ±6 ±7 ±8 ±9 ±10 ±11 ±12 ±13 ±14 ±15 ±16 ±17 ±18 ±19 ±20

Types with respect to A_0, °C

Types with respect to soil composition	Types with respect to moisture (Q_{ph}, kJ m^{-3})	maritime	moderate maritime	moderate continental	continental	rather-highly continental	sharply continental	extremely sharply continental
peaty, fine grained	shallow (167600)	0	7.5	11	13.5	17	21	24
sandy-silty; silty-clayey; more seldom sandy and rudaceous	middle (104750)							
sandy and rudaceous; more seldom sandy-silty and silty-clayey	deep (41900)							

1
2
3
4 — ξ, m

Fig. 11.5. Approximate depths of seasonal thawing (freezing) of ground with the most typical soil conditions at different t_{mean} and A_0.

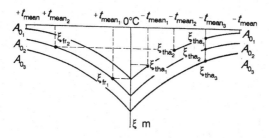

Fig. 11.6. Variation of seasonal freezing ξ_{fr} and thawing ξ_{tha} depth depending on the mean annual temperature t_{mean} and amplitude A_0 of temperature at the ground surface.

increasing ground surface temperature amplitudes, i.e. at $A_3 > A_2 > A_1$ and any fixed value of t (for example, t_3) it is apparent that the depth of thawing $\xi_3 > \xi_2 > \xi_1$. The diagram also shows that the same depth ξ can be developed at quite different combinations of t_{mean} and A_0.

Changes in *lithological composition* of materials leads to changes of their thermal-physical properties: thermal conductivity λ and heat capacity C. As follows from V.A. Kudryavtsev's equation for the calculation of ξ (1.35), this value is proportional to $\sqrt{\lambda}$ and has a more complicated dependence on C, increasing in general with lower C. It is known that finer-grained soils lead to reduced λ. Therefore, with other conditions being equal, the greatest depths of seasonal freezing (thawing) are encountered in rudaceous and coarse-grained materials (for instance, in sands) and the shallowest ones in the fine-grained soils (for instance in clays). Changes in soil humidity affect the ξ value in two ways: firstly through change in thermal-physical properties of soils (λ and C) and, secondly, through the amount of phase transition which, as a rule, has a more substantial impact. The greater the moisture content, the greater the amount of heat consumed for phase transition and the smaller is the depth of seasonal freezing and thawing. Soil composition and moisture content also have a considerable influence over such an indicator of seasonal freezing (thawing) as t_ξ – mean annual temperature at the bottom of this layer, and through this quantity affect ξ – the depth of freezing (thawing). This influence is associated with development of the thermal offset, the value of which is determined by composition and moisture content of the seasonally thawed (seasonally frozen) layers since it is proportional to the difference between square roots of thermal conductivities of the frozen and thawed ground, as well as to the value of the annual heat cycle.

11.3   The influence of landscape-climatic factors on the temperature regime and depth of seasonal freezing and thawing of the ground

The classification of the types of seasonal freezing and thawing elaborated by Kudryavtsev takes account of the influence of the main geological-geographical factors, both individual and in their combination, on the processes of freezing and thawing as well as on the thickness of the seasonally frozen and seasonally thawed layers. One should bear in mind the mutual relationships of all elements of the environment. Thus, for example, variation of thickness or changes of species of vegetation cover will lead to changes of A_0 and t_{mean} parameters and also soil humidity and, perhaps, composition which, in the long run, will manifest itself in change of the depth of seasonal freezing (thawing). Thus, in order to evaluate the effect of one or another natural factor on the depth ξ it is necessary to determine its influence on each of the classification parameters, i.e. A_0, t_{mean}, λ, C, Q_{ph}. It should be especially emphasized that since the parameters t_{mean} and A_0 used in the classification refer respectively to the bottom and top of the layer ξ, their determination will require quantitative data on the influence of surface cover (snow, vegetation, water, ice, etc.), i.e. one should properly treat mean annual temperatures and amplitudes of air temperature fluctuations to gain knowledge of t_{mean} of the ground and A_0 over the soil surface beneath the cover. Let us consider the effect of some of the most important factors of geological-geographical environment on t_{mean} and ξ.

The effect of topography, aspect and steepness of slopes

The position of the site in the topography to a great extent defines the temperature regime of the ground and depth of seasonal freezing (thawing). Thus, as air temperature diminishes with altitude by about $0.4-0.6\,°C$ per $100\,m$, accordingly, the mean annual temperature of the ground is also reduced. This leads to reduced thickness of the seasonally thawing layer and increased thickness of a seasonally frozen layer. Composition of the soils (first of all grain size) varies with altitude as do soil humidity, thickness of snow cover, persistence and species of vegetation, etc.; i.e. there is variation of all the classification indices of the seasonal freezing (thawing) of the soils.

A substantial influence over t_{mean} and ξ is exercised by aspect (orientation of slopes with respect to cardinal points) and steepness of slopes (angle of incidence of solar rays on slopes of different steepness). Mainly, these are effective through the amount of inflowing solar energy absorbed by the surface. The values t_{mean} and A_0 diminish (with other conditions being equal) from southern and south-western slopes towards north-eastern and northern ones. The difference between t_{mean} and A_0 on slopes of southern and northern aspects is mainly determined by the difference in the summer air temperature, since in winter with small inflow of solar energy southern and northern slopes are almost equally cold (provided there is the same snow and vegetation cover). In summer the southern slopes receive much more solar energy. Because of this the depth of thawing ξ_{tha}^n on the slopes of northern aspect is much less than that ξ_{tha}^s on slopes of southern aspect, i.e. always $\xi_{tha}^n < \xi_{tha}^s$, with the exception of the regions of high (polar) latitudes, where $\xi_{tha}^n \approx \xi_{tha}^s$, since there is no solar radiation in winter, and in summer the Sun heats the slopes of all aspects more or less uniformly.

In the case of seasonal freezing, the thickness of the seasonally frozen layer is formed mainly on account of the heat cycle in the winter period, when slopes of different aspects are cooled practically in the same manner, as mentioned above. Therefore, the difference between depths of seasonal freezing on slopes of both southern and northern aspects will be insignificant ($\xi_{fr}^n \approx \xi_{fr}^s$)

The effect of slope steepness over the temperature regime and depth of seasonal thawing (freezing) follows from the different angle of incidence of solar rays and of shadowing, i.e. the different amount of the radiation absorbed by the surface of slopes. Higher temperatures of the ground in the summer period are typical of the slopes perpendicular to solar rays (about 30°) which promotes greater depth of seasonal thawing, while the depth of seasonal freezing remains similar to those on gently sloping hillsides, as there are small differences in their winter cooling. One should bear in mind that the effect of slope steepness (and aspect as well) on temperature regime may be complicated by other factors, for example, uneven distribution of snow cover, differences in vegetation cover, etc., which makes it difficult to distinguish the factors under consideration.

The geographical location of the site and, first of all, its location with respect to oceans is expressed through convective heat exchange of the atmosphere with the lithosphere, forming meridional sectors of climatic continentality. The maximum mean monthly temperature is higher and the minimum is lower with distance from the coast into the continent, i.e. the annual amplitude of monthly temperatures increases. Therefore, with the same mean annual temperature of ground surface in continental regions seasonal freezing and thawing are deeper, as is the penetration of annual temperature fluctuations. However, as a rule, more continental climate involves more severe permafrost conditions, with the lower mean annual temperatures of air and ground. Consequently, the outlines of the permafrost-temperature zones reflect both zonal and sectorial changes of heat- and water-exchange between soil and atmosphere. The southern boundary of the permafrost and, therefore, seasonal thawing, on the Kola Peninsula is confined to zones of tundra, in the north-east of the European part of Russia to the subzone of the northern taiga, in West Siberia to the southern taiga, and in the Zabaykal'ye and Mongolia to permafrost islands, with seasonal thawing encountered in steppes and even semi-deserts.

The influence of snow cover

Snow cover causes substantial changes of the heat exchange between ground and atmosphere. First, the albedo of snow is much higher than that of a bare surface or of vegetation. This leads to reduced absorption of solar energy and lowering of snow surface temperature relative to air temperature. According to meteorological data the mean winter temperature of a snow surface may be $0.5-2\,°C$ lower than the mean winter temperature of the air.

At the same time, snow cover having a low thermal conductivity (varying from 0.12 to $0.46\,W\,m^{-1}\,K^{-1}$ and 5–10 times lower than that of mineral ground) prevents loss of heat by the ground in winter. The soil surface therefore, beneath the snow cover, can have a much higher temperature than that of the air. On the average, increase of snow thickness by $5-15\,cm$ leads to a $1\,°C$ increase of mean annual temperature of the ground. Therefore, with sufficient thickness of snow cover the mean annual temperature of the soil surface can be positive at low mean annual temperatures (to -6 to $-8\,°C$) of the air.

If the snow remains on the soil surface after the air temperatures become positive, it prevents heating of the ground as a considerable portion of the inflowing solar energy, first, is reflected and, secondly, is consumed for snow melting. Melting snow keeps a temperature of zero at the ground surface despite the positive air temperature at this time. This leads to a certain cooling of the ground and lowering of its mean annual temperature.

The influence of snow cover on the temperature regime of the ground is many-faceted. Both the value of the effect and its vector (heating or cooling) are dependent on thickness of the snow cover. Thus, its cooling effect prevails at small thickness of snow due to higher albedo. Then there is the warming effect of snow as a heat insulator. Increased thickness of snow cover (to a certain critical value) leads to an increased cooling effect due to slower melting of snow in summer. For the greater part of the seasonally frozen and permafrost areas the thickness of snow cover is such as to have a warming effect on the underlying ground. Snow cover not only causes an increase of mean annual temperature of the ground surface, but also leads to much reduced amplitude of the soil surface temperatures as compared with that of air temperatures and in some cases may cause changes in soil humidity.

Snow density is an important factor associated with the warming effect of snow. Thus, with snow cover density ρ equalling $75\,\mathrm{kg\,m^{-3}}$ thermal diffusivity K will be $0.36 \times 10^{-3}\,\mathrm{m^2\,h^{-1}}$; with $\rho = 150\,\mathrm{kg\,m^{-3}}$, $K = 0.72\,10^{-3}\,\mathrm{m^2\,h^{-1}}$; with $\rho = 225\,\mathrm{kg\,m^{-3}}$, $K = 1.08 \times 10^{-3}\,\mathrm{m^2\,h^{-1}}$; $\rho = 300\,\mathrm{kg\,m^{-3}}$, $K = 1.44 \times 10^{-3}\,\mathrm{m^2\,h^{-1}}$; with $\rho = 380\,\mathrm{kg\,m^{-3}}$, $K = 1.8 \times 10^{-3}\,\mathrm{m^2\,h^{-1}}$. Loose snow has a greater warming effect on the temperature regime of the ground compared to dense snow due to its low thermal diffusivity and thermal conductivity.

V.A. Kudryavtsev found the relationship between snow cover and values of heat cycles – the amounts of heat going through the soil surface for half-periods of heating and cooling. The bigger the heat cycles of the soil (with other conditions being equal) the more intense is the influence of the snow cover on mean annual temperature and amplitude of surface temperatures. Taking into consideration the patterns of heat cycle variations as determined by geological-geographical factors, one may conclude that the maximum effect of snow cover occurs at a mean annual ground temperature near to $0\,°\mathrm{C}$ (i.e. in the vicinity of the southern boundary of perennially frozen ground) with a maximally continental climate and very wet soils in the seasonally thawing and seasonally freezing layers.

Qualitatively, the effect of snow cover on the depth of seasonal thawing and freezing can be analyzed with the help of graphs as plotted in Fig. 11.7. As snow is a thermal insulating layer, after its removal the amplitude A_0 of surface temperature fluctuation increases and mean annual temperature usually diminishes (with exception of the cases when snow has a cooling effect). Higher A_0 always causes increase of depth of seasonal thawing ξ_{tha} and seasonal freezing ξ_{fr}. Quite different is the effect of t_{mean}: in the area of perennially frozen ground its lowering leads to lower ξ_{tha}, while in the area without permafrost (unfrozen or thawed ground) its lowering leads to greater ξ_{fr}. Thus, in the first case, variation of A_0 and t_{mean} after removal of snow (reduced thickness) is compensated by their different direction and has an insignificant effect at the depth of ξ_{tha} (see Fig. 11.7a), whereas in the second case, the effects are additive and the depth ξ_{fr} changes substantially (see Fig. 11.7b).

The influence of vegetation cover

There are several aspects to this effect on the temperature regime of ground and depth of seasonal freezing and thawing. Vegetation cover causes a change of reflective capacity compared with that of the underlying surface,

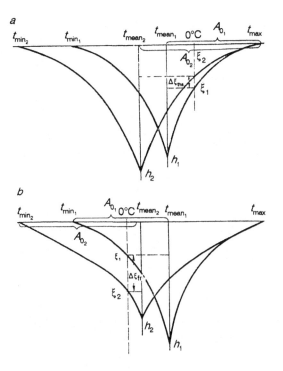

Fig. 11.7. Diagram of the effect of snow cover on depths of seasonal thawing (*a*) and freezing (*b*) of ground according to S.Yu. Parmuzin):t_{mean_1}, A_{0_1}, ξ_1, h_1 – respectively, mean annual temperature of the ground, amplitude of annual temperature fluctuations at soil surface, depth of seasonal thawing or freezing, depth of penetration of annual temperature fluctuations, with snow cover;t_{mean_2}, A_{0_2}, m_2, h_2 – the same characteristics after removal of snow cover; $\Delta\xi_{tha}$, $\Delta\xi_{fr}$ – variation of depths of seasonal thawing and freezing.

absorbs solar energy, causes evaporation of moisture from its whole volume, makes air flow turbulent above the level of biomass development or, on the contrary, causes stagnation of air in it. Conditions of snow accumulation and properties of the snow layer are to a great extent determined by vegetation cover which also has an influence on moisture content and thermal-physical properties of the soil.

Compared with the effect of snow, it is much more difficult to evaluate qualitatively the effect of vegetation cover (as a thermal insulator) on temperature regime and depth of seasonal freezing or seasonal thawing of the ground. This is explained by the fact that vegetation cover insulates the soil both from cooling in winter (as snow) and from heating for the whole summer period. The double effect of these two influences is dependent on duration of summer and winter seasons, continentality of climate, depth of snow cover, moisture content of the underlying soil, etc., i.e. on a number of factors and conditions that determine the role of vegetation cover in heat exchange between the soil surface and atmosphere, on the one hand, and between soil surface and underlying soil and rocks, on the other.

As a first approximation it can be concluded that in the area of perennially frozen ground the influence of vegetation cover on the seasonal thawing depth is greater than that on the depth of freezing in the area of unfrozen or thawed ground (Fig. 11.8). In both cases removal of vegetation cover leads to higher amplitude of annual temperature fluctuations with both lower t_{min} and higher t_{max} (unlike snow that influences only t_{min}). Variations of t_{mean} are dependent on the new values of t_{min} and t_{max}. If $\Delta t_{max} > |\Delta t_{min}|$ it is apparent that t_{mean} increases. In the regions of seasonal thawing (see Fig. 11.8a) both higher A_0 and t_{mean} promote greater values of ξ_{tha} (i.e. the effects are additive). In the area of seasonal freezing (see Fig. 11.8b) greater A_0 leads to greater ξ_{fr} while an increase in t_{mean} would compensate this effect. As a result, ξ_{fr} increases insignificantly which is well seen in Fig. 11.8. It is known, however, that in the northern regions of the permafrost zone where the thickness of the snow cover is not large (0.2–0.3 m) vegetation cover has a warming effect, i.e. it leads to higher mean annual temperature of the ground compared with the regions having no vegetation cover. But this does not lead to deeper thawing as the amplitude of the temperature fluctuation under the vegetation cover is always less than when it is absent.

Forest and shrub vegetation reduces the inflow of solar energy to the surface due to the effects of shadowing, which lead to less warming of the surface in summer compared to bare sites and slower melting of snow. According to the observations on thermal balance obtained by A.V. Pavlov near the town of Yakutsk, Igarka and the village of Syrdakh, the albedo of the forest on the permafrost is less than that of bare sites; the effective radiation balances of forested and unforested regions do not differ substantially year by year and the annual sum of the radiation balance for the forest exceeds that for unforested sites.

The influence of the forest vegetation on temperature regime of the ground is closely connected with the geobotanical zones. The larger the phytomass surface of the forest depending on height, density and closeness of its layers, the less the solar rays penetrate the soil surface. Consequently, with more close crowns in the direction from north southwards the role of forests in the formation of mean annual temperature of the ground changes substantially. In the open woodland of the forest-tundra area and the light forest and shrubs of the northern taiga zone, reduced inflow of radiation at

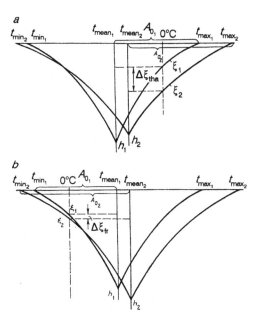

Fig. 11.8. Diagram showing the influence of vegetation cover on depth of seasonal thawing (*a*) and freezing (*b*) of ground (according to S.Yu. Parmuzin): t_{mean_1}, A_{0_1}, ξ_1, h_1 – respectively, mean annual temperature, amplitude of annual temperature fluctuations at the surface, depth of seasonal thawing or freezing, depth of penetration of annual temperature fluctuations, with vegetation cover; t_{mean_2}, A_{0_2}, m_2, h_2 – the same characteristics after removal of vegetation cover; $\Delta\xi_{tha}$, $\Delta\xi_{fr}$ – variations of depth of seasonal thawing and freezing.

the soil surface is compensated by reduced turbulent heat exchange, and under conditions of strong winds more loose and thick snow cover is accumulated compared to woodless regions. As a result, the mean annual temperature of the ground in the northern forest exceeds that of woodless sites.

With closer crowns the incoming radiation is so much reduced that diminished turbulent exchange cannot compensate for it. With weak winds typical of the forested zone (especially in dense, dark coniferous forest) the depth of snow cover is much less than in treeless sites. Therefore, in the central and southern taiga zone and in the area of unfrozen ground in the south of the country, forest serves as a cooling factor. This is confirmed by regular observations in the area of the town of Zagorsk where the mean annual temperature of the soil surface in the coniferous forest is lower by 2 °C than on bare sites. In Central Yakutia the mean annual temperatures of the ground under forest differs from those on bare sites by 1–2 °C. In West Siberia, in the vicinity of the southern boundary of the permafrost zone, islands of frozen ground are confined to mixed and dark coniferous forest with crown closeness of 0.7–0.8.

Grass cover to a lesser degree causes changes in the heat exchange and temperature regime of soil surface and atmosphere. The total thermal effect of grassy vegetation on t_{mean} of the ground can be either warming or cooling, but it does not exceed fractions of a degree. The amplitude of mean monthly temperatures is also reduced insignificantly. Of importance is the influence of soil covers (mossy, moss-lichen, lichen, mossy-peaty) which are thermal insulators, preventing heating of the soil in summer and reducing heat yield from the surface in winter.

A distinctive feature of humid, natural soil covers is a considerable change of their thermal conductivity while passing from the thawed state into the frozen. According to data from field observations thermal conductivity of moss-lichen covers in the thawed state is $0.1–0.7 \, W \, m^{-1} \, K^{-1}$, one third to one half of that in the frozen state. Therefore, the ability of moss covers to retard the entry of heat in the summer period is greater than its ability to restrict the yield of heat in winter by the same order. Thus, the layer of moss 2–3 cm thick reduces the sum of summer temperatures by two-thirds and more. In winter the influence of moss cover on temperature regime of the ground is much less, because of the drastic increase of thermal conductivity. On the average, in winter, temperature beneath moss differs only slightly from that of its surface. Thicker and less water-saturated mossy cover has a greater effect on t_{mean} of the ground. Depending on difference of thermal conductivity of moss cover in thawed and frozen states, duration of summer and winter seasons, thickness of snow, etc., moss covers can have either a warming or cooling effect. One should bear in mind that an increase of mean winter (and minimum) and lowering of mean summer (and maximum) temperature of the soil surface resulting from the heat-insulating effect of mossy covers during a year leads to a sharp reduction of temperature amplitude. Moss covers 15–20 cm thick cause a reduction of temperature amplitude by 5–6 °C and lead to shallower depths, one-half to one-quarter those for seasonal thawing under a bare surface.

The influence of peat cover on the temperature regime of the ground should be dealt with separately. As shown by investigations, mean annual temperature of peat soil is lower than that of mineral soil. In the vicinity of the southern boundary of the permafrost zone, peat cover 0.1 m thick causes reduction of t_{mean} by 0.5–1 °C. Therefore, even with a positive mean annual temperature at the peat surface the underlying ground can be in the frozen state. On the West Siberian and north European plains the southernmost islands of permafrost are, as a rule, confined to peats. Thickness of the frozen strata in the south of the permafrost zone is greater in peat soil than mineral ground. The thermal conductivity coefficient of peat in the thawed state varies within the range of 0.23 to 0.93 W m^{-1} K^{-1} and in the frozen state between 0.93 and 1.28 W m^{-1} K^{-1}. With higher moisture content of the peat, the difference of thermal conductivities of the frozen and thawed peat increases, the ability to prevent the underlying ground from heating in summer becomes greater than the ability to prevent the yield of heat in winter, and the cooling effect increases.

There are a number of calculation schemes and equations for the tentative quantitative estimation of the thermal influence of soil covers (mossy, peaty, sod, snow and others both natural and artificial) that serve as additional heat-insulating layers on the surface of the ground. For example, there is one enabling determination of mean annual air temperature variation $\Delta t_{mean_{air}}$ and of reduction of annual air temperature fluctuation ΔA_{air} owing to different types of covers that exist during warm τ_s or cold τ_w periods or for the whole year T. It is assumed that phase transitions in covers do not occur. In the calculation, the equation of annual sinusoidal (or reduced to sinusoidal) variation of the cover surface temperature (with period T, mean annual temperature t_{mean} and amplitude A_0) is subdivided into two conditional harmonic fluctuations of temperature taking place near 0°C with periods $2\tau_s$ and $2\tau_w$ and amplitudes of summer and winter temperature fluctuation equalling, respectively, A_{air_s} and A_{air_w} (Fig. 11.9). Using the Fourier equation one can determine the reduction of the amplitudes ΔA_s and ΔA_w of the temperature fluctuation for these two conventional sine waves, due to soil cover, respectively for the warm and cold periods of the year:

$$\Delta A_s = A_{air_s}\left(1 - e^{-z\sqrt{\frac{\pi}{K_{th}2\tau_s}}}\right), \quad A_{air_s} = A_{air} + t_{mean_{air}} \tag{11.2}$$

$$\Delta A_w = A_{air_w}\left(1 - e^{-z\sqrt{\frac{\pi}{K_{fr}2\tau_w}}}\right), \quad A_{air_w} = A_{air} - t_{mean_{air}} \tag{11.3}$$

where z is the thickness of cover; K_{th} and K_{fr} are the thermal conductivities of the soil cover under consideration, respectively in the thawed and frozen state; τ_s and τ_w are the duration of positive (warm or summer period) and negative (cold or winter period) temperatures of air. While calculating A_{air_s} and A_{air_w} the sign of the value $t_{mean_{air}}$ is to be borne in mind.

Having knowledge of ΔA_s and ΔA_w it is simple to determine the true

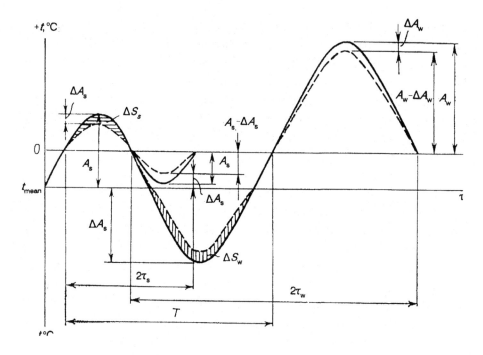

Fig. 11.9. Diagram of temperature variation at the surface of ground cover broken down into two simple harmonic fluctuations with periods equalling the double duration of cold ($2\tau_w$) and warm ($2\tau_s$) periods of a year.

reduced value of annual amplitude ΔA_{air} and the reduction and variation of mean annual air temperature $\Delta t_{mean_{air}}$ caused by soil cover by using the values of area ΔS_s and ΔS_w:

$$\Delta A_{air} = \frac{\Delta A_s \tau_s + \Delta A_w \tau_w}{T} \tag{11.4}$$

$$\Delta t_{mean_{air}} = \frac{\Delta A_s \tau_s - \Delta A_w \tau_w}{T} \cdot \frac{2}{\pi} \tag{11.5}$$

The relationships suggested for the calculation of ΔA_{air} and $\Delta t_{mean_{air}}$ show that soil covers always cause reduction of the amplitude of temperature fluctuation, whereas variation of mean annual temperature $\Delta_{mean_{air}}$ can be both positive (cooling effect) and negative (warming effect) depending on which effect prevails: cooling in summer ($\Delta A_s \tau_s$) or warming in winter ($\Delta A_w \tau_w$). If the cover exists only part of the year (during the warm or cold period), for example, the snow cover period of the year, the equations (11.4) and (11.5) are simplified as the term $\Delta A_s \tau_s$ becomes equal to 0.

By using relationships (11.4) and (11.5) one may transfer from mean annual temperatures $t_{mean_{air}}$ and amplitudes of temperature fluctuation A_{air}

of the air (measured at weather stations at an elevation of 2 m) to the values of t_{mean} and A existing at the ground surface. With this aim in view one should determine the difference in mean monthly (or weekly) temperatures at the height of 2 m and at the ground surface during the warmest month of summer and coldest month of winter, i.e. experimentally find the values of ΔA_s and ΔA_w and, substituting the values τ_s and τ_w into the equations (11.4) and (11.5), one may calculate the values ΔA_{air} and $\Delta t_{mean_{air}}$ at the 2 m interval above the ground surface. In the same manner account should be taken of the influence of slope aspect and steepness on the temperature regime of the surface.

The influence of swamps and surface water

The influence of waterlogging on the temperature regime of ground is to a great extent determined by the general climatic setting and stage of development of swamps. So long as the surface of swamps is partially covered with water, which transmits shortwave radiation and retains long-wave radiation, the mean annual temperature is higher than that of the adjacent sites.

In the course of bog evolution, when there is overgrowth by mosses and heaving of the freezing ground, individual sites are no longer covered with water and sedge-carex species are replaced by mosses and shrubs. The warming effect of snow on these sites is reduced due to the diminished thickness of the snow cover and lower moisture content of the peaty deposits. Mean annual temperatures of the ground of elevated sites are reduced compared to those of low-lying sites. Further drainage of bogs as a result of the surface uplift, leads to gradual dying off of sphagnum and replacement by lichen. Mean annual temperatures of the ground at these sites are, as a rule, much lower than in the surrounding terrain. In the vicinity of the southern boundary of the permafrost zone hillocky peats indicate the occurrence of perennially frozen ground.

Thus, depending on the stage of development the bogs can have either a warming or cooling effect on the permafrost and unfrozen ground.

The temperature regime of non-draining fresh-water bodies is dependent on their depth. If the depth of a lake H_w is greater than the maximum thickness of ice H_i, which, in the most severe conditions, does not exceed 2–2.5 m, then bottom deposits are unfrozen year round. Depending on the size of the lake and the mean annual temperature of the ground on adjacent sites either a through, open talik (if the width of lake exceeds double the thickness of the frozen strata) or a closed talik is formed beneath the water body.

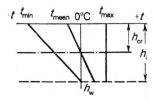

Fig. 11.10. Diagram of distribution of minimum, maximum and mean annual temperatures in a water body.

A water body having a smaller depth than the thickness of ice possible in the given region freezes to the bottom and bottom deposits can have either positive or negative mean annual temperatures. There is a depth of water body (Fig. 11.10) at which the mean annual temperature of bottom deposits will be equal to $0\,^{\circ}\mathrm{C}$. This depth is called by Kudryavtsev a critical value H_{cr}. With a water body of depth H_w less than the critical ($H_w < H_{cr}$) and the mean annual temperature of bottom deposits below $0\,^{\circ}\mathrm{C}$, there is permafrost and only in summer do the deposits thaw to a certain depth, i.e. seasonal thawing of the bottom deposits is observed. With depths of water bodies ranging from the critical H_{cr} to that equalling maximum thickness of ice H_i, i.e. $H_{cr} < H_w < H_i$, mean annual temperature of the surface of bottom deposits will be positive, but in the winter period they will freeze to a certain depth. In this case seasonal freezing of bottom deposits will be observed. The critical depth of water body is mainly determined by climatic characteristics (air temperature and thickness of snow cover). Therefore, the characteristics of the climatic zones determine the critical depths of water bodies. Thus, according to data obtained in West Siberia H_{cr} is 0.2–0.3 m in the vicinity of the southern boundary of the permafrost. Northwards H_{cr} increases steadily reaching 1.6 m in the Yamal and Guydan peninsulas.

If the depth of water body is assumed to be comparable with the maximum thickness of ice H_i, then the temperature regime of bottom deposits can be determined by using Kudryavtsev's scheme (see Fig. 11.10). In this figure t_{min} and t_{mean} designate distribution of respectively, minimum (winter) temperature in the ice cover and the mean annual temperature, and t_{max} is the maximum temperature of the water body in summer, which tentatively is assumed to be uniform for shallow northern lakes (owing to convection mixing). Stemming from this diagram the mean annual temperature at the surface of the bottom deposits (at the depth of the water body (H_w)) is

$$t_{mean} = 1/2[t_{max} + (1 - H_w/H_i)t_{min}] \tag{11.6}$$

The temperature regime of the bottom deposits of saline lakes is different as saline water, being heavier, moves downwards and cools without freezing at negative temperatures. Even in summer the temperature in the bottom of highly saline lakes can be negative. As a result, perennially frozen or cryotic bottom deposits are observed beneath such saline lakes as well as on shallow places off the northern seas coasts, while saline water in the liquid state, having a negative temperature, overlies such deposits.

The influence of convection flows of water and air

The temperature of ground can vary not only by conductive transfer of heat but also by convective transfer by seepage of water or air flows. The inflow of warm or cold air or water flows in the ground can lead to warming or cooling not only due to balancing of heat content between the convection flows and the soil or rock but also through heat released at phase transitions of moisture (freezing – thawing, evaporation – condensation, sublimation – ablimation).

Under natural conditions transfer of heat into ground occurs by infiltration of surface water and, primarily, of precipitation. Intensity of the process is dependent on the amount of infiltrating rainfall, and the temperature, thickness, seepage and thermal-physical properties of the layer of seasonal freezing and thawing. A tentative equation was suggested by Kudryavtsev for quantitative estimation of the increase of mean annual temperature of ground Δt_{mean} at the base of the seasonal freezing (thawing) layer as a result of infiltration of summer precipitation (17):

$$\Delta t_{\text{mean}} = \frac{V t_{\text{ms}} \xi}{\lambda_r T} \tag{11.7}$$

where V is the amount of summer precipitation infiltrating the soil, kg m^{-2}; t_{ms} is mean summer temperature, $°\text{C}$; ξ is depth of seasonal freezing or thawing, m; T is time (year $= 8760\,\text{h}$); λ_r is reduced thermal conductivity equalling the average values (weighted for the year) in the frozen and thawed states, $\text{kJ} (\text{m h} °\text{C})^{-1}$.

According to the calculation data infiltration of the summer precipitation may cause an increase of mean annual temperature of the ground of 1.5–2 °C. This effect will be greatest on sites composed of coarse-grained soil with high hydraulic conductivities. The presence of vegetation cover drastically reduces the infiltration. As a rule, the role of infiltration in the formation of mean annual temperature of the ground at these sites is insignificant and does not exceed 0.1 °C.

An important role in the formation of mean annual temperatures of porous loose and fractured bedrock is played by convection flows of air. In such materials there is a permanent exchange of gases with the atmosphere caused by variations of pressure and air temperatures near the soil surface. The process takes place in the following way: cold atmospheric air replaces warmer and lighter air in the cavities in the material and cools the latter. Such winter ventilation is clearly traced in boreholes and pits and occurs in porous rocks, cooling them intensely for a considerable depth.

In the regions with extensive rudaceous deposits, an additional warming factor is the condensation of water vapour. Thus, in the southern part of the permafrost zone the influence of vapour condensation can increase t_{mean} by as much as $2°C$, in the northern part much less, while it is not observed on high interfluves and in the Arctic zone.

The temperature regime of the ground and the depths of seasonal freezing and thawing vary substantially under conditions of economic development. In large cities specific micro-climate is created with changes in air temperature regime, direction and velocities of wind, evaporation, etc. In the course of development there are substantial changes in vegetation cover, conditions of snow accumulation, drainage or waterlogging of the terrain surface, and artificial water bodies are created. The temperature regime of the ground is much affected by heat issuing from engineering structures. The accelerating tempo of development of the northern and Far East regions leads to deep changes of temperature regime and depths of seasonal freezing and thawing. There is accordingly a requirement for scientifically substantiated prediction of temperature regime and of changes in depths of seasonal freezing and thawing, with a possible need to prepare designs for the purpose-oriented alteration of these characteristics, i.e. management of the processes of seasonal freezing and thawing. Such tasks can be fulfilled only on the base of a thorough knowledge of the processes of seasonal freezing and thawing.

12

Development of the temperature regime and the thickness of the permafrost

12.1   Present-day knowledge of the development of permafrost

Knowledge of the development of permafrost has been arrived at gradually. In the 1930s M.I. Sumgin substantiated the theory of the *degradation* of permafrost (with warming, thawing and northward retreat). Having compared the severe climate of glacial epochs with the warmer present one, Sumgin came to the conclusion that the permafrost formed simultaneously with the ice sheets and had subsequently thawed retreating northward, i.e. it had degraded. He presented data indicating displacement of the southern limit of permafrost northward. However such researchers as S.G. Parkhomenko, P.I. Koloskov, P.N. Kapterev, D.V. Redozubov and others pointed out the fact of new permafrost formation, permafrost temperature decrease and its increase in thickness in other regions of the country. Contrary to Sumgin they followed the alternative theory – that of *aggradation*, i.e. they reasoned that the process of permafrost advance is taking place at the present time. According to their views the permafrost is a result of recent heat exchange (the last 3–5 thousand years) in the system 'atmosphere-lithosphere'. These two points of view were antagonistic for 10–15 years and only V.A. Kudryavtsev's basic works (1953–1963) clarifying the basis for the present theory of permafrost development gave the proper interpretation of these points of view.

Numerous works by Kudryavtsev from 1954 on, showed that the thermal state of the permafrost depends on heat exchange through the Earth's surface, between the atmosphere and the upper layers of the lithosphere, the character of which is defined by:

1) the composition and properties of the soil and rock materials and the processes operating in them;
2) the amount of direct and transformed solar heat arriving at the Earth's surface;

3) the specific features of the Earth's surface absorbing radiant and thermal energy;

4) the amount of heat arriving at the surface from the Earth's interior.

The character and conditions of heat exchange are extremely dynamic. Permafrost strata, their formation, development, transition to the thawed state and all their characteristics (distribution, occurrence, composition, structure and temperature regime) are changing in response to change in the combined natural conditions affecting the course of heat exchange between the atmosphere and the upper layers of the lithosphere. Thus, the dynamics and evolution of the Earth's cryosphere are closely connected with the palaeogeographical and geological evolution of its regions within continents and oceans.

Periodic changes of ground heat exchange through the Earth's surface are responsible for the dynamics of the temperature pattern of the lithosphere's upper layers. In the case of temperature transition through 0 °C the perennial freezing of the subsurface begins. Climatic fluctuations being various in their amplitude and duration propagate differently through the upper layers of the lithosphere. It is known that daily temperature fluctuations extend down for a few tens of centimetres, yearly ones up to 15–25 m, 30–40 year ones to 40–70 m; 300 year ones to 100–150 m and so on. The duration of the temperature fluctuation varies within wide limits from days to a thousand and a hundred thousand years. Therefore the depth of propagation of the heat waves through the ground and thus the thickness of the negative temperature zones will depend on the duration of the fluctuations. Upper boundary conditions for the formation of permafrost and its dynamics in the context of thermal physics are expressed in terms of the following parameters: mean perennial temperature on the permafrost surface t_{mean}^{per}, amplitude of the temperature fluctuations A_{per} on this surface, as well as the period T_{per} of the many years temperature fluctuations.

Temperature behaviour on the permafrost surface for the many-years period is represented by a rather complicated summary curve, which is a combination of temperature fluctuations of shorter periods. To show the complexity of the temperature conditions on the surface one can superpose (as a rough example) only three various-period temperature fluctuations (Fig. 12.1), although the actual temperature curves for the surface are found to be much more complicated. At the same time, as is shown in Fig. 12.1 (curve *IV*), both increase and decrease of the surface temperature, dictated by the fluctuations of the shorter periods, can be observed on the background of a general temperature increase resulting from a 300-year fluctuation.

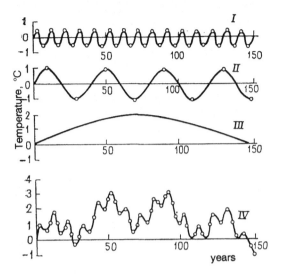

Fig. 12.1. Surface temperature fluctuations with periods and amplitudes (after
V.A. Kudryavtsev): $I - T_1 = 10$ years, $A_1 = 0.5\,°C$; $II - T_2 = 40$ years, $A_2 = 1\,°C$;
$III - T_3 = 300$ years, $A_3 = 2\,°C$; IV – resulting curve at their superposition.

In the upper layers of the lithosphere a multitude of periodic temperature
fluctuations of various periods and amplitudes from daily and yearly to many
years are observed: $T_{per_1} = 11$ years; $T_{per_2} = 40$ years; $T_{per_3} = 300$ years; ...;
$T_{per_n} = 1800$ years and others with periods of ten and a hundred years. Some
of these fluctuations may, it seems, have causes of an astronomical nature,
others arise by periodic changes of the geological environment and geographical situation in the course of the Earth's evolution. Thus, for example, it
is shown in the work by M. Milankovich *Mathematical theory of climatic
fluctuations* that if we use the periodic precessional change of the position of
the Earth's rotational axis and periodic change of its mean albedo value
taking place in this case, as one of the reasons for this process, we can discern
four minima corresponding to four glaciations in the Northern Hemisphere
and three relative maxima between them representing interglacials and the
thermal maximum after glaciation in the Quaternary period on the background of one general wave of cooling of period $T_{per} = n \times 100\,000$ years. At
the present time the majority of researchers assume the existence of many
years' fluctuations of heat exchange conditions on the lithosphere surface
with relatively short period (up to a few hundred years), medium (from a few
hundred to a few thousand years) and long period (from a few thousand years
to a hundred thousand years) as a whole.

As these fluctuations propagate downward through the ground, the damping out (reduction) of the amplitude A_{per} with depth is observed as it is known from Fourier's first law (see §1.1) and phase delay (in time) of the ground temperature fluctuation takes place as is known from Fourier's second law. The shorter the period of the temperature fluctuation T_{per} the sharper is the damping of the amplitude, i.e. according to Fourier's third law the fluctuations extend for a smaller depth. This phenomenon is rather well illustrated in Fig. 12.2. Actually, in layer *I* four temperature fluctuations, the seasonal (one per year) and the subsequent three periods are superimposed. In layer *II* the yearly fluctuations are not traced while the 10-, 40- and 300-year fluctuations are superimposed. In layer *III* only 40- and 300-year temperature fluctuations exist. In layer *IV* there is only the temperature fluctuations of the longest period $T_{per} = 300$ years. At one and the same depth both the rise and the fall of temperature (see Fig. 12.2, curve 4) associated with penetration of either the warm or the cold portion of the integrated temperature wave to this depth can be observed at different moments of time. In other words, degradational and aggradational trends in the development of the permafrost can exist simultaneously at various depths in one and the same place. Such analysis allowed Kudryavtsev to show that such fluctuations can substantially violate, for example, the general rule of the permafrost thickness increasing from the south northward (Fig. 12.3). Suppose that the fluctuations of the period $T_{per}^1 = 10$ years cause permafrost degradation, while the fluctuations of the period $T_{per}^2 = 40$ years and $T_{per}^3 = 300$ years lead to aggradation and degradation respectively, with the effect of 40-year fluctuations being stronger than that of 300-year ones. Based on this fact the degradational trend causing reduction of permafrost thickness will dominate in the first layer as a result of the superimposing of temperature fluctuations of different periods and phases (two degradational effects and one aggradational effect). In the second layer fluctuations of $T_{per}^1 = 10$ years do not occur while the 40-year (aggradational) and 300-year (degradational) ones lead to the aggradation process dominating, i.e. there is increase of permafrost thickness. The fluctuation with the 300-year period provides the degradational trend in the permafrost (decrease of its thickness) and is the only one penetrating into the last (third) layer. Ultimately, instead of the conventionally linear increase of the permafrost thickness from the south northward presented in Fig. 12.3 as a line *AB* there is a rather characteristic curve of the change in permafrost thickness south northward (shown as a broken line). It follows from Fig. 12.3 that in different places we can find alternation of the processes of degradation and aggradation when moving from south northward.

It is necessary to view the processes of permafrost aggradation and

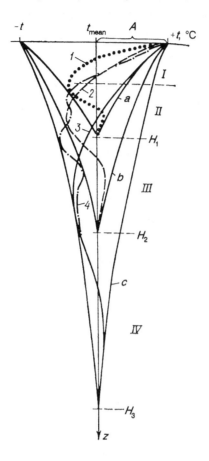

Fig. 12.2. Attenuation of amplitudes with depths depending on the period of the
temperature fluctuations: $1 - T_1 = 10$ years; $2 - T_2 = 40$ years; $3 - T_3 = 300$
years; $4 -$ integrated rock temperature; $a, b, c -$ envelopes of temperature
fluctuations with various periods.

degradation for the specific places and depths and to associate them with the
specific time intervals and periods of fluctuations. Thus, as is shown by
Kudryavtsev, permafrost development is a result of the continuous process
of superimposition of a great number of temperature fluctuations with
various periods and amplitudes on the Earth's surface and their propagation into the ground depending on the whole combination of geological and
geographical factors and conditions.

12.2 The effect of boundary conditions on the permafrost thickness and temperature regime

The depth of the perennial freezing ξ_{per}, the thickness of permafrost
formed as a result of harmonic temperature fluctuations on its surface, is

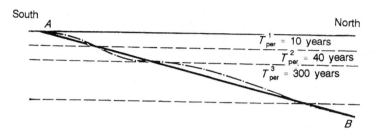

Fig. 12.3. Diagram of possible variations of permafrost thickness from south northward, given the superposition of temperature fluctuations with various periods (after V.A. Kudryavtsev).

defined by the upper $(T_{per}, A_{per}, t_{mean}^{per})$ and lower (g) boundary conditions as well as by specific lithological features: heat conduction λ; heat capacity C and moisture content W of the materials:

$$\xi_{per} = f(A_{mean}^{per}, t_{mean}^{per}, T_{per}, \lambda, C, Q_{ph}, g) \qquad (12.1)$$

It follows from this dependence that the processes controlling permafrost thickness are in many respects analogous but not identical to the processes in seasonal ground freezing. Given the similarity of thermal-physical processes in seasonally and perennially frozen ground, the development of the latter depends on a more complicated interaction of geological and geographical conditions. As follows from the equation (1.35), the depth of perennial freezing appears to be directly proportional to the period of the fluctuation:$\xi_{per} \propto \sqrt{}$, i.e. it increases with the increase of the duration of the fluctuation. This dependence still applies under any change of the other parameters $(Q_{ph}, A_{per}, t_{mean}^{per}, \lambda, C)$. At the same time, as in the case of seasonal ground freezing with $T = 1$ year, the process of the perennial freezing proper lasts usually during a third part of the period ($\tau \approx 1/3\ T$). As an example we can consider the results of calculations (using a hydrointegrator) for perennial freezing of the upper layers of the lithosphere under the following conditions: $t_{mean}^{per} = 0\,°C$, $T_{per} = 100\,000$ years, $A_{per} = 6\,°C$, $g = 0.01\,°C\,m^{-1}$, $\lambda = 2.89\,kJ\,(m\,hr\,°C)^{-1}$, $Q_{ph} = 99,219\,kJ\,m^{-3}$ (Fig. 12.4). The data obtained show that the permafrost thickness being formed on account of 100 000-year fluctuations under the values of thermal physical coefficients and moisture content given comprise from 180–210 m. It takes approximately 33 thousand years to freeze these thicknesses entirely. The rate of freezing, i.e. the rate of advance of the permafrost base, is a few centimetres per year at the beginning of the process, while at the end of freezing it is fractions of

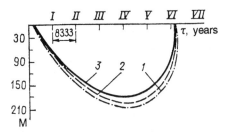

Fig. 12.4. Graph of the variations of depth of ground freezing with time under 100 000-year temperature fluctuations on the surface (after V.A. Kudryavtsev) for (1–3) moisture contents W of 10, 15, 20%, respectively.

centimetres per year. When phase transitions of the water in the ground take place, the freezing rate is two to three orders of magnitude lower than the rate of penetration of the thermal wave into the permafrost where phase transitions are absent. Actually, as mentioned above, the temperature wave of period $T = 300$ years in dry ground reaches a depth of about 180 m, while freezing of moisture-rich rock of the same thickness takes 33 000 years, when the period of fluctuation (T) is equal to 100 000 years.

The effect of the amplitude A_{per} and the mean (for the period T_{per}) temperature (t_{per}) on the depth of perennially frozen ground in the context of thermal physics are similar to their effect on seasonal freezing and thawing.

An increase of amplitude A_{per} of the temperature fluctuation on the surface of a frozen rock unit under a fixed value (for the fluctuation period) of the mean temperature t_{per} causes deeper perennial freezing of the ground and formation of permafrost of greater thickness (Fig. 12.5). The dependence of permafrost thickness on variations of amplitude of the temperature wave, presented in the diagram, was obtained by Kudryavtsev using the formula (1.45) and computer computations, given the following input parameters: $T_{per} = 100 000$ years; $g = 0.01 °C$ m^{-1}; $t_{per} = 1 °C$; $\lambda = 2.89$ kJ (m hr. °C)$^{-1}$; $Q_{ph} = 52 375$ kJ m^{-3}. According to the calculations, the increase by each $2 °C$ of A_{per} in the range of amplitudes from 2–8 °C causes the thickness of permafrost to build up by 43, 28 and 23 m, respectively.

The dependence of the change of permafrost thickness ξ_{per} on the mean temperature t_{per} (for the period of fluctuation) has also been obtained by Kudryavtsev from the calculations (Fig. 12.6), using the same values of input parameters as in Fig. 12.4. The maximum permafrost thickness appeared to be: $\xi_{per} = 130$ m under $t^{per}_{mean_1} = 4 °C$; $\xi_{per} = 90$ m under $t^{per}_{mean_2} = 2 °C$; $\xi_{per} = 45$ m under $t^{per}_{mean_3} = 4 °C'$ all other things being the same. Thus with the increase of mean perennial temperature, for every $2 °C$ in the range of t^{per}_{mean} from 0 to 4 °C, permafrost thickness decreases by 40 and 45 m

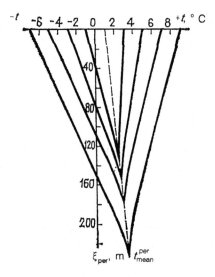

Fig. 12.5. Variation of the permafrost thickness depending on the amplitude of many-year temperature fluctuations on the surface at $t_{mean}^{per} = 1\,°C$ (after V.A. Kudryavtsev).

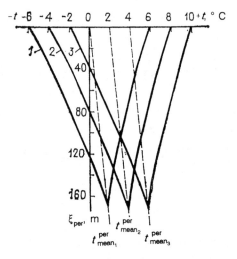

Fig. 12.6. Variation of the permafrost thickness depending on the mean ground temperature for many-year periods of temperature fluctuations on the surface (after V.A. Kudryavtsev): 1–3 – envelopes of the many-year temperature fluctuations at $t_{mean}^{per} = 0;\ 2$ and $4\,°C$, respectively.

respectively.

However the mean temperature t_{mean}^{per} (for the period of fluctuation) at the permafrost surface has an effect not only on the permafrost thickness and its temperature regime but in accordance with the sign (minus or plus) and the

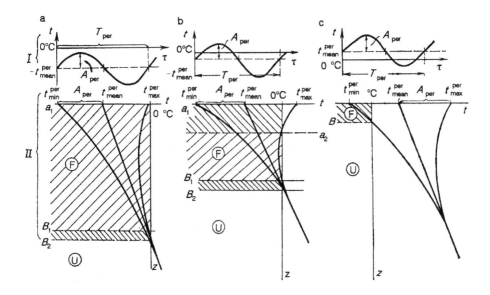

Fig. 12.7. Diagram of periodic variation of permafrost upper and lower
boundaries under periodic temperature changes and various values of mean
(t_{mean}), maximum (t_{max}) and minimum (t_{min}) temperatures on the surface (after V.A.
Kudryavtsev).

value of t_{mean}^{per}, also on the formation of different types of permafrost. Actually, according to Kudryavtsev, three typical cases of permafrost formation
are possible depending on the relationship between the mean (t_{mean}^{per}), the
maximum (t_{max}^{per}) and minimum (t_{min}^{per}) temperatures on the permafrost surface
(Fig. 12.7): 1) $t_{mean}^{per} < 0\,°C$ and $A_{per} < t_{mean}^{per}$; 2) $t_{mean}^{per} < 0\,°C$ and $A_{per} > t_{mean}^{per}$;
3) $t_{mean}^{per} > 0\,°C$, $A_{per} > t_{mean}^{per}$.

In the first case, when $t_{mean}^{per} < 0\,°C$, permafrost existing during the whole
period of fluctuation is formed, with the depth of its base changing in the
range $B_1 - B_2$ (see Fig. 12.7a). Such a type is widespread in severe climatic
conditions of the northern geocryological zone, characterized by thick and
hence rather stable permafrost. In the second case (see Fig. 12.7b) when
$t_{mean}^{per} > 0\,°C > t_{per}$, permafrost is formed with the episodic appearance (during part of the period T) of a layer of multiyear thawing in the range $a_1 - a_2$
and with periodically changing depth of the permafrost base in the range
$B_1 - B_2$. This permafrost is typical of zones having mean temperatures from
0 to $-2\,°C$. In the third case, when $t_{min}^{per} < 0\,°C < t_{per}$, permafrost appears
periodically (during part of the period T_{per}) within the shaded layer (see Fig.
12.7c). This case is typical of the southern zone of permafrost having mean
annual temperatures close to $0\,°C$.

When perennial freezing occurs over a long time, the role of the geothermal flux, and consequently, of the geothermal gradient in the unfrozen ground underlying the frozen, is substantially greater.

The lower boundary conditions are to a large extent responsible for the regime of movement of the permafrost base on account of the relationship between the heat fluxes on each side of the dividing line B_2, that is, on account of the heat flux Q_{th} arriving at this line from the underlying unfrozen ground and of the heat flux Q_{fr} being removed from this line through the frozen ground (Fig. 12.8). The temperature on the dividing line 'unfrozen-frozen ground' proper appears to be fixed and equal to 0 °C. When $Q_{fr} > Q_{th}$, cooling and freezing of the underlying layers proceeds. The boundary of perennial freezing $(B_1 - B_2)$ moves downward in this case. When $Q_{fr} < Q_{th}$, warming and thawing of the overlying frozen ground takes place and the lower boundary of the permafrost B_2 moves upward. When $Q_{fr} = Q_{th}$, temperature conditions at the base of the permafrost will be steady, and the boundary will be fixed. At the same time it is easy to show that if $Q_{fr} = \lambda_{fr}$ grad $t_{fr} = \lambda_{th}$ grad $t_{th} = Q_{th}$, grad $t_{fr} = (\lambda_{th}/\lambda_{fr})$ grad t_{th}, i.e. temperature gradients in the permafrost near its base should be less compared with grad t_{th} in the underlying thawed ground, because most often the values $\lambda_{fr} > \lambda_{th}$.

Thus, the permafrost thickness appears to be essentially dependent on the value of heat flux moving from below upward, i.e. on the geothermal gradient in the underlying thawed ground, g_{th}. The larger is the heat flux moving from below upward, and consequently, the geothermal gradient, the less the permafrost thickness. Estimation of this effect (at $A_{per} = 6\,°C, T^{per}_{mean} = 0\,°C, \lambda = 2.89\,kJ$ (m hr °C)$^{-1}$, $Q_{ph} = 52\,375\,kJ$ m^{-3}, $T_{per} = 100\,000$ years) show that with the increase of geothermal gradient from 0 to 0.03 °C m^{-1}, the permafrost thickness decreases by a factor of approximately two thirds to one half (Fig. 12.8). This rule is followed rather clearly in the part of the permafrost zone where heat flux differentiation is caused primarily by the different age of geological structures. For example, within the Siberian Platform the smallest heat flux values (from 13 to 25 mW m^{-2}) at which the permafrost zone thickness is 800–900 m on the average noted within the most ancient structures of the basement (Anabar anticline). Increase of the heat flux value (up to 40–60 mW m^{-2}) and decrease of the permafrost zone thickness (to 800–600 m) are noted within Mesozoic structures such as the Vilyuy (syncline) and the Predverkhoyansk and Yenisey-Khatangan depressions. A similar dependence of the permafrost thickness on heat fluxes is noted by V.V. Baulin for various structures of the Western Siberian Plate, where q values (in the range between 0.10–0.13 and 0.25–0.30 mW m^{-2}) are in accordance with the permafrost thicknesses of 350–400

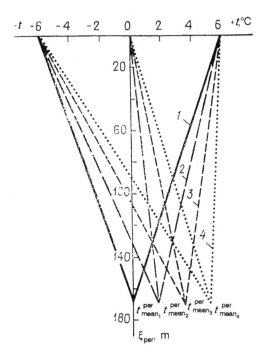

Fig. 12.8. Variations of the thickness of permafrost depending on the value of geothermal gradient at $A_{per} = 6\,°C$ and $t_{mean}^{per} = 0\,°C$ (after V.A. Kudryavtsev): 1–4 – geothermal gradient = 0; 0.01; 0.02 and 0.03 $°C\,m^{-1}$, respectively.

and 135–225 m, respectively.

The highest intensity of the heat fluxes (from 40 to 100 kW m^{-2}) are noted in Russian North-East (regions of active orogenesis) according to V.T. Balobayev's data. It is precisely this situation which is associated with the wide variations in permafrost thickness here (from 150–500 m), with low temperatures (from -6 to $-9\,°C$). Still higher values of heat fluxes are a possibility within the regions of contemporary volcanism (for instance on Kamchatka) where the permafrost can be completely absent as a result.

12.3 Dependence of the permafrost thickness and temperature regime on geological factors and processes

The effect of geological factors and processes on the permafrost thickness and temperature regime is essentially corrected by a heat balance of the upper layers of the lithosphere. This correction can take place as a result of the anisotropy of heat transfer conditions in rock which is dependent on the specific features of their composition, properties and structure, as well as on the redistribution of heat fluxes because of the occurrence of additional heat sources such as active zones of fractures, vapour-thermal springs, water-bearing horizons and zones and areas of vigorous chemical reactions with heat release (for example, oxidation of coal, of sulphides-bearing ores, etc.). Processes associated with heat absorption (for example, adiabatic expansion of gases rising to the surface, etc.) are also developed in the upper layers of the lithosphere. Fundamental changes in the freezing conditions for permafrost occur as a result of neotectonic movements which are responsible, for example, for sea transgressions and regressions, displacement of the water area of large lakes, changes in rate and character of sediment accumulation and spatial expansion and narrowing of denudation areas. These factors and processes affect the permafrost thickness and temperature regime in various ways.

Effect of special lithological features and moisture content of freezing ground

The composition and properties of the permafrost materials, through their moisture content, are responsible for expenditures of heat on moisture phase transition Q_{ph}, and through the thermal properties (λ, C). Increase of the moisture content causes increase of Q_{ph} values and, consequently, decrease of the permafrost thickness. Analysis of the freezing through the years for a periodic, step change of temperature at the surface has shown that differences in permafrost thickness do not exceed 40–50% (Table 12.1). The depths of the permafrost can be changed by a factor of 1.5–2, on account of variations in geothermal gradient.

The depth of the permafrost depends essentially on the thermal conductivity of the freezing soil too. When considering seasonal freezing it was pointed out that its depth is directly proportional to thermal conductivity λ. It is thought also that during permafrost formation the depth of freezing $\xi_{per} \propto \sqrt{\lambda}$. At the same time, disturbance of this dependence under the effect of geothermal flux should be taken into consideration. According to Kudryavtsev's data the value of this disturbance does not exceed 10–15%.

In hard rocks where the values of λ reach 10–12 kJ (m hr°C)$^{-1}$, permafrost thicknesses (all other factors being the same) are approximately 1.4–1.6 times greater than those of soils with $\lambda \approx 4$–5 kJ (m hr°C)$^{-1}$. Therefore similar thicknesses of the permafrost of hard rock and of loose deposits most likely indicate a different age for the formations, i.e. a greater age for the frozen units composed of loose deposits.

The geological structure of a region and particularly the specific features of the occurrence and composition of the freezing materials, their moisture content and heat conductivity, can have a very fundamental effect on the

Table 12.1. *Variation of the permafrost thickness (m) depending on heat of water phase transition Q_{ph} under temperature waves of various amplitudes at the surface A_{per} and the geothermal gradients g (after V.A. Kudryavtsev)*

Q_{ph}, kJ m^{-3}	$A_{per} = 2°C$		$A_{per} = 8°C$	
	$g = 0.01°C$ m^{-1}	$g = 0.03°C$ m^{-1}	$g = 0.01°C$ m^{-1}	$g = 0.03°C$ m^{-1}
20 950	86	47	193	126
62 850	66	41	145	105
104 750	55	37	118	92
146 650	49	34	103	83

formation of permafrost. Thus, given a not very thick cover of loose materials with lower thermal conductivity λ_1 occurring over more heat-conductive crystalline rocks λ_h the maximum permafrost thickness (for the period of development) will be greater than that of uniform loose sedimentary formations. And on the contrary, if the rocks with high values of thermal conductivity λ_h (for example, effusion rocks) occur over sediments with low thermal conductivity λ_1 (for example, clays or silty clays) the thickness of the lithologically two-layered permafrost will be less than that of the uniform profiles with λ_h. Crystalline bedrock, having greater thermal conductivity than loose materials, at the same depth from the surface (for example, at the depth of 100–200 m) always has higher temperature. In this connection the permafrost thickness (in the conditions of a platform) increases with the dip of the bedrock surface as a whole.

Effect of structure on the permafrost thickness

Redistribution of the geothermal flux as a result of inhomogeneity of the structure and thickness of the sedimentary covers significantly affects the thickness of the permafrost. Increased values of the heat flux are observed above the vaults of anticline structures while there is a decrease above the syncline structures. Such distribution of heat fluxes is explained by the fact that where folding occurs, in addition to general vertical upward heat flux additional heat transfer from depressions to vaults of the anticline structures takes place. Additional heat transfer is associated here with differences in thermal resistance of the sedimentary rocks along the bedding and across it. Values of the geothermal gradient in such situations, for example, on the north of the Western Siberian Plate, can be 4–5 times greater than that where the bedding is horizontal, according to V.V. Baulin. The effect is to decrease permafrost thickness in structural vaults by 100–200 m. Thermal convection in water-bearing horizons and water circulation in strata can increase as well as attenuate thermal anisotropy of local structures. Thus the dependence of the permafrost thickness on the structure and thickness of loose sedimentary strata takes various forms.

Effect of the hydrogeological factor

Development of permafrost always takes place in a dynamic thermal interaction with ground water. The several effects of the latter on the depth of perennial freezing is evident in various hydrogeological structures. The effects are associated with the particular conditions of supply, regime and discharge of the water-bearing horizons, typical of these structures. In the general case, the movement of fresh ground water having positive temperatures creates positive anomalies of the heat flux and increases the geothermal gradient below the permafrost. The change of the geothermal flux on account of the seepage of ground water along a bed can be calculated from M.M. Mitnik's formula:

$$t_{ob} = t + \Delta t_w \exp(-nx) \tag{12.2}$$

where Δt_w is the temperature deviation of water entering the bed; x is the distance from the place of water entrance to the place of observation; t is the normal temperature of the bed; t_{ob} is the observed temperature; $n = 2a[m\Delta z(v + \sqrt{v^2 + 4a^2/(m\Delta z))})])]$; Δz is the depth of the bed occurrence; m is the bed thickness; v is the seepage rate; a is the thermal diffusivity of the rock.

Accounting for the upward or downward movement of water can be done by way of adding the molecular heat (conductive) flux and the flux due to the seepage $v_f C \rho \tau$ (convective), where v_f is the seepage rate; $C \rho \tau$ is the heat capacity of water per unit volume. Then when the water is moving upward the geothermal gradient should decrease depending on the rate of water movement, heat capacity, density and thermal conductivity of the water-bearing bed, according to N.A. Ogilvy. It follows that the rate of water exchange plays an important role during the establishment of the permafrost thickness, all other factors being equal. The movement of stagnant ground waters has a minimal effect. High-temperature water associated with deep faults and artesian water rising from great depths have important warming effects on the permafrost. The thickness of permafrost along thick water-saturated zones of faulting is as a rule less than that within adjacent undisturbed rock.

There is a possibility of an abnormally decreased heat flux as a result of horizontal seepage of brines (cryopegs) colder than the surrounding permafrost. Such conditions occur within the Anabar anticline where the maximum thicknesses of permafrost (for the territory of the former USSR) are noted.

Effect of gas reservoirs

The effect of gas reservoirs on the thickness of the permafrost shows itself most often when adiabatic expansion take place, which may lower the rock temperature by up to 5°C. The most favourable conditions for this process occur within the zones of increased jointing, facilitating gas penetration through the joints. This effect is typical, for example, of reservoirs having a bed temperature of 18–21°C.

Under certain conditions, interaction between natural gases and underground waters occurs in association with formation (and conversely, with disintegration) of natural gas hydrates. As a great amount of heat is released during their formation, while the same amount of heat is absorbed during their disintegration, such thermal effects can cause an increase or decrease of the permafrost thickness as well as appropriate changes in temperature regime of the rocks occurring above and below the zones of hydrate formation. It should be noted that the heat of 'water-ice' phase transformations is 334×10^3 J kg^{-1}, while during the formation of natural gas hydrates 520–540×10^3 J kg^{-1} of heat is released, i.e. the thermal effect of the latter process can be highly significant in the course of formation of the permafrost thickness.

13

Taliks and groundwater in the permafrost zone

13.1   The types and formation of taliks in the permafrost zone

Even in the most severe climatic conditions the spatial distribution of permafrost is not universally continuous. Within the permafrost zone the frozen ground can be absent within many sections of river valleys and watersheds, on south-facing slopes, under lakes, at sites of concentrated discharge of groundwater or of its recharge by seepage, volcanic craters, calderas, under some of the present glaciers and above the interior-Earth thermal anomalies associated with oxidation reactions, etc.

Within the region of distribution of continuous permafrost, unfrozen ground occupying small areas and existing continuously for more than a year is termed a *talik*. When the area of unfrozen ground appears to be comparable with or larger than the area of frozen ground the unfrozen ground is termed a *talik zone or massif of unfrozen ground*. According to the relation of taliks and talik zones to the surrounding frozen ground, they can be divided into open (penetrating through the whole frozen stratum) and closed (sometimes they are termed 'pseudotaliks'), penetrating into the frozen stratum to some depth and underlain by permafrost. Often thawed and unfrozen layers, lenses, channels and bodies of other form, bordered at the top, bottom and sides by permafrost are found in geological sections of the permafrost zone. Such formations are termed *inter- and intrapermafrost taliks* in the literature.

Strictly speaking, in addition to taliks proper (frozen earlier, but thawed now), the so-called original taliks, represented by ground not frozen before and existing in the unfrozen state, can be found in nature. For example, it is known that the ground under many rivers of Siberia and the Far East as well as under large lakes of tectonic, glacial and volcanic origin has never been frozen. In conditions of continuous permafrost the taliks are being formed at the present time also. Many of them are formed as a direct effect of human activity within the permafrost zone in regions under development.

Thus, taliks within the permafrost zone can be of various origin. Some of them should be considered as existing continuously in the course of perennial freezing (for many hundred thousand years), the others – as occurring naturally at the present time (after the thermal minimum of the Late Pleistocene) or as a direct effect of human activity.

There exists a great number of talik classifications reflecting in one way or another the nature of occurrence of thawed and unfrozen ground within the permafrost zone. Such kind of elaborations were carried out by N.I. Tolstikhin, I.Ya. Baranov, N.A. Vel'mina, N.N. Romanovskiy, N.A. Nekrasov, S.M. Fotiyev, Ye.S. Sukhodol'skiy and others. As the basis of these classifications, various particular features of general or specific character are brought together under headings such as topographic position, (watershed, valley, and slope taliks), conditions and degree of water saturation (water-bearing, impermeable, or dry), shape and size (rounded, slightly and strongly extended, thawed fissures, craters, etc.), lifetime (stable, unstable, seasonal, perennial), hydrologic peculiarities of the ground (water-absorbing, water-removing, water-conducting, water-containing), the occurrence of water-bearing strata in relation to the permafrost (suprapermafrost, intrapermafrost and subpermafrost), mechanisms of heat transfer (conductive, convective and their combinations), heat sources (solar, ground, surface and atmospheric water energy, the geothermal flux, exothermic reactions, etc.) and others.

The classification by N.N. Romanovskiy according to which taliks are divided into types with respect to the causes of their occurrence is most commonly used. The main causes are: radiation-heat balance on the Earth's surface including heat flow from the interior by both conduction and convection; warming effect of water courses, water reservoirs and glaciers; exothermal oxidation reactions; volcanic activity etc. (Table 13.1). On the basis of these main causes the taliks are subdivided into seven types: radiation-thermal, hydrogenic, hydrogeogenic, glaciogenic, chemogenic, volcanogenic and technogenic (Fig. 13.1). This subdivision of taliks points out the details of their origin. Thus, the *radiation-thermal* type is subdivided into three subtypes: radiation, thermal and pluvial-radiation.

Taliks of the radiation subtype are formed on account of solar energy arriving at the Earth's surface. Positive ground temperatures are maintained within the areas composed of water-impermeable rocks mainly by way of conductive heat transfer without the effect of infiltrating atmospheric precipitation. Such taliks are most widely distributed near the southern limit of the permafrost regions with a great number of sunny days and small amounts of snow (Middle Asia, Southern Siberia, southern part of the Far East).

Table 13.1. *Classification of taliks (after N.N. Romanovskiy using those of I.A. Nekrasov and S.M. Fotiyev)*

Source of heat	Type	Subtype
Exogenous	I Radiation–thermal	Radiation, thermal, pluvial-radiation
	II Hydrogenic (water-thermal)	Shelf, below an estuary, below a lake, below a river bed, near a river bed (flood-plain)
Endogenous	III Hydrogeogenic (underground-thermal)	Subaerial, below a lake
	IV Chemogenic	—
	V Volcanogenic	—
Polygenous	VI Glaciogenic	—
	VII Technogenic	—

Thermal taliks are formed due to the insulating effect of snow (increasing with depth and lower density) causing positive temperatures at the base of the seasonally freezing-thawing layer. This layer is usually represented by slightly water-permeable or water-saturated ground (within swamps) as in taliks of the radiation subtype. This kind of talik is typical of the regions with maritime and moderate continental climate with wind redistribution of

Fig. 13.1. (*opposite*) Position and structure of various types of taliks and their involvement in water exchange: 1–7 – geological structure; (1 – crystalline bedrock; 2 – terrigenous-sedimentary rocks; 3 – extrusives; 4 – intrusions; 5 – loose materials of various origins; 6 – tectonic dislocations; 7 – rocks with increased joints); 8 – pluvial-radiation infiltration open taliks; 9–17 – hydrogenous (underwater) taliks; (9 – stagnant, closed below a lake; 10 – stagnant, open below a lake; 11 – infiltration, closed below a lake; 12 – infiltration, open below a lake; 13 – pressure-seepage, open below a lake; 14 – infiltration, open below a river bed; 15 – pressure-seepage, open below a river bed; 16 – ground-seepage, closed below a river bed; 17 – ground-seepage, open below a river bed); 18–20 – hydrogeogenous taliks (18 – subaerial pressure-seepage, open; 19 – subaerial pressure-seepage, closed; 20 – pressure-seepage, open below a lake); 21–22 – glaciogenic taliks (21 – infiltration open; 22 – pressure-seepage, open); 23–24 – volcanogenic taliks (23 – infiltration open; 24 – pressure-seepage, open); 25–26 – boundaries (25 – of permafrost; 26 – of areas of deep and ultra-deep freezing without groundwater); 27 – direction of groundwater flow; 28 – groundwater icings; 29 – lakes; 30 – glaciers.

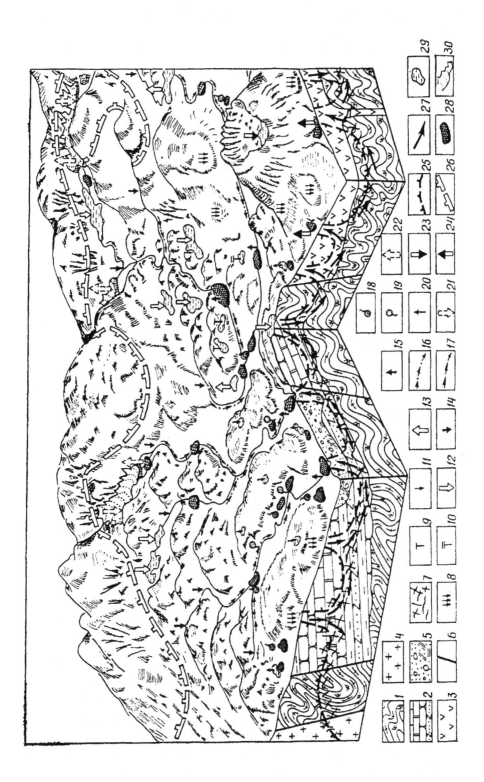

snow (European North of Russia, Western Siberia, mountain regions of Southern Siberia and the Russian Far East).

Taliks of the pluvial-radiation subtype are formed by the heat effect of infiltrating rain-water. Such taliks are typical of low altitude (up to the absolute altitude of 900–1200 m), plane watersheds and gentle slopes composed of Quaternary formations and bedrock having good infiltration properties. They are formed in the southern parts of the permafrost region with discontinuous development of permafrost from the surface, where the amount of summer precipitation exceeds 300–400 mm. The pluvial-radiation taliks occupy the greatest areas (to 50–80% of the area). It is through such taliks that groundwater is replenished from atmospheric precipitation.

Romanovskiy divides the taliks of *hydrogenic type* (underwater taliks), formed under the warming effect of surface waters, into five subtypes: shelf taliks (under the effect of sea water), taliks below an estuary (under the warming effect of river and sea waters), below a lake (owing to the water containment effect), below a river bed (under the influence of the water flow) and flood-plain taliks (under the temporary, periodic effect of flooding waters).

It should be noted that taliks below a river bed and flood-plain taliks are not formed solely by the warming effect of surface waters. Powerful underground streams in the deposits of river beds have a very significant effect. In plan the free cross-sectional area of such taliks often extends beyond the flood-plain and into terrace deposits. Taliks below a lake (see Fig. 13.1) are clearly important in various ways for water exchange in the permafrost. They exist below lakes of various origins and can be both open and closed. Some amount of lake water feeds deep subpermafrost groundwater flow through open taliks of watershed lakes of glacial and tectonic origin. Some portion of this groundwater flow in turn discharges subaquatically through open taliks in lakes or trough valleys situated at lower levels.

It should be stressed that the presence of an open talik beneath most lakes in former river beds and lakes of thermokarst origin means the probability of groundwater recharge or discharge. However if water-impermeable clayey ground predominates in the section of the talik, or slightly permeable water-bearing ground alternates with water-impermeable ground, such taliks are practically free of water or are characterized by a stagnant or slowed-down regime of water exchange.

Glaciogenic taliks are formed by glacier meltwater, the streams of which are concentrated near the glacier base through fissures. Such taliks are found within high mountain systems of Pamir and Tien-Shan as well as below a number of glaciers of Verkhoyan'ye, Chukotka, and Koryakiya which have the most severe climatic and permafrost conditions.

Hydrogenic, chemogenic and volcanogenic types of taliks are classified based on the fact that they exist due to thermal anomalies resulting from processes in the Earth's interior. The origin of these hydrogeothermal anomalies is associated with convective heat transfer by subpermafrost waters under pressure. The conductive heat transfer causes anomalies of geothermal origin (exothermal, heat flux from the Earth's interior, hot gases outlets, volcanic heating).

Taliks of the technogenic type are formed in the course of human activity and develop as taliks of various types in natural conditions. They are formed below artificial water reservoirs, below straight stretches of river beds, under spoil heaps and fills of coal- and sulphide-containing rocks, below areas with the soil-turf layer artificially removed, with disturbance of the radiation-thermal regime existing at the ground surface, etc. The time of formation of technogenous taliks and their areas are different for the regions of continuous and discontinuous permafrost. Thus, in the south of the permafrost zone a small, often rather short-term, change in heat exchange conditions on the ground surface is sufficient for taliks to be formed; to the north long-term changes are necessary because the low-temperature permafrost has a more considerable thermal lag.

The character of taliks, the tendency for their development and the rate of the process depends in many respects on the relations with ground waters and on the type of water exchange. It is the water (if it is considered as an accumulator and carrier of heat from the Earth's interior, especially within the sections below lakes, where it moves vertically), which has most effect on the change in geotemperature pattern. Various types of taliks are subdivided into the following-classes: free of water (water-impermeable or 'sushentsy'), stagnant (water-containing), infiltration (water-absorbing), ground-seepage (water-conducting) and pressure-seepage (water-releasing). Groundwater recharge, movement in the zone of supra- and subpermafrost groundwater flow and pressure discharge occur through the taliks of the various classes.

Open taliks below a river bed can play various roles seasonally. Thus, in summer, when the abundant recharge of groundwater occurs, they exist as pressure-seepage taliks, while as the spring comes they change into infiltration ones within the upper reaches of valleys, as winter depletion of groundwater has taken place.

The nature of groundwater movement, as well as the rate of water exchange and the degree of participation of various taliks in it, depend on the cryogenic change of hydrogeological structures and vary with the different regions of permafrost development. Paradoxically as it may seem, the water exchange and groundwater movement are most active within the region of continuous permafrost. Concentration of the underground flow along the most washed-out rejuvenated tectonic dislocations takes place here. Within the regions of discontinuous permafrost development this groundwater flow becomes less concentrated. The cryogenic part of the geological section and the degree of washout of water-containing rocks in taliks is reduced significantly in this situation. Such a tendency is retained as a whole when moving from the areas of discontinuous cryogenic water-confining strata to massif-island, island and sporadic ones and is rather typical.

13.2   Groundwater of the permafrost regions

In response to repeated changes in the climatic conditions through the Pleistocene and Holocene, which included glaciations and general cooling, fundamental cryogenic transformations occurred within the upper part of the lithosphere. They extended over vast portions (as far as area and depth are concerned) of the Earth's crust, creating quite a specific hydrogeological situation. Groundwater survived only in taliks of various types, below the permafrost base, within the zones of supra- and intrapermafrost groundwater flow and in the form of intrapermafrost water-conducting zones. The rest of the hydrogeological section turned into cryogenic water-confining strata under the effect of great external cooling. A large amount of water was transformed into ice, promoting formation of expanded cryogenic structures as well as ice-rich horizons with numerous ice lenses in the permafrost.

For the territory of the permafrost zone, groundwater is subdivided, with respect to cryogenic water-confining strata, into supra-, intra-, and subpermafrost water, and groundwater between permafrost units. These specific categories of groundwater have a number of inherent features allowing their grouping together for various permafrost conditions and their consideration in a unified context in spite of these hydrogeological differences.

Suprapermafrost water is subdivided into two varieties. These are temporarily existing water of seasonally thawed layers and continuously existing water of closed taliks.

Groundwater between permafrost units is typical of the permafrost zone which has a two-layered structure, and the water is situated between the upper (Holocene or recent) and lower (Pleistocene or 'relict') cryogenic layers. The waters under consideration usually have hydraulic connection with the other (supra- and subpermafrost) groundwater.

Intrapermafrost water is delimited on all sides by frozen ground. It is an uncommon category of groundwater typical of the intensively freezing hydrogeological structures. Such groundwater is excluded from water exchange as a rule.

Subpermafrost water is water found in water-bearing horizons, complexes and zones of jointing, below the base of the frozen strata.

The water of open taliks belongs to a particular category of groundwater within the permafrost zone, connecting all the types of groundwater noted above (except for the intrapermafrost type) into the unified hydraulic system.

Suprapermafrost water of the seasonally thawed layer

This exists exclusively during the summer period. This is pore-fissure, pore and fissure water in Quaternary deposits, in crusts of weathering and in exposed bedrock. Its distribution correlates with the position of the permafrost surface and depends on the specific features of topography. The depth of thawing and the thickness of water-saturated horizons depend on soil composition and they increase from north to south. Waters of the seasonally thawed layer carry the greatest amount of heat within the sections where the flow velocity is great. These are sites at the base of slopes composed of well washed-out water-saturated rudaceous formations with good seepage properties of the ground.

Recharge of water in the seasonally thawed layer is by atmospheric precipitation, vapour condensation, thawing of snow patches etc. 'Golets' ice in rudaceous formations is of great importance for this process. Such ice is formed in spring during the infiltration of water from melted snow into frozen rudaceous materials.

Frozen ground is not always the water-confining horizon for waters of the seasonally thawed layers. They can also be underlain by water-impermeable thawed deposits, giving typical vadose water.

For the summer period the groundwater can be classified with respect to recharge sources such as periodically appear after a rainfall (within watersheds), periodically disappear as a result of long absence of rains (upper and middle parts of slopes) and continuously exist on account of inflow of water from the seasonally thawed layer in areas situated hypsometrically higher (lower parts of slopes, valleys).

The water considered above has mainly low mineralization, and a hydrocarbonate composition with variable cation relationships. It is enriched in oxygen, in humic acids and in organic matter. Near the coasts of northern seas as well as in arid zones (Central Yakutiya, Southern Zabaykal'ye) brackish and saline water is often found in the lower part of the seasonally thawed layer.

This class of water is of no interest as a source for water supply because of short lifetime, variable water regime, predisposition to pollution and unreliability in respect to sanitation as a consequence.

Suprapermafrost water of closed taliks

All the types of groundwater accumulation that persist throughout the year near the permafrost table are included here. This is water in taliks such as the pluvial-radiation type, taliks below or near a river bed, below a lake, as well as pressure-seepage taliks of the subaerial subtype. The thickness of the closed taliks noted above can vary in the range from a few metres to 40–60 m and rarely to 100–120 m, depending on the sources of supply, the character of the profile and its degree of flushing, capacity properties of the water-containing ground and conditions of water exchange. Most of these taliks (except for those situated near a sea coast) are characterized by free water exchange and fresh composition of the groundwater, the recharge of which depends on water of the seasonally thawed layer, atmospheric precipitation and surface water. These sources include soil pore, stratum and fissure water in Quaternary deposits and weathering crusts and zones of jointing in hard and semihard rocks.

Pluvial-radiation taliks are developed mainly in the south of the permafrost zone, with their water coming from rain and snowmelt as well as from condensation of vapour. Wide variations of groundwater levels resulting from the irregularity of precipitation of varying intensity are typical. These waters correspond to atmospheric precipitation and to waters of the seasonally thawed layer in chemical composition and are of low (up to $0.1 \, \mathrm{g} \, \mathrm{l}^{-1}$) mineralization.

Water of taliks below or near a river bed represents hydraulically uniform soil pore water in loose valley deposits, as well as fissure waters in the underlying unfrozen hard and semihard rocks. This category of groundwater is the most widespread within the permafrost region, and is of great practical importance for providing a temporary or regular water supply. The replenishment is by surface water courses maintained by atmospheric precipitation and by water of the seasonally thawed layer as well as by subpermafrost groundwater flow as a result of subaquatic pressure discharge. The water regime is unstable throughout the year. Chemical composition corresponds to that of surface water courses. Continuity of the flow below a river bed is broken in winter. Separate basins appear, cryogenic heads are formed and the quality of the groundwater changes. Where

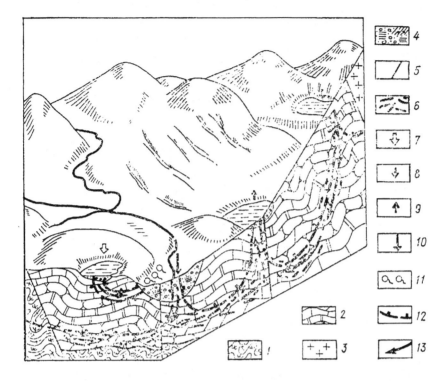

Fig. 13.2. Taliks below a lake within a zone of active water exchange:
1–6 – geological structure (1 – crystalline bedrock; 2 – terrigenous-carbonate
rocks; 3 – intrusive rocks; 4 – loose materials of various origins; 5 – tectonic
dislocations; 6 – rocks with more than usual jointing and karst processes);
7–11 – hydrogenous (underwater) taliks (7 – infiltration, closed below a lake;
8 – infiltration, open below a lake; 9 – pressure-seepage open, below a lake;
10 – infiltration open, below a river bed; 11 – subaerial pressure-seepage closed);
12 – permafrost boundaries; 13 – direction of groundwater flow.

subpermafrost water under pressure discharges into the taliks below a river
bed the groundwater regime is marked by greater stability.

From a regional perspective, the valley taliks' parameters and the cross-
section for seepage flow change, increasing southward and decreasing
northward depending on the permafrost conditions, the intensity of surface
runoff, type and washout degree of the underlying rocks and the presence of
disjunctive dislocations.

Waters of taliks below a lake are formed within various topographic
elements. The chemical composition of these waters is practically the same
as that of water reservoirs. These are pore- , pore-stratum and fissure waters
in the underlying loose and slightly lithified Cenozoic deposits as well as in
bedrock (Fig. 13.2). The thickness of these taliks varies from a few metres to

40–80 m and greater, decreasing from south to north under similar conditions. In freezing lake basins, below alases and below lakes that are freezing or drying-up, groundwater mineralization can be as much as 20–80 g l^{-1} because of the process of cryogenic metamorphization. Water of closed taliks below a lake often has a stagnant regime and limited storage.

Waters of pressure-infiltration closed taliks are rare and situated within the areas of thick (more than 200–300 m) permafrost with rather high mean annual ground temperatures (up to -3 or $-5°C$). Movement of the groundwater of this type occurs within the perennially frozen masses along pipes and channels that are intrapermafrost or situated between two permafrost masses. These are fissure, karst-fissure and vein-fissure waters with an intensive water exchange. The discharge takes place through down-tending sources of groundwater causing formation of small and medium or, rarely, large icings in winter. They are supplied from the atmosphere as well as from surface waters of the seasonally thawed layer. The water is usually fresh. It is often used for supplying installations having a small water consumption.

Groundwater of open taliks

This represents an important link in the system of the existing water exchange between surface water courses, closed taliks, and water of deeper horizons and complexes situated near the permafrost base. There is pore, pore-stratum, fissure, vein-fissure and karst water. These categories are found in practically all these taliks and are subdivided into infiltration and pressure-seepage classes. Downwards or upwards flow of groundwater of this category occurs through the system of thawed fissures and channels in the permafrost. There are differences in specific hydrological features for various subtypes of taliks. Thus, groundwater levels in pluvial-radiation taliks of the infiltration class experience wide, sometimes even spasmodic fluctuations (from 20 to 30 m, or 40 m) in the summer period depending on the precipitation amount and intensity. These fluctuations are gradually damped out away from the talik boundaries. In winter the drop in levels in the upriver infiltration taliks situated below and near a river bed can be as much as tens to a few hundred metres, as the depletion of the surface discharge and of the discharge below the river bed progresses.

Large, or gigantic, icings are usually formed near the pressure-seepage taliks. In large rivers within such places, the warming effect of the subpermafrost groundwater flow being discharged into them leads to the existence of polynias in winter. Along the coasts of northern seas and within the intracontinental regions of the Siberian platform the discharge of highly mineralized water with negative temperatures through taliks below a river bed, produces a cooling effect on the ground.

Subpermafrost groundwater flow

This flow exists near the base of the permafrost and has a significant effect on the cryogenic structures and on the permafrost thickness. There are pore, stratum, fissure, vein-fissure, karst and other types of groundwater. These types are characteristic of the permafrost zone. Formation of the cryogenic water-confining horizons has an effect on the spatial distribution of groundwater flow in various seepage media near the permafrost base. The position of such groundwater results in its definition as subpermafrost, to some extent smoothing over the differences between the separate water-confining horizons, complexes and zones of jointing.

Subpermafrost water can be pressure contacting (in contact with frozen rocks) or non-contacting (separated from frozen rocks by thawed ones) with respect to the permafrost base. The fresh and slightly mineralized water below the permafrost has positive temperatures as a rule, while brines (cryopegs) have negative temperatures. The water differs in chemical composition in accordance with the type of bedrock. For the most part the water has hydrated calcium carbonate perhaps with sodium carbonate, if fresh water, and in salt water sodium chloride perhaps with calcium chloride and sodium and calcium sulphates. Groundwater of the non-contacting type can be under pressure (below the water-impermeable thawed ground) or can have a free water table (below permeable thawed ground). In the course of changes in the climatic situation and repeated raising and lowering of the permafrost base, fissure zones of cryogenic disintegration with higher water content are widely developed and occupy large areas. The total thickness of these layers of cryogenic disintegration near the southern limit of the permafrost zone can be more than 150 m.

The water level regimes of groundwater of the contacting and non-contacting types are different. For the non-contacting type, wide seasonal variations of groundwater levels are noted near the sources of recharge. For the contacting type, the groundwater levels are more stable.

Groundwater between permafrost units and intrapermafrost groundwater

These are typical categories within the permafrost zone and were formed at the closing stages of the cryogenic transformation of the hydro-geological structures.

Groundwater between permafrost units occurs widely within the hydrogeological structures of the European North of Russia, Western Siberia and in some parts of the intermontane basins of the Baikal type, etc. In addition, the water-bearing strata with fresh and brackish waters occur below draining and freezing thermokarst lakes and hollows, as well as at the sites of ancient water bodies.

The closed infiltration taliks recharging groundwater between permafrost units are found within high terraces and interfluve areas of the Lena and Vilyuy basins, Putorana plateau, etc. Such groundwater is situated below certain lakes, small tributaries and pluvial-radiation taliks. Under climatic cooling the channels conducting water flow inside the permafrost can become separated from the surface being transformed into intrapermafrost water-containing cavities which, in the end, freeze.

13.3   Interaction of groundwater with the permafrost and types of cryohydrogeological structures

Permafrost affects the groundwater condition through the mechanism of mutual interaction. The water loses energy to the frozen rock, tending to thaw it, while the frozen rock, on the contrary, tends to freeze the water by taking its heat energy. The permafrost zone currently existing is derived from this complex process proceeding in various directions.

Partial or full freezing of water-bearing horizons, complexes, and zones of jointing has occurred as a result of such interaction. They have been excluded from natural water circulation to a significant depth of the geological section and even entirely in some places. The particular conditions of exchange of the ground, atmospheric and surface waters through the system of taliks discussed above have arisen in this way. Storage capacities of the bedrock have been reduced and the groundwater circulation has narrowed spatially. At the same time separation of the hydrodynamic systems previously interconnected becomes typical of the upper part of the lithosphere. The interrelations in the primordially existing water-bearing horizons, complexes and zones of jointing have been disturbed and the directions and intensity of groundwater flow near the permafrost base have been changed.

Layers of frozen ground (with ice), cryotic (without ice) and cooled (with cryopegs) were formed. In the vertical section there can be one-, two- and even three-layered cryogenic units within the various parts of the lithosphere. At the same time the total thickness of the cryogenic water-confining horizons can vary from the upper few metres in the south of the permafrost zone, to 500–700 m and even more than 1000 m in the most severe conditions of highly dissected mountain structures.

In the hydrogeological structures changed during the course of freezing, abnormally high heads were formed. The initial structures were separated into individual basins with enormously concentrated run-off at times and with extremely restricted water exchange at other times. In the course of the thawing of the hydrogeological structures during warming periods, zones of abnormally low heads developed.

The initially fresh composition of groundwater underwent cryogenic metamorphism as a result of the interaction with the perennially frozen water-confining horizons. The main point is that in the course of transition of water into ice some portion of the dissolved salts precipitates. The content of easily dissolvable components increases in the freezing solution and is pressed out by the growing ice inclusions into the lower part of the section promoting cryogenic concentration. During warmings when degradation of the cryogenic water confining horizons takes place, not all the precipitated salts are redissolved. As a result, the water-bearing systems that are reestablished in the course of thawing have lower salt content (are cryogenically desalted).

Formation of cryogenic water-confining horizons affects the development of salt waters with negative temperature (cryopegs) and it has caused the formation of fissure zones of cryogenic disintegration with higher than usual water content near the permafrost base, thus giving entirely new (cryogenic) hydrogeological structures and accumulation basins (in depressions and fissure zones) and groundwater flow (along valleys and tectonic zones) of a particular kind.

At the same time the existence of groundwater beneath penetrating permafrost actively counteracts the freezing process and this must have an effect on the character of the developing cryogenic section. As this takes place, the spatial involvement of ground in the freezing changes considerably, not only in the wide regional sense but also locally, depending on the groundwater discharge, its heat content and the character of water exchange in the hydrological system. Thus, the very severe freezing conditions in high latitudes and within mountain regions had an effect on the formation of the very thick (up to 800–1000 m) cryogenic water-confining horizons found within Baikal-Stanovoy, Verkhoyansk-Kolyma and other folded regions. Sizable areas here have no significance at all as far as hydrogeology is concerned. The zones of regional exogenic jointing are fully frozen within these areas. The groundwater contained in them before the process of lithosphere cooling began was either carried away or transformed into ice, having an effect on the permafrost ice content. Within the areas situated around the highly elevated rock massifs with entirely frozen zones of jointing excluded from water exchange, there exist places with extremely concentrated groundwater flow along cavities, pores, fissures and veins in the rocks. Such groundwater concentration occurs along dislocations with a break of continuity, along neotectonic depressions and the network of valleys and it changes the temperature conditions of the frozen rocks, their thickness and continuity. At the same time in the immediate neighbourhood one can observe areas of continuous, ultradeep freezing (more than 300–500 m) with vast areas of discontinuous and massive-island development of the cryogenic water-confining horizons and even areas where the permafrost is absent. Such a pattern is typical, for example, of a number of hydrogeological structures within the Baikal-Stanovoy folded region. Rock massifs with ultradeep freezing (along the main watersheds of Severo-Muyskiy, Udokan, Kodar and other ridges) are adjacent here to areas where the permafrost is absent. These areas are highly water-saturated and alternate with permafrost massifs and islands, the thickness of which is one third to one fifth of that in the areas of continuous freezing situated around them.

Salt groundwater and brines assume negative temperatures and are transformed into cryopegs in the course of climatic cooling. Remaining liquid, they affect the intensive rock cooling, increasing the thicknesses of the cryogenic part of the section abnormally because of their cooling ability. Within the Siberian Platform, the Verkhnevilyuysk, Verkhnemarkhinsk and Turukhansk uplifts and Putorana volcanogenic massif, etc., show such distinctive cooling involving cryopegs.

The mutual effect of groundwater and cryogenic water-confining horizons on each other can be shifted sharply in either direction, as a result of natural as well as artificial disturbances in the surface conditions, in spite of the apparent relative equilibrium of this interaction. These disturbances can arise as a result of general climatic changes as well as of geotechnical activity. At the same time, any change in the equilibrium towards warming causes a usually greater and more efficient effect of the groundwater on the permafrost. As a consequence, it very quickly becomes the leading factor in permafrost degradation.

The heat exchange between groundwater and permafrost is revealed in various ways by groundwater flow. If one follows the groundwater flow within the permafrost zone from recharge to discharge, in the most general terms it is obvious that the character of the thermal interaction with the permafrost changes. Where downward flow along the open infiltration taliks occurs, the rocks of the upper part of the section are warmed. Closer to the base of the permafrost the temperatures of groundwater and bedrock equalize and become lower than that of the thawed ground near the lower boundary of the cryogenic water-confining horizons. Within this interval

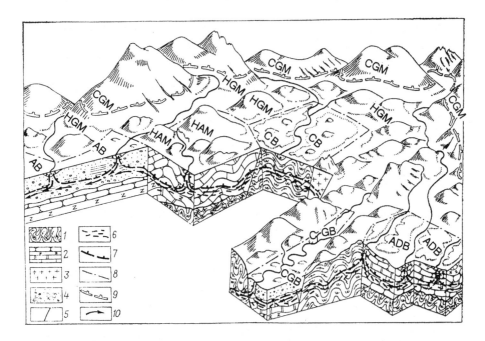

Fig. 13.3. Main types of hydrogeological structures changed by cryogenesis:
1–6 – geological structure (1 – crystalline bedrock; 2 – terrigenous-carbonate and
carbonate-terrigenous rocks; 3 – intrusive rocks; 4 – loose and slightly lithified
materials – of various origins; 5 – tectonic dislocations; 6 – rocks with increased
jointing and karst processes); 7–9 – boundaries (7 – of permafrost; 8 – of
hydrogeological structures of various types; 9 – of areas of deep and ultra-deep
freezing without groundwater); 10 – direction of groundwater flow. *Main types of
cryohydrogeological structures*: CGM – cryogeological massif; HGM –
hydrogeological massif; AB – artesian basin ; HAM – hydrogeological admassif;
ADB – adartesian basin; CB – cryogenic basin of pressure-fissure waters; CGB –
cryogeological basin.

the groundwater flow acts as a cooling factor resulting in the permafrost
thickness increasing. In the course of more or less horizontal subpermafrost
flow the strata are heated by the heat from the Earth's interior. As this flow
is parallel with the isolines of the temperature field, the groundwater exerts a
neutral effect on the thermal state of the ground. As the groundwater flows
toward the places of discharge (upward flow) its temperature is higher than
that of the surrounding frozen material, on which its exerts a warming effect.
As a consequence, the open pressure-seepage taliks remain stable.

Within the permafrost zone as well as within the regions free of permafrost, hydrogeological massifs and artesian basins are separated in the
course of *hydrogeological regionalization* (Fig. 13.3). Protrusions of crystal-

line basement rocks with fissure and vein-fissure types of groundwater occur in the massifs. Artesian basins are basins, depressions and platforms filled by horizontal sedimentary rocks of various age and composition with pore, pore-stratum, stratum-fissure and karst-fissure waters. In addition, the artesian structures of the intermediate type (between massifs and basins), share a number of traits with both of them. For example, the adartesian basins are similar to the artesian ones as far as their geological-structural position is concerned; however, their water-containing rocks are folded into synclines. The adartesian massifs are characterized by the wide development of anticlines and represent positive topographic structures. In either case the fissure and stratum-fissure groundwater accumulations prevail as is the case in the hydrogeological massifs. The flow of the groundwater is centripetal in adartesian basins and centrifugal in adartesian massifs.

The names of the types of hydrogeological structures (probably it is more correct to call them cryohydrogeological or hydrogeocryological structures) should reflect the cryogenic characteristics considered as a result of the process of groundwater interaction with the cryogenic water-confining horizons: for example, hydrogeological structures of continuous ultradeep freezing (more than 300–400 m); of continuous deep freezing (200–300 m); of not very deep freezing (100–150 m); of discontinuous intermittent freezing (frozen massifs comprise 50–75% of the total area); of discontinuous massive-islands (25–50%) and of islands and sporadic freezing (5–25%).

Water-bearing horizons, complexes and zones of jointing are the main taxonomic units in the cryohydrogeological structures. One is forced to characterize them additionally in the context of cryogenic changes that have taken place in them: for example, frozen entirely, dry; frozen partly, across the section; mainly frozen with local water saturation, etc.

Rocks in cryohydrogeological massifs and admassifs are water-saturated near the base of cryogenic water-confining horizons and along taliks of various types. At the same time, separation of a single system of pressurized water in massifs into a number of systems in the basins of river runoff, with localization of recharge sources and formation of cryogenic heads (pressures) takes place. The most deeply frozen parts in such structures are 1500–2000 m higher than the present river beds and they are characterized by the total absence of groundwater in the liquid phase (except for the suprapermafrost water during the summer period). Such structures are termed *cryogeological*.

In the course of freezing of the large Meso-Cenozoic artesian basins, when the permafrost thickness becomes greater than that of the fresh water belt, there are only cryopegs in the lower part. Such structures are termed *cryoartesian basins.* Within the more shallow artesian structures corresponding to Cenozoic superimposed basins, the complete freezing of water-containing rocks of the artesian cover often takes place. This causes transformation of these artesian structures into the so-called *cryogeological basins.* Within these basins the ground waters are contained in faulted rocks of the basement as well as in open and closed infiltration-, pressure seepage- and ground seepage taliks.

In the course of cryogenesis particular hydrogeological structures were formed when groundwater was concentrated near the permafrost base or within the zones of cryogenic disintegration. These are cryogenic basins of pressurized-fissure water. They are characterized by the continuous extent of the permafrost water-confining horizons. The layer of cryogenic disintegration with the pressurized-groundwater does not usually exceed 15–20 m.

A particular type of cryogenic basin appears near the southern limit of the permafrost. Permafrost water-confining horizons in these basins are discontinuous while the thickness and the number of layers of cryogenic disintegration increases by an order of magnitude. Fissure water in the zones of cryogenic disintegration has water tables that change throughout the year.

Water exchange in the far from complete list of hydrogeological structures discussed above and subjected to cryogenesis, is termed 'open', 'closed' or 'partly open'. Recharge and discharge of groundwater in open structures takes place directly within them. Within partly open structures only one process (either recharge or discharge) proceeds, while the missing element of the water exchange cycle is implemented through the nearest structure. In the hydrogeological structures closed by the cryogenic or lithological water-confining horizons the recharge and discharge of the groundwater do not take place inside the structure. They are implemented by groundwater transfer and flow from the basement rocks or from the cover of the nearest structures.

IV

Regional features and evolution of permafrost

14

Permafrost evolution in the Earth's history

14.1   History of the development of permafrost and its distribution on the planet

The evolution of the Earth extends over almost 5 billion years. Its history can be divided into three significant stages which are basically responsible for the present state of the lithosphere. These are the initial or Archaean (2×10^9 years), Proterozoic (on the order of 2×10^9 years) and Phanerozoic (0.6×10^9 years) stages of the development of the geosphere and lithogenesis. During the Archaean stage of the Earth's development the vulcanogenic type of lithogenesis, characterized by accumulation of lava and loose ash material on the sea bottom with a low content of weathered terrigenous sediments, was dominant. The differentiation of the lithogenetic types is likely to have begun only at the end of the Archaean stage, being most apparent in the Proterozoic-Riphean stage with the formation of humid, arid, volcanogenic-sedimentary and glacial or cryogenic types of sedimentary rock. The first certain glacial drift of the continental type is referred to this stage. The expansion of the area of the continents caused not only glaciation and formation of specific deposits of the cryogenic type but also considerable increase of deposition of detrital and dissolved material in oceans (especially of carbonates) with the predominance of exogenous lithogenesis over the volcanogenic-sedimentary.

The cryogenic type of lithogenesis in the Proterozoic is the latest and it becomes more and more important as the recent epoch is approached. An increase in climatic severity and frequency of glaciation supports this. We can see the process clearly beginning from the end of the Mesozoic era (Fig. 14.1), that must have led to the regular and more frequent occurrence of the cryogenic type of lithogenesis. In this manner the irreversibility of the evolution of the type of lithogenesis shows itself in progressive replacement of the volcanogenic-sedimentary type at first by the humid, then by the arid type, and finally we can see a trend toward prevalence of the cryogenic type

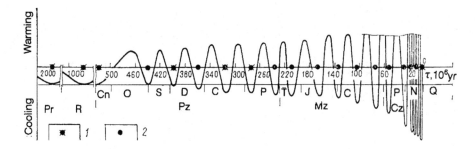

Fig. 14.1. The repeating intense climatic warming-cooling rhythms (glacial epochs) in the post-Archaean history of the Earth (tectono-cryogenic periodicity): 1–2 – glacial epochs (1 – recognized; 2 – supposed).

(10). In this connection we can follow a change to accumulation of more and more polygenetic, polydispersed detrital poorly sorted moisture-rich (ice-rich) deposits of the permafrost type for the most part of mesomictic (quartz, feldspar, hydromica, montmorillonite, etc.) and sand-aleurite (sand, coarse silt) composition.

General trends in planetary development (such as Earth's radiogenic heat decrease, expansion of continental areas and, consequently, climatic severity and increase of continental sedimentation rate) may cause not only cryogenic freezing of the stratigraphic units but also, possibly, the formation of syncryogenic deposits as early as the end of the Proterozoic (Fig. 14.2).

As the cryogenic type of lithogenesis has not been adequately studied yet (with the exception of the glacial drift and syncryogenic rock units) there exist to date only few data for the occurrence of permafrost in the ancient epochs. Ancient continental glaciations are recognized from moraine deposits, from the established phenomena of rocks broken away from the glacier bed and of detritus transportation and reworked deposits from glacier thawing. Consolidated and cemented ancient moraines with a density close to that of rocks of the sandstone type are termed *tillites*. Discovery of such kinds of formations of Precambrian, Ordovician, Silurian, Permian and Carboniferous age in different regions of the Earth indicates in a unique manner the repeated appearance, existence and disappearance of glacial covers and, consequently, of the permafrost.

The relations between the areas of the glacial ice cover and the permafrost can be various and depend on the glacier type, sizes and thickness, climatic severity and continentality, atmospheric moisture content and amount of solid precipitation, etc. Large glacial covers, for example, of Antarctica and Greenland, cause an essential decrease of mean annual temperatures not only over the glacier itself (up to -30 to $-60\,^{\circ}$C), but also over the large

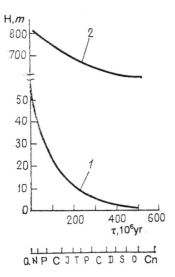

Fig. 14.2. Tentative character of the change in the maximum thickness of (1) syncryogenic and (2) epicryogenic sedimentary layers in the development of the Earth's lithogenesis.

adjacent territories (on account of cold and dry glacier winds). Thus, large ice covers have as a rule rather wide adjacent periglacial zones within which the low temperature permafrost exists. At the same time the permafrost can be lacking under ice covers of great thickness because of the pressure of the ice mass or because of the increased geothermal flux. In any case the thickness of frozen ground under glaciers will probably be less than that within the areas free of ice. In accordance with this it can be expected that permafrost degradation (its decrease of thickness) will be observed under the advancing glaciers and that the severity of permafrost conditions will increase on the ground after the recession of the glacier. On this basis we can conclude that the glacial cover and the permafrost will develop asynchronously, i.e. the maxima of areal development of the glaciation and of the permafrost will not be in phase. However in any case the occurrence of large ice covers indicates that permafrost also occurred, developing over larger areas than the glaciers themselves.

On the basis of the greatest glaciations determined with the help of deposits, we can recognize five time intervals (periods) in the Earth's history, when the permafrost should have been widespread on the planet (Fig. 14.3). These are early Proterozoic (2.4–2.1 billion years BP), Late (Riphean) Proterozoic (1–0.6 billion years BP), Palaeozoic (460–420 million years BP), Late Palaeozoic (330–320 million years BP) and Late Cenozoic (25–0 million years BP). The places where tillites were found and, consequently, the regions

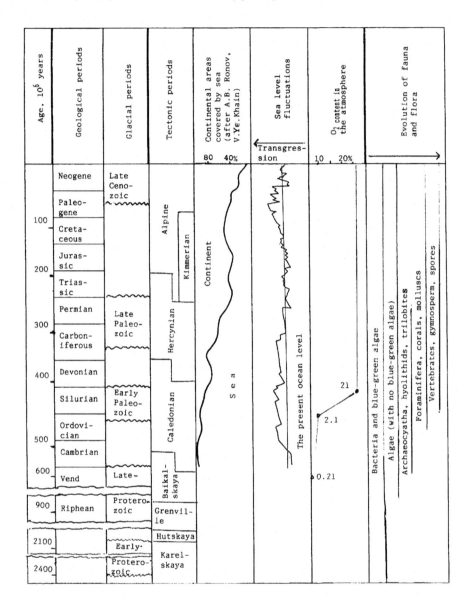

Fig.14.3. Diagram of the correlation of glacial periods with tectonic epochs and the development of continental and sea areas, O_2 content in the atmosphere and evolution of fauna and flora.

of glacial and permafrost development in those periods are scattered and now separated from each other by great distances. Thus, for example, traces of simultaneous glaciation during the Carboniferous-Permian period are found in Central and South Africa, in Australia, Antarctica and on the

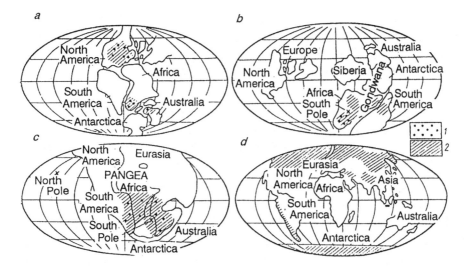

Fig. 14.4. The proposed ice sheets and occurrence of permafrost on the continents during the Earth's history: *a* – Early Proterozoic glaciation (after G. Yang, N.M. Chumakov); *b* – Early Palaeozoic glaciation (after B. John; S.A. Ushakov and N.A. Yasamanov); c – Late Palaeozoic glaciation (after A. Wegener; B. John, S.A. Ushakov and N.A. Yasamanov); *d* – Late Cenozoic glaciation (after B. John); 1 – glaciation traces; 2 – areas of the supposed ice sheets and occurrence of permafrost.

Arabian peninsula, in the Indian mountain regions and in South America. We can explain this taking into account continental drift and the arrangement of continents relative to each other in ancient epochs, reconstructed in the context of plate tectonics over the period of the Earth's development. Fig. 14.4 shows the suggested position of the continents over the different periods of the Earth's geological development and the regions of the assumed development of permafrost, determined on the basis of the traces of the ancient continental glaciation that have been found. The data on the ancient continental glaciations and on their characteristics required for such geological reconstructions and cited below, are taken from different literature sources (E. Derbyshire, B. John, S.A. Ushakov, N.A. Yasamanov, N.M. Chumakov, etc.).

The most ancient – *Early Proterozoic* – glacial period (2.4–2.1 billion years BP) in the Earth's history is associated with the discovery of sedimentary rocks of great thickness (up to a few hundred metres) which are considered to have a glacial origin. They were found in the region of Lake Huron in Canada, in the eastern part of South Africa and in the northwestern part of Australia. Such stratigraphic units have great similarity to moraine and consist of debris of various shapes and sizes with fine-grained material, and can be characterized as typical tillites. At the same time bedrock surfaces and debris with glacier striations were found in the Huron deposits. The three regions of the Early Proterozoic glaciations mentioned above are widely separated now. According to recent reconstructions of the position of the continents in the past, the giant 'Proterozoic supercontinent' is supposed to have existed with two main centres of glaciation: in North America and in South Africa-Western Australia. Possibly it was precisely those regions with which large areas of the permafrost were associated in Early Proterozoic (see Fig. 14.4a).

The *Late Proterozoic* glacial period (0.95–0.6 billion years BP) as well as the Early Proterozoic one appear to have comprised three (if not more) individual glacial stages: about 600, 750 and 950–900 million years ago. The formation of the ice sheets took place in this case on the background of the break-up of the 'Proterozoic supercontinent' and of its rapid dispersal. A number of American researchers reason that this glaciation was the most wide-spread in the Earth's history because the ice sheets could be formed in high, middle and even low latitudes. Traces are found within practically all the present continents. The glacial drift sections were studied best of all in Scotland where tillite formations are characteristically up to 870 m in thickness and accommodate more than forty 'mixtite' horizons of undetermined origin, which can be interpreted as tillites. The mixtites are sedimentary rocks having a mixture of fine particles, pebbles and larger-sized debris which usually alternate with sandstones, aleurites and other layered and sedimentary rocks. This fact points to their formation in conditions of a shallow water shelf. Generally speaking, the Late Proterozoic tillite formations are characterized by glacial-marine cycling, indicating possible advances and recessions of ice sheets on large areas of shallow water shelf. Large polygonal fissures, filled with sandstones and conglomerates, have been found in the top of some mixtite horizons. These can be interpreted as fissure-polygonal formations (sand, and earth wedges and possibly ice wedge pseudomorphs) common to the zone of recent permafrost development and periglacial regions.

The Late Proterozoic glacial drift could, obviously, have formed under the effect of mountain glaciers; however, the majority of researchers believe the widespread continental ice sheets made the greatest contribution. At the same time the development of those glaciations in the Late Proterozoic is likely to have been associated with orogenesis followed by the rapid movement of the ancient continents which were fractured as a result. Because of this, these different continents moved towards high and middle latitudes in succession.

The *Early Palaeozoic* (Late Ordovician) glacial period (460–430 million years ago) has been identified by researchers in the past 15–20 years following the work of petroleum geologists in Western Africa and the Sahara. They have found firm evidence of a large continental glaciation. Sand composition is one of the main features of the Ordovician glacial drift. Sandstones with high porosity are oil reservoir rocks. Tillite formations in this case often contain giant boulders with traces of glacial striation as well as boulders and pebbles in sandstones and more rarely in aleurites. The data for the Sahara suggest the existence of not less than three glacial advances and recessions in the Ordovician. At the same time a number of sea transgressions and regressions are noted in the Late Ordovician all over the Earth. They could be associated with the alternation of interglacial and glacial periods.

By the Late Ordovician time (compared with Late Proterozoic) the fundamental change in the position of the continental plates had taken place (see Fig. 14.4b). The ancient analogues of North America and Europe in the west, were separated from each other and the supercontinent named Gondwana included the present South America, Africa, Antarctica and Australia. The South Pole was located in the region of the modern Sahara at that time, therefore the Late Ordovician glacial drift and permafrost were developed on the territory of the present Africa, Saudi Arabia and South America simultaneously (see Fig. 14.4b). There is also scattered evidence of Upper Ordovician boulder horizon outcrops (which are thought to be glacial drift by some researchers) in Scotland, Spain and France.

The joining together of several different ice caps could only take place during the main stages of glaciation, on the large and likely rather plane territory of the Gondwana continent. There probably existed vast areas with sharply continental climate and low negative temperatures between the several ice caps with a consequent formation of a great thickness of frozen ground. The traces of the proposed Silurian and Devonian glacial formations were found in a number of regions on the east of South America and on the west coast of South Africa. However these data need further support and are unlikely to indicate the existence of large glaciations. They most likely are the traces of mountain glaciers that have survived under favourable conditions after thawing of large continental glacial covers on plains.

The *Late Palaeozoic* (the Great Permian-Carboniferous) glacial period lasted 50 million years (310–260 million years ago). In this period ice sheets, permafrost, shelf glaciers and icebergs were wide-spread. It was the time of the 'assembly' of all the continents joined together in the large supercontinent Pangea (see Fig. 14.4c) with a consequent fundamental rearrangement of the ocean currents. Plate tectonics and orogenesis were intensive in Permian-Carboniferous time. When the plate margins collided with each other new mountain ridges were formed (Ural Mountains, Hercynian mountains of Europe, etc.). The formation of large mountain chains was accompanied by widespread volcanic eruptions, lava flows, and thermal ejections into the atmosphere along the plate margins, causing a lowering of atmospheric transmissivity and decrease in the amount of solar energy arriving at the Earth's surface. At the same time the gigantic concentration of continents would have led to snow accumulations and glaciers, causing an important increase of the amount of reflected solar radiation (at least in the region of the South Pole and in high mountain regions even near the Equator). All this would contribute to cooling of the ground, global temperature lowering and development of the greatest glacial period, as far as duration and extent is concerned. The Late Palaeozoic glaciation in the Southern Hemisphere was of continental type. The centre of that glaciation is likely to have been located in East Antarctica-South Africa, with the glaciers moving from there toward Australia, Hindustan and South America. In the Northern Hemisphere, which was for the most part oceanic, traces of the glaciation have been found only in a limited number of regions in the north-east of Siberia.

Within the greater part of Gondwana (southern part of Pangea) glacial and periglacial conditions prevailed for many million years and vast areas of polar desert covered with tundra vegetation extended in front of the ice margin. In the *Permian-Carboniferous* glacial period the permafrost is likely to have occupied the largest areas in the Earth's history in high as well as in middle latitudes. These areas increased and decreased repeatedly. The maximal glacial extension probably took place about 280 million years ago. The Permian-Carboniferous glacial period is likely to have had several glacial epochs, asynchronous within different continents (see Fig. 14.4c). During the main glaciations the separate thick ice sheets are likely to have joined together forming a single giant cover twice as large as the present Antarctica. Characteristically the ice sheet of Gondwana was partly situated in the sea, since glacial-marine deposits have been found in the regions of the Permian-Carboniferous ice sheets. Aggradation and degradation of the ice sheets was accompanied by sea-level fluctuations of 150–200 m during glacials and interglacials.

The most dramatic evidence for the Permian-Carboniferous glaciation has been found in the southern part of South America. Glacial formations up to 1600 m in thickness and including unstratified tillites, bedrock surfaces with glacial striations, trough valleys, stratified tillites with glacial-marine deposits, cut pebbles, osers, etc. have been preserved in Brazil and Argentina. It is supposed that the glacials covering Brazil and Uruguay were moving from a centre situated in South Africa. Such a supposition is partly based on the fact that the Brazilian tillites contain erratics of quartzites, dolomites and cherts unknown in those countries but typical of the Namibian tillites where they are often found in bed deposits. The glacial drift in South Africa and Madagascar (of up to 1000 m in thickness) is mainly represented by three types: homogeneous tillites of continental origin; stratified tillites with marine fossils; and ribbon-layered formations with sporadic inclusions of rudaceous deposits. Evidence for the glaciations in the Southern Hemisphere has been found also in southern Australia, Antarctica, India and Pakistan.

In the Northern Hemisphere the mixtites of Massachusetts, USA, formed probably under local mountain conditions, as well as the glacial-marine deposits in the north-eastern part of Siberia (for example, in the Omolon river basin at the Kolyma river inflow), are evidence of the existence of the Permian-Carboniferous glaciation.

In Antarctica the *Cenozoic* glacial period (generally 25–0 million years ago) is likely to have begun more than 30 million years ago, and about 8–12 million years ago the high mountain ridges of Alaska were subject to glaciation. According to S.A. Ushakov, the intensive spreading (drift) of continents took place latitudinally. In the Palaeocene the Greenland plate was separated from Eurasia and from the North-American continent and Australia was split off from eastern Antarctica. In the Oligocene epoch the collision and joining of India and Eurasia took place, causing the formation of a number of large mountain systems in Central Asia. The distribution of the continents in accordance with their present position probably occurred in the Miocene. During the Cenozoic era cooling developed nonuniformly because of the alternation of warmings and coolings which occurred at particular time intervals. It is supposed that in the Eocene the Earth's generalized mean temperature was higher than at present* by 9–10 °C and at the end of Oligocene by only 4–5 °C, with the temperature decrease being more intensive in the Southern Hemisphere, and probably caused by the appearance of glaciers in the subpolar regions.

In the period when the deep strait was formed between Antarctica and Australia (in the middle of the Oligocene epoch), the Antarctic circumpolar current developed with the consequent decrease of influx of middle latitude

---

*According to M.I. Budyko (1), the recent temperature is about +15 °C.

waters and the strengthening of cyclonic activity. Associated with this was the origin of the ice sheet in Antarctica moving slowly toward the South Pole region. The Antarctic glaciation in turn caused a sharp increase of the surface albedo and greater cooling of this continent and, as a consequence, further lowering of the Earth's mean temperature. The glaciation in Antarctica is likely to have originated in the Transantarctic Mountains and Gamburtsev Mountains as early as in the Oligocene (38–26 million years ago).

The glacial cover existing now is likely to have begun to form in early Miocene (25–20 million years ago).

The earliest evidence for glaciation in the Northern Hemisphere has been found in the high mountains of south Alaska where the glacial drift alternates with lavas. According to potassium-argon dating the ice sheet existed from the middle of the Miocene (about 10 million years ago).

The onset of the ice cover in the Arctic Ocean is probably comparable with the formation of the Antarctic ice sheet as far as the effect on the Earth's climate is concerned. Beginning from that time progressive cooling took place on continents on the background of rhythmic climatic changes. There were several large glaciations in the Middle and Late Pleistocene such as the Oka, Dniepr (Samarov), Valday (in the European North of Russia), Zyryan and Sartan (in Siberia). The widespread development of the permafrost is associated with these rhythmic epochs of cooling.

14.2   Reasons for the development and evolution of the permafrost in the Earth's history

The origin, development and disappearance of the glacial covers at different stages of the Earth's development depend on the heat exchange level at its surface. Continuity of the development of the geological and geographical environment in the course of the development of the planet causes corresponding changes in the development of the heat exchange between the atmosphere, the hydrosphere and the lithosphere. When the heat exchange leads to negative temperatures on the Earth's surface, glaciation and ground freezing develops. The reasons for the glaciation and permafrost formation are various. By convention, they are divided into external (space or astronomic) and internal (Earth's or planetary) ones.

Scientists usually consider the external reasons causing decrease of solar energy arriving at the planet's surface to be variations of the astronomic factors in the solar system and outside, i.e. effects having an extraterrestrial origin. They assign factors causing some change in the amount of solar radiation arriving at the surface of the atmosphere (solar constant) to the first group. Thus, for example, it is often pointed out that it is possible for the

Solar system and the whole galaxy to cross dust particle accumulations causing a decrease of solar radiation and of radiant energy from the other stars arriving at the Earth. Because our galaxy rotates in space with a rotation period of about 200–250 million years, with the Sun in addition being displaced upward and downward relative to the galaxy plane, it can regularly cross such sections causing the global cooling of the longest periods, which may be responsible for the largest glacial periods in the Earth's history. The majority of researchers consider the variability of the solar radiation as the main reason, among others, for rhythmical climatic changes. It is supposed that there exist cycles of solar activity with periods from 250–300 million years to 200–400 thousand years. Cyclic changes of the solar constant with a recurrence period of 11, 22 and 44 years have been established with certainty. The results of modelling show (although not always uniquely) that a 2–5% reduction of the solar constant can lead not only to great cooling but also to the increase of ice cover.

The second group of factors of astronomical origin is the most debated, including variations of the Earth's axial tilt to the plane of its orbit, variations of the Earth's orbit eccentricity (the degree of orbital deflection from a circle) and displacement of the equinox point with the orbital movement of the Earth (gyroscopic wobbling of the Earth's rotational axis). Change in the Earth's rotation rate can also occur. These changes of the orbital parameters, superimposed on one another, determine the tilt of the Sun's rays incident on the Earth, and consequently, the intensity of the heating of the Earth's surface. As a result climatic cooling and warming epochs alternate on the Earth.

Internal reasons causing glaciation and formation of permafrost are associated with the irreversibility of the development of the atmosphere, lithosphere and the Earth's interior as well as with the interconnected periodic and episodic variations of the factors of the geologic-geographical environment.

The irreversibility of the development of our planet is most conspicuous first of all in the decrease of the Earth's radiogenic heat flux, in progressive continental expansion on the planet surface and in variations of the carbon dioxide and oxygen content in the atmosphere. Thus, for example, many researchers consider that climatic changes on the Earth's surface are associated with the process of the Earth's cooling down and with variations in the carbon dioxide and oxygen content of the atmosphere. This is connected with the fact that CO_2 transmits shortwave radiation but absorbs longwave radiation, creating the 'greenhouse effect'. A number of climatologists connect the possibility of the global temperature increase on our planet with the rise of the atmospheric content of CO_2 liberated by mineral fuel combustion and this 'greenhouse effect'. However during the Earth's history as a whole, the proportion of CO_2 in the atmosphere decreased steadily and the proportion of oxygen grew nonuniformly (from 1% in the Late Proterozoic to 10% in the Silurian) reaching the present level (21%) in the Late Cenozoic. The progressive continental expansion in the Phanerozoic (see Fig. 14.3) is considered to be the third factor in the climatic development of the planet which would have provided the increase in climatic continentality and more intensive cooling in the lithosphere, i.e. the climatic severity increase, cooling and resultant decrease of the surface temperature on our planet. Similarly the irreversible unidirectional changes in the atmosphere, lithosphere and in the Earth's interior are observed beginning in the Proterozoic, and these together have led to a process of persistent climatic cooling and the decrease of the mean planetary temperature.

Continental drift, size of the continents and their position in relation to the Earth's poles are considered to be internal reasons contributing to repeated glaciation and permafrost development. Formation and breaking up of the continents are certain to have affected the planetary climate as a whole, especially during their movement toward middle and high latitudes. A comparatively cold but moist climate when the amount of winter precipitation is more than that melting in the summer, is favourable for glaciation. Having begun to grow, the glacier stimulates dramatic cooling through the sharp increase of albedo. Growing in size, it can cause the surrounding air temperature to fall by 10–20 °C. Therefore the growing ice cover occupies an increasingly large area and causes further climatic severity. Consequently the formation of thick ice sheets creates a source of continued cooling, i.e. the global refrigerators on our planet. The Antarctic and Greenland ice sheets currently serve as such refrigerators.

Orogenesis caused by the colllision of plates of the lithosphere is closely associated with continental drift. In this case developing mountains can rise higher than the snow line where the snow remains throughout the year. The decrease of air temperature and the continuous accumulation of solid precipitation at such heights results in glacier formation. The main local centres of glaciation grow steadily, subsequently covering valleys and plains and forming the ice sheets.

In the literature there are many attempts to correlate large-scale cooling and glacial epochs with the epochs of extensive sea regressions. In a number of cases such correlation exists (see Fig. 14.3) and is explained by the fact that during the regression period as the sea retreats and the continental area grows, the continentality of the Earth's climate increases and cooling begins in the middle and high latitudes. However, it is thought that regressions could have been the consequence of formation of large ice sheets because gigantic masses of water were taken from the seas with a consequent essential sea-level lowering. Thus, for example, according to numerous estimations the World ocean level was 120 m below the present during the Late Pleistocene glaciation.

It is obvious that the oceanic and atmospheric water and air mass transfers, especially the great heat transfer from the tropical regions to the polar ones and vice versa, also had a large effect on the formation of the Earth's climate. Thus the continental drift (break up and assembly of the continents) essentially rearranged the circulation of the oceanic waters and caused the heat redistribution on the planet with a consequent development or termination of glaciation.

Individually, however, none of the separate external or internal reasons for the planetary climatic cooling could cause the formation of the ice sheets and the permafrost on the continents. The origination of the main glacial periods in the Earth's history becomes possible with the 'favourable' coincidence of a few factors which can mutually strengthen each other. Factors having an origin extra-terrestrially (long-period changes of solar activity or periodical passage of the Earth through physical nonuniformities of various kinds in our galaxy space) as well as global factors having a terrestrial origin, showing themselves at long time intervals (dynamics of lithospheric plates, jointing and break-up of large continents, positive tectonic movements, etc.), can coincide. Nevertheless factors having a terrestrial origin are likely to dominate in the development of each of the five major glacial periods.

Shorter term (medium periodic) rhythms of cooling and warming lasting a few million years and leading to a retreat or full disappearance of ice sheets or permafrost, followed by their increase in area as well as in thickness, are distinguished within each of the glacial periods discussed which last not less than tens of millions years. The reasons for the medium period and especially short period coolings could be lesser-order tectonic processes, redistribution of continents and sea areas, regressions, specific circulational currents of ocean waters, etc. It should be stressed that formation of permafrost on the planet is associated not only with the periods of ice sheet development. Within separate zones of the continents there could be conditions when glaciers could not develop because of shortage of precipitation, even though freezing of the upper layers of the lithosphere occurred. Thus, for instance, the cold and dry climate of Siberia allowed the formation of thick frozen ground in the Pleistocene but did not lead to development of an ice sheet.

Possibly even in the Mesozoic, which is characterized by warm and mild climate, the permafrost could concurrently be formed and persist in high mountains (in accordance with altitudinal geocryological zonation). It should be taken into consideration in this case that the general irreversible trend of the Earth's development, showing itself in progressive cooling and temperature lowering, should favour more frequent development of permafrost on our planet.

14.3   The history of the geocryological development and the main stages of permafrost formation on the territory of the former USSR in the Late Cenozoic

The evolution of natural conditions in the Late Cenozoic caused by global cooling of the Earth's climate led to the formation of the permafrost in the north, north-east and within high-elevation mountain systems on the south of the former USSR which continues to the present day. The history of the cryogenic development of the north-eastern part of Eurasia in the Late Cenozoic recorded in stratigraphy has not been studied uniformly and adequately as a whole. The initial period of the study of geocryological history is associated with investigations by A.I. Popov in the Taymyr Peninsula, P.A. Shumskiy, B.N. Dostovalov in Central Yakutia, and V.N.Saks, V.V. Baulin, N.S. Danilova, G.I. Lazukov in Western Siberia. Ye.M. Katasonov, N.N. Romanovskiy, Yu.A. Lavrushin, B.I. Vtyurin, T.P. Kuznetsova, T.N. Kaplina, A.V. Sher, A.A. Arkhangelov, S.V. Tomirdiaro and other researchers have made a great contribution in the study of the geocryological history of the Russian North-East, as have and A.A. Velichko, S.E. Sukhodolskiy, I.D. Danilov, N.G. Oberman and others in the study of European Russia.

Reliability and comprehensibility of the palaeogeocryological reconstructions depend on completeness of the geocryological sections and on the study of the palaeogeographical and palaeoclimatological surroundings correlated with these sections. Sections revealing Cenozoic rock units, formed during a large time interval and suitable for visual investigation and sampling, are the key to reconstructing the history of the regional geocryological development.

In this case a method for studying the permafrost 'footprints' buried in layers of the Neogene and Quaternary deposits is central when studying the history. Originally formed ground veins (ground wedges, pseudomorphs of ice veins, parallel-layered ground wedges), ground involutions (cryoturbations), ice bodies of different shapes and sizes, various relationships between the lithological bedding and the layers with high iron content, peat inclusions and ice streaks, shear and displacement of the layers observed in the exposures and in drill cores at the present time, constitute these 'footprints'. Such 'footprints' are of great variety and when occurring *in situ* give much information on the types of sedimentation and freezing.

This method of perception of the history of the permafrost is based on the principle of actualism, allowing the drawing of parallels between the present natural conditions and the cryogenic phenomena existing now and the related buried ancient forms. The analysis of such present-day forms gives us some idea of the conditions existing during the time of their formation. The essential features for the analysis are the occurrence of ground layers and ice streaks in the profile, their succession and the character of transformations of one layer into another.

Methods of Quaternary geology play an important role and thus the study of fossil animal bones and vegetation remains and inclusions (fragments of roots, branches and trunks, inclusions of lignites, coals, etc.) is also important in confirming the stratigraphy and origins of the frozen layers.

Palynologic data obtained as a result of spore and pollen analysis of samples from the sections under study allow one to estimate the climatic conditions for vegetation during the accumulation of the various horizons. Analysis of diatoms allows assessment of the moisture conditions and origin (marine or continental) of sediments.

Determination of the cryogenic age and of the type of freezing is possible only when studying the syncryogenic strata, the cryogenic and geological ages being the same. In this case the cryogenic features should testify to the absence of thawing during the warm periods, after freezing, i.e. these features should be the primary features. Otherwise these 'foot prints' will show the effects of repeated freezing and, consequently, the epigenetic cryogenic origin of the stratum. In the past two or three decades the ^{14}C dating of organic inclusions as well as palaeotemperature reconstructions based on the isotope ratio $^{18}O/^{16}O$ have played a great role in reconstructing the geological history in the Late Cenozoic.

The history of geocryological development (because of incomplete study in different regions of the permafrost zone) is not unambiguous and very often the same materials are treated differently by researchers. Weaknesses in the absolute dating and palaeotemperature methods, as well as ambiguity in the interpretation of palynospectra and of the other kinds of inclusions in deposits showing evidence of syngenesis or of subsequent thawing and new freezing, may be a factor. During different periods of study this situation caused underestimation as well as overestimation of the cryogenic age of the same layers and of the whole section, making it difficult to correlate palaeogeocryological materials in different regions. It must be added that because of the great rate of disintegration of ice-saturated materials in the exposures that are 'key' for the syncryogenic sections it is not always possible to repeat and to improve their description and sampling. For the epicryogenic strata the main evidence for the long duration of the permafrost and of the main stages of its development is the cryogenic structure across the section and the thickness of the whole permafrost layer, i.e. the depth of the 0°C isotherm. We can see the manifestation of cooling and warming periods in the epicryogenic bedrock in the formation of the cryogenic disintegration zones near the permafrost base and by the expansion of primary fissures in the ground as a result of water repeatedly freezing in them. The groundwater pressure deficit appearing as a result of incomplete filling by water of fissures in the thawing ground indicates thawing taking place at the base of the permafrost. We can recognize the refreezing of thawed frozen ground by the incomplete filling of fissures by ice. The distribution of ice content and the thickness of ice streaks proceeding downwards give much information about the conditions of freezing in loose epicryogenic materials. The combination of the features as a whole allows firm reconstruction of the history of the freezing of the stratum.

Nonuniformity of the evidence of permafrost over the territory and the ambiguity of the interpretation of the conditions of accumulation and of freezing of the cryogenic strata, as well as a weak understanding of the current regional geocryological environment, have led different scientists to separate and describe different stages and numbers of stages of geocryological development, giving each of them their own age range and assessment of the cryogenic role in the formation of the present features of the permafrost zone. The present description has been made in accordance with the four stages of geological and geocryological development of the Eurasian continent in the Cenozoic, distinguished by T.N. Kaplina, as well as with the four stages (with a detailed breakdown of the Holocene) distinguished by V.V. Baulin, N.S. Danilova, A.A. Velichko, etc.

At the beginning of the Cenozoic the Antarctic continent occupied its south-pole position and because of reduced solar radiation began to cool from the surface. The circumpolar oceanic currents sweeping the Antarctic continent were closed off with time and no longer picked up the heat from middle and equatorial latitudes, with a consequent sharp cooling of the atmosphere, lowering of the snow line and glacier formation on the elevated areas of the continent. The subsequent growing and joining of glaciers in the Middle and Late Miocene produced the great ice sheet of Antarctica and associated with this the lowering of world ocean levels by 55 m and more at the boundary between the Miocene and Pliocene. An increase in icebergs began to contribute to the world ocean cooling and caused the formation of the permafrost zone of the Earth's Northern Hemisphere in the Pliocene.

It should be remembered that along with the great land mass of the former USSR from west to east and with the transfer of Atlantic air masses dominating in the Cenozoic the cold periods were longer in the eastern part of the country and shorter in the western part. This spatial pattern was compensated by the latitudinal zonation of the solar radiation complicated by the altitudinal zonation in mountain regions. Therefore the longest cold periods existed in the north-east, the shortest ones in the south-west of the country. This tendency has persisted and caused the great variations in the duration or time intervals of the natural conditions during the stages of cryogenic development described (Fig. 14.5).

The **first stage** of geocryological development refers to the *Pliocene-Early Pleistocene* ($N_3 - Q_I$) and is the longest one (from 2.4–1.9 to 0.9–0.73 million years ago). The general cooling of the Earth's Northern Hemisphere climate is linked to the Early Pliocene (Fig. 14.6) when the polar basin glaciation, which has existed more or less uninterruptedly for the past 3 million years, began to develop because of the change in the warm Atlantic current and of the global cooling of the oceans. Thus, according to Ch. Emiliany's data, the bottom water temperature in subpolar regions of the Earth was the following on the background of a general cooling trend: in Middle Oligocene (30–35 million years ago) it was $+10.4\,°C$; in Early and Middle Miocene (10–20 million years ago) it was $+7\,°C$; in the Late Pliocene (3 million years ago) it was $+2.2\,°C$. At the boundary between Miocene and Pliocene the ocean level fell by a few hundred metres on the north-east of the Asian continent, and as a result of the Polar basin regression the whole Arctic shelf became a continent from which perennial freezing of the ground began, expanding southward. The great regression had a global character and was associated with the Antarctic glaciation as well as with the tectonic uplift of continents and the tectonic subsidence of the ocean bottoms.

The data on the oxygen $\delta^{18}O$ isotope ratio in ice columns from Antarctica and Greenland reported by A.S. Monin and Yu.A. Shishkov (Fig. 14.7) allow us to follow the stages of the Earth's climatic cooling in the Cenozoic and the stages of the accumulation of thick ice sheets as well as the associated ground freezing. It has been determined with the help of K-Ar dating that the Greenland ice sheet, with the ice thickness as much as 3 km and more during the glacial maxima, was formed about 3 million years ago in the Pliocene.

The initial reasons for glaciations were not only the global climatic

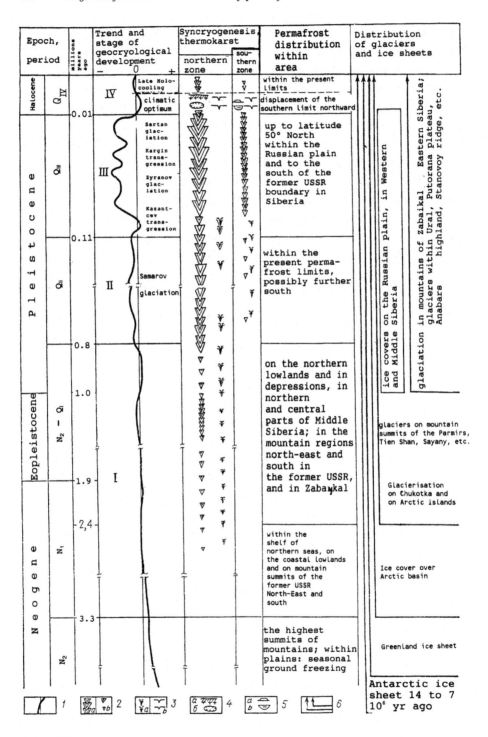

(Fig. 14.5.)

changes but also the particular combinations of regional natural conditions. The main among them are regional uplift as a result of tectonic movement and the increase in solid precipitation. Given the limited precipitation during the periods of cooling, very deep ground freezing took place.

In the northern half of the former USSR the freezing began first of all on high peaks of mountain ridges with the formation of glaciers and then within the high elevation mountains of the middle and southern latitudes. It has been established using palynologic data, that as early as in the Eocene within the mountains of north-eastern Asia the dark-coniferous taiga (which is now typical of the northern zone of unfrozen ground) was displaced by larch forests with shrubs and mosses; mountain tundras appeared on the mountain summits for the first time. In the early Pliocene the north of the continent was occupied by coniferous forests undemanding of the soil-climatic conditions, with swampy areas, and at that time the permafrost had already appeared on the plains of the North-East of Russia. In the places where the continental accumulation of fine-grained sediments prevailed, the traces of cryogenesis have survived to the present.

The most ancient indications of the syncryogenic ground freezing, observed in the form of ground wedges and pseudomorphs of ice wedges, have been found in the Russian North-East in the Kolyma river basin (the Begunov suite and the lower bench of Oler suite) and on the Krestovka river (Kutuyakh suite) the sediments of which have been assigned by A.V. Sher, T.N. Kaplina, A.A. Arkhangelov to the Pliocene (2.4–1.9 million years ago). We can observe from one to four layers of pseudomorphs in these suites of sediments – associated by Kaplina with cooling-warming cycles and the consequent permafrost aggradations and degradations. The sizes of the pseudomorphs of ice wedges suggest what we might call severe conditions during the formation of syncryogenic lacustrine-alluvial sediments, with the ice wedges on the north of Chukotka being assigned by Arkhangelov to the Eopleistocene.

In *Early Pleistocene* such traces of perennial freezing have been studied in Western Siberia at latitude 60° N in alluvial and lacustrine deposits of the Irtysh river valley and in Central Yakutiya. The upper terraces of the

Fig. 14.5. (*opposite*) Diagram of the development of permafrost in the former USSR, in Late Cenozoic: 1 – curve for the climatic fluctuations (coolings and warmings); 2 – accumulation of syncryogenic deposits with (*a*) large and (*b*) small ice wedges; 3 – formation of (*a*) pseudomorphs of ice wedges and of (*b*) hillocky terrain as a result of ice thawing out; 4 – (*a*) increase of seasonal thawing layer and (*b*) regional development of thermokarst processes; 5 – regional development of (*a*) peatlands and of (*b*) swamps; 6 – time of existence of glaciers and ice sheets.

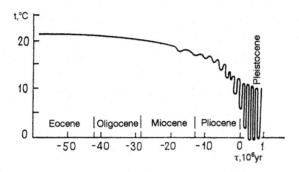

Fig. 14.6. Diagram of mean annual temperature fluctuations in the Cenozoic in Central Europe (after A.S. Monin and Yu.A. Shishkov).

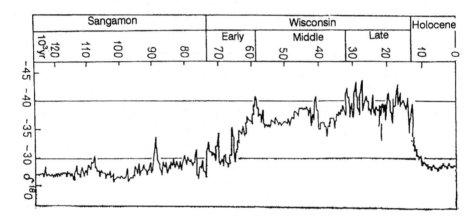

Fig. 14.7. Profile of oxygen $\delta^{18}0$ isotope ratio in an ice core from Camp Century – Greenland ice sheet (after A.S. Monin and V.A. Shishkov).

Lena river and its tributaries consists of deposits with ice wedges of up to 10–20 m and more in thickness, which can be attributed to syngenetic freezing in the early Pleistocene, according to T.P. Kuznetsova. Studies in the Russian North-East of pseudomorphs of ice wedges, indicate their syncryogenesis in the period of accumulation of the Pliocene-Early Pleistocene deposits of the Oler and Khroma suites. In the Oler suite, for example, an increase of size of the pseudomorphs of ice wedges and a decrease of distance between them has been revealed in the vertical section, pointing to continuous lowering of the ground temperatures and to increase of the amplitude of temperature fluctuations on the ground surface during the period of ice wedge growth.

The cold continental climate over the territory of the Siberian Platform at that time contributed to epicryogenic ground freezing over great areas.

Cyclic climatic fluctuations with relatively insignificant warming periods throughout the Quaternary period did not interrupt this pattern, because during most of the Pleistocene the climate was colder than at present.

The present permafrost thicknesses exceed 500–700 m in the Russian North-East and in Zabaykal'ye and 1000–1500 m on the Siberian Platform, and these depths lead us to assume that in these regions the epicryogenic frozen strata underwent freezing from the Early Pleistocene and, locally, possibly beginning from the Late Pleistocene. As the degradation periods associated with climatic warming were not so significant and long as the periods of cooling (Fig. 14.5), the thawed ground refroze entirely during the succeeding aggradation period. When the tendency towards increase of duration of the cooling periods and towards decrease of the warming periods took place, the degradation process was damped out and the aggradation process was amplified with time, as well as northward (and with height in mountains). The intensity of these processes varied over the region depending on the specific features of the natural surroundings. Therefore in the north of Siberia there was the possibility of the ground remaining frozen from the Early Pleistocene but mainly in epicryogenic ice-rich soils and bedrock. The syncryogenic deposits always produced relief during their formation and thawed in the warming epochs. They were reformed in the thermokarst during epigenesis with further formation of pseudomorphs in ice wedges.

As a whole the data presented here point to severe geocryological conditions throughout a stage when the ground temperature was not more than $-3°C$ and the southern limit of permafrost was situated at latitude 60° N and just south of it including Taimyr and the northern and middle part of the Middle Siberian Upland (Fig. 14.8).

The **second stage** (from 0.9–0.73 to 0.15–0.11 million years ago) refers to the Middle Pleistocene (Q_{II}), which became the key stage in the permafrost formation process. During this stage the aggradation-degradation cycles took place mainly on a background of negative ground temperatures. Therefore the development of the cryogenic strata in the north of Siberia in the Middle Pleistocene has been continued practically without interruption up to the present time. In T.N. Kaplina's opinion the syncryogenic deposits formed in the Middle Pleistocene on the northern plains have persisted locally, to the present.

The geological (and geocryological) age of the deposits of this stage was determined in the lower course of the Kolyma, by dating of fossil rodent bones by A.V. Sher, T.N. Kaplina and others, from layers in the section of the syncryogenic strata that have persisted until the present. On the north-

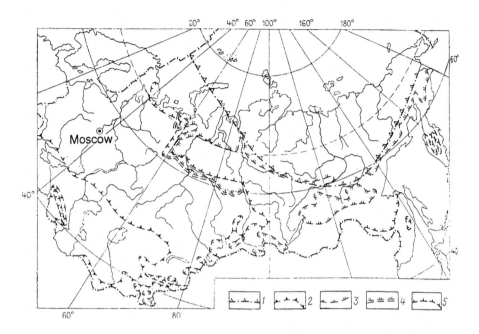

Fig. 14.8. Possible permafrost distribution limits in Cenozoic: 1 – in
Neogene-Early Pleistocene; 2 – in Middle Pleistocene; 3 – during the Holocene
climatic optimum; 4 – during the Holocene climatic optimum and in Late
Holocene (at the depth of 70–200 m from the surface); 5 – at the present time.

ern plains of Yakutiya in all the studied sections of such age, ice-rich strata
with thick ice veins occurring at various depths have been found.

Analysis of the permafrost traces in sediments of the Middle Pleistocene
in other regions show that the permafrost was developed in this period not
only within the present limits but also significantly south of them (see Fig.
14.8). Within the European plain the pseudomorphs of ice wedges and
ground veins are reported by A.A. Velichko, A.B. Bogutskiy and others to
latitude 50° N. In Western Siberia and Northern Kazakstan such traces
have been observed to latitudes 49–50° N, and in the south of Central
Siberia and in Zabaykal'ye, to the boundary of the former USSR. In T.N.
Kaplina's opinion such wide permafrost distribution in the Middle Pleistocene points to much more severe geocryological conditions at that time
compared to the present. Only late in the period and for up to forty
thousand years (Kazantsev interglacial period) was the climate close to the
present one. On the general background of these severe geocryological
conditions in the Middle Pleistocene, periods of warming (for example,
Tobol and Messov-Shirtin time in Western Siberia) are recognized by
palaeogeographers and geocryologists. In N.S. Danilova's opinion the ex-

pansion of northern taiga and forest-tundra northward and thermokarst development with accumulation of lacustrine and bog deposits, are to be associated with those periods.

The **third stage** of the permafrost development is the *Late Pleistocene* (Q_{III}), from 150–110 to 11–10 thousand years ago. This stage is characterized by sharp and strong climatic cooling (see Fig. 14.5). Wide development of surface ice sheets and permafrost over the whole region of the former USSR is associated with this cooling. This period has been termed in literature the *Glacial period*. The upper Pleistocene deposits in the North and North-East of the country are relief-forming and have been studied in a great number of key sections, with dating of fossil vegetation and bones and by ^{14}C. They have inclusions of thick ice wedges, 50–80 m in vertical extent and of width 3–6 m. In the Northern Hemisphere the Late Pleistocene is divided into four periods according to the degree of climatic severity.

The beginning of the Late Pleistocene, *Kazantsev time* (Q^1_{III}) with a duration of about 30 thousand years, is described by V.V. Baulin and others as the time of the Kazantsev shallow sea transgression covering the vast territories of Western Siberia (according to G. I. Lazukov up to latitude 65° N). The Salekhard marine plain in Yamal and Gydan remained un-flooded. Syncryogenic littoral-marine and alluvial deposits of Kazantcevsky age containing syngenetic ice wedges have been described by G.I. Dubikov. Pseudomorphs of ice wedges occurring in latitudes more than 3–4° further south than the regions of present-day ice wedge formation, are found in deposits from the beginning of the Late Pleistocene in Western and Central Siberia. This points to more severe climatic conditions than the present.

The *Zyryan* (60–50 thousand years ago according to N.V. Kind) and *Sartan* (27–15 thousand years ago) periods are the coldest in this stage, with the ice sheets being at a maximum and the perennial ground freezing taking place under a temperature of 5–10 °C below that of the present. Therefore it is thought that the Zyryan-Kargin-Sartan period ($Q^2_{III} - Q^4_{III}$) lasting about 70 thousand years is the period of the maximum buildup of permafrost thickness and its maximum extension southward over most of the former USSR. Traces of the severely perennially frozen ground (pseudomorphs of ice wedges, hillocky terrain, cryoturbations, etc.) were followed by A.A. Velichko over the vast area of the Russian plain extending to 48–49° N. In Western Siberia such traces have been found at 52° N. On the south of Taz peninsula at the latitude of the Arctic Circle, the vertical dimension of the pseudomorphs in such deposits reaches 12 m with width up to 2 m, with the underlying sand deposits carrying some evidence of the previous syngenesis. Under present conditions only irregular frost fracturing takes place in peat terrain at this latitude. North of the Arctic Circle the Upper Pleistocene deposits of the III – I alluvial terraces have syngenetic sand and ice wedges which have persisted from the time of their accumulation.

In the north of Central Siberia, within the northern plains and in the Meso-Cenozoic depressions of Eastern Siberia, the accumulation of syncryogenic ice-saturated horizons continued during that period, with wedge ice of great size (up to 80 m and more) – the 'ice' complex (see §9.2) which still exists in relics of the 'yedoma' suite which form relief. Typical of this time in the North-East of Russia, was a sharp decrease of forest area and its replacement with a particular kind of tundra-steppe vegetation having no analogue at the present time. In mountain regions the mountain tundra and deserts were widespread. Under the severe conditions of the *Sartan* period (27–15 thousand years ago) as a result of marine transgression, the freezing of the shelf and cryopeg formation took place. According to T.N. Kaplina's estimation the mean annual temperatures possibly reached $-25\,°C$ in that time, i.e. the temperatures were $10\,°C$ lower than at present. This extensive region of the country was occupied at that time by continuous permafrost of great thickness, 100–200 m greater than at present, containing thick syngenetic ice wedges in the Middle-Late Pleistocene deposits and epigenetic ice wedges in the deposits formed earlier – with minimal development of thermokarst and taliks below water courses, as well as of frost weathering in bed rock and eolian processes on sandy areas of the drained shelf and tundra-steppes. The long persistence and remarkable severity of freezing conditions led to ice wedges of great thickness, having no analogues previously, which are observed up to the present in the Upper Pleistocene deposits within the northern plains.

The occurrence of the 'second' buried permafrost layer separated from the surface by thawed ground (see Fig. 14.8) in Western Siberia and on the European North-East of Russia points to the great thickness of frozen ground in the Late Pleistocene. In Western Siberia the base of the relict layer of permafrost has been found at the present time at a depth of 300–400 m in latitude 59–60° N, i.e. to the south of the Ob river's latitudinal course and south of the present occurrence of permafrost islands. The existence of the Barents Sea and Kara ice covers and deep ground freezing in Central and Eastern Siberia where the absence of great glaciations, due to the deficit of atmospheric moisture and precipitation, caused the formation of a thick cryogenic stratum, are considered to be associated with that time. By the end of the Late Pleistocene in the central part of the Siberian Platform this thick cryogenic stratum had reached the maximum thickness (of 1000–

1600 m) on the Earth as far as platform conditions are concerned and it is represented by a thick negative temperature layer with salt waters and brines overlain by the permafrost layer with ice in fissures. Severe geocryological conditions extended over most of the former USSR territory at the end of Late Pleistocene and the southern limit of permafrost was situated at 47–49° N on the south-west excluding only the Black Sea coast and plains of Central Asia, and beyond the boundary of the former USSR to the southeast (see Fig. 14.8). After the Sartan minimum the global climatic warming is apparent, most intensively in middle latitudes of the European part of the former USSR and in Western Siberia.

The **fourth stage** of the permafrost development is the *Holocene* one (Q_{IV}). The climatic optimum Q^2_{IV}) taking place from 10–9 to 4.5–3 thousand years ago, is shown very well within the whole territory of the former USSR and a great amount of palaeogeographical and palaeocryolithological evidence shows this was the main event at this stage. On the Russian plain the ice sheets disappeared, in the Russian North-East, in Zabaykal'ye and in other permafrost regions, the areas of mountain-valley glaciers decreased, the soil-vegetation zones were displaced northward, etc. The thawed layer occurring above the layer of Pleistocene permafrost on the European northeast and in Western Siberia is one of the important relics of the permafrost degradation in the Holocene. Temperature curves with low gradient or without any gradient in cryogenic strata in Western Siberia and in other regions are evidence for the geocryological conditions becoming milder in the Holocene optimum.

The permafrost distribution by the end of the climatic optimum was such that (see Fig. 14.8) permafrost 200 m in thickness, lay roughly along the present southern limit of the permafrost. The estimations by V.V. Baulin and others show that thawing from the surface of bedrock at a temperature higher than -3 °C, could involve 100–250 m and that some part of this (approximately 100 m) could become frozen again in the period of the Late Holocene cooling. By and large it can be considered that the ground temperature was 3–4 °C higher during the climatic optimum than at the present.

In Eastern Siberia the period of the Holocene climatic optimum is likely to have been 9.5–8 thousand years ago. Regional development of thermokarst and formation of lacustrine-thermokarst plains and swamp deposits took place within the northern plains composed of ice-saturated ground in that period.

The Late Holocene (Q^4_{IV}) was marked universally by cooling and by the aggradational development of the permafrost zone as a whole (see Figs. 14.5

and 14.8) reflected in its present state. Thus the permafrost up to 50–100 m and more in thickness of Late Holocene cryogenic age was formed in thawed ground in the southern part of the permafrost zone; the layers thawing in the optimum were partly or completely frozen again; the mean annual permafrost temperature became 1–2 °C lower; the southern limit of permafrost was displaced to the south by a few degrees of latitude; the ice wedges of 1.5–2 m began to be formed and to grow within the peat lands of the European North of Russia and in Western Siberia. Within the present southern permafrost zone perennial and seasonal ground heaving is actively taking place with hillocky peatlands being formed. Within the northern permafrost zone new fairly thin syncryogenic deposits are being formed on river and lake flood-plains; frost fracturing is active on flat ground; icing formation is developing widely as a result of river and stream beds freezing up; 'golets' ice is being formed in rock streams and in coarse block deposits.

Thus the permafrost strata show various cryogenic ages (times of onset of the perennial freezing) across the section through the territory of the former USSR. Thus it was reported by N.S. Danilova that within the northern part of the permafrost zone the younger strata (as far as the cryogenic age is concerned) are situated near the permafrost base. These are the frozen rocks of the Late Pleistocene cryogenic age, because the maximum cold pulse is associated with that period of geological development of the Earth's permafrost zone in the Cenozoic. Under the alternating conditions of sedimentation and freezing (in the southern geocryological zone, in the regions of sea transgressions and ice sheets etc.) complex strata (as far as their cryogenic age is concerned) are being formed. In syncryogenic deposits the lower part of a section is always of more ancient cryogenic age than the upper one. However the cryogenic age of the permafrost base in the epicryogenic part can be the same as in the syncryogenic.

15

Zonal and regional features of present-day geocryological conditions in the territories of the former USSR

15.1 Distribution of permafrost and spatial variations of its mean annual temperature

The great extent of the territories of the former USSR from west to east and from north to south, the complexity of orographic, geological, tectonic and hydrogeological structures, of landscape and of climate, together with the changes in the late Cenozoic, are responsible for the variety of geocryological conditions in their different parts.

The zonal distribution of permafrost in the territory of the former USSR is associated with two main circumstances:

1) the history of the cryogenic development of the upper layers of the Earth's crust in Neogene-Quaternary time during which the stage was set for the present geocryological conditions;
2) the present conditions of heat exchange at the surface and in the ground. The mean annual ground temperature for the particular period, including for short-period fluctuations, indicates the level of heat-exchange for every climatic sequence.

The present heat-exchange conditions under which the mean annual ground temperatures are being formed, are defined by the following basic factors and conditions:

1) the dependence on latitude of the incoming solar radiation, upon which depends the radiation balance at the Earth's surface;
2) the effect of the Atlantic, Arctic and Pacific oceans on the atmospheric circulation, which gives the regional nature of climate;
3) the distance between the territory and sea basins and the position of the territory within the continent, this being responsible for the climatic continentality, i.e. for the amplitude of the annual fluctuation of temperature of the air and of the Earth's surface;

4) the orographic conditions which control the normal distribution of heat exchange with altitude in middle and high mountain regions and the associated inverse distribution within mountain plateaus;
5) the geological-tectonic conditions, with which ground composition, structure and properties, and the value of the geothermal flux and gradient of temperature in the permafrost are associated;
6) the hydrogeological conditions existing as a result of interaction of the permafrost and groundwater;
7) the landscape-climatic conditions (vegetation, swampiness, peatiness, microtopography, etc.).

The combination of these factors and conditions results in great variability in distribution and mutual arrangement of the perennially frozen and unfrozen ground and in the ground temperature regime. The distribution of permafrost, the spatial relationship between the permafrost and unfrozen ground, as well as the spatial changes of the mean annual ground temperature, show a close connection with the landscape-climate zones. Therefore we should study the geocryological environment of large territories in relation to the associated natural and geocryological regions. Within each region the pattern of the permafrost distribution and temperature regime is similar and differs from that of the adjoining regions. Such a regional or zonal approach allows study and comparison of the main geocryological trends within the whole territory of the former USSR.

We can appreciate visually the regular change of the mean annual ground temperature from south northward and from west eastward and with altitude in mountain regions, from the Geocryological Maps of the former USSR at the scale of general survey (1:2 500 000 – 1:25 000 000). In these the

Fig. 15.1. (*opposite*) Map of the permafrost region and mean annual ground temperatures (°C) in the former USSR. *Subaerial and subglacial permafrost regions.* The distribution of permafrost in the southern geocryological zone: 1 – sporadic permafrost (from +2 to −0.5°C); 2 – permafrost islands and massive-islands (from +2 to −2°C); 3 – discontinuous permafrost (from +0.5 to −2°C). *The northern geocryological zone with continuous permafrost:* 4 – from −0.5 to −3°C; 5 – from −0.5 to −5°C; 6 – from −2 to −5°C; 7 – from −4 to −9°C; 8 – from −7 to −9°C; 9 – from −9 to −11°C; 10 – > −11°C; 11 – from −11 to −13°C. *The submarine part of the permafrost region:* 12 – permafrost islands and discontinuous permafrost with cryopegs (from 0.5 to −2°C); 13 & 14 – continuous permafrost with cryopegs (13 – from −0.5 to −2°C; 14 – from −1 to −3°C). *Limits:* 15 – southern limit of occurrence of permafrost; 16 – limits of geotemperature zones – in subaereal (*a*) and submarine parts (*b*); 17 – of northern and southern geocryological zones.

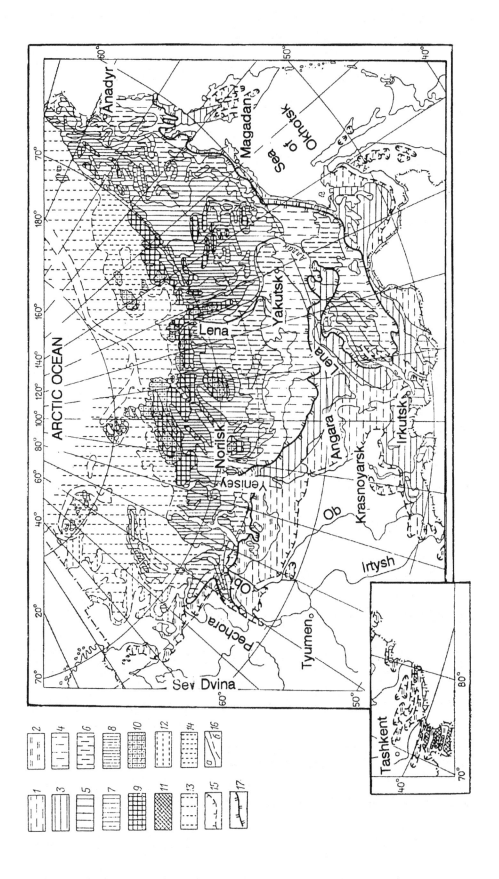

geotemperature zones are shown with temperatures to an accuracy of 1 to 2°C. We can see from the map (Fig. 15.1) the essential south-eastward departure of the identical geotemperature zones which are situated 2–3° further south in Western Siberia and as much as 6–8 °further south on Prilenskoye plateau and in Southern Yakutiya as compared with the regions within the European part of the former USSR, as a result of regional climate and the continentality of the heat exchange at the Earth's surface. At the present time the distribution of the permafrost and its temperature regime within the former USSR are such that the severity of geocryological conditions increases from south northward and north-eastward and with altitude in mountain regions. An azonal heat exchange appears locally superimposed on this background, with a combination of regional natural factors playing a greater role than that of latitude. The valleys of large rivers such as the Ob, Yenisey, Lena and others situated meridionally, and large lakes, do not freeze to the bottom and large sand masses of eolian, fluvioglacial, alluvial and other origin are geocryologically azonal. On the former USSR territory as a whole the mean annual ground temperatures vary over a wide range from +21°C in unfrozen ground of the subtropical zone to −15°C and lower in the permafrost of the Arctic zone and in high mountains. All of the territory of the former USSR is divided into two main regions: of permafrost with seasonal thawing from the surface (the permafrost region) and of soil with seasonal freezing of soil horizons and of the ground below (seasonal freezing region). The conventional dividing line separating these regions drawn around the southern permafrost islands in general maps is termed the *southern limit of the permafrost region.* To the north of the southern limit (and in mountains higher than where it is situated) the area of the permafrost islands and their numbers increase and the mean annual ground temperatures are lower. In connection with the latitudinal variation of the solar radiation and under the effect of the Atlantic warmth, the permafrost limit is generally directed from the northwest (Kola Peninsula) south-eastward (up to Yenisei river), being turned south-westward and westward in the neighbourhood of the city of Krasnoyarsk and then passing through the foothills of the mountain system of the southern former USSR.

The relationships of unfrozen to frozen ground and the spatial changes are closely associated with and depend on the ground temperature regime. The most southern permafrost islands are a few ten to a few hundred square metres in area, and can develop periodically in the unfrozen ground within two geotemperature zones adjacent to the permafrost region. In these zones the ground is characterized by rather low (from +3 to 0°C) positive mean annual temperatures. Within them permafrost can occur under undisturbed conditions during cooling, but only in peatlands and in clayey-silt deposits which are free of snow cover or heavily shaded. In areas of geotechnical disturbance (rock dumps, quarries etc.) permafrost may develop even without any cooling. Within the permafrost region the southern and northern geocryological zones are distinguished depending on the distribution of permafrost and its peculiarities, mean annual temperatures and thickness.

The southern geocryological zone

This zone is characterized by the discontinuous distribution of the permafrost, and its aerial extent and continuity which increase northward from 25% and less in the area of the sporadic permafrost to 25–75% in the areas of massive permafrost and islands and to 75–90% in the areas of the incomplete permafrost (Table 15.1).

In the southern zone the dynamic nature of the permafrost and thawed ground, especially near the southern limit, is associated with mean annual ground temperature fluctuations from year to year of as much as 1–2 °C and even more in years of extreme weather conditions. Such changes can cause new permafrost to form or may cause a separation of the permafrost from the seasonally frozen layer. Investigation of materials and estimations show that such effects occur most often in locations with mean annual temperatures ranging from +2 to −2 °C. The permafrost in the southern geocryological zone is for the most part of late Holocene cryogenic age. New permafrost forms in the regions of intensive change in natural conditions. In addition to frozen islands and masses, thawed ground (radiation-thermal taliks) occurs and is widespread in all topographic elements, under conditions where the summer ground heat balance exceeds the winter one.

The smallest proportion of permafrost to unfrozen ground is typical of the (sub)zone of *sporadic permafrost and permafrost islands* adjacent to the region of unfrozen and seasonally freezing ground. The prevailing mean annual temperatures in the unfrozen and frozen parts within this subzone range from +2 to −0.5 °C (Table 15.1). Under these conditions the permafrost islands exist only under favourable combinations of permafrost-forming factors such as surface shading, full or partial absence of snow cover, high peat content of the deposits, wide development of silty-clay surface-water-confining strata, etc. Thus within the flat areas of this subzone the appearance of the first permafrost islands is associated with peat frost mounds, with shaded peaty clay-rich silt, covered with mosses, and with the bottoms of valleys of streams and watercourses, overgrown with thick coniferous forests. Within the flat watersheds the permafrost can be

Table 15.1. *Distribution of the perennially and seasonally frozen (thawed and unfrozen) ground on territory of former USSR*

Region (with respect to presence or absence of frozen ground)	Zone	Subzone		area occupied by permafrost within the zone %	mean annual ground temperature °C	Type of seasonal thawing of permafrost and seasonal freezing of ground (with respect to the stability of the process)
		with respect to specific features of the perennially frozen, thawed and unfrozen ground distribution				
I) Of seasonal and perennially frozen ground – permafrost region	Northern geocryological	Continuous (northern part of the zone)		>95	<−10 −5 to −10 −3 to −5	Arctic and polar Stable Stable for a long time
		Mainly continuous (southern part of the zone)		90–95	−0.5 to −3	Transitional, semi-transitional and stable for a long time
	Southern geocryological	Discontinuous Massive-islands Islands and sporadic		75–90 25–75 <25	+0.5 to −2 +1 to −1 +2 to −0.5	Transitional and semi-transitional
II) Of unfrozen and seasonally frozen ground – seasonal freezing region	Northern seasonally frozen	Continuous thawed and unfrozen ground		Fractions of percent — — —	0.5 to 2 2 to 5 5 to 10 10 to 15	The same Stable for a long time Stable Southern
	Southern seasonally frozen and frozen for short periods	Continuous unfrozen ground		—	>15	Subtropical

associated with elevated bogs. Within foot-hills and 'golets' regions the first permafrost areas still occur at low elevations and appear on the shaded, moss-covered slopes blanketed with clay-silts and rock-debris. The individual permafrost islands may be situated at the foot of a terrace where snow patches lie.

The permafrost in this subzone is characterized as a whole by high negative mean annual temperatures rarely lower than $-1\,°C$ with the most commonly encountered mean temperatures (t_{mean}) varying from 0 to $-0.5\,°C$. The unfrozen ground dominates within all the topographic elements, especially within watersheds and on river terraces. Within the areas most favourable for summer precipitation infiltration, snow accumulations, solar insolation, etc., the mean annual temperatures can be significantly higher and reach $+4$ or $+5\,°C$.

In the (sub)zone of *distribution of massive islands* the frozen ground is characterized by an increase in number and size of islands and masses while the thawed ground is characterized by decrease of the area of open, radiation-thermal taliks. This (sub)zone is situated to the north of the previous one with plains and highlands and higher than the previous one in mountain regions. The lower mean annual temperatures of frozen as well as of unfrozen ground are typical. The prevailing temperature varies from $+1$ to $-1\,°C$. At the same time the mean annual ground temperature (t_{mean}) can reach $+2\,°C$ on the south-facing dry slopes (comprising not more than 25% of the area) and only $-2\,°C$ within the peaty north-facing slopes. The permafrost is widespread within the swampy terrain with hillocky peatlands in the boreal-tundra zone of the European part of the former USSR.

In the Northern and subpolar Urals the small permafrost islands give way to larger permafrost masses as the topographic relief and slope steepness increase. From north southwards their hypsometric position rises from 200 to 1000 m. The permafrost masses on the west-facing slopes, being more moist and snowy, are situated 100 m higher in the subpolar Urals and 500 m higher in the Northern Urals, compared with the corresponding east-facing slopes. In the Central Urals the perennially frozen soil masses are situated only on the mountain 'golets' summits, when their altitude exceeds 1000 m.

Within the Western Siberian plain the permafrost is developed in the form of large and small masses of epicryogenic Holocene peatlands characterized by temperatures to -1 and $-2\,°C$. Such permafrost masses are distributed at latitudes in this subzone in which unfrozen soils also occur in the form of islands and masses, and the zone itself extends almost up to the polar circle (see Fig. 15.1). To the east of the Yenisei river, permafrost islands/masses are typical of the lower areas within the Tunguska plateau occupied by light coniferous taiga, and in the Predbaikalskaya plain. To the east of Lake Baikal such permafrost is developed on the steppes of the Uda-Selenga highland as well as on the peaty and swampy Amur-Zeya plain occupied by sparse coniferous-taiga forests.

There is no clear dividing line between the first two geotemperature subzones because of the fact that change in surroundings from south northward is gradual and consequently gradual change in geocryological conditions is typical of the plains. Within the mountain regions permafrost areas increase more sharply because of lowering of the mean annual ground temperatures, with a mean gradient of 0.4–0.6 °C for every 100 m increase in altitude, and in connection with the sharp change of topographic and landscape-climatic conditions across the altitudinal belts.

The unfrozen ground within this zone occupies from 50 to 25% of the area, and is composed of radiation-thermal and underwater taliks the mean temperature of which is rather low (up to $+1$ C°, more seldom $+2$ °C. When the natural conditions are disturbed they become thermodynamically unstable and can become frozen within some areas.

The (sub)zone of *discontinuous distribution* of permafrost is the most severe within the southern geocryological zone. The mean annual permafrost temperatures here vary from 0 to -2 °C and lower still within 25% of the territory. Unfrozen ground occurs only under favourable conditions in the form of islands and, rarely, masses, and decreases in area from south northward from 20 to 5%. It is represented by open taliks of radiation-thermal and water-thermal origin up to $+1$ °C, more seldom $+2$ °C in temperature, and occurring within certain limits in areas of sand floodplains and terraces.

The permafrost of this subzone is widespread in the north of the boreal tundra of the European part of the former USSR. In Western Siberia it is represented by large masses composed of mixed soils and peats. To the east of the Yenisei river on the interstream area between the Nizhnaya Tunguska and Podkamennaya Tunguska rivers the permafrost of this zone extends south of latitude 60° N, and also occupies low plateaus and highlands within the western part of South Yakutiya and the Aldan-Tympton interstream area. The relationship between permafrost and unfrozen materials in Cambrian carbonate deposits and Jurassic sandstones depends on the warming effect of summer precipitation infiltrating through the karst cavities and in crystalline rocks through the water permeable fracture zones. The mean annual temperature increase due to heat from precipitation infiltration can reach 2–3 °C. As a result, unfrozen masses and linear zones occur among the surrounding high-temperature (from 0 to -2 °C) permafrost on the watershed surfaces of plateaus and highlands composed of fractured rocks. As a rule pine-larch forest, growing on areas with dry sandy silty gravels and loose detritus, is the indicator of absence of permafrost or of the deep seasonal thawing of the ground.

In regions of loose Quaternary deposits open and closed taliks exist under rivers and lakes in sands of eolian, alluvial and fluvioglacial origin, and in flooded lowland swamps. A narrowing of this geotemperature subzone eastward within the Central Siberian highland, compared to the first two subzones, is the noteworthy feature. This is connected with the change from gently rolling, wavy, slightly dissected topography to an elevated surface with diabase intrusions, sharply dissected by deep valleys and affected by glaciation. The latter features were evidently also the limit of a widespread thawing from the surface of permafrost during the climatic optimum of the Holocene. At the present time the combined impact of these factors has brought about a sharp latitudinal transition from intermittent to continuous distribution of permafrost and to a sharp reduction of the areas of unfrozen ground in this region.

The zone of continuous distribution

In this zone of the permafrost (the northern geocryological zone) there is disturbance of the continuity by open taliks developed locally under large river beds, deep lakes and in the areas of ground water discharge. Taliks of radiation-thermal origin are developed only under particularly favourable conditions for their existence, being represented for the most part by closed taliks which occur more often in the southern subzone of the northern geocryological zone where temperatures down to $-2\,°C$, more seldom to $-3\,°C$, are found. As a rule they are developed on weak Jurassic sandstones as well as on alluvial and eolian sands. The permafrost of the northern geocryological zone as a whole is characterized by the mean annual temperatures varying from 0 to $-15\,°C$ and lower, decreasing regularly from the southern margin of the zone northward and north-eastward in accordance with landscape-climatic and altitudinal zonality. On plains the intensity of the lowering of the mean annual ground temperature is associated with the landscape and lithological-moisture features of the areas, in highland regions with the continental climate (development of temperature inversions, deep depressions and valleys characterized by complex air-exchange conditions compared with those in the surrounding interstream areas.

The mean annual ground temperatures in the south of the northern geocryological zone are characterized by high negative values (for the most part from −1 to −3°C) at present. Such permafrost temperatures are observed in Western Siberia in the latitude of the Arctic circle. They exist also to the north of this latitude along large river valleys in accordance with the distribution of the boreal-tundra landscape. The high temperature continuous permafrost is widespread on the right bank of Yenisey river, in Zabaykal'ye and in Southern Yakutiya, in the south of the Russian Far East (see Fig. 15.1). Its existence depends on the warming effect of snow cover and summer precipitation, on a favourable radiational regime, good drainage and relatively low ice content of the deposits which repeatedly thaw from the surface and freeze again.

To the north of this subzone the mean annual ground temperatures decrease gradually. Thus they decrease from −2 to −7°C (in the latitude of Anabar massif) within vast areas of the Central Siberian highland. The lowest ground temperatures are established on the northern plains of Western Siberia (from −5 to −9°C), of Central Siberia (from −9 to −13°C), of Eastern Siberia (from −9 to −15°C) and on the coasts of the Arctic islands (from −5 to −15°C). Within the northern geocryological zone higher mean annual ground temperatures (than the zonal ones) are established in alluvial deposits of river valleys within all the geotemperature subzones. Within mountain systems of the southern part of Siberia, Central Asia and the Caucasus owing to the high incoming solar radiation at the surface in southern latitudes, continuous permafrost is developed only at high altitudes: from 2700 m in Gornyy Altay, from 3000–3500 m in Dzungarskiy Alatau, from 3500–4500 m in Tien Shan and the Pamirs and from 3000–3500 m in the Caucasus. As the topographic relief increases, the mean annual ground temperatures in mountain massifs decrease from valleys to watersheds, reaching −15°C and lower on high summits. Within the mountain slopes covered with mobile tills and deposits of large blocks with 'golets' ice the lowest temperatures occur on the steep north-facing slopes. The decrease in mean annual ground temperatures at comparable altitudes is greater with the increase of the topographic dissection and steepness because of the extent of cooling of the steep slope forms and increase in drainage of the hard and semihard rock masses. Glaciers increase the temperature of ground under them leading to glacial talik formation within firn zones and decrease it down to the mean annual air temperature value within the ice zones, thus modifying the permafrost distribution and its temperature regime.

Thus regular decrease of the permafrost temperatures takes place within similar topographic elements over the whole territory of the continuous permafrost. In the case of different ground composition the lowest temperatures are formed in cohesive and peaty ground. Local increase of the zonal temperature is possible under favourable conditions for the heat exchange in loose clastic deposits; however, the general latitudinal zonation holds with the temperatures decreasing northward.

15.2   Structure of the permafrost and spatial variability of its thickness

The permafrost is defined as the part of the Earth's crust in which the earth materials have negative temperature, regardless of the presence or phase composition of water in it. The structure of the permafrost, its thickness and distribution reflect the combined result of the whole history of its formation and the dynamics of perennial ground freezing from the end of the Neogene up to the present time. The permafrost thickness within the territory of the former USSR varies greatly, from the first metres to 1500 m and more depth, and is distinguished by continuous and discontinuous (interrupted) distribution over the area and in section. The continuity of the permafrost over the area (or its discontinuity) depends on the conditions of the ground temperature regime established in the layer of annual temperature fluctuations, on the distribution of open taliks, and on its thickness and structure. In accordance with these features it is subdivided into the northern and southern geocryological zones of continuous and discontinuous (sporadic to interrupted) permafrost. The discontinuity of the permafrost over the area (see Table 15.1) is associated with the development of open and closed taliks of different genetic types (see Chapter 13).

The distribution of the permafrost across the region depends on the geological tectonic structure, on the conditions of perennial ground freezing and on the interaction of the ground with ground waters and essentially on the history of geocryological development in the Late Cenozoic. As a result, continuously frozen strata (through the whole depth) as well as two frozen layers separated (vertically) by a layer of thawed material, have been formed in various geologic-structural, hydrogeological, topographic and landscape-climatic conditions. With respect to geological structure the permafrost zone can be represented by loose deposits of Cenozoic (mainly Quaternary) age and by geologic formations of the PreCenozoic age.

With respect to the type of cryogenesis and structure of the permafrost, syncryogenic and epicryogenic sediments can be characterized as having low ice content, average ice content or as being ice-rich. With respect to cryogenic age (the onset of perennial freezing) the strata of the permafrost region are Pliocene-Pleistocene (not differentiated), Late Pleistocene, Late Holocene and recent materials.

In the cryogenic structure of the permafrost profile there may be several

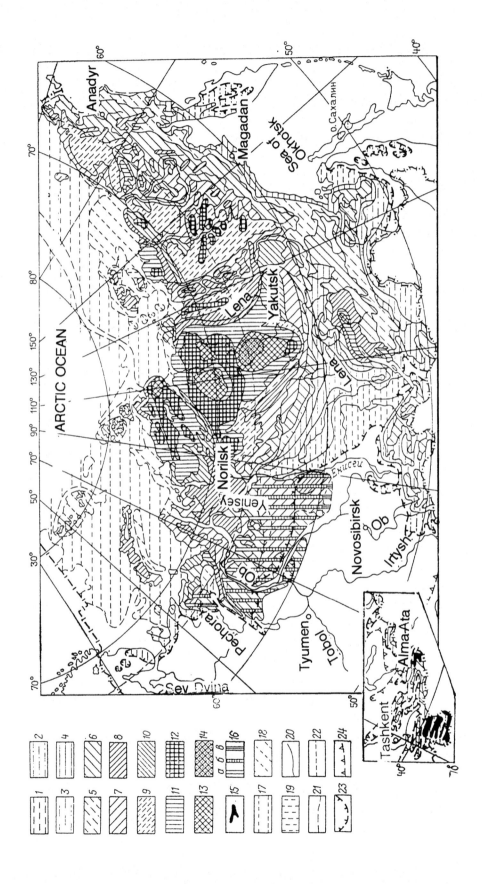

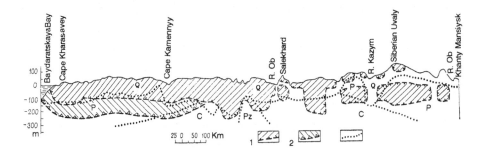

Fig. 15.3. The structure of the West Siberian permafrost zone along a meridional profile (after G.I. Dubikov): 1 – permafrost; 2 – ground below 0°C with cryopegs; 3 – stratigraphic boundaries between: Quaternary sand-clay layer Q, Palaeogene clay P, Cretaceous clay and sand-aleurolite C, and Palaeozoic crystalline materials Pz.

strata modified by the presence and content of ice, salt waters, brines and gases in them. These strata are as follows: 1) frozen and cryotic, i.e. with and without ice; 2) cooled below 0°C, containing salt groundwater and brines (cryopegs) and gas hydrates; 3) relict (Pleistocene) permafrost, occurring beneath the surface and overlain by thawed and perennially frozen (Late Holocene) soil and recently frozen ground.

Strata with negative temperatures containing ice, salt water and brines are termed *cryogenic* strata after S.M. Fotiyev. The variations in permafrost thickness depend mainly on the existing geocryological latitudinal variations (or the altitudinal zonation in mountains) and on the peculiarities of the development of the permafrost zone in the Late Cenozoic.

The formation of the permafrost is considered to have occurred at various times from many hundred thousand years ago (about 2 million or more years ago) to some years ago. Thus the first formation of permafrost on the north of Eurasia was Late Pliocene and Early Pleistocene. With respect to the occurrence and conditions of formation in the upper part of the Earth's

Fig. 15.2. (*opposite*) The map of the thickness of the permafrost of the former USSR. *Southern, subaerial and subglacial, geocryological zone:* 1 – 0–15 m; 2 – 0–5 m; 3 – 0–50 m; 4 – 0–100 m. *Northern, subaerial and subglacial, geocryological zone:* 5 – 100–200 m; 6 – 100–300 m; 7 – 200–400 m; 8 – 300–500 m; 9 – 300–700 m; 10 – 400–600 m; 11 – 400–700 m; 12 – 500–900 m; 13 – 700–1100 m; 14 – 900–1500 m; 15 – 100–1000 m; 16 – relict permafrost of 100-200 m thickness occurring at depths of up to 100 m (*a*); 100-200 m (*b*); > 200 m (*c*); from the surface. *Submarine part of the permafrost:* 17 – 0–100 m; 18 – 100–300 m; 19 – 200–400 m. *Limits of permafrost:* 20 – limits of subaerial and subglacial parts; 21 – of submarine permafrost; 22 – of relict permafrost; 23 – of present day permafrost; 24 – of possible distribution of the permafrost in the Pleistocene.

crust the permafrost region is subdivided into the subaerial part of the continental permafrost, the submarine permafrost part, situated under the Arctic basin and basins of adjacent seas, and the subglacial part of the permafrost under glaciers. Permafrost thickness and the types of its cryogenic structure within the territory of the former USSR are shown on the map drawn at the 1:25 million scale (Fig. 15.2).

The subaerial permafrost occupies the continental part of the territory of the former USSR and is divided into southern and northern geocryological zones with respect to areal distribution, thickness, cryogenic structure and cryogenic age.

Within the southern zone the cryogenic thickness from the surface is represented by one layer only of perennially frozen strata commonly up to 100 m in thickness. Changes in thickness depend on the latitudinal variations of heat exchange on plains and highlands and on the altitudinal variations of heat exchange in mountains. The permafrost strata are underlain by unfrozen strata with hydrological conditions in accordance with the hydrogeological structure.

The thickness (varying from 0 to 1500 m and more) and the structure of the cryogenic rocks in the northern geocryological zone depend to a lesser degree on the present conditions of heat exchange, being dictated instead by the history of geocryological development and depending essentially on the type of tectonic and hydrogeological structure and on topography. Thus the permafrost is represented within the mountain and ancient shield regions mainly by one permafrost layer, that is, by perennially frozen materials underlain by unfrozen water-bearing strata.

The permafrost layer (from 25 to 500 m in thickness) within ancient platforms, young plates and on the coast of the Polar basin is underlain by cooled rocks, with cryopegs. There are two permafrost layers within the Western Siberian plate, on the north-east of the Russian Platform and in the intermontane depressions of Pribaykal'ye, these regions having been subjected to deep thawing from the surface during the climatic optimum of the Holocene (Fig. 15.3). The upper layer of the permafrost is of Late Holocene age, and the lower one is of Late Pleistocene age. The lower layer of relict frozen strata is usually associated with sand-clay and clayey-silty sand materials showing large phase transitions of water.

The cryogenic development of the Eurasian continent in the Pleistocene and Holocene retained the zonal distribution of thickness of permafrost on young plates and partly on ancient platforms, and was responsible for the more complex pattern of its distribution over the vast areas of the mountain regions of the Russian North-East.

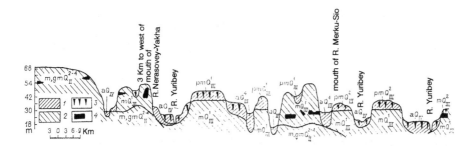

Fig. 15.4. Geocryological section of Gydan peninsula (after G.I. Dubikov): 1 – syncryogenic unit of Upper Quaternary proluvial-marine (pmQ^1_{III} and pmQ^2_{III}), recent and Upper Quaternary alluvial (aQ_{IV}, aQ^4_{III}) deposits; 2 – epicryogenic unit of Upper Quaternary marine (mQ^1_{III}) and Middle Quaternary marine and glacial-marine (m, gm Q^{2-4}_{II}) deposits; 3–4 – ice wedges in syncryogenic and sheet ice in epicryogenic deposits, respectively.

The latitudinal zonation of the permafrost thickness is best-defined in the *southern geocryological zone* which includes sporadic permafrost (up to 15–25 m in thickness), massive islands (up to 25–50 m in thickness) and discontinuous permafrost (up to 100 m in thickness). The permafrost may extend over limited areas (less than 25% of the total area) and may be more or less than the thicknesses given by a factor of 0.5–1.5 or even 2. Such a decreased permafrost thickness can occur only under favourable conditions azonal for the given area, while an increase of 50–100 m in thickness can occur in the areas where joining of Late Holocene permafrost is possible (as far as its cryogenic age is concerned) with frozen layers surviving from the Late Pleistocene. The permafrost within the southern geocryological zone as a whole is mainly of Late Holocene and recent cryogenic age correlated with the perennial freezing of the ground after the Holocene climatic optimum (see Fig. 14.8). The layer of relict (Pleistocene) permafrost occurring at some depth (70–200 m and more) from the surface has been retained within the southern geocryological zone owing to dynamics of climate in the Pleistocene and Holocene, sea transgressions and regressions and glaciations under particularly favourable conditions (unconsolidated ice-rich Mesozoic and Cenozoic material). Such relict permafrost is found at the present time even to the south of the Arctic circle on the north-east of the Russian plain, from the latitude 60° N to the Arctic circle within the Western Siberian plain (see Fig. 15.3) with the thickness varying, characteristically, from 300 to 100 m and less. In the places where the permafrost of Holocene cryogenic age is developed above it the 'two-layered' permafrost occurs. On the north of the Russian plain and within the Western Siberian plate, cryopegs cooled below 0°C are developed below the relict permafrost in marine deposits filling depressions between dome structures. This cooled layer is 50 m and more in thickness. Relict permafrost, not very thick (to 100 m), is developed also at depths of 40-80 m within the Upper Angara and Barguzin depressions of Pribaikal'ye.

Within the *northern geocryological zone* the permafrost of the Late Holocene and of Pleistocene cryogenic age is mainly joined, forming cryogenic units of great thickness: to 300-500 m in Western Siberia, to 1000–1500–m within the Siberian Platform and to 300-800 m and more on the mountain ridges and to 200–400 m in depressions of the North-East of Russia. Within this cryogenic unit open and closed taliks of the underwater and hydrogeogenic types are developed due to the heat effect of aquifers, streams and ground water. Radiation-thermal and infiltration taliks situated within sand masses, for the most part of the closed type, are developed locally.

Two types of permafrost are developed from the surface within low plains on platforms and young plates. These are: 1) ice-rich epicryogenic (marine, glacial) loose deposits with occurrences of sheet ice (Fig. 15.4); 2) ice-rich syncryogenic (lacustrine, alluvial, colluvial, biogenic, etc.) loose deposits, with thick syncryogenic ice veins up to 40–80 m – the 'ice' and 'yedoma' complexes. Syn- and epicryogenic permafrost of the loose cover is underlain by a layer of epicryogenic hard and semihard bedrocks at various depths.

There exists rather clear-cut geocryological zonation in the distribution of permafrost thickness within the Western Siberian plate, on the North-East of European Russia and within the plains part of the Siberian Platform. It is disturbed only by altitudinal geocryological zonation in mountainous regions. At the same time the subzone with the permafrost layer of 300–500 m in thickness (300 m in valleys; 500 m on watershed areas) is typical of highland surfaces having an absolute altitude of about 400 m within the Siberian Platform.

Such thickness of the permafrost layer is typical of the hyperzone extending from the left bank of the latitudinal flow of the Lena river to the latitudinal flow of the Anabar river. The permafrost, being 300–400 m in thickness, is mainly the continuation of this zone in Western Siberia. Within the more ancient high altitude topography the permafrost increases in thickness to 500 m. The permafrost thickness on the north of the Siberian Platform and on Taymyr peninsula reaches 400–600 m. So the latitudinal zonation of the level of heat exchange during the Pleistocene and Holocene manifests itself in this way.

The essential disruption of the zonal permafrost distribution is the layer of cooled rocks with cryopegs. This is seen most clearly within the Siberian

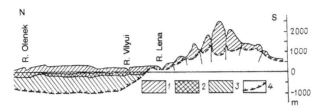

Fig. 15.5. Profile along meridian of the Central Siberia and Zabaykal'ye permafrost zone: 1–3 – epicryogenic units of hard and semihard rocks (1 – with ice in fractures freezing in the period $N_2 - Q_{IV}^{3-4}$; 2 – with ice in fractures and with saline waters at negative temperature freezing in the period $Q_I - Q_{III}$; 3 – with negative temperature brines, freezing in the period $Q_I - Q_{III}$); 4 – bottom of permafrost.

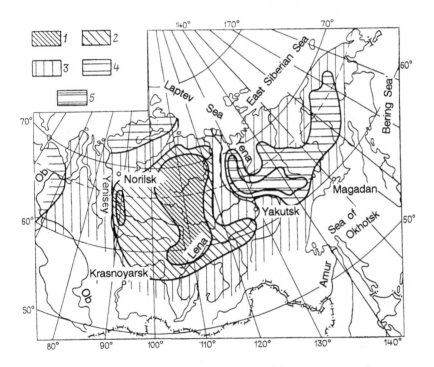

Fig. 15.6. Schematic map of heat flux of the North-East of Russia (after V.T. Balobayev): *Heat flux*, W m^{-2}; 1 – < 0.02; 2 – 0.02–0.04; 3 – 0.04–0.06; 4 – 0.06–0.08; 5 – > 0.08.

Platform. The permafrost layer and the layer of cooled rocks (below 0 °C) with cryopegs has been revealed in bore-holes in the upper reaches of the Markha river (central part of the Siberian Platform) extending to a depth of 1500 m (Fig. 15.5). Such deep freezing is associated with directional cooling of the Earth's crust in the Pliocene (see Fig. 14.2), with the increase of Arctic

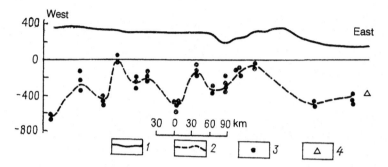

Fig. 15.7. Schematic profile along the Olyenyok-Zhigansk line (after V.S. Yakupov):1 – topography; 2 – lower limit of the permafrost; 3 – vertical electric sounding; 4 – geothermal data.

Ocean ice cover throughout the Pleistocene, with the increase of climatic continentality in connection with this and the sharp decrease of atmospheric precipitation. Periodic climatic warming probably did not cause the permafrost to thaw while the low mean annual temperatures (more than 10°C lower than the present ones) contributed only to a limited temperature increase within their negative values, in the opinion of T.N. Kaplin. As a result the great permafrost thickness formed on the platforms of the Earth's continents was comparable with the permafrost thickness found only within the highest mountain ridges today.

The inertia of negative temperatures in the interior of the Siberian Platform is associated with the weak tectonic activity of this platform in the Cenozoic with a consequent small intensity of geothermal flux (Fig. 15.6). As a result at the present time small gradients of rock temperature of the order of 0.5–1°C per 100 m (0.1–0.5°C per 100 m in the zone of the greatest thickness) are typical of this platform. Such gradients in the permafrost contribute to the development of rather high mean annual ground temperatures also. The analysis of these temperatures with due regard to information on the nature of the permafrost layer shows the discrepancy between the present conditions of heat exchange and the thickness of the permafrost. This fact most likely reflects the ancient Pliocene-Pleistocene cryogenic age of this unit.

Within the areas where the Precambrian basement elevation is the aquiclude for the water-bearing horizons of the strata of the Siberian platform, the permafrost zone is mainly represented by one layer of epicryogenic Palaeozoic and Mesozoic rocks. The permafrost is 700–800 m in thickness according to the data from a bore-hole situated on the Anadyr-Olenyok interstream area. At the same time great variations in permafrost thickness, as much as 300–500 m (Fig. 15.7), are noted in the eastern part of the Siberian Platform from geophysical data from the cross-section Olenek-Zhigansk. Such variations in thickness are associated with the effect of tectonic structures of different order and with the presence of fractured zones with various rock structures and composition, with their gas and water permeability, etc., in the opinion of V.S. Yakupov.

By and large the essential distinctive geocryological feature of the Siberian Platform, which is part of the Eurasian continent of low mobility, is that its northern half has the most continuous permafrost in area and the thickest in section of the continental subaerial type and, as opposed to the other great megastructures, it does not have open taliks with positive temperatures. These taliks are developed only under the bed of the Vilyi river, in the region of the Central Yakutiya depression filled with thick Cenozoic and Mesozoic rocks and under some sections of the Lena river bed within the eastern margin of the platform.

The enhancing of geocryological altitudinal zonation by geocryological latitudinal zonation is seen most clearly on the high lavas of the Putorana plateau, in the Anabar highlands and within the Byrranga mountains in Taymyr. As a result the thickness of the permafrost in the upper belt of the Putorana plateau at altitudes of 1500 m and higher and on the summits of the Anabarskoye highlands at altitudes of 900–1000 m, reaches 1000 m and more. The epicryogenic perennially frozen Archean, Proterozoic, Palaeozoic and Mesozoic rocks contain ice along the tectonic dislocations, the number of which decreases with depth. Within the lower structures the permafrost is underlain by cooled rocks, containing cryopegs, cooled below 0 °C (see Fig. 15.3). Open taliks are developed in the Western part of the platform under large lakes of glacial-tectonic origin (Lama, Keta, Khantayskoye, Dyupkun, Vivi, etc.).

The permafrost has a multi-layered structure (see Fig. 15.3) on the low lacustrine-alluvial, fluvioglacial and marine plains of the North Siberian lowland, including the northern part of the Central Siberian highland and the Yenisey-Khatanga depression.

The upper layer consisting of ice-rich syncryogenic Late Pleistocene lacustrine-alluvial deposits up to 50–70 m in thickness and originally frozen glacial drift, is developed patchily in this region on account of the particular topographic and sedimentation conditions. It is underlain by epicryogenic deposits of fluvioglacial, glacial, marine, eolian and other origins, and having a Pleistocene cryogenic age, varying from a few tens to 200 m in thickness. The third layer is composed of epicryogenic Jurassic and Cretaceous sedimentary deposits of a cryogenic age varying from Early to Late

Pleistocene. The fourth layer is represented by rocks cooled below 0 °C with cryopegs and with gas hydrates in dome structures. The cryogenic structure is defined merely by the layers of epicryogenic frozen unconsolidated bedrock within the area of surficial marine, littoral-marine and fluvioglacial surficial deposits. Unfrozen layers from 20–80 m in thickness have been found at depths of 30–60 m within the areas of marine transgression along the valleys (for example along Pyasina river) where there are epicryogenic sand masses interbedded with sandy-silt and clays.

The layered structure of the permafrost in mountain regions depends on geological-structural features, topographic variability and intensity of water-exchange in the stratum. The upper part of the profile (to 100–300 m) in mountain regions usually consists of permafrost containing ice along fractures. Ice does not fill the structures completely and the separate blocks of rock are frosted below this upper part. The thickness of the permafrost in mountain regions increases with altitude and dissection because of better drainage and lower temperatures.

With the same height, degree of dissection and general orientation toward the Sun, the thickness of the permafrost within the mountain regions is subject to geocryological latitudinal zonation, that is, it increases from south northward. The geocryological latitudinal zonation becomes more important with increase in altitude. Thus for example the base of the permafrost in the Pamirs and Tien Shan is situated at an altitude of 3500–4000 m while in the ridges of the northern part of Russia it goes down to 300–500 m below sea-level.

The discontinuity of the permafrost in the valleys which follow tectonic dislocations, is the distinctive characteristic of the orogenic region of the Russian North-East.

The permafrost from 300–500 m in thickness, constituting the so-called hyperzone of the low mountain ridges which extend to the east of the Lena river between the Arctic and Pacific oceans, is the most widespread within this part of the territory. The extensive formation of this permafrost is caused by a combination of increase of mountain relief and height from north southward, while the Arctic cooling effect weakens in the same direction, with the surface radiation balance increasing at the same altitudes. At the same time the thickness of the permafrost in the region of intensive neotectonic movements is more variable than that within low-movement regions because of the deep permafrost of mountain ridges and shallow freezing within river valleys with water-bearing taliks typical of the former.

The permafrost is represented mainly by three layers in the large intermontane basins in the Russian North-East. The upper layer consists of Late Pleistocene syncryogenic ice-rich sediments of variable, mainly lacustrine alluvial origin with thick ice wedges. The middle layer is represented by Neogene-Pleistocene epicryogenic materials of fluvioglacial, glacial, lacustrine alluvial and other origins and has less ice content; the lower layer is composed of epicryogenic hard and semihard Mesozoic and Palaeozoic rocks. Loose syncryogenic deposits can occur immediately above the epicryogenic bedrock within shallow depressions.

Northern coastal lowlands such as Yana-Kolymskaya, Srednekolymskaya and Abyiskaya consist mainly of polygenetic multi-layered permafrost (as far as wedge ice is concerned) occurring on epicryogenic Neogene-Pleistocene loose materials or on strata cooled below 0 °C, with cryopegs and with gas hydrates in individual structures. The upper permafrost layer gives topographic expression and is represented by a complex of ice-rich syncryogenic, polygenetic deposits of Late Pleistocene age with thick ice wedges. Ice-rich syncryogenic alas and epicryogenic lacustrine deposits of various cryogenic ages (from late Pleistocene to recent) are included in the upper part of this unit. The middle layer is composed mainly of epicryogenic Pliocene-Early Pleistocene loose deposits with moderate ice content. The lower layer of the permafrost is represented by epicryogenic hard and semihard rocks of the Mesozoic cover with ice inclusions along tectonic dislocations. The lowest layer of the permafrost consists of rocks cooled below 0 °C, with cryopegs.

The Zabaykal'ye folded mountain system occupying a mid-latitude part of the region of the former USSR between latitude 50 and 60° N is distinguished by the broadest range of geological conditions. This territory, situated within the region of sharply continental climate, lacking sea basins, with latitudinally oriented mountain ridges and peaks of high absolute altitudes (from 2500 to 3000 m) and a long history of geocryological development including numerous glaciations in the colder periods and thawing during the shorter warm periods, has led to a complex combination of geocryological latitudinal and altitudinal zonation. At the present time the thickness of permafrost on ridges and peaks reaches 1000 m, sharply decreasing toward deep valleys and intermontane basins and often thinning in their bottoms because of the warming effect of ground water. The permafrost is one-layered in mountainous areas. It is composed of epicryogenic bedrock of Pleistocene and Late Holocene cryogenic age with ice along tectonically fractured zones overlain by a 5 m layer of coarse boulder deposits on watersheds, and rock streams on slopes cemented with 'golets' ice. The permafrost is overlain by thawed water-rich rocks in which water exchange is very intensive in connection with the high seismicity of the region. This circumstance contributes to a sharp decrease of the permafrost thickness in valleys formed along tectonic fractures. The cryogenic section in valleys is mainly two-layered. The upper layer is represented by alluvial, fluvioglacial, slope wash and other ice-poor epicryogenic or thawed, relatively thin layers (to 25 m, more seldom 50 m). Taliks situated under large river beds and along numerous fractures with thermal mineralized ground water outlets are typical of the valleys. The lower layer is represented by epicryogenic hard and semihard frozen rocks to 50-100 m in thickness.

A two-layered permafrost structure is typical of the Verkhneangarskaya and Barguzinskaya depressions.

The low mountain and highland regions of Zabaykal'ye are characterized by temperature inversions accompanied by decrease of snow cover in valleys, development of marshes, fine-grained deposits, development of sparse forest landscapes, etc. Therefore the thinnest permafrost (about 50–150 m) is typical of the lower watersheds. The thickest is typical of valleys (100–200 m) and high watersheds (200–300 m). Within the low mountain belt and in the south-eastern highland area of Zabaykal'ye with steppe landscapes, the permafrost is developed in depressions and in valleys and is characterized by a one-layered structure. It is represented by epicryogenic loose deposits of various origin, of Late Holocene cryogenic age and not more than 50–100 m in thickness decreasing to 50–25 m southward.

The thickness of the permafrost of epicryogenic hard and semihard rocks is in agreement with the altitudinal geocryological subzones within mountain systems to the south of the former USSR (Sayany, Altay, Tien Shan, the Pamirs, the Caucasus). The hypsometric level of these zones rises from the east westward. The greatest permafrost thicknesses are typical of the periglacial area and range from the first metres near the foothills (Sayany, Altay) and, at altitudes of 2000 m and higher (Tien Shan, the Pamirs), to 1000 m and more under the summits of the high ridges and the highest peaks.

The thickness of the permafrost follows a general pattern within the territory of the former USSR as a whole, which appears throughout the history of geocryological development: a continuity in distribution with area and depth increasing from the south northward, from the regions with oceanic climates to the regions with continental climates, from low mountains to high ones, from tectonically movable seismically active structures to structures of low mobility in recent time, and from young structures to more ancient ones as far as their cryogenic age is concerned.

Submarine permafrost

This is situated within the Polar basin and includes oceanic and shelf areas. The oceanic permafrost includes the greater part of the Arctic basin depression but is absent within the areas affected by branches of the North Atlantic current. It is represented only by cooled sea water-saturated materials in the upper ten metres, with temperatures to $-0.7°C$. The shelf permafrost was formed as a result of the submersion of the permafrost below sea-level in the Holocene. This permafrost was formed on the continent during the Late Pleistocene epoch by regression of the sea and possibly as a result of depression of the ground surface under a glacier. The upper ice-rich horizons of epi- and syncryogenic deposits were subject to marine action and the temperature of bottom deposits which ranged from -9 to $-13°C$ and lower, increased to -1.9 to $-0.7°C$; the ground ice was partly melted and replaced by salt water on account of the marine transgression. As a result the cooled materials include relict layers and lenses of frozen material degrading from the top and from the bottom (because of the Earth's heat) and which were formed under the sea water at the isobaths of 60–10 m. The discontinuity of relict frozen layers increases and the thickness decreases from the coast towards the sea basin. Seasonal freezing and thawing occurs on shoals and on the shelf in shallow waters (with isobaths of less than 10 m); littoral-marine syncryogenic deposits are formed locally. Cryopegs are developed in the lower layers within shallow water parts of the shelf, in deltas and on the elevated bottom area. Near the mouths of large rivers the permafrost is sporadic or is absent entirely.

Subglacial permafrost

This is situated under cold glaciers, the bottom temperature of which is below $0°C$. In the case of the base of the glacier lying above sea-level the permafrost is represented mainly by frozen ground; however, it is represented by cooled material with cryopegs when situated under ice shelves. The thickness of the subglacial permafrost varies from a few metres to 500 m and more depending on the glacier basal temperature, its thickness and the geothermal flux. Under warm glaciers with an ice-firn mass that has existed for a long time there is a possibility of unfrozen materials with positive temperatures (but close to $0°C$) holding fresh and slightly saline waters, giving way at depth to rocks below $0°C$ with cryopegs.

Within the high mountains of the Caucasus, the Pamirs and Tien Shan, the glaciers exert a cooling effect where firn masses are absent in the near-surface portion of the glacier, as well as having a warming effect when firn is formed on icesheds and ice caps. The latter occurs in glacier beds

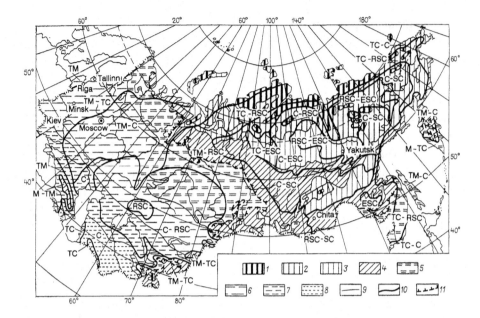

Fig. 15.8. Map of types of seasonal ground thawing and freezing in the former USSR. *Types of seasonal thawing with respect to mean annual temperature,*°C: 1 – arctic and polar (below − 10°C); 2 – stable (− 5 to − 10°C); 3 – stable for a long time and semitransitional (− 1 to − 5°C); 4 – transitional and semitransitional (+ 2 to − 2°C); 5 – type of seasonal thawing and freezing – semi-transitional and stable for a long time (1 – 5°C). *Types of seasonal freezing with respect to the mean annual temperature,* °C: 6 – stable (5–10°C); 7 – southern (10–15°C); 8 – subtropical (15–20°C and higher). *Types of seasonal freezing and thawing with respect to amplitude of temperature fluctuations on the surface,*°C: M – marine (less than 7.5°C); TM – temperate-marine (7.5–11°C); TC – temperate-continental (11–13.5°C); C – continental (13.5–17°C); RSC – rather sharply continental (17–21°C); SC – sharply continental (21–24°C); ESC – extremely sharply continental (higher than 24°C); 9 – boundary of seasonal ground thawing and freezing with respect to the mean annual temperature; 10 – the same with respect to the amplitude of temperature fluctuations on the surface; 11 – southern limit of permafrost.

when the ice is flowing down the slope and when it is thawing from the bottom in the course of sliding. Therefore the mean annual ground temperatures under the moving glaciers can be positive or close to 0°C (negative) while the temperatures below glaciers of low mobility and temperatures of rocks adjacent to glaciers can decrease to − 10°C, − 15°C and lower. Taliks of the closed type occur under small thawing glaciers surrounded by low temperature ground with great thickness of permafrost.

15.3   Distribution of main types of seasonal ground thawing and freezing

The main distribution pattern of different types of seasonal ground thawing and freezing on the territory of the former USSR is shown on the general map-diagram (Fig. 15.8) on which we can follow in succession the zonal-regional changes depending on the stability of process (on t_{mean}) and on the climatic type at the surface (on A_0). Within every type there exists (almost without exception) areas with azonal conditions of freezing and thawing, typical of more northern or more southern regions because of regional variations of natural conditions. Only the most northern and the most southern parts of the country are characterized by the smallest range of seasonal thawing and freezing depths because the differentiation of the natural conditions in those parts does not cause any marked changes in geocryological environment.

In spite of the fact that considerable variations of the permafrost conditions are noted within one region or even within a part of a region, especially in areas with permafrost, the distribution of various types of seasonal ground freezing and thawing follows the zonal and regional pattern within the predominant landscapes over the former Soviet Union as a whole.

Seasonal thawing of ground

Zonal-belt as well as sectorial changes of mean annual temperature t_{mean}, and as a consequence spatial differentiation of various types of seasonal thawing, are typical of the vast areas with permafrost. The successive change of seasonal thawing types from Arctic and Polar to semi-transitional and transitional takes place in connection with a regular increase of mean annual temperature t_{mean} from north southward (within plain topography) and from summits to valleys (in mountain systems) as a whole. Increase of amplitude eastward and with lower altitude in mountains is typical of the permafrost areas. Thus a successive change in the type of seasonal thawing takes place with respect to continentality (through A_0) from west eastward from the temperate-maritime type in the north-western part of the European Russia to the extremely sharply continental type in the valleys of Central Yakutiya, Eastern Siberia and Zabaykal'ye.

The range of thickness of the seasonally thawing layer varies from the first centimetres to 3–5 m and more. The increase in the depth of thaw takes place in accordance with increase of summer heat flux through the ground surface, southward as well as eastward and with lower altitudes in mountain regions as a whole. The maximum values of ξ_{th} occur in the zone of discontinuous permafrost where transitional and semi-transitional types of seasonal thawing of surface deposits prevail. The thaw depth for low moisture content coarse-grained soils in conditions of increased continentality can reach 4-5 m. The minimal thawing depths occur in moisture-rich peaty grounds, all other factors being similar.

A brief description is given below of the main types of seasonal thawing within the former USSR.

Arctic and polar types of seasonal thawing occur where the mean annual temperature is below −10°C, and are observed within European Russia only in the zone of Arctic and ice deserts of the archipelago of Franz Josef Land (Zemlya Frantsa Iosifa) and on the high (800–1000 m) summits of the islands of Novaya Zemlya. Within the Asian part of Russia, they occur on the islands of Severnaya Zemlya, on the northern coastal lowlands and in the most elevated parts of the mountain systems of Central and Eastern Siberia and the Russian Far East. These types occur in the south of the country in the nival zone of the high mountains of the Pamirs and Tien Shan. The temperate-marine and temperate-continental types of seasonal thawing prevail on the Arctic islands, coastal lowlands and on the summits of the Middle Asia mountains. Continental and rather sharply continental types of seasonal thawing are typical of the belt of Arctic mountain deserts and high mountain of Eastern Yakutiya. The depth of thaw of ground within these regions is not very great (from the first centimetres to a few tens of centimetres). Only within high mountains of alpine type can the summer thawing thickness reach 1 m and more on coarse-grained soils in sunny situations. The ground thawing can be absent entirely in extremely cold years.

Stable and long-term stable types of seasonal thawing of ground the mean annual temperature of which varies from −2 to −5°C, or −10°C, are developed in the north-eastern European part of Russia, in the high mountains of the Northern and Polar Urals, on the archipelago of Novaya Zemlya, to the north of the Arctic circle on the West Siberian lowland, in the northern and central parts of Central Siberia, practically over the whole territory of Eastern Siberia and the Russian Far East as well as in the mountain part of Zabaykal'ye. The extent of these types increases steadily from west eastward and south-eastward. The amplitudes of the temperature fluctuations at the soil surface caused by growth of the climatic continentality with distance away from the Atlantic Ocean increase in the same direction. Thus in the North-East of European Russia and in Western Siberia temperate-maritime and temperate-continental types of seasonal ground thawing prevail. Continental and rather sharply continental types of thawing occur under natural conditions in Central and Western Siberia. Formation of the extremely sharply continental type is a possibility in valleys with

scarce or absent vegetation cover. Thus while the thawing depth of moisture-rich clay-silt deposits with a mean annual temperature of $-5\,^\circ$C does
not exceed 1 m in the European part, the summer thawing depth in Middle
Siberia can reach 2 m under the same conditions.

The depth of summer thawing within these regions varies greatly depending on surface and ground-moisture conditions. Thus coarse-debris soils
can thaw to a depth of 3 m in Zabaykal'ye under conditions of continental
and sharply continental types of heat exchange at the ground surface.

Transitional and semi-transitional types of seasonal thawing are widespread in the region of discontinuous permafrost, the mean annual temperature of which varies from 0 to 2 °C. The territory under review includes the
northern and central parts of the Kola Peninsula, Malozemel'skaya Tundra,
central and southern parts of Bol'shezemel'skaya Tundra, the greater part of
the West Siberian lowland, practically all the southern part of Central
Siberia, the middle-altitude mountain zones of Zabaykal'ye, the Russian
Far East and Kamchatka. These types of seasonal ground thawing and
freezing are developed also on the most elevated areas of Sakhalin and the
Sikhote-Alin mountains, and prevail in the middle-height mountains of
Sayany, Altay and in the southern mountains of Middle Asia and in the high
mountains of the Caucasus.

The unstable character of the ground thermal regime, the presence of
short-term and discontinuous permafrost, the frequent change of seasonal
ground thawing and freezing types over the area and of the maximum values
of seasonal thawing and freezing depths of soils being observed, are typical
of the territories with such types of seasonal thawing. Thus the thawing
depth of coarse-grained deposits is 2 m on the soil surface within Kola
Peninsula in the conditions of a temperate-continental climate. The seasonal thawing depth increases to 3 m in dry sands within Western Siberia
where a rather sharply continental type of thawing occurs. Within mountain
regions of Central Siberia and Zabaikalye under conditions of rather sharply and sharply continental regimes of temperature fluctuations at the
ground surface, the thickness of the seasonally thawing layer in coarse-
grained materials can reach 3–4 m. The seasonally frozen layer is 5–6 m in
thickness within the same region on the areas without permafrost.

Seasonal freezing of ground

Annual seasonal freezing of the upper horizons of soils and ground
takes place practically within the whole region of ground situated south of
the permafrost region. According to available data the thickness of the
seasonally freezing layer in coarse-grained materials does not exceed 5–6 m,

while the seasonally freezing layer in clay ground does not exceed 3–4 m; the thickness of the seasonally frozen layer in peat does not exceed 1–1.5 m within the former USSR. The lowest values of winter freezing are observed in the southern regions and extend only to the first centimetres.

The distribution of types of seasonal ground freezing outside the permafrost zone according to their mean annual temperatures has a mainly latitudinal pattern. Some tendency for eastward displacement on the European territory as well as in Western Siberia is caused by the general increase in continentality in this direction.

Transitional, semi-transitional and long-term stable types of seasonal freezing of ground with mean annual temperatures not exceeding 5 °C are developed on the Kola Peninsula, in Karelia, in the northern part of the Russian plain, in the foot-hills of the Urals Preduralye, in the Middle and Southern Urals as well as in central and southern parts of the West-Siberian plain. In the east of the country the seasonally freezing ground of these types is widespread on the southern foot-hills of the Sredinnyy ridge and the southern part of Kamchatka as well as on the Kurile Islands, in northern and central parts of Sakhalin and in Primorye.

Within the European territory the amplitude of temperature fluctuations at the ground surface corresponds to temperate- maritime and temperate-continental types of seasonal freezing as a whole, with the general tendency for continentality to increase eastward. The freezing depth increases in the same direction, varying, in dry sands, from 2.2 to 2.5 m with mean annual temperature of $+1$ °C. In similar deposits to the south when the ground temperature increases to $+5$ °C, the depth decreases to 1–1.5 m.

Some increase of freezing depth (to 3–3.5 m) is noted in Western Siberia southwards within this zone. It is associated with increase in the amplitude of the ground surface temperature fluctuation and decrease of the moisture content in the freezing layers. The same situation is observed within the European territory of the country southwards although the increment in thickness of the seasonally frozen layer does not exceed 0.5 m.

The stable type of seasonal freezing of the soil occurs throughout nearly all the European part of the former USSR, the most southern part of the West-Siberian Plain, Kazakhstan, the southern part of Sakhalin and valley territories of Primorye. A certain successive change in the freezing types from west eastward and south-eastward is noted in connection with climatic continentality. Thus the temperate-maritime and temperate- continental types of seasonal ground freezing prevail in the European territory while the continental and especially continental types prevail on the north of the Caspian region and on the foot-hills of the Southern Urals and in Kazakhstan. Within the latter (mainly in Kazakh melkosopochnik) the maximum thickness of the seasonally frozen layer reaches 2–2.5 m in conditions of stable, high temperature (7–9 °C).

The southern type of seasonal freezing where the mean annual ground temperature varies from 10 to 15 °C is developed in Trans-Carpathians, within the Black Sea lowland, in the Crimea, Pre-Caucasus and Trans-Caucasus on the Caspian lowland and in the central part of Central Asia. The depth of seasonal ground freezing varies from a few centimetres to 0.1–0.3 m within the European part in conditions of temperate-maritime and temperate-continental types of climate. As far as the Caspian region and Central Asia are concerned, the depth of winter freezing can reach 1 m under the conditions of a continental and extremely continental-temperature regime.

The subtropical type of seasonal ground freezing occurs where the mean ground temperature exceeds 15 °C and is found on the southern coast of the Crimea, in Trans-Caucasus (Kura lowland) and in the southern desert zone of Central Asia. Regular seasonal ground freezing is absent over the greater part of these territories. Only a short-term or daily (nightly) manifestation of the process is noted.

16

Principles and methods for regional geocryological investigations

16.1   Geocryological survey as the basis for regional investigation of the seasonally and perennially freezing zones

Purposes and problems

Geocryological conditions and their spatial variability are dictated by the features of the natural environment (such as geological and tectonic structure, hydrogeological and thermal-physical conditions, landscape-climatic surroundings and topography) and by the history of the development of the permafrost in the Cenozoic. Geocryological conditions in the regions of the seasonally and perennially freezing zones are studied following geocryological surveys. The development of principles and methods for such surveys are to a great extent associated with the work of V.A. Kudryavtsev.

The *geocryological survey* represents by his definition a system of field, laboratory and office work the purpose of which is: to study *the principles of formation and development of seasonally and perennially frozen ground* and low-temperature geological processes as a function of the natural conditions, their change in the Pleistocene and Holocene, as well as a result of the economic development of the region; to compile geocryological maps and to make geocryological forecasts on this basis as well as to work out the measures to control the permafrost processes and to protect the environment.

For purposes of the geocryological survey we can examine the following matters (15):

1) the principles of the distribution of seasonally and perennially frozen ground over the area depending on the variability of the geological-geographical environment and on the details of heat exchange at the Earth's surface;

2) the cryogenic structures (occurrence and condition of the cryogenic

strata and of the layered structure of the permafrost across the section) as they depend on geological structure, neotectonics, the effects of surface and groundwater, and on climate dynamics in the Cenozoic;

3) the composition and properties of frozen, freezing and thawing materials of the geological-genetic complexes and formations in the various landscape-climatic conditions;

4) the cryological features of frozen strata (cryogenic texture, macroinclusions of ice and ice content of Quaternary deposits and Prequaternary bedrock) and their dependence on the composition, genesis, age and neotectonic development, type of freezing and dynamics of the permafrost process;

5) the characteristics of the temperature regime at the soil surface, at the base of the seasonal and perennial freezing or thawing ground and at the level of zero annual amplitude, on the basis of analysis of their dependence on the individual natural factors;

6) the dynamics of the depths of seasonal and perennial freezing and thawing and their dependence on the system of natural environmental factors;

7) the spatial change of the thickness of the permafrost over the area and across the section in connection with the existing natural conditions and the history of regional geocryological development;

8) the characteristics of the formation and development of taliks, relating to their genesis, distribution and manifestation;

9) the characteristics of the cryogenic processes and phenomena, their distribution and development;

10) the nature of the interaction between the permafrost and the groundwater;

11) the engineering-geocryological conditions and engineering- geological severity of the terrain in relation to its economic development and geocryological forecast;

12) the experience from construction and other kinds of development with regard to the type of engineering structures, the character of the development and the dynamics of the permafrost process;

13) the history of the permafrost as a function of the dynamics of climate, geological history of the region and nature of human activity.

All the questions listed above or at least some part of them can be resolved depending on the scale of the survey and its aim within the area under investigation. These questions must be considered when making up a pro-gram for a geocryological survey.

Methodology

The methodological basis for geocryological surveys representing the scientific background, has been developed and formulated by V.A. Kudryavtsev (see Introduction, §5). In V.A. Kudryavtsev's opinion the most important conclusion is that the main object for study is the earth materials themselves. It is impossible to establish the principles of ground temperature regime and all the other permafrost characteristics without studying the materials. It is the seasonally and perennially frozen layers which should be the main object of research when studying geocryological conditions.

With allowance made for this fact the process of geocryological survey is based on the following premises:

1) the main result of a geocryological survey is the establishment of geocryological conditions through the study of the geocryological characteristics in their interaction with the natural surroundings, their change over the territory and the history of development;

2) variations in the geocryological conditions over the territory de-pend on variations of natural conditions, structure, and type of tectonic structure, geothermal and hydrogeological conditions and topography, landscape-climatic environment and the history of cryogenic development of the territory;

3) a change of one of the cryoformative factors from one region to another causes a change of geocryological conditions;

4) environmental change and short-term climatic change (3,11,33-years) appear first of all in the change of seasonal freezing and thawing depths, mean annual ground temperatures and cryogenic structure of the layer of annual temperature fluctuation, in taliks and areas of formation of new permafrost as well as in the change of cryogenic processes and character of the phenomena;

5) a change of climatic and geological-structural surroundings in dif-ferent periods of geological development causes change of all the components of the geocryological conditions on the surface and in the layer of temperature fluctuation as well as across the whole layer of permafrost;

6) the study of the present geocryological conditions involves the study of two groups of geocryological characteristics: heat exchange at the ground surface and in the layer of annual temperature fluctu-

ation for the upper boundary conditions; and heat exchange at the permafrost base and in the permafrost itself for the lower boundary conditions;

7) reconstruction of the history of geocryological development explaining the present cryological structure and the possible changes in the upper 20 m of the ground-cryogenic layer is possible only when analyzing the history of the whole complex of development of the natural environment;

8) the study of geocryological conditions in the course of the survey is carried out during preliminary, field and office work periods with application of a complex of special permafrost, geological-geophysical, geomorphological, climatic, laboratory and other methods of investigation as well as aerial and satellite photography in various spectral bands;

9) the study of multi-factor dependence of every geocryological characteristic on the cryoforming factors of the natural environment. Informativeness and reliability of a geocryological survey results when it is conducted in combination with engineering-geological and hydrogeological surveys, using scientifically the same methods of integrated research.

The key landscape method

The main methodological provisions of the survey can be realized with the help of the key landscape method. The principle of this method consists of the following. In the first stage the typologic *landscape* (geological-geographical) of the region is subdivided in accordance with factors and conditions dictating the formation and existence of the particular types and variations of the seasonally and perennially frozen ground. Then the *key areas*, typical of the separated types of regions (microregions), are selected on the basis of the general map and the factual data collected in the process of earlier research. Integrated investigations on a wide range of problems are solved with the help of different methods centred on these key areas during the field period and individual, general and regional geocryological characteristics are studied. Similarities and differences between the different microregions are established through traverses planned on the landscape regionalization map and with the help of aerial and satellite photographs. In conclusion the geocryological map is compiled by extension of the previously obtained geocryological characteristics to the identical types of regions similar to those studied in the key areas and along survey routes. In this case the landscape method of partitioning of the territory becomes the landscape-indicative method of mapping when mapping the ground temperature regime, seasonal freezing and thawing and cryogenic structure of the layer of annual temperature fluctuation.

Landscape regionalization

This is carried out to reveal the similarities and distinctions between areas under investigation with respect to the complex of natural features and associated permafrost conditions. It is commonly accepted that such combinations of natural complex elements as topography and the rocks, soils, vegetation, microtopography and microclimatic features comprising it form persistent genetic homogeneous areas on the Earth's surface. The topographic elements – microregions or types of terrain – represent such areas. The separation into microregions should be based on the geological structure and geomorphological topographic elements, characterized by uniform origin of rocks, relief (height, steepness and orientation of slopes) and conditions of heat exchange at the ground surface. Within the particular topographic elements the subsequent division of the area is carried out on the basis of more detailed uniformity – of cryoformative conditions (composition and moisture content, type of vegetation cover, swampiness, peatiness, ground surface microtopography, etc.).

Selection of the features of landscape regionalization under the particular zonal-regional conditions should be made taking into consideration the clearness of boundaries of the natural complexes, while the gradations should be justified by the degree of effect of each cryoforming factor on the geocryological environment of the area.

Regional features of the area under study and the scale of the survey introduce new criteria for the selection of features and regionalization gradations. Thus it is necessary to analyze the structural-geological and geographical conditions and to prepare a classification of regional features before the landscape regionalization mapping. Depending on the survey scale, topographic elements and their parts differing from each other by their complexity and area can be separated as microregions. Thus when the small-scale survey (1 : 100 000 – 1 : 500 000) is conducted, the large topographic forms characterized by uniformity of development and geological structure are mainly separated as microregions. These are, for example, watersheds having particular heights of their ground surface and degree of dissection; watershed slopes to river valleys subdivided into 2–3 gradations according to their steepness; river valley elements with various surface conditions (degree of forest cover and swampiness, etc.). When the medium-scale surveys (1 : 25 000 – 1 : 50 000) are conducted the areas are subdivided into a number of smaller ones depending possibly on the soil composition and properties as well as on more detailed consideration of the other natural factors typical of the areas being subdivided. Thus, for instance, within watersheds homogeneous in geological structure, the areas with different combinations of soil beds and composition, microtopographic features, different degree of swampiness, different vegetation cover, etc. can be distinguished.

When the large-scale survey (1 : 2 000 – 1 : 10 000) is conducted and the whole area is situated within one or two topographic elements, for example, in river valleys or on one particular watershed or its slopes, etc., such details of the components of the natural complex components which are considered too small when conducting the small-scale and medium-scale investigations, become more and more important as the regionalization of the landscape is pursued further. Thus, for instance, when a large-scale survey is conducted, areas are distinguished by the prevailing composition of the deposits and their moisture content and also by lithologic features of particular sections as well as by moisture distribution across them. Thus, for example, the areas are separated by degree of forest cover depending on the density of a stand of trees and on degree of surface shading etc.

The key areas

During the field work period it is profitable and economically attractive to carry out geocryological studies not with the same detail over the whole territory but to concentrate them within particular, smaller areas, susceptible to the use of all the methods of investigation. These key areas should include the most typical widespread microregions as well as anomalous areas occurring locally. They are subdivided into general and special-purpose areas according to the thrust of the investigations. Within the general-purpose key areas the principle conditions of the permafrost, typical of the region, are studied. Within the special-purpose key areas the particular thematic questions of a regional and methodological character are solved, including the problem of determination of geocryological relationships.

When the small- and medium-scale surveys are conducted it is profitable to select key areas, not for every microregion individually but to cover a few microregions simultaneously. This allows not only the establishing of geocryological relationships for each microregion, but also the tracing of their spatial variability.

When a large-scale survey is conducted the special problems associated with the particular economic development of the region are dealt with.

In practice the sizes of the general-purpose key areas vary from 1 to

10 km^2 in the small-scale survey and from 0.2 to 1 km^2 in the large-scale survey depending on the complexity of the geocryological conditions; the sizes of the special-purpose areas vary from 0.2 to 2 km^2 and from 0.1 to 0.3 km^2, correspondingly. The number of key areas depends on the complexity of zonal and regional terrain conditions, i.e. on the number of landscape types, each of which should be studied within not less than two key areas. The use of aerial and satellite photographs is of great importance for their selection, with their representation of larger territories and with the possibility of the preliminary integrated interpretation of natural conditions and cryogenic formations. In an effort to use geological surveying and prospecting data more completely and rationally it is profitable that key areas for small- and medium-scale surveys should be located on the site of mineral deposits, that is, where there has been construction and a dense network of bore-holes, shafts, trenches and pits.

The number of key areas (sites) is specified by the stage of the project and by the permafrost and other engineering-geological features considered when the large-scale survey is carried out.

Exploratory traverses

These are carried out when conducting geocryological surveys to study microregions distinguished on the landscape regionalization map and to ensure the proper extrapolation of the factual data on the key areas to the whole territory under investigation. Therefore the main problems for exploratory investigation are the following: 1) to follow geological, geomorphological, and geobotanic boundaries to investigate the changes in natural conditions 2) to determine the depths of seasonal freezing and thawing and the changes depending on the combination of particular factors of the natural environment of each microregion; 3) to study and to map cryogenic and other geological formations; 4) to sample surface waters and the upper water-bearing horizons (suprapermafrost water and vadose water); 5) to map and to sample groundwater outlets and icings; 6) to study the cryolithologic section of the genetic types of Quaternary deposits in pits, clearings and exposures; 7) to study dislocations and breaks of continuity and jointing in exposures and their expression at the surface; 8) to study the experience being gained from construction and other kinds of economic development of the region.

Successful solution of the problems outlined above with high efficiency and on short notice depends first of all on the correct selection of field traverses. When selecting the traverse route it is necessary to follow these rules: 1) the routes should cross all the types of main areas being distinguished in the course of landscape regionalization; 2) the routes should be planned transversely to river valleys with examination of watersheds as well as along them, to study the elements of the valley complex; 3) the density of the route network and distance between observation stations depend on complexity of the area (region). At the same time the greatest number of observation stations per square kilometre should be located on the key areas and sites of detailed work; 4) in the course of surveying the necessary corrections should be introduced into the plan for exploratory investigations.

It is advisable to use topographic maps of 1 : 100 000 scale and geological maps of the same or larger scale for small-scale exploratory surveys. When medium- and large-scale surveys are conducted it is profitable to use maps of the surveying or larger scales. In any case the explorations should be carried out with the map of landscape regionalization and aerial photographs interpreted beforehand. Along with black and white and spectral band aerial photographs it is desirable to use satellite photographs.

Scales for geocryological surveys

By analogy with engineering geological surveys the geocryological surveys are divided into the small scale (1 : 500 000 – 1 : 100 000), the medium-scale (1 : 50 000 – 1 : 25 000), the large-scale (1 : 10 000 – 1 : 5000) and the detailed (1 : 2000 and larger).

Geocryological field surveys are not carried out at smaller than 1 : 500 000 scale while the geocryological maps are compiled in the course of office work by gathering and generalizing the published and field data of geocryological, engineering-geological and other investigations coordinated with the zonal-regional geocryological background. The latter can be obtained from the geocryological maps of smaller than surveying scale and using the geocryological maps of 1 : 2 500 000 scale for small-scale and general maps.

The small-scale geocryological survey is usually made for large areas with the goal of geocryological, hydrogeological and engineering-geocryological assessment of the region for its potential economic development. Such surveys should be conducted by the regional geological boards of the country, first of all on mineral-rich areas and by other organizations and departments for the purposes of industrial and agricultural development of the individual regions. The selection of the particular survey scale (from 1 : 500 000 to 1 : 100 000) depends on the complexity of natural conditions, the level of knowledge of the region, the specific thrust of the investigations and the stage of development of the region.

The medium-scale geocryological survey is the target one and is conducted on comparatively small areas that are to be economically developed in a particular way. Such surveys are conducted at the stage of pre-design documentation – technical and economic substantiation and technical-economic estimation – of a project and in some cases depend on the stage of substantiation of a project according to the type of construction and complexity of natural conditions. The medium-scale survey results are optimal when these surveys are conducted on the background of small-scale ones, i.e. when the predetermined geocryological elements are improved and worked out in detail.

The large-scale surveys (1:2000 – 1:10000) are, to a still greater extent than the medium-scale surveys, specialized for a particular type of construction. These surveys are usually conducted at the stage of technical development of a project and elaboration of designs and are supplemented with investigations for each specific structure with its particular technological and construction characteristics.

When conducting the surveys at any scale uniform procedures must be followed, consisting of the study of geocryological features based on geocryological mapping, forecasts of changes in the permafrost environment and the control of cryogenic processes, in an effort to exploit the geological surroundings harmoniously. At the same time it is obvious that the small-scale survey gives an indication of geocryological conditions of large areas in connection with natural macro- and meso-conditions. The medium-scale survey is conducted for the purpose of studying smaller regions while the large-scale surveys are conducted to study small sites and sections. The small-scale geocryological surveys, as well as all the other types of the small-scale geological survey, are the basis for carrying out the larger-scale investigations. Only on the background of knowledge of characteristics of the permafrost formation and development over large regions can any special manifestation of these characteristics on small sites and particular areas be understood correctly. Thus the detailed geocryological investigations on small areas without wide general study of the permafrost conditions will always be less effective than a combination of surveys at different scales. The great role of regional investigations including the small-scale (and in some cases the medium-scale) survey follows from this.

16.2   The methods and carrying-out of geocryological surveys

Carrying out geocryological surveys at any scale includes three stages: preliminary, field and office work.

The preliminary stage begins with the drawing up of the work program, laying out the main problems of the survey to be conducted. For its realization field observations and published materials such as topographic maps, aerial photographs at large and small scale (black and white, colour and spectral range) and satellite photographs (when the small-scale survey is conducted), geological and tectonic reports and maps, descriptions of geological sections and exposures, geomorphological maps and Quaternary deposit maps, geobotanical maps, climatic and hydrogeological data are gathered and studied. The study of earlier geocryological, hydrogeological, engineering-geological and geophysical investigations showing the character of the geocryological surroundings, plays a special role. At the same time the reliability and representativeness of the materials of the previous investigations carried out during various periods of time, at various scales and usually using various methods, are established.

On the basis of this analysis and preliminary interpretation of aerial photos and satellite photos (for the small-scale survey) the preliminary (geological-geographical) map of the landscape region is compiled. Then the factual data characterizing distribution, structure, occurrence, thickness, temperature regime of the permafrost, etc. are entered on the landscape map. All this allows one to get an idea of geocryological conditions and of the level of knowledge about the area, to determine the possible interval of change of geocryological characteristics, the program of field-work necessary and the most rational methods and degree of complexity of investigations as well as the plan for, and the problems of, reconnaissance, aerial-visual and correlation (control) exploration during the preliminary period.

In the absence of such data it is necessary to study general maps to identify the regional geocryological background and to get an idea about the geocryological conditions of the territory under study. The preliminary assessment of the effect of natural factors on the ground temperature regime contributes to this.

On the basis of the maps obtained and materials studied the key areas are considered and the requirements for drilling and sampling, for geophysical observations and for observations in the laboratory and under special conditions, and other types of work within key areas are drawn up. A study of experience with development is also drawn up; the extent of ground sampling, the volume of surface and ground waters and the content and extent of calculation work using a computer are determined.

The main problems of the *field work period* are to study the dependence of each geocryological characteristic on components of the natural environment and to determine general and regional characteristics of the formation and spatial variability of geocryological conditions. The main study is carried out within key areas and then it is supplemented with the help of exploratory investigations using aerial photos, geocryological, geological, geomorphological, geobotanic, microclimatic and geophysical methods for special purposes, etc. As the materials are obtained the preliminary map of the landscape regions is improved, columnar sections are made from bore holes, exposures and mine openings, field geocryological profiles are drawn and calculations for revealing the interdependence of factors in the natural environment and geocryological characteristics are performed. All the data are plotted on the map of record or on the map of landscape regionalization. The latter is more rational because the map of regionalization takes geocryological content into account and becomes the basis for geocryological mapping.

Within the key areas geological formations and distribution of geological-genetic rock complexes, their composition and properties are studied. Ice content and study of cryogenic structure, revealing syn- and epicryogenic ground freezing are a major concern. In addition, ground temperature measurements and observations of seasonal freezing and thawing depth are carried out, the permafrost thickness is determined, and the regime and temperature of groundwater in bore holes and in springs are studied. Water sampling for chemical and spectral analyses, and sampling of rocks with disrupted and undisrupted textures (including frozen rock massifs) for laboratory studies of their composition, thermal physical, physical-mechanical and permeability properties are performed. Observations of special regimes, thermal-moisture field dynamics and of the development of geocryological processes are carried out.

Studies of experience with construction, with thermal amelioration of the ground and of the efficiency of environmental protection measures are a major preoccupation. Disturbance of natural conditions and consequences of the associated processes are thoroughly studied and charted. All these measures help to improve the calculation of forecasts for the conditions.

The result of exploratory and detailed investigations within key areas are used for the regional classification of seasonally and perennially frozen ground, of taliks and exogenous geological processes and phenomena and of groundwater. Then on the basis of these classifications and geocryological sections the geocryological maps on a larger scale than that of the survey are compiled for the key areas.

The field work period is completed with preparation of the geocryological maps of the whole area under study on the basis of the landscape regionalization maps for key areas. These maps are supplemented with summary tables showing characteristics of permafrost, taliks and the seasonally thawed layer, profiles, plots representing the dependence of certain characteristics on each factor of the natural environment and other explanatory materials.

Special methods are used to get specific geocryological information during the field work period. Thus, for example, the palaeopermafrost analysis and palaeopermafrost reconstruction method is used for studying the formation, evolution and dynamics of the permafrost. The method of permafrost facies (cryolithological) analysis combined with the geological survey facilitates study of the Quaternary deposits in the permafrost parts. The landscape indicators help to establish the dynamic series of landscapes associated with development of specific complexes of geocryological processes.

During the *office work period* researchers complete the processing and analysis of field and laboratory investigations and of the observations of regime, classify the permafrost-geological sections, use laboratory data to establish characteristics of the composition and properties of the permafrost of different origin and age and determine the absolute cryogenic age of syn- and epicryogenic materials, and make clear the genesis and special features of talik zones, and the development of various cryogenic phenomena. Using computer calculations they make clear the effect of each factor of the natural environment on the mean annual temperature and depth of seasonal ground thawing and freezing, assess the interaction of the permafrost with groundwater and the spatial variability of the permafrost thickness, and make the natural-historical geocryological forecasts. Then they improve regional classifications of permafrost and taliks, compile the final geocryological maps for natural and predicted conditions, perform final geocryological (engineering-geological) regionalization of the territory and give the region-to-region description of geocryological conditions. In conclusion the engineering-geocryological assessment of the territory is performed and recommendations for deliberate changes of natural conditions for optimizing the regime of the geocryological environment are prepared. The results of the investigations, their substantiation and scientific conclusions are given in the scientific report including a text, maps, graphical and textual data supplements. The text of the report should include also the methods of map compilation and analysis, permafrost forecasting, engineering-geological assessment of the area and recommended measures for the control of permafrost processes.

The investigation techniques

Regional geocryological investigations techniques can be combined into three groups each of which include particular types of work.

I. *Determination and measurement of various indices of geocryological characteristics* (factual data): 1) mining-drilling; 2) geophysical; 3) experimental-instrumental; 4) photography (ground-based as well as aerial and satellite); 5) visual and aerovisual observations (description, drawing).

II. *Primary processing and interpretation of geocryological data:* 1) statistical (analysis of the law of distribution of exponents); 2) graphical (plots, diagrams, histograms, etc.); 3) tabulation; 4) interpretation; 5) logical, physical-mathematical correlations; 6) statistical simulation (trend analysis, cartographic-statistical, regression, correlation, factor analyses, etc.).

III. *Establishment of natural characteristics:* 1) comparison, conjugate analysis (through natural components and conditions) of the territory and landscape indicators; 2) systematization and classification (preparation of particular and general regional diagrams and classifications); 3) charting and integrated regionalization; 4) mathematical and physical simulation.

The main aims of the drilling work are: study of permafrost parameters (composition, ground temperature, thickness, occurrence, cryogenic structure, ice content, etc.), sampling of frozen and thawed materials for laboratory analyses; thermometric and hydrogeological observations, integrated well hole logging. The specific nature of the permafrost study imposes particular demands on the design and technology of drilling. Thus *bore-hole drilling* is profitably carried out with air blowing of the bottom of the hole, providing for the smallest disruption of the natural ground temperature regime the most complete conservation of the permafrost core and the most exact determination of the permafrost thickness and of the depth of the water-bearing horizon.

Geocryological test holes should be equipped with moisture-insulated observation pipes. Temperature observations should be carried out after the natural temperature field is restored. According to various investigations the time of such restoration varies from 10 days to a few years depending on the hole depth, its construction, type and time of drilling. Geocryological bore holes are divided by their depth into two groups: mapping, through the layer of annual temperature fluctuations (10–25 m); and key geothermal holes through the whole permafrost. In the course of survey the mapping holes are positioned at two to three points within each type of landscape (type of area, microregion). Key holes are positioned usually taking into account changes in the geological structure of the territory that are causing regional change of the permafrost thickness. Supporting bore holes are used also to study the interactions between the permafrost and the groundwater.

Geophysical methods are used to study temperature fields, the state (frozen or thawed), composition, structure and mode of occurrence of various genetic ground types in section and in plan. For these purposes thermometric, electrical, acoustic and nuclear logging of holes, electrical profiling and vertical electrical soundings are carried out, infra-red, thermomicrowave, radiolocation and aerial photography are conducted. Particular attention is given to the method of measurement of parameters of physical fields. The widest experience is gained in this regard when using the integrated bore hole logging and direct current (electrical) prospecting methods. The geophysical techniques and particular geophysical investigations used in the course of the permafrost survey depend on a number of factors, mainly: 1) special features of geological and permafrost structure of the territory under study; 2) the scale of researches; 3) the end use of the survey being conducted.

Thus, for example, in the course of engineering-geocryological and hydrogeocryological survey of hard-to-reach poorly known areas in mountain regions with discontinuous permafrost the most productive are electrical prospecting techniques such as electrical profiling (EP) and vertical electrical sounding (VES).

Geophysical techniques used in small-scale surveys present the most complete characterization of, overall, the given region (or of some parts of this region such as topographic elements and associated landscapes) with values of such parameters as permafrost thickness, ice content and its most typical variability in the section, thickness of sedimentary deposits, microinclusions and ice bodies, thickness of the fractured zone in rock massifs with close crystal bounds, degree of fissuring of rocks within these zones, talik zones with the outlines and extent of thawed materials and water saturation. The specific aims in this case (revealing the most typical background values of the parameters under study) lie in the investigation method: first of all VES points are installed in the best-known typical undisrupted conditions, and only then over the rest of the area. The VES curves obtained within the typical areas are interpreted first, the special features of which in the permafrost section allow quantitative interpretation on the basis of parametric VES in the best understood holes. In the second stage, the remaining VES curves are interpreted. However, in this case the data on specific resistance of the thawed and frozen sections obtained at the first stage of quantitative interpretation are used as key ones.

The results of the interpretation of the small-scale survey are presented in the form of a section as a rule. Typical of this section are generalized data, the values of which are calculated by averaging all the special determinations of these characteristics made using the electrical prospecting techniques.

The geophysical investigations in the course of the large-scale surveys are designed to reveal the inter-landscape variability of the values of the features studied, with the consequent necessity of substantially increasing the density of observational points. The large scale of the investigations require also detailed study of the whole territory, including objects with steeply falling boundaries. For these purposes the range of geophysical techniques used is substantially extended. Various types of logging, of refraction-seismic exploration (RSE), of vertical seismic profiling (VSP); with laboratory measurements of specific resistance and elastic wave velocities (SR) are used, as well as acoustic and penetrational logging to study composition, moisture, density, porosity and ice content of sandy, clay-rich, coarse silty and rudaceous grounds and also the composition, state and properties of bedrock (porosity and degree of fracturing, static elastic modulus, modulus of strain, of short-term uniaxial compression resistance, the stress state coefficient, etc.).

To study such geocryological phenomena as karst, thermokarst, icing, occurrence of ice bodies in frost mounds and elsewhere, ground veins, pseudomorphs of ice wedges etc., various kinds of logging, electrical profiling, VES, resistivity measurements in bore holes and water basins, RSE, SR, gravimetry, VSP are used.

Specific presentation of the geophysical investigations in the course of the large-scale survey, except for the detailed geoelectrical sections constructed with the help of particular profiles, is by the compilation of special-purpose maps (for example, the map of various depths important for a particular kind of construction within the area under study).

16.3   Classification and regionalization in the course of geocryological survey

Classifications in geocryology systematize research data and reflect cause and effect linkage (in some cases, time linkage) between phenomena and their characteristics, properties and processes.

In the late 1960s, V.A. Kudryavtsev and others had systematized and developed the main features of the classification of natural factors and permafrost for the purposes of geocryological survey and mapping on the basis of the theory of permafrost formation and development. The features were divided into three groups. The natural factors responsible for the conditions of formation of seasonally and perennially frozen ground constituted the first group. These are geological and geomorphological structures of the area, lithologic-genetic peculiarities of the freezing deposits, and hydrogeological, geobotanic and climatic conditions.

The second group of features characterizes heat exchange at the ground surface and inside the ground unit whereby the permafrost is classified with respect to the latitude and climatic continentality; to the mean annual temperatures of ground and duration of temperature fluctuations at the surface with which the developed thickness of permafrost is associated; and to the value of ice content and cryogenic structure.

The features by which permafrost is subdivided with respect to its specific peculiarities constitute the third group. These are the distribution, structure and particular geocryological characteristics. For example, the permafrost is subdivided with respect to its distribution into continuous permafrost, discontinuous permafrost, permafrost islands and sporadic permafrost; with respect to mode of occurrence in profiles, merging into the seasonally thawed layer, or occurring at great depth from the surface and being separated from the seasonally frozen layer as relict frozen strata etc.; with respect to dynamics into degrading across the whole thickness, or with aggradation from the top and degradation from the bottom; aggradation across the whole thickness, etc.

Classification features of all three groups are mutually related: the features of the first group predetermine the features of the second group, while taken together, they predetermine the features of the third group. The features of the first group and partly of the second group are used for landscape regionalization of the area in the course of geocryological research; the features of the second group and of the third (partly of the first) are used to compile geocryological maps. When compiling maps for the particular regions researchers select from the general system of features those which are typical, are responsible for their peculiarities and are adequate for the aims of the investigations. With respect to these features the regional permafrost classifications are drawn up and in accordance with them legends for the maps are worked out. A good example of such a classification is the regional classification of frozen Quaternary deposits of the north of Western Siberia (Table 16.1).

Of great scientific and practical importance are classifications of frozen ground with respect to cryolithogenetic features as a guide to the ground composition, structure and properties. In this regard the general tabulation for subdivision of cryolithogenesis put forward by the author has considerable promise. In this tabulation the physical-geographical conditions of the permafrost formation (radiation-thermal and water balances), tectonic, geologic-geomorphological conditions, character of seasonal or perennial

Table 16.1. *Detail of the engineering-geocryological classification of Quaternary permafrost: the example of Western Siberian region*

Type (with respect to occurrence in geomorphological stages, topographic elements)	Subtype (with respect to genesis of deposits)	Class (with respect to material composition)	Subclasses (with respect to cryogenic structure of syngenetic and epigenetic frozen soils)	Group (with respect to mean annual temperature of ground, °C)	Subgroup (with respect to frozen thickness, m)	Kind (with respect to mode of occurrence in section)	Variation (with respect to special features of distribution in plan)
Permafrost composing Salekhard interstream plain	Glacial marine	Fine grained (sandy-silty and silty clay)	Epigenetic with network and layered network structure ($W_{tot} > W_{com}$)	High temperature (from −1 to −2)	Medium (50–100)	Homogeneous	Discontinuous
				Low temperature (from −3 to −5)	Great (>100)	the same	Continuous
	Marine	Fine-grained (silty-sand, clay, sandy-silt)	Syngenetic with layered structure ($W_{tot} \gg W_{com}$)	the same	the same	the same	the same
Permafrost composing Kargino terrace	Alluvial	Course grained (sands)	Epigenetic with massive structure ($W_{tot} > W_{cr}$)	Transitional (from 0 to −1)	Small (20–50)	Not merging	Island permafrost
	Marine	Fine grained (sandy-silty, silty-sandy, clay)	The same with network structure ($W_r < W_{tot} < W_{com}$)	High temperature (from −1 to −2)	Medium (50–100)	Homogeneous	Discontinuous and island permafrost

freezing of deposits and forms of lithogenetic process are taken into account. This schematic allows classification of frozen ground as a function of particular lithogenetic processes and gives a lead to the systematization of cryogenic rocks not only of the Earth but also within the Solar system as a whole (10).

Classifications of frozen rocks put forward by I.Ya. Baranov, Ye.M. Katasonov, B. I. Vtyurin and Sh. Sh. Gasanov are united by a common genetic approach to subdivision of the frozen layers with respect to the type of their freezing (epigenetic or syngenetic). Classifications put forward by V. A. Zubakov and I.D. Danilov use, respectively, formations and facies as approaches to the subdivision of the frozen materials. A.I. Popov in his classification considers the frozen materials to be a result of cryolithogenesis. He subdivided them with respect to the content of ice as a mineral. There exist now a great number of classifications of frozen materials in which they are subdivided with regard to particular gradations of one or another characteristic (for example, mean annual temperature, thickness and ice content), as applied to the problems under investigation. The indicators can be genetic, spatial, temporal or quantitative.

Aligning of the classification criteria in a ranking scale is carried out every time on the basis of establishment of cause and effect linkage: to identify the first (lower) ranks we should select the features amost commonly associated with features of the next higher subdivision. These rules of multistep classifications (by the greater number of features) should be abided by from step to step. Each step can be a simple one-row classification within which the recognized subdivisions of the subject are equal in rank. Such particular classifications form the basis for the integrated Geocryological Map of the former USSR at the 1 : 2 500 000 scale edited by the author.

When using the relatively independent groups of features (for example, regional and zonal) cross matrix classifications are often constructed.

Regionalization in the course of geocryological investigations is closely related to the construction and usage of the classifications. At the present time three types of regionalization, typological, individual and individual-typological, are widely used. *Typological* regionalization refers to ground types or landscape types similar in one or a few features and selected for the purpose of regionalization, which are set aside and separated from land without such features (or particular gradations of indicators). In the course of *individual* regionalization the adjacent units and regions being relatively similar as far as some fundamental feature is concerned, are unified and separated from the parts without this feature. In the course of *individual-typological* regionalization the units of territory are first isolated individually with respect to one important feature and then are subdivided into types of ground each having common features.

Terrain regionalization with respect to the end use is multifunctional in geocryology. Regionalization is performed in the first place to study and analyze the variability of natural complexes (landscape regionalization) in which the perennially frozen ground exists; secondly to follow spatial variability of geocryological characteristics or to reveal common geocryological features (geocryological regionalization proper); and thirdly to assess the area for the solution of practical problems. And if the selection of regionalization features for the first purpose is conditioned by structural-geological and geographical features and by geocryological characteristics, the selection of features for the assessment regionalization is defined by the specific engineering use of the geocryological environment (by types of economic development) and by the particular engineering-geocryological conditions. In all cases certain requirements and limitations are imposed on the selection of regionalization features and their gradations by the scale of map.

16.4   Regionalization in geocryological mapping

Geocryological maps are compiled to study geocryological conditions in two cases: at the stage of completion of a geocryological survey at any scale, and when summarizing the geocryological and other materials at the general scales (without conducting the geocryological surveys).

The procedure of compilation of geocryological maps at surveying scales is based on the integrated study of two groups of cryoforming factors and conditions: 1) zonal landscape-climatic and 2) regional structural-geological, hydrogeological and orographic. In accordance with the surveying procedure the maps are compiled first of all for well-studied 'key areas' and then on the basis of landscape regionalization for less-studied areas situated between them. The maps are the main result of regional integrated surveys or thematic study of geocryological conditions in the course of surveys within the permafrost as well as within the area of seasonal ground freezing.

Geocryological maps at surveying scales compiled as a result of field investigations are always included – as along with permafrost characteristics the geocryological structure of the territory is one of the main components. The geocryological characteristics revealed and studied within the key areas compiled with the help of various techniques of investigation are distributed in the course of compilation of geocryological maps on the basis of landscape microregionalization which takes into account topographic elements and relationships between them, rock composition, moisture content and cryogenic structure, ground microtopography, swampiness, vegetation, insolation conditions and microclimate. As this takes place, the analyses of natural conditions of the areas and distribution of geocryological characteristics within them should be made using aerial and satellite photographs.

With respect to interrelations between the factors of the natural environment and the geocryological characteristics established in the course of the survey, a set of geocryological maps at the surveying scale (or larger than surveying scale) is compiled on the basis of geocryological classifications of seasonally and perennially frozen ground. For example, on the map of geocryological conditions a single geocryological characteristic (for instance, mean annual ground temperatures) or a few (for instance, mean annual ground temperatures, permafrost thickness and taliks) can be presented by appropriate graphics reflecting their spatial variability.

The main method for presenting the geocryological conditions on maps is separate mapping of each geocryological characteristic and each permafrost-forming component of the natural environment over all the topographic elements within the map. Such an approach allows one to get information on one or another parameter of the geocryological situation with the help of the map and also to reveal the spatial variation and association with natural environmental factors and to make a preliminary engineering-geocryological forecast, on the basis of these interrelations, for the regions open for economic development.

Geocryological maps are accompanied by geocryological sections on which, in addition to geological structure and rock composition the permafrost thickness and distribution, temperature field below the level of zero annual amplitude (in the form of lines of equal temperatures), moisture and ice content in a profile in depth with indication of the type of cryogenesis (syn- and epigenetic), etc. are shown. The content of a map should be in agreement with its name and function. The main mapping means – colour – showing clearly the spatial variation of the major (for the purpose of investigation) geocryological characteristic in the region is used for the main content of the map. Characteristics secondary in importance (for the given map) are mapped with the help of hatching and symbols.

Field maps are mainly compiled as particular work maps at the surveying or larger than surveying scale, i.e. reflect only one characteristic of the natural environment (for example, topographic elements or Quaternary deposits) or one geocryological characteristic (for example, distribution of thawed and frozen ground). Maps of observational material, Quaternary deposits, vegetation, and swampiness, topographic elements, cryogenic phenomena and technological formations, etc. are such maps. The maps completing the geocryological survey or constituting those for a report are compiled during the office work period. The compilation of the geocryological maps at all surveying scales is fundamentally unified but has its special features.

The large-scale maps

The large-scale geocryological maps are compiled mainly on the basis of data obtained within each area under study and are designed for the solution of concrete engineering-geological problems. Therefore the large-scale geocryological map is always engineering-geocryological while its main content is associated with geocryolithogenetic peculiarities of area deposits (structure, composition, cryogenic structures and ice content of frozen ground, their properties, stability of temperature regime to change of natural conditions in the course of development, etc.).

Separation of the areas with various cryolithologic peculiarities is associated with the detail of the landscape microregionalization when each topographic element is divided into a few landscape areas (microregions) with respect to natural characteristics.

For every portion of the map the following characteristics are shown throughout its area: 1) mean annual ground temperatures with gradations of 0.5 and 1°C; 2) depths of seasonal thawing and freezing (mean, with gradations of 0.5 m and extreme values); 3) thicknesses of permafrost and taliks (within the southern geocryological zone – with gradations of 10–25 m; within the northern one – of 50–100 m); 4) cryogenic phenomena in their specific areas of development. Such mapping is conducted against the geological background – shown on the map at the level of suites, subsuites and facies with particular outlining of their boundaries and reflection of their physico-mechanical and thermal-physical properties in the legend. In addition, such maps must reflect 1) areas of thawed and frozen ground within their natural boundaries which, on the maps of smaller scale, are shown by symbols for permafrost islands and discontinuous permafrost distribution; 2) the depths of seasonal ground thawing for warm years and of seasonal freezing for cold ones along with the mean perennial depths; 3) intensity and stage of development (initial, mature, dying out) of cryogenic phenomena within the areas of their distribution; 4) areas of periodic separation of permafrost from the layer of seasonal freezing. For the purpose of engineering-geological assessment it is necessary to show the value of ground heaving in the course of freezing for every cryolithogenetic complex under the natural conditions as well as its change in the course of the planned economic development. Every area must be assessed also with respect to possible development of the thermokarst process when icy ground layers (or pure ice) occur near the surface; and with respect to development of thermal erosion etc.

According to the complexity of the natural terrain conditions, the information necessary for solving the engineering-geological problems can be presented on one integrated map or on a few maps. Usually a set of maps includes cryolithological and geocryological maps, a map of seasonal ground freezing and thawing types, maps assessing development of cryogenic processes, etc.

The importance of engineering-geocryological maps becomes greater during the period of structural design as well as during the period of post-construction investigation, provided the mapping was carried out with small intervals of gradation of the geocryological characteristics. The engineering-geocryological large-scale map must be supplemented with forecast maps for the construction period and with advice how to control the geocryological situation and how to improve it.

The medium-scale maps

As well as the large-scale maps, medium-scale geocryological maps are compiled for the purpose of the particular type of economic development but can include larger areas under study, especially for the purposes of hydro-technical constructions.

Maps at such scales are optimal for reflection of the geocryological conditions within the boundaries of all the topographic elements taking into account their landscape specific features and cryogenic structure (Fig. 16.1) and do not require additional partitioning of geomorphological elements with respect to landscape-lithological microconditions. The special features of the medium-scale map are : 1) agreement between boundaries of geomorphological elements and their typical geocryological characteristics; 2) the necessity to reflect quantitative parameters on the map and in sections not only for engineering-geocryological conditions but also for hydrogeological ones; 3) reflection of deeper investigation of the interaction between the permafrost and the groundwater.

The slightly less detail of mapping on the medium-scale maps compared with the large-scale ones can be illustrated by the following example. On the large-scale maps within the swampy ground of flood-plains, terraces, watersheds and other elements the chains of frost mounds, individual frozen peatlands and thawed intermound depressions are shown. On the medium-scale maps such an area (and other ones with close alternation of small patches of frozen and thawed ground) can be shown only by the symbol for

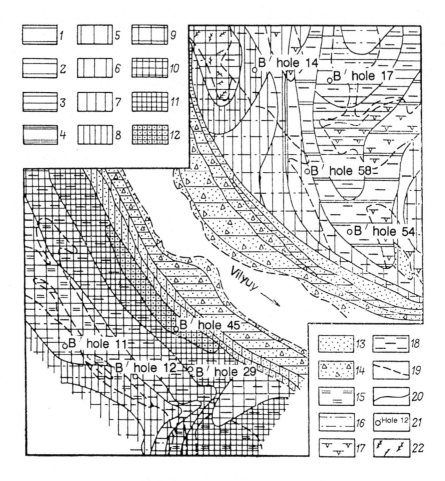

Fig. 16.1. The medium scale geotemperature map pattern: *Mean annual ground temperatures*, °C: 1 – from 0 to −0.5°C; 2 – from −0.5 to −1°C; 3 – from −1 to −1.5°C; 4 – from −1.5 to −2°C; 5 – from −2 to −2.5°C; 6 – from −2.5 to −3.7°C; 7 – from −3 to −3.5°C; 8 – from −3.5 to −4°C; 9 – from −4 to −5°C; 10 – from −5 to −6°C; 11 – from −6 to −7°C; 12 – below −7°C. *Materials*: 13 – medium-grained sands; 14 – boulder-pebble deposits; 15 – sandy-silty-clay with rock waste of diabase and sand; 16 – sandy-silty with rock debris of diabase; 17 – 18 – silty-clay-rich material (17 – with rock debris of tuffas and diabase; 18 – with rock debris of diabase); 19–20 – boundaries (19 – of lithologic variations; 20 – of areas with various temperatures); 21 – key bore holes; 22 – bedrock precipices.

permafrost islands, with a description of the proportions of thawed and frozen ground in the legend.

The set of medium-scale maps completing the geocryological survey usually include: a geocryological map, a map of ground freezing and thawing types, engineering-geocryological and hydrogeocryological maps compiled for natural conditions (for the period of investigations) and a forecast geocryological map with regard to the general plan of construction and the means for protecting the natural environment.

The small-scale maps

These geocryological maps are always integrated as they are compiled at the pre-design stages of investigations and must provide answers to problems on the many aspects of development of an area and on environmental protection. Their problem is to reflect the complex of geocryological, hydrogeological and engineering-geocryological conditions over large areas, often undeveloped but having potential for exploration and exploitation of mineral resources, for linear construction, etc.

Field maps are compiled in the course of small-scale surveys first of all for the key areas at a scale 2–3 times larger than that of the survey. The purpose of compiling working maps for key areas is to make clear the main geocryological conditions in each microregion (particular geocryological characteristics) and their variation over the area from one microregion to another (the general geocryological characteristics). Preparation of the integrated geocryological maps for the whole region is based on the maps of the areas and on the analyses of relationships between natural environmental factors and geocryological characteristics.

As the regions under small-scale mapping are large and often complex as far as their geocryological structure is concerned, it is profitable to compile the integrated maps separately for each characteristic taking into account every particular prepared map in succession. The final integration of all the geocryological characteristics and permafrost-forming factors of the natural environment on one or a few maps reflects regional geocryological characteristics.

The small-scale integrated surveys as a whole are completed with the preparation of a set of integrated maps such as a geocryological map (map of permafrost distribution and mean annual ground temperatures, of permafrost thickness and structure, of cryogenic phenomena and taliks); a map of seasonal ground freezing and thawing types; an engineering-geocryological map; and a hydrogeocryological map. The set can include a map of Quaternary deposits, a cryolithological map and geocryological maps for assessment (by stability and variation of cryogenic phenomena and cryogenic conditions as a whole), etc. Such a set of maps facilitates the preliminary assessment of competing areas to be developed, to select the most favourable one as far as engineering-geocryological or hydrogeocryological conditions are concerned. Overall the wide integration and great depth of investigations in the course of the small-scale geocryological mapping are based also on the geocryological development of the region in the Late Cenozoic and during each of its stages.

Forecast and assessment maps

Compilation of such maps at the large and medium scales is dictated mainly by the need to solve various practical problems.

The aim of the compilation of forecast geocryological maps is mapping of such geocryological characteristics which can develop in the course of the proposed development, or through the natural dynamics of climate. Their basis is the geocryological forecasting through all the key parameters of the geocryological environment and the existing map of currently researched geocryological conditions. Lines of demarcation on a forecast geocryological map differ essentially from the map of the existing situation only within the areas of intensive change of natural conditions.

The aim of the assessment maps is the reflection of engineering-geocryological information obtained in the course of survey in a form suitable for use when making projections and when solving other practical problems. Selection of the features for the assessment depends on the specific engineering use of the natural environment, the geocryological forecast and engineering-geocryological characteristics.

The forecast and assessment maps and the map of existing geocryological conditions are the basis for compilation of a map showing tolerance of natural conditions to technological changes or a map for environmental protection. This map includes the subdivision of terrain with respect to degree of inertia (or degree of reaction) to geocryological change following various disruptions, and recommendation on trends in the designated geocryological conditions.

General mapping and regionalization

Up to now the geocryological maps covering the whole territory of the former USSR have been compiled at the general scales (1 : 40 000 000 – 1 : 5 000 000) mainly because of rather limited data.

The first maps of the permafrost regions (beginning from the end of the eighteenth century up to the 1940s were very simple as far as their content and illustration were concerned and included only one or two characteristics represented schematically: at first only the conventional line of the southern limit of permafrost, then conventional lines of mean annual ground temperature with values of -1, -3, -5 and $-10\,°C$. In spite of their simplicity, the content of those maps was rather progressive for that time (maps of A.F. Middendorf, G. Wild, A.V. L'vov, L.A. Yachevskiy, V.B. Shostakovich, M.I. Sumgin and V.F. Tumel') as they gave the idea of zonal permafrost variation. In the 1950s as observational material was accumulated and the science of permafrost developed for separate regions and then for the whole permafrost zone of the former USSR, there appear the permafrost maps by V.A. Kudryavtsev, I.Ya. Baranov, I.A. Nekrasov, A.I. Popov, A.I. Kalabin, P.I. Melnikov, and others at the 1:40 000 000 – 1:20 000 000 scales. In the 1960s I.Ya. Baranov compiled the first geocryological map of the former USSR at the 1:10 000 000 scale, which generalized in somewhat schematic form the data on seasonally and perennially frozen ground, the mean annual temperatures and on physical-geographical phenomena accumulated by that time. Then Baranov published in 1977 the new geocryological map of the USSR at the 1:5 000 000 scale, which was substantially more detailed with respect to the concept of types of cryogenesis and the main characteristics of seasonally and perennially frozen ground. In those years geocryological maps were published, at the general scale, varying in content and in detail, for regions showing promise for economic development. Thus a great stride forward in geocryological mapping is a set of three maps at the 1:1 500 000 scale, compiled for the region of permafrost within the Western Siberian Plate published in 1976 (edited by Ye.M. Sergeyev) and a set of larger scale maps published in 1986 (edited by V.T. Trofimov).

A great number of maps at small and general scales was compiled for particular regions of the permafrost under intensive development: for Southern Yakutiya, Western Siberia, southern part of Central and Eastern Siberia and Zabaykal'ye.

In 1986 the Geocryological Map of the former USSR at the 1:4 000 000 scale, edited by A.I. Popov, was published. The main content of this map was the distribution of syn- and epigenetic types of frozen strata, the cryolithological characteristics of which were presented in columns typical of the permafrost sections. In the early 1970s the Department of Geocryology (Geological Faculty, Moscow State Unuversity) began the integrated small-, medium- and large-scale geocryological surveys within the large and various lands of the permafrost regions and this was completed with the 1986 compilation of the new geocryological maps for the whole territory of the former USSR at the 1:2 500 000 scale, including the Geotemperature Map of the USSR, the map of the permafrost thickness, and the Geocryological Map of the USSR. Geocryological characteristics on all the maps cited above have been correlated (as far as the scale permitted) with the homogeneous natural conditions within each of the areas considered.

The Geocryological Map of the former USSR at the 1 : 2 500 000 scale

This integrated map is compiled on a geologic-genetic and formational basis and reflects the relationships between the perennially frozen strata and the geological structure and the composition of bedrock, the landscape-climatic conditions and the topography, as well as the neotectonic, hydrogeological and deep geothermal conditions, by means of separate mapping of the main geocryological characteristics within the topographic elements and groups of elements.

Such a method allows representation of the regular change of geocryological conditions from south northward (in connection with the latitudinal zonation of heat exchange on the surface), from west eastward (in connection with variations of sectorial structure and continentality of climate), from low topographic elements to high altitude ones and from slightly sloping topographic elements to steep ones (in connection with altitudinal zonation of heat exchange and with topography), from stable tectonic structures to mobile ones (in connection with geothermal flux increase), and from young deposits to more ancient ones, etc.

Analysis and correlation of geocryological characteristics within each area with the permafrost-forming factors of the natural environment, were carried out on the basis of the method of key landscape maps. Maps of large-, medium- and small-scale geocryological surveys and thematic investigations were used as key ones.

The main content of the map includes the distribution of frozen and unfrozen ground, spatial variations of their mean annual temperatures (see Fig. 15.2), thickness and structure of the permafrost, the relationship between frozen, cooled and relict layers over the territory and in section (see Fig. 15.1), permafrost-geological phenomena and taliks as well as geological structure of the territory reflecting the real composition of permafrost, its cryogenic structure and ice content which are shown on the map through geologic-genetic complexes of Quaternary deposits and geological formations of pre-Quaternary rocks. The type of freezing, macro-ice content and cryogenic structures are also shown.

Comprehensive representation of the geocryological characteristics, especially within the southern zone, facilitates the map's use as the basis for regional geocryological forecasting after taking into account the general construction procedures (deforestation, removal of moss-turfed soils, partial removal of snow, etc.).

The content of each geocryological map at the 1 : 2 500 000 scale is supplemented and generalized by general maps at the 1 : 25 000 000 scale: 1) of the distribution of frozen and unfrozen ground; 2) of the mean annual ground temperatures; 3) of permafrost thickness; 4) of cryogenic age of the ground and type of cryogenesis; 5) of hydrogeocryology; and 6) of the engineering-geocryological regional characteristics of the former USSR territory.

The Geocryological Map of the former USSR at the 1 : 2 500 000 scale can be used together with much specialized data extending the characteristics and assessment of geocryological conditions and their monitoring as the new natural data arrive. For this purpose we can use a computer, the memory of which can store and analyze all the information about each element of relief recognized on the map in digitized form.

V

Rational use of frozen ground and environmental protection in the course of economic development of the permafrost regions

17

The effect of different types of development on the natural geocryological environment

17.1 The basic principles of rational use of frozen ground in the course of the economic development of the permafrost regions

The intense pace of economic development of the territories within the permafrost region continues to increase steadily. Under the effect of different types of development all or some of the components of the natural environment, including the geocryological conditions, can change, resulting in transformations of the natural complex as a whole. 'The geological environment' is defined as an essential constituent of the natural environment. We address it in several aspects as far as various types of economic activity are concerned: 1) as the engineering-geological environment in which different structures are developed and operate; 2) as a source of mineral resources; 3) as the geological environment, the most important component of the natural complex as an animal and human habitat.

Engineering geocryology as a branch of geocryology studies the freezing ground of the Earth's crust as an environment for human life and activity. Among the main problems of engineering geocryology is the engineering-geological background for design, construction and operation of different engineering structures and undertakings within the permafrost region. The aim is to provide and select the most reliable and economic means of development of an area.

One of the main features of design, construction and operation of engineering structures within the permafrost regions is the necessity to take into account and to regulate heat exchange between the ground, the construction and the environment. Change of ground thermal and moisture conditions in the course of economic development especially in connection with the temperature going through 0 °C, causes changes of ground composition, of structural properties as well as of strength, bearing capacity and compressibility, of heaving and shrinkage stresses and deformations in freezing and thawing ground, of workability within the permafrost zone as far as excavation work and mining are concerned, of intensity of thermal erosion, icing, thermokarst, solifluction and other cryogenic processes and phenomena which can turn some terrains into badlands.

Observational data show that the mean annual increase of length of ravines developing in frozen ground reaches tens and even hundreds of metres and that hundreds of cubic metres of thawed soil are removed from one running metre of the northern sea coasts (the height of the shore cliff is about 10 m). Solifluction processes can cause creep of deposits on slopes over a distance of some tens of metres (when the viscous-flow strains in thawing soils take place). Settlements due to warming under constructions can reach tens of centimetres and more. Thermokarst processes lead to ground subsidence and to paludification of large areas. Change of the depth of seasonal freezing and thawing and consequent change of groundwater regime often cause activation of icing processes. Thus, for example, formation of 60 icings with a total area of 107 km^2 was observed on one section of a road in Central Yakutiya during one winter season. And finally, activation of frost heaving processes appears not only in the uplift of the ground surface but also in the increase of differential heaving on the surface, as a result of technological change of ground temperature regime.

Thus there exist specific conditions for construction or for any other economic activity within the permafrost regions. Therefore attempts to apply the standard methods and techniques for construction usable outside the permafrost zone to the regions where frozen ground is widespread often lead to inadequate and sometimes even to catastrophic consequences and almost always to unnecessary labour, material and input of time. Thus P.D. Bondarev, A.I. Dementyev and other researchers inspecting buildings constructed within the permafrost region on frozen ground in the city of Vorkuta and in Vorkuta District found that about 80% of the buildings had unallowable deformations (Fig. 17.1). About 30% of all the stone buildings inspected had catastrophic deformations and needed overhaul. Among 1230 buildings inspected in Yakutsk, Chita, Vorkuta and the Buryat Republic around 63% turned out to have deformations (Fig. 17.1). About 30% of all the dwelling-houses on the Arctic coast are deformed. Losses through the overhaul and reconstruction of damaged buildings are as much as 10% of their total cost. The main reason for deformations are permafrost thaw settlement or ground heaving during freezing.

According to TsNIIS* data the extent of railway sections disturbed by heaving processes is about one third of the total extent of the deformed

*Tsentral'nyy Nauchno-Issledovatel'skiy Institut Transportnogo Stroitel'stva [Central Research Institute for Transport Construction]

Fig. 17.1. Building deformations as a consequence of nonuniform thaw settlement of perennially frozen ground under the foundation, Vorkuta (photo by Ye.M. Chuvilin).

sections. About 90% of all the resources allotted for repairing the railway bed are used for control of heaving processes. Highway and railway construction cause considerable icing formation. Thus the quantity of icings on some sections of the BAM railway has increased 50-70%. The cost of icing-prevention structures within the Tynda-Urgal section alone is as much as 5 million roubles [1989].

Investigations show that an unjustified approach to the designing of a number of dams in the rivers of Magadanskaya Oblast' caused thawing in the foundations of the crests of the dams, formation of talik zones inside the dams and water permeating through them.

Analysis of reasons for a negative state of buildings and structures within the permafrost regions, show that from 15 to 30% of emergencies are associated with mistakes made in the preparation of the engineering-geological support for building sites, i.e. in the characterization of the geo-cryological conditions. A substantial percentage of deformations is associated with mistakes in design and with violation of operating conditions. Only the correct expert consideration of geocryological conditions of the territory and the justified selection of the design will make it possible to prevent unforeseen deformations of buildings and structures and to provide for their reliable operation. There are numerous examples of successful development of buildings and structures on permafrost. These buildings are supplied with central heating, hot water and have all modern conveniences. An example is Norilsk at 69° N in an area of extensive permafrost, and Mirnyy, a centre of the diamond mining industry, Vorkuta and many other cities are also good examples (Fig. 17.2).

Four types of economic development are usually recognized in engineering geocryology, characterized by specific effects on permafrost and the geocryological surroundings:

1) regional development of large areas within the permafrost zones connected with profound changes of natural conditions (construction of large water storages and hydroelectric water-power stations, destruction of extensive forests and forestation, drainage of swamps, etc.);

2) economic development of the permafrost regions through different types of construction (civil, industrial, highway, hydro-technical constructions, etc.);

3) development for the mining industry and of underground constructions in the permafrost regions;

4) agrobiological types of development (development and amelioration for the purposes of agriculture).

It is obvious that effective development is impossible without considering technological changes of the geocryological conditions, without special measures intended to control geocryological processes so as to prevent, to eliminate or to limit any dangerous consequences. In this connection it is necessary to carry out a complex of scientific work including: a) study of the existing geocryological situation of the area to be developed; b) study of the possible technological impact on the geocryological environment; c) forecasting of changes in the geocryological conditions associated with this impact; d) elaboration of measures aimed at environmental protection. All this scientific work listed above should be carried out before the period of active development of the region and the beginning of capital construction.

The main aim of the study of the existing geocryological situation is to

Fig. 17.2. An example of building construction keeping the ground in the perennially frozen state during the period of operation, Vorkuta (photo by Ye.M. Chuvilin).

establish the main features of the distribution, formation and dynamics of seasonally frozen ground and permafrost and of the geocryological processes, to compile geocryological maps and to model natural geosystems for the purposes of geocryological forecasts. At the same time, in A.V. Kudryavtsev's opinion, it is not sufficient only to note the existing geocryological situation when studying geocryological conditions but it is necessary also to find out the nature of its formation and development, to establish and to assess the role of the particular natural factors in formation of the temperature regime and other geocryological characteristics of the upper ground layers. Then, knowing the character of the natural complex as a whole during the proposed development of the area we can predict the possible change in geocryological conditions. The geocryological survey procedure is the best way to carry out this kind of scientific research (see Chapter 16).

Technological impact causes various changes in geocryological conditions such as increase or decrease of mean annual ground temperature, and of seasonal or perennial ground thawing or freezing. And it is necessary to work out the systematization (typifying) of technological impacts to establish the direction and extent of the effect of economic development on the

permafrost conditions. Most of the proposed schemes for classifying technological impact are based on the assessment and separation of deliberate
impacts from the spontaneous ones. Thus, for example, spontaneous climatic changes which can take place over large areas as a result of a number of
ecological disturbances have assumed great importance for the harmonious
exploitation and protection of the environment in the permafrost regions.
The applied impacts are usually divided into mechanical, physical, chemical,
biological and others according to their nature, with the first being specified
loads, the others being specified impacts.

The consequences of technological loads and impacts depend on the
duration and the size of the area where they occur. According to duration
they can be *continuous*, as determined by the length of time the newly
created technological landscape has existed in the designated state and
regime or, for example, by the time of the operation of the construction;
temporal – taking place for a number of years (for example, during the period
of survey, preparation for construction and construction itself) and *pulsed*
with a duration of not more than one season (for instance, a single modification of snow thickness and density, consolidation and deformation of moss-
shrub cover, release of water on the surface, etc.). As a rule pulsed impacts do
not cause changes of the permafrost as a whole but lead usually to a change
of seasonal thawing depth. Temporal and especially continuous impacts can
cause a change of ground heat state at greater depths even leading to the
complete thawing of the permafrost.

The thickness of the technologically changed permafrost depends strictly
on the size of the area affected. Now it is generally agreed that there exist
point, linear and areal disturbances of natural landscapes during the development of an area.

A number of processes caused by technological loads and impacts are
planned and regulated by engineering design. These processes include, for
instance, processes of formation of 'thaw basins' under buildings and structures, with heat release and thaw settlements being permissible within
certain limits by design. These processes are contrasted with newly appearing, unplanned weakly regulated ones. Among these processes are, for
example, thawing of permafrost within building sites caused by blackening
of the surface in the regions of coal mining. Attendant disturbances often
represent secondary change following the initial disturbances (being an
immediate result of engineering undertakings) as cause and effect. It has
been suggested reversible, irreversible and destructive disturbances be dis-
tinguised according to the character of the response of the geological
environment.

The consequences of technological impact on the natural environment can be shown in a change of landscape-climatic conditions, geodynamic state of the topographic elements, geocryological surroundings, geophysical characteristics, engineering-geological, hydrogeological and other conditions. Depending on the type of economic activity the contribution of every above-enumerated impact will change.

Thus, for example, the consequences of hydrotechnical construction can show themselves in the change of geodynamic, hydrogeological and geothermic conditions and topography. Specifically, this is a redistribution of large water volumes on the Earth's surface, an elevation of groundwater level with paludification, water sedimentation, ground subsidence, abrasion and thermal abrasion and the formation of new talik zones in water storages and the beds of large water courses, together with the development of karst and thermokarst processes, slides, etc.

The consequences of mineral exploitation can take the form of changes of topography, and of the geodynamic, cryological, hydrogeological and engineering-geological features of an area. All mining causes depletion of mineral resources and a change of the geostatic field of the Earth as a result of the development of working cavities. Besides, underground mining is often connected with ground subsidence, followed in most cases by activation of thermokarst processes with formation of a new hydrological situation and depletion of groundwater resources, with topography transformations and rock waste heaps. Opencast mining is associated with topographic transformations, deep depressions and radical change of the hydrodynamic system, with weathering and breakup of outcropping rocks, slope slides, heaving of pit bottoms and slopes and suffusion. Placer mining is associated with changes of river valley topography, river bed alluvium and mud deposition, and decrease of discharge below the bed and, as a result, freezing of taliks below the river bed.

Ground-based construction (civil and industrial construction, linear construction) is mainly connected with changes of the geothermal, geochemical, hydrogeological, engineering-geological and geophysical environment. And under the influence of engineering construction the temperature as well as the ratio between volumes of frozen and unfrozen ground change, exogenetic geocryological processes (such as thermokarst, icing formation, heaving, solifluction, etc.) become more active, ground water regime changes, chemical water pollution and ground salinization take place and induced physical fields appear.

The character and intensity of technological impacts on permafrost associated with construction of foundations can be controlled. Undesirable consequences can be avoided if techniques for construction and the construction sites are selected adequately and if proper protective measures are applied.

Thus every type of construction or economic development of the terrain causes its own disturbances and technological loads, defining the strategy for engineering-geocryological research and for the particular forecasts. To make up a classification scheme for technological impacts on the environment it is necessary to study experience gained in the course of the development of other areas and to carry out special observations of regime using ground stations and repeated aerial photography.

The quantitative and qualitative characteristics of the change of permafrost conditions under the effect of different types of impact within different areas can be given or geocryological forecasts can be made after performing the geocryological investigations and knowing the character of technological impacts.

The principles and methods for rational use of the geocryological environment of the permafrost areas are selected and developed on the basis of geocryological investigations and geocryological forecast data. The following are the general principles:

> – 'free use' without any limitations by geocryological surroundings;
> – conservation of the natural heat state of the ground (frozen or unfrozen) using measures for limiting the change of ground temperature regime and the engineering-geocryological processes;
> – allowing change of the natural state of the ground (perennial thawing, or more severe temperature regime and new formation of frozen ground) using measures for the control of the processes to avoid negative consequences.

The main ways of achieving these are associated with:

a) limitations on the area of development (a given type of development is allowed only within particular areas);
b) limitations on the construction technology;
c) control of permafrost processes to create the required engineering-geocryological situation;
d) abandonment of the development of a given area (method of prohibition).

Recommendations on the principles and methods of development of the permafrost regions, in accordance with rational siting of industrial installations and with technically and economically efficient designs for protection of the terrain, form the basis for the rational use of the permafrost geological environment.

17.2 Regional environmental change in the course of development of extensive areas within the permafrost zone

Engineering work can cause disturbance of the geological-geographical environment over large areas and regions and even on the Earth as a whole. The formulation and control of many problems of geocryological investigations associated with regional environmental disturbance is very topical in the context of harmonious exploitation and problems of environmental management in recent years. The effect on the geocryological environment of such projects as creation of water reservoirs in rivers in the permafrost zone, of urban agglomerations and the intensive development of large areas within oil and gas fields can exceed the possible changes of the geocryological conditions due to the natural changes of climate and evolution of the environment.

Thus air temperature in large northern cities (such as Norilsk, Anadyr, Magadan and others) is 2–3 °C higher in summer and 4–5 °C higher in winter compared to that in the adjacent region. This is associated with direct heat release, albedo changes connected with dust, asphalt, abundance of concrete etc. Under further enlargement of cities, for example, up to the sizes of some Japanese and American centres, their heat loads are correlated with disturbances at the atmospheric scale according to some data. The mean global temperature will not change but the redistribution of heat energy over large areas due to the disturbance of global atmospheric circulation can lead to a change in the geocryological environment.

Atmospheric pollution over the country adjacent to large cities has a great effect on the vegetation cover, which is one of the leading temperature-forming factors. Thus there are data showing that atmospheric emissions of sulphur, nitrogen, sulphur dioxide, nickel and other metals typical, for example, at Norilsk cause formation of concentric belts with different degrees of disturbance of the ecosystem. The zone of complete destruction or significant disturbance of moss-lichen cover and degradation of the tree-shrub community can be 4–10 km in width and up to 1000 km^2 in area. The initial stage of vegetation cover disturbance is noted over areas of 40–60 thousand km^2. The destruction and change of species composition of vegetation are followed by a change in conditions of snow retention, infiltration and runoff, soil and ground chemical composition over large areas. These inevitably cause changes of mean annual temperature of the ground, of the soil properties and the depth of seasonal thawing and of the cryogenic processes.

Oil and gas production in Western Siberia is also accompanied by important changes of the natural environment. Lowering of reservoir pressure causes ground consolidation and progressive settlement of the ground surface. In one of the fields of the Shainskaya group this settlement has reached 56 cm, while under Western Siberian conditions, ground surface settlement of even 0.5 m causes significant paludification. Consequently it may be suggested that if during the next 10–12 years the reservoir pressure continues to fall the subsidence will reach 1.5 m and the land will be swamped completely. To the south of Sibirskiye Yvaly where the permafrost is associated with peat lands situated 0.7–1 m above the surface of the swamps, the settlement of the ground will lead to complete thawing of the permafrost.

In the early 1980s a project to divert some part of the northern rivers' runoff to southern parts of the country was under discussion. Realization of that project could cause fundamental changes in the geocryological situation over large areas of Western Siberia. One of the variants of that project considered, for instance, taking a water volume of 4–5 km^3 from the intake of the lakes of Karelia and diverting it through the Mariynskaya Canal system into the Volga basin. This could cause the water level of the lakes to fall by 2.8–3 m and artificial drainage of more than 60 thousand km^2 of ground which are currently swamp. The geocryological consequences of this would be an increase in depth of seasonal freezing and formation of short-term permafrost. Furthermore calculations show that evaporation values could decrease by 0.4 km^3 year^{-1} in connection with the reduction of surface of lakes and swamps. As a result mean air temperature as well as ground temperature would rise by 1.6–1.8 °C; climatic continentality would increase. In the southern regions, on the other hand, evaporation would increase through water discharge. This could cause summer air temperature lowering by 2–2.5 °C, i.e. the climate would be less continental.

Analysis of this effect of the planned project shows that in the West-Siberian region decrease of runoff in the Ob' river will cause a change of the amount of heat carried by the Ob' river into Obskaya Guba [Ob' bay]. This situation should have an important impact on ice regime of the Ob' estuary and lead to lower summer air temperatures on the adjacent land. Besides, shallowing of the Ob' river and the beds of its tributaries and lowering of the Ob' bay water level will lead to new permafrost in shallow areas and on exposed sand bars. Changes of ice regime and permafrost formation in the Ob' river and estuary bottom will greatly complicate navigation which is of very short duration now and very intricate because of the small depths and numerous shallows.

In the southern regions of Western Siberia where the vast flood plains of the river Ob' and its tributaries are unfrozen, a decrease of the duration of the flooding period will lead to their perennial freezing and the gradual replacement of meadow vegetation, which is widespread here, by mossy swamps. In the zone situated along the Ob' river and its tributaries, the depth of seasonal thawing will increase as a result of drainage and consequently the thermoerosion processes and thermokarst processes will become more active where there is ice-rich ground and wedge ice.

The proposed diversion of some part of the Ob' river runoff to the south included the creation of reservoirs, which can fundamentally change the microclimate and cause thawing of frozen ground over large areas in the vicinity of the southern limit of the permafrost regions.

The examples cited above show that one of the main problems of regional economic planning is to forecast the change of geocryological conditions over large areas. Such forecasts should be based on a comprehensive study of possible regional change of climatic and natural-ecological conditions and is a very complex integrated problem.

17.3   Economic development of the permafrost regions with various kinds of construction

Economic development is always integrated and multifarious and includes construction and operation: 1) of civil and industrial installations (residential, social, municipal, factory buildings and structures); 2) of linear structures (railroads and highways, pipelines, underground lines, power lines; 3) of airfields; 4) of hydrotechnical structures. The special features of construction and operation of buildings within the permafrost regions will be discussed in the next chapter. This chapter is concerned with the other kinds of construction.

Road and railway building

Considerable attention has been focused recently on roads and railways under construction within the permafrost regions. These are the BAM railroad (3145 km in extent), the Surgut-Urengoy-Yamburg railroad, the railroad and motor road being designed for the Yamal peninsula, etc. *Three alternative principles* are used when designing highways (Fig. 17.3). *The first principle* is preservation of perennially frozen ground in the base of a roadbed during the whole period of operation by means of raising the permafrost table up to an embankment base. *The second principle* is based

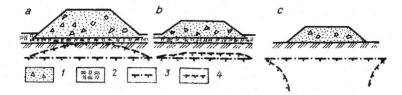

Fig. 17.3. Examples of use of principles of highway and railway design and construction within the permafrost zone: *a* – with creation of a permafrost core in the base of an embankment (the first principle); *b* – with partial permafrost thawing (the second principle); *c* – with prethawing of the permafrost (the third principle); 1 – fill; 2 – moss-vegetation layer; 3 and 4 – surface of permafrost under natural conditions and in the period of operation, respectively.

on the partial thawing of the perennially frozen ground of the base by an amount determined through calculation from the allowable deformations of a roadbed. *The third principle* assumes thawing of frozen ground prior to the beginning of construction with soil drainage under the highway itself and within the strip along the roadway.

Design on the first principle is conducted within areas of low-temperature perennially frozen ground prone to great subsidence, when the thawing of the ground can cause impermissible deformations and destruction of the pavement. The highway is designed with the embankments composed of uncemented clastic materials with retention of the moss-turf cover in the original undisturbed state at the base of the embankment and along the whole roadway. To retain the moss-turf cover it is recommended to build an embankment 'away from you'. Machines being used for construction move on the filled embankment in this case and the soil is placed directly on the moss cover. To reduce the embankment thickness, thermal insulation inter-layers of peat, compacted moss and slag, etc. are placed in the base of a roadbed. Using thermally insulating foam plastic shields with rather high strength give good results. These shields were used successfully in the course of construction of certain sections within the Urengoy-Yamburg and BAM railroads.

Design *on the first principle* can be carried out using the base soil prefreez-ing method. This method was suggested by B.I. Popov, N.F. Savko and others in the course of roadway building in Western Siberia, towards the southern limit of the permafrost. It was impossible to use water-saturated soils from swampy ground with thick peat layers as a base for motor roads. Removal of snow and vegetation cover of the alignment of a road had caused freezing of the upper layer of peat with formation of a thin permafrost interlayer within 2–3 years which was used as a base for the embankment.

Design *following the second principle* is usually carried out for an embankment composed of clay-rich and sandy soils with the moisture content below the plastic limit and showing only slight subsidence in the course of thawing. The moss-turf cover is not removed in this case, at the base of the embankment. Design *on the third principle* is carried out mainly in easily drained soils. It is used chiefly when preliminary thawing of the permafrost, drainage of roadway strip and strengthening of base soils as a result of their pre-construction settlement is possible. It is necessary in this case that the roadway strip should be cleared from forest, scrub, the moss cover should be removed within the strip completely, and the drainage ditches should be made no less than a year ahead of the beginning of work.

Design of railroads and motor roads within the permafrost zone is carried out mainly on fill. Cuts are allowed mainly within the areas with favourable geocryological conditions (hard rock debris and ice-poor gravel) and hydro-geological conditions (absence of suprapermafrost waters). If it is necessary to make a cut within the area of fine ice-rich soils the cut should be designed only with a guaranteed thermal insulation of slopes, replacement of ice-rich soils and assured drainage from the cut.

The differential heaving caused by nonuniformity of soils and of moisture and freezing conditions, presents the most severe hazard for a roadbed. The main measures to prevent and reduce the heaving are the following: replacement of soils, placement of coarse-grained layers (to raise the freezing intensity and to interrupt the paths of water migration); drainage of the areas adjacent to the roadbed; use of thermal-insulation layers (to decrease the depth of freezing); soil salinization (lowering of freezing temperature) etc.

Roadway construction is adversely affected by more intense icing processes. The main reasons are the change of natural ground runoff conditions in the course of construction of cuts, a rising of the permafrost table within embankments constructed on the first principle; inadequate arrangement of surface run-off around constructions associated with water (pipes, small bridges); and change (spreading) of water of rivers and streams in the course of bridge construction. Temporary and permanent snow and soil barriers, ties and board fences, water-diversion ditches, waterproof screens and permafrost belts the effect of which is based on interception of subsurface flows and artificial displacement of icings uphill (Fig. 17.4), serve as measures for icing control. In addition to these methods of icing control, drainage of the area and deepening of stream beds, etc. is carried out. Recently the method of prevention of icing formations with the help of groundwater extraction by pumping from wells has been introduced. When making cuts, cutting slopes

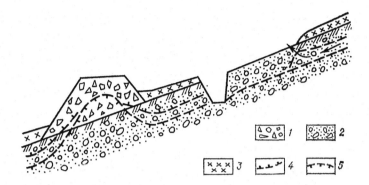

Fig. 17.4. Diagram of one of the structures for use in the permafrost belt for control of icings due to suprapermafrost water: 1–2 – soils (1 – fill; 2 – high seepage 3 – snow; 4–5 – boundaries (4 – of soil freezing from the surface; 5 – of permafrost). Arrows show the direction of flow in the soil.

or when there is inadequate arrangement of water-carrying constructions or features, an intensification of the thermal erosion processes is a possibility, with thermal erosion control being a serious problem making it necessary to do much excavation to remove gullies.

Trunk pipelines

These are now an integral part of the landscape within the permafrost regions and extend over hundreds and thousands of kilometres, passing through different geocryological zones. Pipe laying is performed by various means. At the same time the temperature regime of the product being pumped changes greatly along the pipeline route as a rule. All the above dictates specific geocryological research in the planning of the pipeline.

The construction features and temperature regime of pipelines depend on the character of the product being carried. Thus oil pipelines and water pipelines operate under positive temperature, with the minimal temperature being $+5$ to $+10\,°C$ for oil, because under lower temperature the oil becomes thick, paraffin plugs are formed and the oil becomes unsuitable for transportation. Gas pipelines can have positive as well as negative temperatures.

The pipelines are divided with respect to their position in relation to the ground surface, into those underground, those placed inside an embankment or those without piling-up of soil (exposed) and those elevated aboveground (Fig. 17.5). When passing through a water course, underwaterlaying is used. The pipelines laid under ground exert the greatest thermal

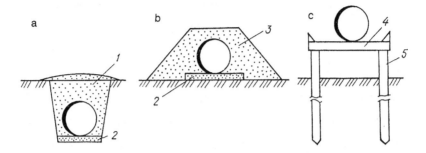

Fig. 17.5. Methods of pipe laying: *a* – underground; *b* – in an embankment;
c – above surface; 1 – earth infill; 2 – sandy pad; 3 – fill; 4 – collar beam; 5 – pile.

effect because the pipes (especially those large in diameter), widely used in trunk pipelines, are laid below the maximum depth of seasonal thawing inside the permafrost. The smallest heat effect on the permafrost is observed with above-ground pipelines.

There can be four main variants of combinations of mean annual temperature t_{mean}, maximal t_{max} and minimum t_{min} gas temperatures characterized by rather different thermal effects on the permafrost (Fig. 17.6), the particular formation of a seasonal or perennial annulus of freezing or thawing depending on the temperature regime of the pumped product (for example, of gas, having the widest range of temperature variations in the course of its transportation). All the illustrated variants of the temperature regime of a pipeline are a possibility within one route. This situation prevents the use of a single pipeline-laying method. The gas has a constant temperature of $+30$ to $+40\,°C$ when leaving the compressor station. Under the interaction of a pipeline with soil or air the temperature is progressively lowered along the pipeline route approaching the temperature of the natural environment. In this connection the thawing of perennially frozen ground around the pipeline is a possibility within the initial section of the pipeline route, while at a distance 100–150 km away from the compressor station seasonal or perennial freezing is a possibility. The Medvezhye-Nadym-Punga pipeline is characterized, for example, by such a regime.

The main undesirable cryogenic processes from laying of underground pipelines with positive temperature are formation of a thawing annulus around the pipeline, ground settlements, paludification and thermokarst development. In addition to this, thermal erosion often develops along the trenches. Under negative gas temperature when a freezing annulus is formed around the pipeline, heaving is a possibility.

Laying pipe inside embankments poses a problem of ensuring the stabil-

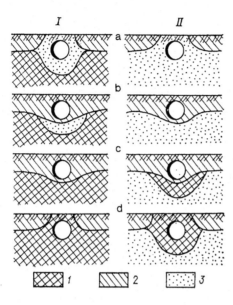

Fig. 17.6. Diagrams of heat interaction between underground pipelines and soils with respect to relations between mean annual temperature t_{mean}, minimum t_{min} and maximum t_{max} gas temperatures: $a - t_{mean} > 0\,°C$, $t_{min} > 0\,°C$; $b - t_{mean} > 0\,°C$; $t_{min} < 0\,°C$; $c - t_{mean} < 0\,°C$, $t_{max} > 0\,°C$; $d - t_{mean} < 0\,°C$, $t_{max} < 0\,°C$; $1 - 3 - $ soil (1 – perennially frozen; 2 – seasonally frozen; 3 – thawed); *I, II* – regions of permafrost and non-permafrost ground respectively.

ity of the embankment itself. Experience gained from operation of the Western Siberian pipelines has shown that native soils are not suitable in practice for construction because within two to three years after the beginning of operation the embankment is disturbed and soil is washed out exposing the pipeline. In addition the extended embankment changes the conditions of surface runoff, contributes to paludification of the right-of-way and of the adjacent terrain and to thermokarst development. When the pipe is laid above ground the main problem is ensuring the stability of the pile trestles against the effect of frost heaving processes developing in the seasonally frozen layer. If pipeline weight is insignificant it is necessary to put piles deeper in order to 'spread out' the tangential forces of pile heaving.

Water pipelines and conduits

These are rather expensive constructions within the permafrost zone. At the same time it is necessary to heat the water for the 6–8 winter months to keep it unfrozen. To select the requirements for laying of these pipes, thermal calculations are made resulting in determination of such

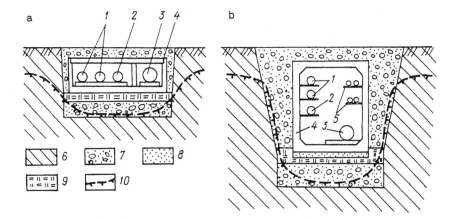

Fig. 17.7. Diagram of underground service lines laid in inaccessible (*a*) and accessible (*b*) ventilated ducts: 1 – heating system; 2 – water pipeline; 3 – sewer; 4 – reinforced concrete section; – electric cables; 6–7 – soils (6 – natural base; 7 – fill, non-frost susceptible); 8 – sandy padding; 9 – sand-gravel-clay building mixture; 10 – permafrost surface.

characteristics as temperature variations along the water pipeline, its heat losses, the zone of thawing around pipes, etc.

Construction practice has established the following means of laying service lines: a) above-ground laying; b) laying inside an embankment; c) underground laying in trenches; d) underground laying in ducts with or without ventilation, in which all the service lines can be laid together (Fig. 17.7). Whatever means is selected, thermal insulation is required. All the means of laying call for use of thermal insulation. Inside the ventilated ducts the air temperature is close to the outside air temperature in summer time. Slight thawing of frozen parts takes place in this period in the base of the duct. In winter the negative air temperature is maintained because of the natural ventilation inside the duct resulting in freezing of that soil which was thawed in summer. In such a manner the perennially frozen ground occurring below the prepared base will be in the frozen state at all times.

Power lines

Having no heat release, these change geocryological conditions to a smaller degree than the other linear structures (motor- and rail roads and pipelines). Importance is attached to calculations of stability of power line supports in power line design. The supports are placed on pile, post and bedplate foundations designed as a rule on the principle of preserving the frozen ground. Stability and strain characteristics of frozen ground determined with regard to possible change of soil temperature regime within the line right-of-way are used in the course of calculations for mechanical stability of the supports. When determining the stability characteristics of frozen soils it is necessary to take into account not only static loads (the trestle and foundation weight) but also dynamic loads caused by the effect of gusts on the supports. In addition, calculations are made of the effect of heaving forces on the foundations of the supports.

Airfield pavements

These consist of artificial pavement and an artificial base overlying the natural ground. The artificial pavement is the uppermost relatively thin and strong layer taking the main impact of loads. The artificial base is composed of a layer or of a few layers of crushed rock, gravel, and sand pretreated with chemical binding materials. It serves to redistribute the wheel stresses over a larger area and their transmission to the natural base, i.e. to the upper layers of ground levelled and compacted artificially (Fig. 17.8). Airfield pavements should correspond to the following specifications: 1) strength and durability; 2) plane surface and wear resistance; 3) roughness of surface necessary for wheels to grip the pavement well; 4) absence of dust; 5) water resistance and tolerance for climatic effects.

To select the pavement in the design of the airfield, the results of strength and stability calculations for constructions and their foundations are used. To do this requires also knowing the strength and deformation characteristics of the ground in the natural base. Usually airfield construction within the permafrost regions is carried out variously, with protection of frozen ground in the base from thawing; with thawing of frozen ground in the course of construction and operation; or with prethawing and improvements of ground properties in the base. The principles for selection of one or another method of airfield design and construction and the procedure for stability calculations for the base of the airfield pavements, are similar to those in roadway construction. The specific character of airfield pavement design compared to that of roadbeds lies in the heavy demand on their stability.

Hydrotechnical structures

The experience in construction of such structures in the permafrost regions is not very great, but includes Vilyuy, Khantayka, Mamakan, Kolyma and Zeya water-power stations, Arkagala National Regional Electric Power Station, and dams in the Irelyakh, Pevek and other rivers. Within the permafrost regions, the most widespread are plain earth and rock-fill

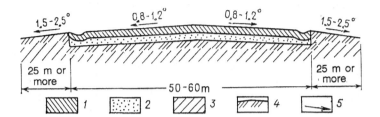

Fig. 17.8. Airfield pavement structure: 1 – pavement; 2–3 – foundations
(2 – constructed; 3 – natural); 4 – vegetation layer; 5 – direction and degree of
slope of surface.

dams, more rarely concrete dams are used. Plain earth dams represent an
embankment of trapezoidal cross-section. Their stability results from great
weight. The greater part of rock-fill dams consists of rock fill, i.e. highly
moisture-permeable material. To reduce water head special-purpose anti-
seepage devices such as shields, diaphragms and cores are used.

Design and construction of hydrotechnical structures are performed on
either of two principles: 1) protection of perennially frozen ground in the
base from thawing during the whole period of construction and operation;
2) progressive thawing of the permafrost in the course of the construction
and operation of the structure and its use as a base after thawing (Fig. 17.9).
Construction according to the first principle is performed usually in the
conditions when highly ice-rich soils with considerable thaw consolidation
are used as a base. When keeping the ground in the frozen state the frozen
curtain (the core of a dam) is arranged to guarantee the stability and to
prevent water seepage through the dam body. Until recently the freezing
systems in which cold atmospheric air was used as a heat-transfer medium
were used widely to freeze the core of a dam. Cold air circulation was forced
through coaxial columns placed in holes drilled from the crest of the dam or
through horizontal air offtakes inside the body of a dam. However, experience tells us that the cost of such systems is high and the reliability is not
sufficient because the columns and the air offtakes are gradually filled with
ice and clearing them is a tedious procedure. Freezing systems with natural
circulation of a heat-transfer agent (self-regulated cooling installation) are
used with increasing frequency today.

Dams designed for progressive thawing of permafrost are allowed when it
is impossible to keep a base in the frozen state or when it is not economically
profitable. When designing dams on the principle of thawing of the base
soils, provision should be made for smooth, gradual thaw settlement of soils,
not exceeding the allowable values. At the same time the soils in the base
must not have high water permeability after thawing.

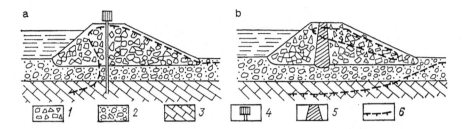

Fig. 17.9. Design of plain earth and rock-fill dams on the first (*a*) and second (*b*) principles: 1 – body of dam; 2–3 – bases (2 – natural, permeable; 3 – impermeable, hard); 4 – cooling installations; 5 – anti-seepage screen; 6 – permafrost surface.

The location of the water storage is a major factor in the design of the hydrotechnical structure. When the engineering-geocryological assessment of the site of the proposed water storage is made it is as well to bear in mind that permafrost degradation will take place in the course of its filling with water and future existence. If the ground is hard, faulted rocks, the fissures of which are cemented by ice, or carbonate rocks with karst cavities filled with ice, or rudaceous materials which are a water confining stratum when in a frozen state can, on thawing, cause water leakage from the reservoir. In addition it is necessary to predict the possibility of thermal abrasion modifying the reservoir banks and to justify the measures for protecting the area from flooding.

Ice and permafrost as building materials

Climatic conditions within most of the former USSR, especially its northern regions, are favourable to wide use of ice, snow and permafrost as building materials. Thus it is possible to use them for constructing seasonal and perennial storehouses, shelters, ice roads and temporary bridges, dams, temporary piers, etc. Underground structures such as storages, coolers, laboratories, etc. are constructed in permafrost. The advantage over the normal underground structure lies in the fact that it can be used without supporting walls and a roof of rock or with light, protective support. Ice lining is also often used.

When designing the ice, snow and permafrost structures it is necessary to select constructional methods providing stability for the structures as a whole if deformations of elements of these structures take place. At the same time, it is not permissible that elements experience bending stresses (beams, plane ceilings) and tensile stresses. It is necessary to follow such constructional principles that all the elements of a structure are subject to compression. In such cases repeated deformations cause only bulging of vaults, height decrease and increase of thickness of columns and walls, without the structure being destroyed. The value of allowable deformations for the period of operation of a structure which will not cause any emergency is fed into the design calculations.

The most widespread constructions made from ice and snow are ice storages. All the basic elements of these constructions are just ice; ceilings are made in the form of adjacent semicircular arches joined at the top with continuous ice plates, with the thickness of the vaults reaching 2 m. The ice storage is protected from thawing by an insulation layer of sawdust, slag and other thermally insulating materials on the outside. During the period of use a temperature varying from 0 to −10°C is maintained inside the storage. In winter latent heat of water freezing is used to prevent too much chilling inside the storage, by spraying the floor and walls with water. In summer a temperature varying from 0 to −2°C is maintained. Ice-salt cooling may be used. It is possible to construct heated rooms inside the ice and snow structures with the help of light thermal insulational screens and cold air ventilation between these screens and the main structures.

Experiments on ice and ice-ground islands and platforms constructed for placing of bore holes within the shelf area of Arctic seas, have been carried out in Russia recently. Experience with such construction in the USA and Canada has shown great promise for their use in prospecting and extraction of oil and gas from the shelf mineral deposits. Various methods are used for platform build-up: layer-by-layer build-up, build-up by water spraying, cooling with the help of self-regulating installations; and forced cooling with the help of various cold-carriers.

17.4 Development in the permafrost regions for the mining industry and underground engineering

The following kinds of work are associated with industrial mining needs: 1) trench and open cast mining; 2) construction of embankments, dumps, tailings dumps; 3) construction clearing of ground and building sites; 4) tunnelling, sinking of shafts and corridors; 5) construction of underground industrial structures (coolers, gas storage, etc.); 6) construction of prospecting and production oil and gas wells.

Sites for the mining industry depend mainly on the location of the mineral deposits. Therefore the problem for the engineering-geocryological studies is the rather limited choice of sites for construction and installations. Within the areas of the location of the mineral deposits, the existing geocryological conditions are assessed, forecasts of their change in connection with construction and operations and the development of the mineral deposits are drawn up, with a plan of measures for control of the permafrost processes.

Trench and open cast mining

This is usually carried out in two ways: explosive and mechanical. The latter applies to 30% of the frozen ground under development in Russia. The main difficulty connected with mechanical mining is the resistance of frozen ground to cutting. This characteristic depends on composition, ice content and temperature of the frozen ground (Fig. 17.10). When the method of mechanical mining is selected it should be taken into account that frozen ground has its lowest resistance at discontinuities. Therefore it is profitable to use the procedure of mining based on ground spalling. When mechanical characteristics of frozen grounds are determined, the values of their instantaneous strength are used. Mechanical mining is used as a rule within areas of fine-grained soils. Gravel-pebble rudaceous materials are mined either with explosives or by using the prethawing method.

Storage of mine wastes

This occurs in refuse dumps, rock spoil heaps, tailing dumps (tailings is the term denoting waste from enrichment – loose material composed of various size fractions of crushed minerals) and is a problem of great concern for the mining industry. As tailings represent a fluidified mass saturated with chemical reagents of enrichment, the construction of dikes is the biggest challenge in the course of construction of tailings heaps. Heaps are designed with preservation of a frozen base or with thawing of the soil. In the latter case measures should be taken to prevent seepage of wastes through the dike or through the base. The problem of stability of heaps arises with waste from mines, open cuts, trenches in tailings etc. The height of the heaps of some mining plants reaches a few hundred metres. Thus there exist waste heaps up to 100 m in height in the city of Norilsk. Stability of the heaps depends substantially on the varying temperature of the heap's material throughout a year. It is necessary to know the strength properties of materials making up the slopes (frozen or thawed) and the temperature regime to calculate the slope stability. The stability of heaps, pits, and trench walls (the safe angle of slope and critical mass of the heap) is calculated on the basis of the strength characteristics (angle of internal friction, cohesion, etc.) of the materials, obtained in the course of investigations.

Open cast placer mining

This involves disposal of overburden (peats), with mining and flushing of ore bearing ground (sands). Placer deposits are represented as a rule

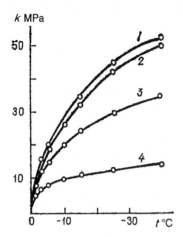

Fig. 17.10. Temperature dependence of specific cutting resistance k:
1 – sandy-silty material; 2 – sandy-silty clay; 3 – clay; 4 – sand.

by gravel-pebble deposits. Mechanical mining is difficult in the frozen state and therefore they are usually subject to prethawing.

According to G.Z. Perl'shteyn the methods of thawing frozen ground used in the course of placer mining include: 1) snow retention (snow thickness increase) during the cold period and its mechanical removal at the beginning of the warm period; 2) removal of vegetation and the upper layer of soil; 3) arrangement of shallow (20–30 cm) warming ponds in summer; 4) flooding of the area with water for a winter period and creation of an ice-air system; 5) injection thawing; 6) steam thawing; 7) electric heating; 8) thermochemical methods. All the methods are used also in the course of preconstructional thawing of frozen ground (see Chapter 18). In addition to the methods cited above, layer-by-layer sprinkling and 'infiltration'-drainage thawing, soil salinization and the formation of artificial 'sushenets' (drained land) are usually used.

When *layer-by-layer* thawing is used the soil is mechanically removed from the surface to a depth of 10–20 cm, as it thaws naturally, resulting in exposure of the frozen soil. Then the natural (radiational-heat) thawing of the next layer to the same depth occurs and it is removed also. This manipulation is repeated during the whole summer period intensifying the thawing process. It is possible to thaw a placer up to 10–15 m in thickness in such a manner during the warm period.

Sprinkling thawing is used for preparation of areas of a placer where the material is highly permeable in a thawed state. The method consists in spraying water with the help of sprinkling installations placed in a grid (Fig. 17.11a). To intensify the process, film coatings reduce the expenditure of

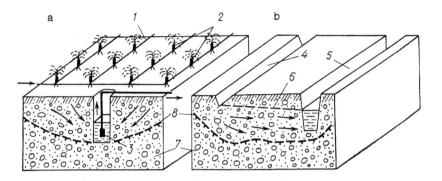

Fig. 17.11. Diagram of thawing of permafrost by sprinkling (*a*) and
seepage-drainage (*b*): 1 – distributing pipes; 2 – sprinkler heads; 3 – drain well
with pump; 4–5 – ditches (4 – irrigation; 5 – drainage); 6 – groundwater level;
7 – gravel-pebble soils; 8 – permafrost surface.

heat in evaporation and losses through turbulent heat exchange, and salt
solutions allow transfer from the frozen state to the thawed under negative
temperatures.

Seepage-drainage thawing of frozen ground involves transfer of heat from
a horizontal seepage flow developed in the thawed layer under the effect of a
difference in water levels between a sprinkler and a drain (see Fig. 17.11b).
On account of the small temperature gradients of the seepage flow this
method is usable only for thawing soils with seepage coefficients higher than
$40 \, \text{m day}^{-1}$.

Gravelly pebble-rich soil with small sand content can be prepared for
year-round mining by *dewatering*. The main point of this method is that at
moisture contents less than 2–3% these materials do not possess under
negative temperature the particular properties of frozen ground and are
similar, in mechanical mining, to thawed materials. Seasonally thawed
ground artificially dewatered in the summer period is often termed
'sushenets'.

Underground workings involve shafts and tunnels. One considers the
special properties of frozen ground within the permafrost zone, and as well,
in regions of unfrozen ground (for subways), because the method of artificial
freezing is often used in sinking or driving in weak water-saturated and fluid
earth. Driving is performed under the protective enclosure of frozen ground
having higher strength than before and low permeability. Once the driving
is finished and the continuous support is installed, there is no need for the
ground enclosure to be frozen and the ground is thawed. Thermal-technical
and statistical calculations for the frozen ground around workings are
performed at the design stage. Thermal-technical calculations dictate the

selection of measures to cool or to freeze the ground and involve determining the temperature field around the underground structures. With the help of statistical calculations it is possible to assess stability and to determine safe dimensions. It should be taken into account that frozen ground can, with time, develop plastic creep strains many times higher than the momentary (elastic) deformations. Therefore when stability calculations for ground are performed it is necessary to start with rheology theory and to carry out the calculations for ground strength and strains with regard to the time factor.

Underground industrial structures (storages, plants, electric power stations, shelters, etc.) within the permafrost regions are, with respect to the thermal regime, divided into structures with positive and negative operational temperatures. The former are usually sited on hard rock (bedrock) although other sufficiently strong and stable ground can be used. In the narrow underground coolers of adit or vertical shaft type the temperature is usually maintained by natural cold reserves inside the permafrost and does not go below $-10\,^{\circ}$C. When the thermal-technical calculations are carried out, the heat remaining inside the structure and the heat coming in through air exchange between the structure and the atmosphere are taken into consideration. Winter ventilation or artificial ventilation and adequate thermal insulation in summer enrich the 'cold reserves'.

There are underground structures the inside temperature of which is lower than the natural temperature of frozen ground. High-powered coolers having a temperature below -20 to $-30\,^{\circ}$C and storage of condensed gas having a temperature below $-180\,^{\circ}$C, are in this category. Structures of such type demand use of additional natural or artificial cooling. In the course of their operation development of a number of processes in the ground associated with the low temperatures is inevitable. These are cracking (increasing the gas permeability of the containing frozen ground); ice sublimation (causing falling and collapse of ground walls); ablimation and icing of walls of underground workings, etc.

Drilling of prospecting and production oil and gas wells

Within the permafrost zone, such activities cause warming of the ground adjacent to the well and its thawing. Experience shows that use of water as a flushing fluid, as is the practice outside the permafrost zone, is practically impossible within the permafrost because it causes further warming and thawing of the ground, formation of cavities around the wellhead and distortion of the wellhead. Therefore special-purpose flushing fluids which do not freeze under negative temperature have been developed. In addition, it is profitable to use air as a working medium. Fixing of casing pipes (strings) is no less complicated a problem.

Operation of oil and gas wells is also associated with heat release. Gas and crude oil in pools have temperatures varying from 20–60°C while discharge from the wells may run as high as hundreds of tons per day. If there are no countermeasures in the course of well construction, one might expect permafrost thawing around the well and ground subsidence around the wellhead (Fig. 17.12) and increase of loads on casing strings with time. Ground thawing can be prevented with the help of well bore zone cooling using circulation of special-purpose coolants (the active way) and with reliable thermal insulation of the well bore (the passive way). In the first case two concentric jigs are positioned inside the permafrost layer with the space between these jigs and the production string filled with air or nitrogen (Fig. 17.13). Thus the problems requiring geocryological investigation in well design are the following: calculation of temperature patterns around the well for the period of sinking and operation, and the assessment of the possibility of cavity formation as a result of thawing of ice-rich ground and ice.

17.5   Types of agrobiological development in the permafrost regions

Geocryological research in the course of development of agriculture is becoming more and more important. It is associated not only with the permafrost regions but also with those with deep seasonal freezing: the non-fertile zone of European Russia, Western Siberia, the Chitinskaya Oblast' the BAM zone, etc. For the present the problem of using only small areas for agriculture is considered here. One of the reasons for poor harvests as well as for their unreliability lies in inadequate warming of the ground.

According to A.M. Shul'gin the main critical thermal conditions of soils are the following: 1) the sum of active temperatures; 2) minimum temperatures of the soil; 3) depths of seasonal freezing and thawing; 4) duration of seasonal freezing and of thawing. The sum of active temperatures of a soil at the depth of 10–20 cm is the main factor responsible for development of vegetation. By this term is meant the sum of mean daily temperatures over the period when the soil temperatures are above 10°C. Knowing the sum of active temperatures of the soil one can establish the possibility of development of one or another kind of agriculture. As seen from the diagram (Fig. 17.14) the sum of active temperatures at the depth of 20 cm is 1200–2000°C over the considerable area constituting the zone of risk for agriculture. According to the values of minimum soil temperatures, of seasonal freezing depths and duration, the greater part of agricultural lands in the former

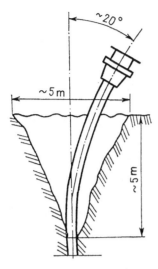

Fig. 17.12. Formation of thaw subsidence cone as a result of thawing of permafrost around an operating gas well (according to data from field observations).

USSR lie in the zone of risk agriculture where it is necessary to solve the problems of improving the climate of the soil and carrying out thermal reclamation.

The main problems of soil thermal reclamation are the following: a) control of deep winter freezing of a soil in the cold period; b) inducing rapid heating of the ground in the warm period. In addition it is necessary to provide the most favourable temperature regime for plants in winter time in the region of winter crop development. At a temperature higher than -3 °C at 5 cm depth, rotting and damage of winter plants take place, and at temperatures below − 15 °C freezing of these plants occurs.

Prevention of deep soil freezing is carried out by snow retention and snow accumulation, with arrangement of shelter belts (protection forests) which can lead to increase of snow thickness on fields of up to 45–50 cm compared to 20 cm within open areas, or with the help of shelter belt farming (measures based on snow retention using belts of plants that do not fall in winter such as sunflower, corn and buckwheat). Snow ploughing is often used for this purpose. When snow ploughing is carried out the snow is pushed up with ploughs into banks of 0.5–0.6 m height, spaced at 10–15 m intervals from each other and oriented at right angles to the wind. Autumn soil ripping contributes to increase of evaporation surface, causes soil drainage, decrease of thermal conductivity, creation of supplementary thermal-insulational cover and decrease of thickness of the freezing layer.

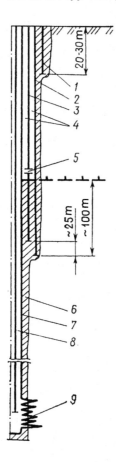

Fig. 17.13. Structure of an operating gas well with thermal insulation of well bore: 1 – guide pipe; 2–3 – jigs (2 – outside; 3 – inside); 4 – thermally insulating air space; 5 – circulation hole; 6 – cement stone; 7 – operating casing string; 8 – casing string; 9 – perforation interval.

Within the zones of excess moisture the snow-removal method, i.e. removal of snow from fields with snow ploughs and cultivators, is widely used in the spring period. When the snow is removed the mean daily air temperature should be higher than the temperature of the upper horizon of a soil. Adequate removal of the snow, in European Russia, advances the beginning of thawing by 10–12 days and reduces the duration of the seasonal freezing of the ground by 7–10 days.

Rolling of soils in the spring period causes consolidation of the upper layer of the soil and increase of thermal conductivity, contributing to better warming and more intensive thawing. Observations show that rolling of soil can raise its temperature by 4.5 °C at the depth of 20 cm during the first

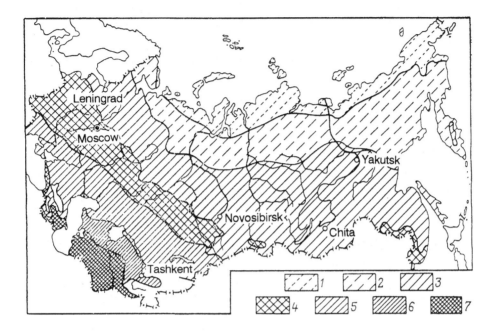

Fig. 17.14. Sum of active soil temperatures (°C) at depth of 20 cm (after V.N. Dimo): 1 – <400°C; 2 – 400 to 1200°C; 3 – 1200 to 2000°C; 4 – 2000 to 2800°C; 5 – 2800 to 4400°C; 6 – 4400 to 5200°C; 7 – 5200 to 6800°C.

summer months. This method is most efficient with dry soils and less efficient for moist ones. The good point of this method is the fact that the rolling has an anti-erosion effect and causes moisture retention as well. Mulching of surface (both complete and interrow covering of the soil with sawdust, straw, peat, manure, soot as well as with paper, emulsified asphalt, etc.) is performed in order to raise or to lower the soil temperature.

In the Nechernozyomnaya (non-fertile) zone, ridge and row sowings are used for fast soil heating. Ridges and rows increase the surface of heating by 20–25%, raising summer temperature on rows by 2–5°C compared to a horizontal surface.

In northern regions where the common methods for increasing the provision of heat do not perform very well, covering of the soils with translucent polyethylene and nylon films is used. These films allow the passage of shortwave solar radiation and hold back up to 90% of longwave radiation from the soil surface, retarding heat and moisture exchange, which results in the warming-up of the surface under these films by 7–10°C (compared with open areas) and in thawing depth increasing. In the north the methods of thermal hydroreclamation, in particular, basin irrigation, are also used. In this case late in May to early in June, the fields are flooded with a water layer 25–40 cm thick for 6–10 days. After water removal the active temperatures (higher than 10 °C) penetrate into the soil 20–25 days earlier and 15–20 cm deeper than under natural conditions.

A group of Magadan scientists under the direction of S.V. Tomirdiaro has developed recommendations for the creation of meadows on bottoms of drained thermokarst lakes. Investigations have shown that bottom deposits of thermokarst lakes are of high productivity. The main point of the method consists of artificial drying of thermokarst lakes, the creation of a drainage system for some dewatering of bottom deposits and in sowing crop grasses. With the observance of proper rules for the usage of such meadows one can provide heavy grass productivity as a feeding base for cattle breeding for 5–6 years, and more seldom for 10 years, and then grass vegetation is replaced by moss-lichen and shrub.

18

Ensuring the stability of engineering structures in the permafrost regions

18.1   Principles of construction on permafrost (bases and foundations)

For construction outside the permafrost regions it is usually believed that all the load from a structure is transmitted through the base of the foundation to bearing ground while ground in contact with the sides of a foundation only occasionally carries a vertical load (as with piles and deep foundations). The interaction between foundations and ground is assessed differently in the regions of deep seasonal freezing and of permafrost. The load is transmitted to ground here through all the ground surfaces in contact with the foundation. This is associated with the fact that adfreezing of the foundation surface to the ground occurs with the result that tangential and normal stresses are transmitted from the ground to the foundation and from the foundation to the ground. The value of the transmitted stresses is then limited by the strength of adfreezing.

The direction of the stresses arising in the interaction between a foundation and the ground can change with time and depends upon the layer (seasonally or perennially freezing) in which the foundation is situated. The main types of embedding of foundations are shown in Fig. 18.1. Thus, within the layer of seasonal freezing (or thawing), tangential stresses transmitted from ground to the foundation develop during the part of a year in which the layer freezes and heaves, and are directed upwards . In response to adfreezing with the side of the foundation within the layer of perennially frozen ground, the frozen ground takes some portion of the load from the foundation not only through its base but also throughout the side surface. At the same time the side adfreezing of the lower part of the foundation with the perennially frozen ground layers increases foundation resistance to heaving developing in the course of freezing of a seasonally thawed layer. The foundation can be stable or can move (heave or settle) as a result of the combined effect of normal and tangential forces as well as heaving forces (and sometimes of frictional forces). Nonuniform movement of foundations

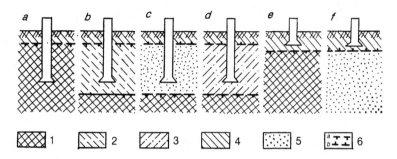

Fig. 18.1. Diagram of foundations (*a, b* – in the permafrost; *c, d* – in closed taliks; *e, f* – in seasonally thawing and seasonally freezing ground, respectively):
1 – 5 – ground (1 – in the frozen state throughout the period of construction
2 – artificially frozen before construction or during the period of operation,
3 – artificially thawed before construction or during the period of operation,
4 – subjected to seasonal thawing and freezing, 5 – in thawed state throughout the period of operation of the structure); 6 – the upper surface of the permafrost (*a*) and the base of the seasonally frozen ground (*b*).

upward or downward is the main reason for deformation of buildings and constructions. Embedding of foundations below the base of the seasonally freezing or seasonally thawing layer can result in the effect of annual heaving processes and ground settlement being exerted only on some part of the side surface. Foundations embedded entirely in the layer of seasonal freezing-thawing experience heaving and settlement not only through the side surface but also through the base of the foundation.

Measures for ensuring the preservation and durability of structures and required operational qualities must be considered in design and construction on permafrost. This is achieved by selecting the optimal design features of the structure, the type of foundations, methods for enhancing the structural properties of bearing ground and by regulating the heat interaction between structures and bases. It is conventional to call a set of these measures or some part of them the method for ensuring the engineered structure's stability. By convention all these methods are unified into two large groups termed the principles of use of frozen ground as a foundation, according to *Building Norms and Regulations II*-18-76.

Principle I: perennially frozen ground is used in foundations in the frozen state, being preserved during the period of construction and throughout the given period of operation of the structure. This includes the following methods for ensuring the structures stability: (a) by keeping the ground base frozen; (b) by limiting the thawing of the frozen ground base; (c) by preliminary freezing of the ground base; (d) by freezing the ground base in the course of construction and operation.

Principle II: the perennially frozen ground is used in the thawed state (being allowed to thaw in the course of operation of the structure or with thawing to the design depth before construction). This includes the following methods for ensuring structural stability: (a) by adapting the above-foundation structure to permit differential settlement of the foundation with allowable thawing of the permafrost in the course of operation (a constructive method); (b) by prethawing of the permafrost; (c) by stabilizing the initial position of the upper surface of the permafrost.

Selection of the principle of construction on permafrost and of the method for its control is based, on the one hand, on comprehensive study of the geocryological conditions of the site taking into account possible change in the course of construction and operation of the structures and, on the other hand, on complete consideration of the features of the construction (sizes of foundations, structural materials being used and the service life of the structures), and of the mode of operation (with heat release or without heat release, using a wet or dry process, etc.). Data on the availability and transportation conditions of structural materials, on power supplies for construction and on the seasonal dependence of the work are important. Let us consider the methods for realization of Principle I of construction in the permafrost zone.

A method for ensuring structural stability by keeping the ground frozen or by cooling it

This method began to be used widely once N.A. Tsytovich had developed a procedure of thermal-physical and strength calculations for perennially frozen ground in foundations in 1928. The method is based on complete removal of the heat released by a building or a structure, i.e. on conservation of the existing temperature regime of perennially frozen ground, or its cooling to a lower than natural temperature. Ventilated cellars under floors, which had received wide recognition in practice due to the ease of construction and the reliability in operation, are the most widespread constructions for maintaining heat removal from the structure. From 1928 on, a great number of engineering structures in which the foundation was kept frozen were constructed in the North and the Russian North-East (cities of Vorkuta, Noril'sk, Yakutsk, Magadan, etc.). Half a century of experience of continuous operation has justified that this method of construction is thriving. Later on, it has been used widely by foreign specialists in the course of construction in Alaska and in the north of Canada. Basements of one of several types (Fig. 18.2a,b,c) are usually used in this case. These are: 1) open, i.e. a space open from all sides under a building

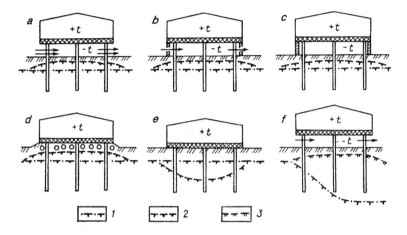

Fig. 18.2. Structures of buildings according to the first principle of using the ground as the base: *a–c* – cellars (*a* – ventilated open, *b* – ventilated enclosure with some air holes, *c* – with cold ground floor), *d–f* – building constructions (*d* – on fill with cooling pipes in the base, *e* – with limited thawing of the permafrost, *f* – with preliminary ground freezing); *1–3* – boundaries of the permafrost (1 – under natural conditions, 2 – under the conditions developing in the course of operation of the structure, 3 – after preliminary freezing).

or under a structure ventilated with air throughout the year (both in winter and in summer); 2) closed, with air holes, by way of cellar space enclosed in a socle and by making ventilation openings in it such that the air required for keeping the base frozen or for its cooling circulates; 3) with a cold socle without air holes being constructed – as a rule, in regions with strong winds where the holes would be plugged with snow in winter.

Special heat engineering calculations based on the assumption of complete heat removal from structures, are carried out to determine the height of ventilated and non-ventilated cellars. The procedure of such calculations put forward by N.A. Tsytovich was developed by N.I. Saltykov, G.V. Porkhayev and others. Experience suggests that the height of cellars can vary from 0.5 to 1.8–2.0 m depending on the severity of the geocryological conditions and size of structures. In the North airtight multi-layered floors with heat insulation are used to reduce heat losses through the floor above the cellar.

If natural ventilation turns out to be inadequate according to the results of the heat engineering calculations, forced ventilation or artificial cooling of the ground is initiated. One such structural variant is shown in Fig. 18.2d.

A method for ensuring structural stability by thawing of perennially frozen ground

With this method perennially frozen ground is allowed to thaw to a depth determined by heat engineering calculations. Foundations are put deeper into permafrost, below the stationary thaw basin (see Fig. 18.2e). One can gain some reduction in the foundation construction cost by partial or complete abandonment of a cooling system. However some increase of the mean annual temperature of frozen base ground takes place with thaw basin formation resulting in a bearing capacity decrease. In order for the bearing capacity decrease to be compensated the foundations must be put still deeper into the ground resulting in a rise in the cost of these foundations. In addition, the method requires very close control of the position of the upper boundary of the permafrost in the thaw basin.

A method for ensuring the stability of the structure by preliminary freezing of underlying ground

In the regions with permafrost islands it is rather difficult to site a building entirely within a frozen mass. In this case preliminary freezing of an unfrozen ground layer is carried out (see Fig. 18.2f) and then it is kept frozen with the help of a cooling system. The essential drawbacks of this method are its high cost and the extending of the construction time by the period of freezing, covering usually 3–6 months. Therefore this method has not been used widely in practice in civil and industrial engineering. However it is used widely in hydrotechnical construction: during the construction of dams within the permafrost zone for creating the counter-seepage frozen ground core in their body; for driving tunnels in ground with high water pressures and in the course of mine operations in regions of unfrozen ground.

A method for ensuring structural stability using ground freezing in the course of construction and operation

This method is used usually in regions where frozen ground of the unconnected type occurs where the permafrost table is separated from the layer of seasonal freezing. The base of the foundation is laid in the layer of thawed ground underlain by the permafrost. After construction of the foundations cooling systems are put into the permafrost at a depth of 0.5–1.0 m, installed adjacent to each part of the foundations and then construction of the above-foundation structure begins. Freezing of the ground around the cooling systems causes an increase in its volume, with great horizontal pressures developing in the unfrozen ground between the cooling systems. Expansive forces fix the foundation preventing it from moving upward under the effect of heaving forces. With the passage of time complete ground freezing with retention of foundation stability occurs.

Types of foundations

In the course of foundation construction according to Principle I, shallow (surface) posts and piles are used (Fig. 18.3). Shallow foundations are used mainly for light-weight, one-storey, most commonly wooden buildings. They are placed in shallow excavations within the layer of seasonal thawing or directly on the ground surface. Operation of the structures with these types of foundations is accompanied by deformations with a pulsating pattern in the course of seasonal freezing and ground heaving as well as in the course of seasonal thawing and settlement. Pressure on a pile in a pile foundation is balanced by counter forces of the frozen ground on the pile face and by forces of the adfreezing of the ground and the pile side surface. If the permafrost strength does not ensure the stability of the structure on piles, post foundations are used, with the shoes being put into the base of a pile (see Fig. 18.3 *II*).

To fabricate the piles, reinforced concrete, wood or metal is used. Placing of single piles, pairs of piles and pile clusters has to be considered in the design. Piles may be placed by boring, and by lowering into position and by boring and driving (see Fig. 18.3 *III*). In the first case piles are installed in a preliminarily bored hole of diameter greater than the pile diameter. The space between the pile and the hole wall is filled with mud or mortar which freezes later. The bored-driven piles are driven into holes of diameter less than the pile diameter. This method can be used when the mean annual temperature is near to $0\,^{\circ}\mathrm{C}$. The lowered piles are installed in preheated and prethawed ground.

As noted above, the second Principle of construction within the permafrost zone includes several methods for ensuring structural stability.

A method for ensuring structural stability by adapting the above-ground structures to differential settlements under permissible permafrost thawing in the course of operation

This is a constructive method for designing and construction, and is the oldest method in foundation engineering in the North. When construction using this method is carried out a thaw basin is formed below buildings and structures (Fig. 18.4a) as the result of heat release from them and ground settlement occurs. This settlement is differential as a rule because of lithological nonuniformity of the ground, nonuniformity of heat release from the building and of load distribution in it and because thaw occurs

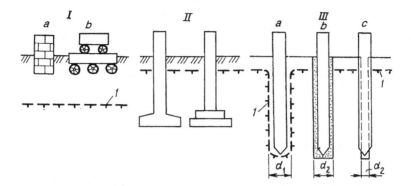

Fig. 18.3. The main types of foundations used in the course of construction on the permafrost: *I* – shallow surface (*a* – brick and concrete, *b* – wooden); *II* – posts; *III* – piles (*a* – lowered, *b* – bored-lowered, *c* – bored-driven); d_1; d_2 – diameters of the thawing zone and of the bore hole, respectively; 1 – the upper surface of the permafrost.

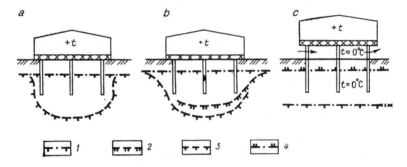

Fig. 18.4. Foundation structure following the second principle of using the ground as a base: *a* – with permissible thawing of the permafrost during the course of operation; *b* – with prethawing of the permafrost; *c* – using the method of stabilizing the permafrost table at its original position; 1–4 – boundaries (1 – of the permafrost in natural conditions, 2 – after the prethawing, 3 – formed in the course of operation, 4 – of the seasonally freezing layer).

under the centre and edges of a building. This method can only be used if the consequences of thawing are taken into account completely and if measures are undertaken for ensuring the adaptation of foundations and structures to the deformations which will take place in the course of thawing. Therefore when construction on thawing ground is carried out, there is a need to maintain as slow and uniform thawing of the ground as possible, this being a factor in the uniformity of settlement. If rapid settlement is allowed, the building construction will, probably, have no time to adapt resulting in unacceptable deformations in the structure causing an emergency.

Foundation designs and calculations with permissible thawing of perennially frozen ground must include: 1) determination of the depth of thawing of the perennially frozen ground for various times during operations, 2) determination of the bearing capacity of the thawing ground, 3) determination of the differential settlements of the thawing ground, 4) determination of size and type of foundation, 5) testing of foundation for heaving.

Two different *foundation structures* are mainly used on thawing ground: the rigid and the flexible. The former is not sensitive to differential settlements because of its rigidity. Raft slab foundations, a system of rigid concrete strip foundations and waffle foundations able to withstand nonuniform thaw settlements of the base are used for greater rigidity. To gain rigidity in structures, perimeter beams of reinforced concrete or steel are used when bracing building walls, load-carrying structures and pile foundations. Buildings covering an extensive area and their foundations are often divided into separate rigid sections (with settlement joints) each of which is reinforced by a grade beam, i.e. the building is constructed on the basis of a rigid structural model for separate blocks able to move independently of one another. The flexible structure model ensures the structure accommodating differential settlement. It is based on the principle of possible realignment of deformed elements of the construction. Articulated joints of elements and building sections are used to achieve this. Stresses due to warping are ruled out when this principle is applied so the buildings in practice follow the settlements. The flexible structure model of construction presents a way of treating displacements in the buildings. In this case preliminary provision is made for jacks to be installed in a cellar with the help of which constructions are realigned.

Following up the state of buildings and structures constructed by the method of allowing for differential settlements shows that many of them are being deformed in the course of operation. Evidently this method can only be used on ground that is practically incompressible in the course of thawing with a settlement value of not more than $0.5\,\mathrm{cm\,m^{-1}}$.

A method for guaranteeing the structure stability by permafrost prethawing

This method (see Fig. 18.4b) has been known from the 1930s but justification for using construction with prethawing was given only in 1958 by V.F. Zhukov. From this time, widespread use of this method occurred. It is most profitable to use the prethawing method in the following cases: 1) when ice-rich ground occurs in a fairly thin layer (to 7–10 m) on little-compressible monolithic hard rocks; 2) when heaved rudaceous soils have compacted rapidly in the course of prethawing and have sufficient bearing capacity; 3) on sandy soils with the rate of contraction being practically in line with the rate of thawing. Prethawing in clay soils also works well, however, only when they alternate with well-filtered sandy and rudaceous interlayers favourable for rapid drainage, or when their consolidation, stabilization and strengthening are accelerated artificially, simultaneously with the thawing.

Thawing to the depth of the whole calculated thaw basin is the best way for preparing the foundation base. However it is very expensive and not used now. Prethawing is carried out usually to a lesser depth (some 60% of the steady-state thaw basin depth) thus representing 60–80% of the settlement possible with thawing of the whole basin. Settlement during further thawing (in the period of construction and operation) proceeds very slowly and does not cause important deformations. Ground prethawing is carried out using thermal methods, or water and thermal amelioration.

After prethawing the soils have as a rule high macroporosity, excess water saturation and are in a poorly consolidated state and unsuitable for construction of foundations. Such soil requires preliminary preparation such as ameliorative drainage, consolidation and strengthening before construction starts. Design and construction on prethawed and stabilized soils are carried out according to their condition without considering their frozen state. The types of foundations are similar to those used in the course of construction on unfrozen weak soils.

A method for ensuring the stability of structures by stabilization of the permafrost table at its original position

This method was developed by L.N. Khrustalev and others in 1967 for developing the areas with permafrost of the unconnected type. The method implies that the upper permafrost boundary is maintained throughout the period of operation at the initially determined level. Construction involves using ventilated cellars with the foundations situated between the permafrost table and the base of the seasonally thawing layer (see Fig. 18.4c). Stability of the permafrost table can be attained only when there is no temperature gradient in the ground above the permafrost table. This condition is achieved when the mean annual ground temperature is 0°C at the depth reached by seasonal freezing. Thus, maintaining a mean annual temperature near 0°C in the ventilated cellar is required for this method. This is achieved with the help of a temperature regulating system in the cellar using air holes that open and close, by using seasonal self-cooling installations, etc. This method of stabilization has been used in civil and industrial engineering in the city of Vorkuta for many years.

18.2   Methods of amelioration of frozen ground for foundations

By amelioration (improvement) is meant a purposeful change of mean annual temperature, structure and properties of frozen, freezing and thawing ground and of the thermal-moisture regime in the direction required for solving the particular practical problems. Thermal, water-thermal, mechanical, physical-chemical and chemical amelioration is used. Thermal amelioration consists of the artificial lowering or raising of ground temperature using various heat sources. Heat transfer is performed in ground by conduction. In water-thermal amelioration the heat transferred by convection and conduction in the course of water injection, percolation or irrigation is used. Mechanical methods of amelioration are used for changing the ground properties and include: ground substitution, consolidation, deconsolidation, drainage, etc. Physical-chemical and chemical amelioration modes consist either of using the heat being released or absorbed in the course of chemical reactions or in changing the state of aggregation of the soil by injecting various chemical reactants and by using the electroosmosis effect.

Methods of thermal and water-thermal ground amelioration involving artificial change in the permafrost temperature, freezing of thawed ground and thawing of frozen ground before or during construction have received the widest acceptance.

Deliberate cooling of frozen ground and freezing of unfrozen ground

This is carried out in various ways. Methods using natural cooling (systematic snow removal in winter, shading and protection of ground surface using heat-insulating material in summer, construction of ventilated cellars, piles allowing natural ventilation into the foundation structure in the cold period, etc.) are examples. The advisability of using one or another method is shown by thermal engineering calculations.

To accelerate the process, the artificial cooling is effected by pumping air into boreholes driven through the whole thickness of the zone being cooled or into freezing pipes, with special air-supplying installations, sunk into these boreholes (Fig. 18.5). Heat removal from such a pipe ranges from 130–140 W m^{-2} according to S.M. Filipovskiy's data. Cooling (freezing) depth is as much as 17–20 m.

The drawback of this method is the necessity for expenditure of energy and for permanent maintenance personnel for frequent inspection of the freezing pipes to remove hoar-frost and ice plugs in them.

Ground cooling (freezing) using ammonia-brine installations guaranteeing heat removal of up to 300 W m^{-2} from the pipe is the most efficient.

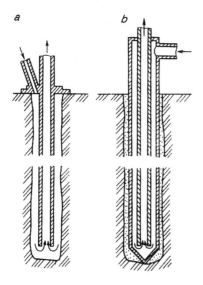

Fig. 18.5. Air cooling installations: *a* – in bore holes; *b* – hollow piles.

However such cooling installations are a rather expensive undertaking and require regular maintenance in the course of operation. Therefore they are used only in a few cases in hydrotechnical engineering and in mining, and to eliminate accidental damage to structures caused by permafrost thawing, and for preliminary foundation freezing when a reduction of the construction period is required, as well as for deep shaft sinking and tunnel driving in highly water-saturated materials. The depth of artificial ground freezing for shaft sinking achieved at present is more than 600 m.

A method of ground cooling using coolant circulation through pipes laid close to building foundations has also been adopted. The design with such a method usually includes ventilated spaces below the structure as well. Using such a complex, labour-consuming method for preservation of the frozen state of the ground is justified in construction of very sensitive structures as well as of buildings and structures with high heat release.

Self-regulated, seasonally acting cooling installations or *thermal piles* are efficient in operation and economical. Thermal piles were developed in the 1960s by S.I. Gapeyev in the former USSR and by E.A. Long in the USA almost simultaneously. The principle of their work is based on the closed convection of a liquid coolant such as kerosene (Gapeyev's thermal piles) or of a vapour-liquid coolant (Long's thermal piles) such as propane, ammonia, or freon. Convection is caused by a difference between densities of the coolant in the above-surface and subsurface sections of the installation (Fig. 18.6). During the cold period the temperature of the upper part of the pile is

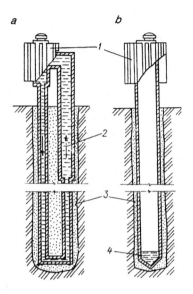

Fig. 18.6. Constructions of the self-regulated seasonally acting cooling installations: *a* – liquid (Gapeyev's system); *b* – vapour-liquid (Long's system); 1 – heat exchanger; 2 – coolant (kerosene); 3 – backfill; 4 – coolant (propane).

lower than the ground temperature at the depth of the lower part. Under these conditions, evaporating propane (in Long's thermal piles) moves upward, condenses on the pile walls and falls to the pile base as a liquid, having a temperature the same as that of the air temperature, which then cools the surrounding ground. Liquid thermal piles work on the same principle. Cold kerosene moving downwards displaces upwards the less dense and warm kerosene thus cooling the ground.

Numerous modifications of liquid and vapour-liquid cooling installations with high cooling qualities have been developed and introduced but all these structures work on the same principle.

Liquid and vapour-liquid systems represent practically automatic permafrost cooling installations dispensing with the need for maintenance. They can be used independently or be installed in the body of pile foundations, therefore they are often termed thermal piles. Mean cold output of the liquid thermal piles varies in the range from 25–50 W m^{-2}, and for the vapour-liquid ones from 60–100 W m^{-2}. The depth of cooling achieved in practice by liquid thermal piles is 25 m and by vapour-liquid ones, 50 m. This method for cooling (freezing) can be used in conditions of discontinuous permafrost or sporadic permafrost for preconstruction ground freezing, and, using the method of permissible freezing in the course of construction and operation, for increasing the ground bearing capacity and for maintaining the prescribed temperature of perennially frozen foundations for buildings and structures. Self-regulated cooling installations are in common use in hydrotechnical construction to create and to keep the impermeable cores of earth dams frozen, and in linear structures when erecting piers, pipelines, power transmission lines and trestles, for permafrost curtain formation, etc.

One of the drawbacks of liquid and vapour-liquid thermal piles lies in the fact that in the summer period, when the thermal piles do not function, an important increase of ground temperature is noted around them. This drawback is eliminated using devices (zeroters), which accumulate cold in winter. Zeroters comprise an envelope filled with a liquid (antifreeze) freezing at temperatures below zero, which is mounted near a thermal pile. With the advent of winter the thermal pile begins to work, cooling (freezing) the ground and the antifreeze. In summer the temperature of the zeroter remains equal to the temperature of the melting antifreeze, i.e. it remains negative providing the prescribed bearing capacity of the base.

Preconstruction thawing of permafrost

Such a method is widely used in foundation engineering in the permafrost zone when construction on the second principle is performed.

The methods of permafrost thawing are of two kinds: using natural heat and using artificial heat sources.

The methods using *natural heat* are employed when enough time is available. The field of their application is limited mainly to southern zones of the permafrost regions with high mean annual ground temperature. To increase the depth of seasonal thawing the following measures are used: 1) snow retention (increase of thickness), that is, ground surface warmth retention in the cold period; 2) snow removal from the surface in spring to use solar radiation; 3) removal of surface vegetation and the upper layer of ground, surface blackening, or the use of transparent polyvinylchloride film; 4) drainage of the construction site; 5) arrangement of 20–30 cm deep heat-collecting pools in summer to increase ground warm-up; 6) foundation pits filled with water for the winter period (if ground thawing properties are not impaired, providing, for example, the ground is sandy and rudaceous), keeping the ground unfrozen and promoting growth of the thaw basin.

Other variants of natural heat use for prethawing of ground based on controlling the radiation balance are also a possibility. Experience shows that such methods are of limited application because of their inefficiency. The methods using *artificial heat sources* are far more productive. In addition to the methods of sprinkling and infiltration-thawing discussed above, methods of thawing with the help of cold and hot water, thawing with the help of steam, warming with alternating electric current of permafrost, and thermochemical methods of thawing are widely used.

A method of thawing with water involves thawing of permafrost through boreholes into which the water is injected (Fig. 18.7a) The boreholes are spaced at intervals of 3–5 m depending on the ground percolation capacity. They are placed into the ground with the help of drilling rigs. It is advisable to use this method for ground with a percolation coefficient of more than 0.01 m day^{-1}. This method is rather expensive, but it enables preconstruction thawing to be carried out within 10–12 days to a depth of more than 10 m (up to 25 m) when the construction is on rudaceous soils, especially in warm permafrost.

Thawing by steam is performed with the help of steam injectors (see Fig. 18.7b), by the open method (when the steam is released at the end of a pipe into ground) and by the closed method (when the steam is circulating within the pipe). In practice the open method is used more often. Steam injectors are sunk into the ground by gravity or are forced slightly by hammering. The use of steam makes it possible to thaw the frozen ground to a depth of 10 m.

Electrical thawing of the permafrost is the most industrial method and is used for any ground, when suitable power sources are available. The technology of thawing using electrical heating is based on the Joule-Lenz law. There are two ways of using the equipment: as electrodes and as electric heaters. In the first case alternating three-phase current at 380 V is used. The electrodes in the form of special steel probes are installed at a given depth and arranged in parallel rows spaced at 2–2.5 m (see Fig. 18.7c). The method is advisable mainly for clay-rich soil. The expenditure of energy ranges from 60–80 kW m^{-3}. In the second case electrolytic or resistance heaters are used to warm up the ground (see Fig. 18.7d,e). The power supply to the electric heaters is alternating current at 20–30 V. Electrolytic heaters are usually arranged in a grid pattern 3 $\times$ 2 m in size, while the resistance ones are arranged in a 6 $\times$ 5 m grid pattern. The most efficient is the combined use of the equipment: as electric heaters (electrolytic and resistance) at the initial stage of thawing and then as electrodes using alternating current of high voltage. Combination of these two methods makes it possible to increase the rate of thawing of the permafrost.

The thermochemical method of permafrost thawing consists of using thermochemical reactions as heat sources. Thermit, a humidified quicklime is used for these purposes. Temperatures of up to a few hundred degrees are developed as a result of slaking, causing thawing of the frozen ground. In addition, the quicklime increases in volume sharply when slaked resulting in the consolidation of the thawing ground.

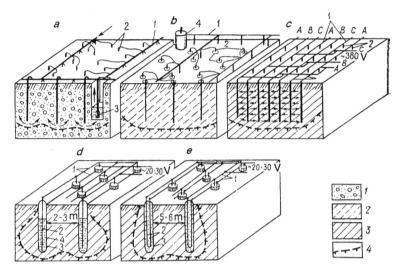

Fig. 18.7. Methods for preconstruction thawing of ground: *a* – using injection points (1 – distribution tube, 2 – injectors, 3 – drain well with pump); *b* – using steam injectors (1 – steam line, 2 – steam injectors, 3 – protective hood, 4 – steam boiler); *c* – using three-phase alternating current (1 – electric conductor, 2 – electrodes distributed by phases *A*, *B*, *C*); *d* – using electrolytic heaters (1 – electric conductor, 2–3 – electrodes inner and outer, respectively, 4 – electrolyte); *e* – using resistance heaters (1 – electric conductor, 2 – electrode, 3 – bore hole); 1–3 – ground (1 – gravel-pebble , 2 – sandy-silty, 3 – sandy-silty-clay); 4 – permafrost boundary.

Consolidation and strengthening of the thawed ground

The methods used for these purposes are identical with those for unfrozen ground. Let us enumerate the main ones.

Mechanical densification is performed with the help of tamping, vibratory compaction and sometimes surface active agents are added to the ground. Tamping is used for damp ground (if the degree of water saturation is less than 0.7%). Clay-rich ground is compacted to a depth of 1.5–2 m by tamping. With vibratory methods clay-rich ground is densified to 10 m depth.

Consolidation by loading, dewatering with the help of ground water table lowering, etc. are physical methods. The most efficient method for clay-rich ground is *vertical sand drains* which accelerates consolidation several tens of times. Vertical drainage is carried out with special sand drains of diameter 40–50 cm placed 2–3 m apart, which consist of bore holes filled with sand. The method of ground consolidation using lowering of the groundwater table is based on the increase of ground weight in the course of the temporary lowering of the water table.

Chemical consolidation and strengthening of soils is performed by injecting chemical solutions into the ground, with these solutions forming a firm foundation in the course of solidification. Chemical methods are suitable for rather water-permeable ground.

The electrochemical (physico-chemical) method consists of electro-osmotic dewatering. In the course of electro-osmosis in saturated colloidal clay-rich and fine soils, pore water movement towards the cathode is initiated and the effective pressure on the ground skeleton near the anode increases, contributing, on the one hand, to ground dewatering and on the other, to consolidation of its skeleton. The method is used in fine-grained soils.

Electro-osmotic consolidation is carried out with direct current. The consolidated ground mass has a few electrode loops (Fig. 18.8). Electrodes of the exterior loop are cathodes and of the interior, anodes. Electrodes serve as vertical drains simultaneously. Chemical admixtures such as saturated solution of quicklime with 7% calcium chloride content are required for uniform ground stabilization by electro-osmosis. This method is often used simultaneously with the use of strengthening agents, for example, with waterglass injection. Electro-osmotic dewatering is used as a rule following ground electro-thawing with the help of the same electrodes but with a modified loop.

Protection of the thawed ground layer against seasonal freezing

This measure is an important part of ground amelioration, because excavation of the thawing and thawed ground, and construction, are seldom completed during one season. Deep winter freezing causes additional expenditure on thawing the seasonally frozen layer and reduces the time for work, after the extended ground thawing. To decrease the depth of freezing or to exclude freezing, such measures as artificially increasing snow thickness, provision of heat-insulating covers, flooding with water, etc. are carried out. Local materials such as moss, pine needles, sawdust or artificial synthetic heat insulators such as foam plastic, polystyrene can be used as heat-insulating covers. Complete protection of the ground against seasonal freezing is achieved by flooding the areas with water to a depth somewhat greater than the depth of freezing in natural water bodies.

18.3   Principles of foundation design and selection of type of foundation for construction on permafrost

As shown in Chapter 8, the physical-mechanical properties of permafrost depend essentially on its temperature. They change drastically when the ground passes from the frozen into the thawed state. Therefore structural stability can be maintained only if the appropriate temperature

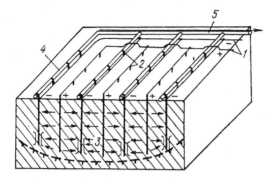

Fig. 18.8. Diagram of electro-osmotic consolidation: 1 – electric conductor; 2 – electrodes; 3 – filters; 4 – connecting hose; 5 – spillway.

regime for the supporting ground is set up. In addition, changes in the temperature regime of the permafrost can cause development or activation of unfavourable cryogenic geological processes such as heaving, thermokarst or icing formation, frost fracturing, thermal erosion, etc. And finally, the heat losses from the structure depending on the ground thermal state, affect the operational expenditures. Thus selection of the type of foundations, calculation of their area and the depth to which they are laid begins with determination of the ground temperature field.

The thermal physics of frozen ground is closely associated with its mechanics; therefore the combined consideration of three fields (temperature, moisture and stress), that is, of the so-called *interconnected problem*, is most appropriate. However as these problems are very complex, they may only be defined qualitatively. In addition, in the majority of cases the temperature field depends only slightly on the stressed state of the ground under the pressures that occur in practice. This makes possible consideration of the thermal physics problems without regard to mechanics, i.e. consideration of the problems as *not interconnected*. For example, first of all the ground temperature pattern for the particular point in time is determined and then the strength parameters are estimated.

Analytical determination of the ground temperature field is also associated with a number of complexities. Within the locality under construction it is necessary to take into account the thermal effects of buildings, structures and heating systems on each other; to estimate the change with time of the ground temperature field, and to take account of conductive as well as of convective heat transfer by water movement. Such problems belong in the class of non-steady-state, multidimensional problems of the Stefan type. Their analytical solution is obtained only for particular cases.

As a rule, the main simplifications of the problem of the ground temperature field in the locality under construction amount to the following: 1) water moving in the ground and, consequently, heat transfer by the water, is assumed to be absent; 2) phase transitions are assumed to proceed at the boundary of the thawing ground under buildings, at a temperature t of $-0°C$; 3) heat transfer in the ground (frozen and thawed) is assumed governed by the Fourier transfer equation. Given these prerequisites, the quantitative assessment of the thermal effect of buildings and structures on the permafrost temperature regime has been obtained by many authors and presented in the works of G.V. Porkhaev, L.N. Khrustalev and others. In a general form the non-steady-state two- and three-dimensional heat conduction problems are solved by numerical methods using computers. When designing structures according to the *first principle* one carries out the heat engineering calculations with the aim of: (a) justification of the measures selected to guarantee retention of the frozen state of the ground (calculations for ventilated cellars, cold ground storages, cooling pipes and ducts, cooling devices, etc.); (b) determination of the design seasonal thawing depth of the ground; (c) calculation of the design permafrost temperatures at various depths for determining the strength characteristics of the frozen ground. When designing according to the *second principle* one determines with the thermal engineering calculations (a) the depth of seasonal freezing; (b) the depth to perennially frozen ground under buildings and structures.

After determination of the thermal state of the ground, strength and deformation characteristics of the soils are calculated in order to ensure the operational reliability of buildings and structures.

Whether the first or second principle of ground use as a foundation is followed, when the structure is erected the calculation for the foundations are carried out from two limiting states: 1) from bearing capacity (the first group of limiting states); 2) from strains (the second group of limiting states). In addition, foundations are tested for stability with respect to frost heaving. Calculations from the first group of limiting states are performed to determine the capacity of the ground to carry the structural load. The purpose of the calculations from the second group of limiting states is to determine the permissible ground strains without causing a loss of structural stability.

The procedure for calculating the stability of foundations is outlined in *Building Norms and Regulations II-18-76 'Bases and foundations on permafrost'*. The fundamental tenets for assuring ground support for foundations designed on the principle of *keeping the ground frozen* are of the greater interest, because when design is according to the second principle, foundation design is the same as for unfrozen ground.

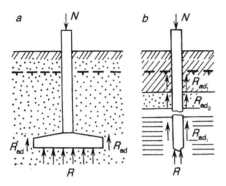

Fig. 18.9. For the calculation of bearing capacity of posts (*a*) and piles (*b*) for foundations in the permafrost.

Stability of the foundation is calculated from *bearing capacity* using the following formula:

$$N \leqslant \Phi k_{\mathrm{r}}$$ (18.1)

where N is the design load on the base; Φ is bearing capacity of the base; k_{r} is the factor of safety (used in accordance with the class of structure with respect to the *Norms*).

The design load N on the foundation base includes the weight of the structure above the foundation with its effective load and the weight of the foundation proper (from the level of a cellar ceiling to the base of the foundation).

Bearing capacity Φ of the base includes the shearing strength of the frozen ground along the adfreezing surface of any part of the foundation frozen into the permafrost and the resistance of the permafrost to pressure normal to the foundation base (Fig. 18.9). For post and pile foundations:

$$\Phi = m\left(RF + \sum_{i=n}^{n} R_{\mathrm{ad}_i} \cdot F_{\mathrm{ad}_i} \right)$$ (18.2)

where m is the coefficient depending on the type and structure of the foundation, varying in the range from 1.2 to 1.1 according to the *Building Norms*; R_{ad_i} is the design shear strength of the *i*-th layer along the adfreezing surface of the foundation; R is the design resistance of the frozen ground to normal pressure; F_{ad_i} is the area of the *i*-th layer of the side surface; F is the area of the foundation base.

Values of R_{ad} and R are set equal to the values of long-term shearing and compressive strengths and depend on ground composition, ice content and temperature at the depth at which the foundation is laid.

The permafrost possesses plastic properties, i.e. the capacity to deform under the effect of long-term loading, especially at temperatures close to 0°C. A stability assessment for deformations is performed, to establish that:

$$S \leqslant S_{per} \tag{18.3}$$

where S_{per} is the limiting permissible deformation of the base and structure determined in accordance with the type of the structure and the particular features of the foundation and structure; S is the value of the deformation of the ground base depending on the pressure around the base of the foundation and the composition and properties of the frozen ground (strain modulus or compressibility factor).

The conditions (18.1–18.3) are responsible for the stability of the foundation base depending on the strength and strain properties of the permafrost. However during the period of seasonal ground freezing the foundation is subjected to heaving forces directed vertically upward and tending to lift, or to 'pull out' the foundation from the perennially frozen ground. Two types of heaving forces, normal and tangential, are distinguished. Normal forces are exerted normally to a foundation base situated in the layer of seasonal freezing or thawing. They must be considered in the design of shallow foundations (see Fig. 18.1).

Forces of heaving resulting from the adfreezing of the frost-susceptible ground to the side surface of the foundation are exerted at a tangent to its surface. These forces are considered in the design of the basic types of foundations, the depth of which must be greater than the normative depth for seasonal ground thawing or freezing in the area, according to *Building Norms and Regulations II-18-76*. The normative depth of the seasonal ground thawing is defined as the greatest depth observed over a period of no less than 10 years, at sites free of vegetation and peat cover, and where the snow cover disappears in spring. This depth is taken to be equal to the mean depth of maximum annual seasonal freezing in accordance with observational data for the period of no less than 10 years, for open ground, free of snow, with the groundwater level situated below the depth of the seasonal freezing.

Foundation tolerance for the effect of heaving forces is calculated, on the basis that the heaving forces against the foundation should be smaller than those keeping the foundation from being heaved (Fig. 18.10):

$$(P_{heav} \cdot F) - N \leqslant \left(\frac{m}{k_r}\right) \cdot P \tag{18.4}$$

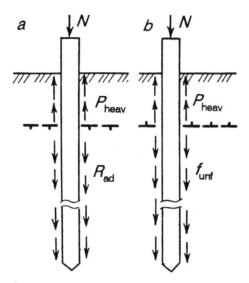

Fig. 18.10. For the calculation of heaving forces in (*a*) permafrost and (*b*) non-permafrost ground.

where P_{heav} is the specific tangential force of heaving; F is the area of side surface of the foundation situated in the layer of seasonal freezing or thawing; N is the design load on the foundation including structure and foundation pressures; P is the design value of the force keeping the foundation from being heaved; m is a coefficient of work conditions equal to 1 according to the *Building Norms and Regulations*; k_r is the safety factor taken to be equal to 1.1.

In the case when a foundation is situated in perennially frozen ground (see Fig. 18.10), the value P is defined by the shearing strength, R_{ad}, of the frozen ground along the adfreezing surface and by the area of vertical adfreezing surface F_{ad} equal to the product of the foundation periphery and the depth of placement in the perennially frozen ground:

$$P = R_{ad}F_{ad} \tag{18.5}$$

If the foundation is situated within the *unfrozen seasonally freezing ground*, the forces keeping it from being heaved are defined by friction resistance of the unfrozen ground along the friction surface f_{unf} and by the area of the friction surface F_{fri}:

$$P = f_{unf}F_{fri} \tag{18.6}$$

The specific tangential forces of heaving depend on composition, moisture content, depth of ground freezing and the material of the foundation involved in the adfreezing. Their value varies from 0.05 to 0.13 MPa. When the depth of seasonal freezing is greater than that of the placing of the foundation, the normal forces of heaving are engaged and their value is added to the left part of the equation (18.4). This places fundamentally more stringent requirements upon the foundation structure, because the value of the normal forces of heaving is significantly (sometimes an order of magnitude) greater than that of the tangential forces.

It follows from the formulae (18.2) and (18.4) that the bearing capacity of the base and the forces keeping the foundation from being heaved depend on composition, properties of ground and, in addition, are in direct proportion to the area of the foundation side surface situated in the permafrost. It follows herefrom that one will be able to fulfil the conditions (18.2) and (18.4) and justify the selection of optimal sizes for foundations, ensuring structural stability, by changing the foundation area, increasing or decreasing the depth of its placement into the frozen ground and using various constructive solutions.

The depth of seasonal thawing on which the total value of heaving forces depends is also susceptible to control. By achieving its reduction and using types of foundation having minimal area of adfreezing with the seasonally thawed ground layer, one can decrease the values of heaving forces by limiting the placement depth of the foundation.

However, it is not always possible to achieve stability of the structure under design with the help of constructive solutions only and these solutions may turn out to be economically unattractive. In these cases there is a need to use methods aimed at changing the properties of the ground.

18.4   Normative documents for engineering design and construction in the permafrost regions

Engineering geological surveys, design work and construction are carried out in conformity with documented norms. In accordance with the *System of Norms for Construction* adopted by the USSR State Construction Committee in 1983, they are subdivided into All-Union, department and republic normative documents.

The All-Union normative documents are obligatory for all enterprises, organizations and amalgamations subordinate to the All-Union, departmental and republic bodies, carrying out work in a particular industry. Thus the *All-Union Construction Norms and Regulations* is *obligatory* for all the organizations carrying out construction as well as preconstruction work such as engineering survey and design, independently of their position and departmental ranking. The documents of the former USSR Ministry of

Geology are obligatory for all organizations carrying out exploration. The All-Union documents of the other ministries and departments are of a similar character.

The *Construction Norms and Regulations* dictate the main points for performing works. This document contains the main fundamental requirements and regulations, dictates the main survey and design problems, establishes the procedure for their solving, dictates the types and the character of work, establishes requirements for work organization and the operation of buildings and structures. The Building Regulations are documents having the character of a directive so all the engineering staff of surveying and designing organizations must be acquainted with their requirements.

Clearly it is very difficult to consider all the main survey and design problems as applied to various kinds of construction in one document. Therefore various departmental *Building Regulations* are issued for industrial, civil, hydrotechnical, highway and other kinds of engineering. The main documents regulating survey and design work in the permafrost zone are *Building Norms and Regulations II-18-76 'Bases and foundations on permafrost'*, and *Building Norms and Regulations 2.02.07-87 'Engineering survey for construction'. The Building Norms and Regulations 2.05.02-85 'Automobile roads'*, the *Building Norms and Regulations 2.05.08-85 'Airfields'* and other documents touch on designing in the permafrost zone to some extent as well.

The departmental (branch) *Building Regulations* refer to normative documents. Contrary to the All-Union normative documents they regulate the work specific to a given ministry or department. Clearly the departmental normative documents must not be in contradiction with the All-Union ones. They can merely develop and supplement the latter, reflecting the construction typical for each ministry or department. The former USSR ministries and departments approved departmental normative documents in accordance with the rights given them in agreement with Gosstroy of the former USSR. They are obligatory for all the organizations, institutions and enterprises of the ministry (department) which has approved these documents. These documents can be obligatory for organizations of other ministries and departments if they are approved by the relevant ministry or department.

The republic *Building Regulations* establish requirements for design, engineering survey, construction and operation of buildings and structures taking into consideration the peculiar conditions (natural, economic, etc.) of the republic. These documents are approved by the Union Republic Gosstroy or by other republic bodies in keeping with the rights given them. The republican normative documents are obligatory for all the organizations, institutions and enterprises carrying out design work and construction for structures situated on the territory of the given republic independently of their departmental subordination. These documents must not be in contradiction with the All-Union normative documents nor duplicate them.

Normative documents contain requirements having a direct relationship only to the subject under standardization. They include enumeration of fundamental tenets necessary for carrying out given kinds of work such as survey and design work.

Many kinds of work which are a part of survey and design can be carried out by various methods. Each method is used within particular limits and has its own accuracy and cost depending on numerous natural and engineering factors. Therefore in the majority of cases the *Building Norms and Regulations* must not simply regulate the use of one or another method. Thus, for example, the *Building Norms and Regulations* 2.02.07-87 indicate that in the course of survey it is necessary to make ground temperature observations in bore holes, to determine depths of ground freezing and thawing using various methods. In an effort to unify work performance by one or another method, the normative documents have been prepared regulating work carried out by the unified procedure using equipment of the same kind as far as its main parameters are concerned. Such documents are termed standards in the State standardization system.

Standards in the former USSR are divided into the following categories: former USSR State standards; departmental standards; republican standards; standards for enterprises. Only the State standards are used in engineering geological survey and design practice.

The standards are divided into the following types: for technical conditions, parameters, types, marks, structures and sizes, for specifications, test methods, rules, markings, packing, transportation and storage etc. The standards for test methods and for transportation and storage of samples are used in engineering geocryological survey.

To develop normative documents scientific research institutes and designing organizations develop special handbooks (guides). The whole text of the standard document is retained in these guides. Not being the normative documents they develop the main points of the relevant *Building Regulations*, departmental *Building Regulations* or republic *Building Regulations* in detail, contain examples and calculation algorithms, textual, tabulated and graphical data and other reference sources necessary for survey, design work and construction. The main purpose of these handbooks is to help specialists to understand and to learn the normative document. After the name of the handbook the name and the imprint of the normative document for which it has been developed are given. To develop a single document, one or a few handbooks can be issued depending on its content.

In addition to normative documents and handbooks, various ministries and departments publish a great number of documents of an advisory and procedural character (recommendations, directions as to methods, etc.). They are prepared for subsidiary and individual kinds of work for which there is insufficient experience in practice, with respect to methods of performance, methods of design, testing or determination. The main purpose of these documents is to help in promoting the results of scientific research being carried out by research and design organizations, into survey and design practice as soon as possible. Individual points of the recommendations and methods can be included in the normative documents after evaluation.

19

Engineering geology in support of design, construction and operation of structures in the permafrost regions

19.1 Engineering-geological survey in the permafrost regions

Engineering-geological and geocryological investigations carried out to help in the design of various structures are termed *engineering-geological surveys*. The first practical work in the world on the problem of geocryological investigations for construction purposes was that by V.N. Yanovskiy, *Methods of investigation of permafrost for engineering construction purposes*, published in the 1950s in which experience in the development of the European North and Siberia in the 1930s and 1940s was outlined. At the present time engineering-geological surveys are standardized. The basic document on the problems of surveys in the region of unfrozen as well as perennially frozen ground is the *Building Code and Regulations (SNiP)* 1.02.07-87 '*Engineering surveys for construction purposes*'.

According to the *Building Code* 1.02.07-87 the main problem for engineering-geological surveys is the integrated study of the engineering-geological conditions within the locality where construction is planned. This includes the topographical, geomorphological, seismic and hydrogeological conditions, geological structures, composition, state and properties of the soil, geological processes and phenomena and changes in conditions following construction, in order to get necessary and sufficient information to justify structural designs taking into consideration the harmonious exploitation and protection of the geological environment, as well as to make forecasts concerning changes in the engineering-geological conditions in the course of construction and operation of enterprises, buildings and structures. The engineering-geological survey is closely associated with and precedes the design of the particular structures. This dictates the specific study of the engineering-geological conditions of the locality selected for construction purposes. Clearly the engineering-geological support for the design of hydroelectric stations, underground structures, motor and rail roads, industrial and civil structures requires specific integrated geocryological investigations. Therefore the main points and the details of an engineering-geological survey must depend on the particular type of construction.

The requirements for the individual kinds of engineering-geological work, are determined by the number of design work stages, the complexity of engineering-geological conditions, and the character and degree of importance of the buildings and structures being planned.

Normative documents used today provide for a survey in three main stages. By a stage is meant the finished portion of the particular engineering-geological survey distinguished by the degree of detail and the procedure for the work depending on the number of structural design stages. In line with the established design procedure the engineering survey is performed to develop: 1) *predesign documentation – technical and economic substantiation (TES) and technical-economic estimation (TEE)* for construction, for enlargement, modernization and reequipment of buildings, buildings and structures in service, of overall plans for cities, settlements and populated rural areas; of projects of cities planning industrial zones (regions), of projects of planning in detail; of structural models and overall plans for industrial centres; 2) *designs* (contractor's designs) for plant, buildings and structures; 3) *service forms and records* of undertakings, buildings and structures. The name of the survey stages accord with the design work stages. Therein lies the succession and interrelation of engineering survey and design work as a single process.

The engineering-geological (engineering-geocryological) survey is performed under a program developed by surveyors (engineering geologists, geocryologists, hydrogeologists) on the basis of the *technical assignment* given by the designing organization. The successful survey has a number of requirements imposed by the technical assignment in accordance with particular features of the structures under design and their characteristics. These details are different at various design work stages. Thus at a predesign stage the technical assignment must contain the following data: the character of construction (new construction, modernization, enlargement, reequipment) and of the enterprises under design, the importance of the buildings and structures and the effect of the structures under design on the environment. When industrial and civil construction is carried out it is necessary to have a schematic of the overall plan for the site representing the most economical arrangement of the structures under design taking into account technological and other particular features of their mutual arrangement.

In the course of survey at the design stage the technical assignment must contain data on the justification of the principle of using ground as foundation base, on the proposed types of foundations, on the laying depths of subsurface portions of buildings and structures and on the interaction of the structures under design with the geological environment.

The technical engineering survey assignment for developing working documentation must include data on which principle is accepted for using ground as a foundation base, on the technological function and heat regime of the structure under operation, on the way of laying service lines and on the measures providing the accepted principle of construction and control of the cryogenic processes.

Study and assessment of the terrain (for alternative sites for route sections) is to justify selection of the most favourable site for the particular structure to be located, constructed and to operate, taking into consideration protection of the structure against dangerous geological processes and protection of the geological environment. At this stage the principle of using perennially frozen ground as a foundation base (by keeping it frozen or by thawing it) and the estimated cost of construction are determined, and the area under survey is two times or more greater than that of the particular construction site. Sometimes not one site but a set of competing sites, which can be widely separated, is considered.

The survey at this stage includes the basic work in the field, laboratory and office. The work includes collection, analysis and classification of materials from the preceding survey, materials from the geological survey, and engineering-geological and hydrogeological work and other data on natural conditions within the region under design. So, the landscape is classified with preliminary compilation of a landscape map and the engineering-geocryological conditions are noted on this base, using results of interpretation of air photos.

The field work includes reconnaissance, geocryological study of the area to improve and extend the accumulated data, to give a comparative assessment of the complexity of geocryological conditions within the area under investigation, to establish the distribution and development of dangerous geological processes, to reveal changes in geocryological conditions within the developed terrain, as well as analysis of construction experience. In the course of the reconnaissance, field observations, geophysical investigations, drilling with soil sampling, groundwater sampling and laboratory investigations of the samples are carried out.

In poorly known regions where geological investigations have not been performed earlier and for which construction experience is absent, it is necessary to carry out the survey of sites at the 1 : 25 000 – 1 : 10 000 scale and of linear structures at the 1 : 50 000 – 1 : 25 000 scale.

During the office work period the report on the survey is prepared, in which natural and engineering-geocryological conditions within the territory are described. An essential requirement for the report is a forecast of changes in the geocryological situation under natural conditions and in the course of development of the site. The report is supplemented with the map of landscape types, a geocryological map with the terrain classified by construction conditions and with engineering-geocryological sections, graphs and tables of the permafrost and ground ice properties. When incorporated in the technical assignment the report is supplemented with maps of types of seasonal ground freezing and thawing, ground ice content, permafrost thickness and cryogenic processes and phenomena.

Comparative assessment of the construction sites and of the routes of linear structures and alternatives, with justification of the selection of the best and determination of the construction principle to be followed, are the result of work at the predesign stages of the engineering geological survey.

The engineering-geological survey within the selected site (design stage) must provide the materials necessary and sufficient to justify the particular site for the buildings and structures, to select the approach to their construction, to determine measures for engineering protection of the ground, for geological environmental protection and for developing the designs for construction and organization of the work. For these purposes an integrated study of the engineering-geological conditions within the site or route under construction is carried out. This investigation includes field, laboratory and office work. The engineering geological survey and experimental field studies of ground composition and properties are the main methods for field investigations. The survey is carried out at a larger scale compared with that being used at the predesign stage, 1 : 5000 – 1 : 2000 for industrial and civil engineering structures as a rule and 1 : 10 000 – 1 : 5000 for the routes of linear structures.

Beginning from the stage of the survey, permanent observations including those of permafrost temperature dynamics, of cryogenic processes, of seasonal freezing and thawing dynamics, and of changes of properties of thawing and freezing ground are considered in the design according to the normative documents.

In the course of office analysis of survey materials, the geocryological assessment of the site with forecasts of the changes in permafrost conditions in connection with construction are made, recommendations are developed on the arrangement of the constructions under design, on their base and foundation design, and measures for controlling the cryogenic process are considered. Compilation of a report is the result of the survey in the same manner as at the first stage. The report must contain data on geocryological conditions within the site for construction, results of field and permanent investigations of ground properties and improved recommendations on the selection of the particular principle for use of the ground as a base.

The engineering-geological survey task for particular building sites (the stage of working plans) consists of the determination of geocryological conditions and of the state and properties of sites for particular buildings and structures, necessary to design foundations according to the technical approach accepted. Engineering-geocryological mapping is not carried out at this stage. Surveys include engineering-geological (engineering-geocryological) prospecting consisting of drilling, geophysical investigations, pilot field investigations (testing of frozen ground by a heated probe, of ground pillars for shearing, of ground by static loads on piles, etc.) within the limits of buildings and structures. Fixed observations initiated at the second stage are continued.

In addition to overall geocryological characterization of the site the data necessary for foundation design are included in the report on the survey results. When the permafrost is used as a foundation base on the first principle, the design mean annual ground temperatures at depth in the geological section, the design values of pressure on the permafrost, of permafrost shearing strength, results of field testing, data on frost susceptibility of the seasonally thawing layer and results of permanent observations are presented. When the permafrost is used as a base on the second principle, the physical and thermophysical characteristics for heat engineering calculations for thawed and thawing soils at various depths within the thawing zone under the structure are presented, as well as the design strengths of the soils after being thawed, data on frost susceptibility of the seasonally frozen layer and the results of field and permanent observations.

The concepts presented are common to all kinds of on-land industrial and civil engineering. Developments such as hydrotechnical construction, underground construction and agrobiological utilization of the permafrost areas require additional survey. The content and volume of this work are regulated by normative documents of departmental ministries. In some cases a specially designed survey is carried out for an individual design, according to the particular technical requirements.

19.2   Forecasting change in the geocryological conditions in the course of development

According to V.A. Kudryavtsev (18) *geocryological forecast* refers to the scientific prediction of changes of geocryological conditions which will occur in the future, either in the context of natural circumstances or in the course of economic development.

In line with this definition two types of forecast, the evolutional (often designated natural-historical) and the technological are distinguished. The first includes forecasting the changes in permafrost characteristics and of the processes within it under the effect of the natural dynamics of climate, of neotectonics, of world ocean levels, of denudation and sedimentation processes, of ice covers and of hydrogeological, hydrological and geobotanical conditions. It also includes forecasting of the reverse effect of the geocryological characteristics on the components of the geosystem.

Technological forecasting includes assessment of changes of geocryological conditions under the effect of various technological disruptions of the natural system (beginning with changes of the local landscape and geocryological conditions and ending with global natural transformations such as creation of water reservoirs, industrial air, surface and groundwater pollution, changes in the temperature regime, etc.). Engineering-geocryological forecasting is one of the most important divisions of the technological forecast. The engineering-geocryological forecast is made to solve such practical problems as selection of construction sites and routes for linear structures, assessment of arrangement of structures and the selection of principles for their construction, selection of methods for highway location and of mining methods and a number of other questions necessary for design, construction and operation of various structures. On the basis of geocryological forecasting measures are recommended for eliminating or limiting the consequences of disruption of the balance of natural geosystems, dangerous to structures and the environment, and methods for permafrost control are developed. At the same time in line with the character of technological impact a forecast is made of changes in ground temperatures, of the amount of permafrost freezing and thawing and of changes in its composition, structure and properties and the possibility for the activation and initiation of new geocryological processes, and the time for their stabilization, and of conditions causing their progressive development. In all cases the engineering-geocryological forecast must take into account the results of the evolutional one. The character and formulation of the problems depend on regional features of the terrain and on the detail of the investigation (geocryological survey scale) and its aims.

The main objective, problems and methods of the forecast at various stages of engineering geocryological investigations are presented in Table 19.1. We can see that when the forecast is made at small and medium scales the assessment of sensitivity and stability of the permafrost to technological

Table 19.1. *Stages, problems and methods of geocryological forecast*

Stages of engineering geological investigations	The aim of cryological forecasting	The main tasks for the cryological forecast	The methods of forecasting
Stage I – Master plans for development of national economy on the basis of engineering-geological investigations of natural regions and zones at general scales (1:2 500 000–1:1 000 000)	The assessment of the influence of the geocryological situation on natural resources	1. Forecast of the natural dynamics of the geocryological conditions. 2. Assessment of the sensitivity of the geocryological situation to the technological impact. 3. Assessment of the possibility of controlling the 'cryogenic processes' for effective utilisation of natural resources.	1. The method of analogies on the basis of experience of development. 2. Logical and mathematical simulation. 3. Expert assessment.
Stage II – The diversified development of individual regions on the basis of national engineering-geological survey at the 1:500 000 (1:100 000) scale	To guarantee the rational use (optimizing, arrangement and type of construction included in the industrial complex)	1. Assessment of the stability of the geocryological situation and of permissible technological impacts. 2. Assessment of the reverse effect of changes of the geocryological environment on landscape development. 3. Forecast of the efficiency of measures to ensure the reliability of the construction and environmental protection.	1. The method of analogies on the basis of the study of construction experience. 2. Mathematical and physical simulation (including observations of regime. 3. Extrapolation. 4. Classification of permafrost and cryogenic processes.

Table 19.1. (*Continued*)

Stages of engineering geological investigations	The aim of cryological forecasting	The main tasks for the cryological forecast	The methods of forecasting
Stage III – Engineering-geological suport for the design of industrial and other structures on the basis of the engineering-geological survey at the scale of 1:50 000 to 1:2 000	Engineering-geocryological assessment of building sites and routes	1. Forecast of the influence of the engineering preparation on the geocryological environment. 2. Forecast of the thermal and mechanical interaction between the structures and ground. 3. Forecast of the geocryological environment in connection with use of ground amelioration methods and landscape recultivation.	1. Mathematical and physical simulation. 2. Method of analogies on the basis of study of construction experience.

impact is one of the main results. In this case by 'sensitivity' is meant the response of the geosystem to the impact and the degree of its change, while by 'stability' is meant the capacity of the geosystem to resist the impact without any change in its state and structure, i.e. without such changes of the components of natural complexes and interactions between them which could cause impermissible deformations of structures or irreversible deterioration of the ecological situation. In essence, sensitivity and stability are closely related concepts and are specified by the 'value' of changes in geocryological characteristics. However in the latter case some restriction on the limits of permissible changes is assumed depending on the practical problem under solution. Clearly, the stability of geosystems can be assessed from various aspects, with respect to the various kinds of structures, as they respond to changes in ground conditions, properties and in the processes underway. The same changes in geocryological conditions can be dangerous or not depending on the kind of structure. And if the stability of geosystems is defined by permissible technological impact, not causing dangerous changes in engineering-geological conditions, they can also be essentially different depending on the kind of structure (for example, permissible impacts in the course of airfield construction differ from permissible changes in the course of highway construction and, especially, in the course of industrial construction). At the same time for the same kind of structures the permissible impacts depend on the initial natural situation. Thus, the concept of 'stability' is largely a matter of convention and in each particular case one should note the aspect of its application.

The sensitivity of the geosystem (of the natural complex) does not depend on technological impact – being the property of the system, the capacity to respond to the impact. Therefore this concept is more distinctive, it is always characterized by the degree (value) of changes of individual (or several) geocryological characteristics under the individual (or a set of) impacts.

Technological geocryological forecasts are subdivided with respect to impact of construction on the natural environment into: a) general forecasts, providing for the assessment of the changes in components of the natural complex without taking into consideration the thermal and mechanical impact of the structure itself; b) engineering forecasts, which include the assessment of the results of the direct impact of the structures on the geocryological situation.

When the *general geocryological forecast* is made the possible changes in the permafrost conditions as a result of such processes as disturbance of vegetation cover or of snow accumulation conditions, peat removal and replacement of soil, grading, change of surface and ground runoff, arrangement of artificial cover, grass plots, trees and shrubbery plantations, etc. are described. The necessity for the general geocryological forecast when the engineering problems associated with economic development of the permafrost zone are being solved, stems from the fact that heat releasing and heat absorbing structures under any kind of construction occupy only a part of the area of the disturbed terrain. Thus, according to G.V. Porkhayev's and V.K. Shchelokov's data the density of buildings in northern cities of the former USSR accounts for 13–40%, the rest of the area of the cities is occupied by roads, streets, squares, public gardens, etc. According to [All-Union Research Institute of Hydrogeology and Engineering Geology] VSEGINGEO'S data heat releasing structures occupy only 9% of the area of Western Siberian gas fields; within the rest of the area changes in natural conditions in the course of construction take place. In addition there are numerous cases in practice when there exists a time lag between the beginning of the construction and putting the structure into operation. In such cases the results of the general forecast serve as initial data for making the engineering geocryological forecast.

When the *engineering forecast* is made the particular features of the thermal and mechanical impact of structures on the permafrost are assessed.

Various methods are used to assess the technological impact on the geocryological situation. Among these methods those of mathematical modelling have received wide recognition. However experience shows that efficiency and reliability of the forecasts (as far as the whole range of problems is concerned) increases when several methods are used, such as physical and mathematical simulation methods, methods of analogues, analysis of construction experience and extrapolations and classifications.

When making the engineering-geocryological forecast using one method or a series of methods, one must use the universal method consisting of successive studies of the geocryological conditions with the assessment of the role and effect of each natural environmental factor and typification of technological impact, on the basis of the requirements of the projected engineering structure and with the background of experience. The forecast is made as a result of solving a set of problems to assess the possible changes in engineering-geocryological characteristics and the efficiency of amelioration measures intended to guarantee the reliability of the structure and to control the dangerous engineering-geological processes.

Depending on the duration of time for which the geocryological conditions are predicted, the forecast is subdivided into short-, long- and ultralong-term. The short-term forecast is made for a period of up to 10 years and characterizes changes in geocryological conditions under the effect of short-term (3–11 years) climatic fluctuations, excavation, construction work and operation of structures during the first years, when changes of the state and characteristics of the ground, especially near the surface (in the layer of seasonal thawing and in the layer of annual ground heat storage), proceed most intensively and can be dangerous for the structures. The long-term forecast is made for the period from 10 to 100 years and characterizes changes in geocryological and engineering-geological characteristics corresponding to the new steady-state temperature and moisture regime of the ground. The ultralong-term forecast is made for especially important structures, the period being longer than 100 years. It is used mainly to assess changes in geocryological conditions under the effect of natural environment dynamics or of regional and global transformations causing changes in ground thermal conditions over large areas, as well as to assess the long-term effect of structures on the geocryological situation.

It was shown in Chapter 17 that the various kinds of construction have different effects on the environment. Therefore the range of problems of forecasting the ground temperature regime will have particular features for each kind of construction.

In the course of *urban construction* the ground temperature regime is formed under the effect of many factors which are arbitrarily divided by G. V. Porkhayev and V. K. Shchelokov into three groups: common, local and particular. Among the common factors are components of external heat and mass exchange within the construction area such as radiation balance, turbulent heat exchange, heat expenditure on evaporation and condensation of moisture at the ground surface. Among the local factors causing change in ground temperature regime within a comparatively small area are the thermal effects of buildings, structures and service lines. Among the particular factors are those typical of particular places only. For one region these may be hydrogeological features, for another these may be atmospheric circulation conditions, etc. Forecast of changes in geocryological conditions within the locality where construction is taking place must be made taking into consideration variations in the enumerated factors: common and particular factors are taken into consideration in general forecasts, local ones in engineering forecasts.

The variability in heat sources and flows within the ground causes complex effects on the ground temperature regime. Thus, for example, in parallel with formation of thawing zones under heat-releasing buildings and structures, lowering of mean annual ground temperature and freezing can occur within an area where taliks existed before construction.

The problems of technological geocryological forecasting in the course of *linear construction* include assessment of changes in ground temperature regime within the route, beyond the area under the direct effect of the linear structure (general forecast) and within the area under heat and mechanical effects of the structure (engineering forecast). In the latter case the necessary parameters of embankments, excavations and artificial structures (such as heights and material for the embankment, the necessity for thermal insulation layers, the depths of excavations, etc.) for rail roads and highways are established for the purpose of selecting the appropriate principle of using ground as the roadbed base. For pipelines the thickness and dynamics of thawing or freezing annuli are assessed, providing the means for selecting the appropriate manner of laying, the technological regime of operation and particular construction features.

The main problems of forecasting in the course of *hydrotechnical construction* are to facilitate the selection of the design of a dam body (in the frozen or in the thawed state); the design of cooling systems required for keeping the core of a dam frozen; and the estimation of the dynamics of the many-year ground thawing under water reservoirs. Assessing the intensity of abrasion of banks and of changes in natural conditions (including geo-cryological) within the territory adjacent to the reservoir is of great import-ance when the hydrotechnical construction is carried out, because very large water masses cause fundamental changes in microclimatic conditions and their effect can extend a considerable distance from the water storage body.

When *subsurface construction* is carried out the most important problem for forecasting is assessment of the ground temperature regime to select the safest and most economical way of mining, sinking of shafts, tunnelling or pile-driving.

Technological geocryological forecasting for the purposes of *agrobiologi-cal development* includes the problem of providing the optimal soil climate, and the assessment of ground temperature regime and of depths of seasonal freezing and thawing, as well as the assessment of the duration of positive and negative temperatures at the soil surface and at various depths.

Forecasts of changes in the permafrost *properties* as well as in other geocryological characteristics are based on the knowledge of their nature as discussed in Chapter 8.

Formulation of the problems in the forecast of changes of properties depends first of all on the forecast of changes of ground temperature regime. Two cases are considered here: 1) when the mean annual ground tempera-ture changes (increases or decreases) remaining negative, when changes of the proportions of unfrozen water and ice in the soil occur; 2) when the sign of the mean annual temperature changes and changes in state of soil aggregation, composition and structure occur. Forecasts of changes in soil properties can be made in the first case from the results of field and laboratory tests conducted in the negative temperature range at various densities and moisture contents, and with various soil structures, taking into consideration possible changes in conditions in the course of construction and operation of buildings. In the second case when change of the sign of the mean annual temperature at the soil surface causes perennial thawing there is a need to assess the change of thermal-physical properties, to determine thaw settlement, shearing strength and other properties of thawing soils. If new permafrost formation takes place the problem arises of determining changes in the thermal-physical and mechanical properties of thawing soils depending on the new steady-state regime. This concerns possible changes in density, the total moisture of the frozen soil being assessed based on its composition and initial density (that is, existing at the moment of freezing) and moisture content.

Forecasts of *cryogenic geological processes* are based on the establishment of relationships between the nature of each process and the factors responsible for the process. Methods for forecasting quantitative cryogenic processes as a whole are developed in less detail than those for forecasts of temperature regime and soil properties; this is associated with the complexity of the physical models of the processes – with the necessity to take into consideration when the forecast is made, the great number of parameters responsible, and with the complexity of mathematical simulation of the cryogenic processes.

The greatest amount of work is devoted to methods for assessment of the *soil frost susceptibility index*, the quantitative characteristic of the soil's heaving value, and of nonuniformity of the heaving processes over the area. These are obtained using calculation methods put forward by N.A. Puzakov, I.A. Zolotarev, V.O. Orlov, G.M. Fel'dman, E.D. Yershov and others. The problems of quantitative forecasting of natural and technological thermokarst are considered in papers by G.M. Fel'dman, Yu.L. Shur, V.V. Lovchuk and M.S. Krass among others, in various detail. The forecasting of the possibility of solifluction processes developing can be made using the approximate formulae put forward by L.A. Zhigarev and V.S. Savel'yev. Methods for the forecasting of thermal erosion are considered in papers by E.D. Yershov, D.V. Malinovskiy and V.K. Dan'ko.

19.3   Principles and methods of the control of cryogenic processes

When development is undertaken in the permafrost regions one is repeatedly forced to solve the problems associated with the necessity of changing the geocryological conditions in the direction favourable for the national economy. Defining the problems of the control of the cryogenic process on its own requires knowledge of the natural conditions and of the current geocryological situation which will be subject to artificial change at a later time.

A combination of work carried out to control the cryogenic processes is performed in a certain order. *Study of the natural (including geocryological) conditions of the region under development is fundamental and obligatory.* The study of geocryological conditions is based on revealing the particular and general principles of formation and development of seasonally and perennially frozen ground and its characteristics depending on the combination of natural conditions and on each element of the geological-geographical environment. This problem represents the immediate object of geocryological survey and is solved successfully in the course of surveying. Therefore clearly the geocryological survey must be a component part of the integrated investigations for the control of cryogenic processes and must precede all other kinds of investigations. This stage ends with the compilation of geocryological maps.

The next stage of work includes *description and typification* of the projected technological loads and impacts in the course of the development, and the revealing of the associated changes in natural conditions. This stage of investigations is carried out to get the starting data for the geocryological forecast.

Then, the *geocryological forecast* is made on the basis of the revealed characteristics in the formation of the geocryological conditions and consideration of technological impacts. Possible changes of mean annual ground temperature, of soil composition and properties, of the depths of seasonal and perennial thawing and freezing, and of the character of cryogenic processes are assessed in the geocryological forecast.

The problems of the next stage of investigations include revealing the areas *within which geocryological conditions can develop which do not meet the requirements that might be imposed on them.* In the design of specific buildings and constructions these can be areas within which there is no possibility of implementing the accepted principle of construction. For example, in the course of commercial mineral development there are localities within which the practical way of extraction cannot be applied. Within these areas desired changes of the geocryological conditions (of permafrost distribution and conditions of occurrence, of temperature regime, of seasonal thawing and freezing depths, of composition, structure and properties of seasonally and perennially freezing ground, and of cryogenic processes) are made clear by the results of the forecast.

Such a desired change of individual characteristics of frozen ground can be carried out using various methods of amelioration. Therefore it is necessary to select the most acceptable, rational ways in each case. For this purpose it is necessary first of all to study the methods of water-thermal amelioration and to justify the possibility of use of one or another method for directed change of the natural situation in the given region to create the desired permafrost conditions. It is necessary that the estimate of the economic efficiency of the measures used should take into consideration the time allotted for them to be carried out. It should be noted that each method, in parallel with its main (direct) effect on the individual elements of natural surroundings, causes changes of other elements of the geological surroundings to a greater or lesser extent. In some cases this can cause undesirable changes in the geocryological conditions. The results of this stage are elaboration of *the measures aimed at purposeful change of the permafrost conditions*, and regionally, separation of the areas characterized by similar permafrost conditions within which the particular methods (or set of methods) for the control of cryogenic process must be used.

At the final stage of investigations, each of the distinct areas of the region under amelioration is characterized with respect to the artificially created geological-geographical and geocryological conditions which will occur after amelioration. *The post-amelioration geocryological maps are compiled.*

A set of measures aimed at the purposeful change of geocryological conditions is performed in the following order.

1. Study of the geocryological conditions in the course of the geocryological survey.
2. Characterization and typification of technological impacts and loads.
3. Forecast of changes in geocryological conditions.
4. Substantiation of the need for designed changes of geocryological conditions.
5. Elaboration of the project measures aimed at designed changes of geocryological conditions.
6. Compilation of post-amelioration maps.

There exists now a great number of methods for control of cryogenic processes. Some of the methods of frozen ground amelioration widely used in the course of economic development in the permafrost regions were considered above. V. A. Kudryavtsev and E. D. Yershov have developed a classification of methods for changing geocryological conditions, in an effort to straighten out the existing methods. This classification allows selection of the most efficient measures as far as any kind of economic development of the permafrost region is concerned. The classification is based on consideration of the thermal-physical aspects as well as of the geological-geographical aspects of the cryogenic process to be controlled. All the existing complex of methods is subdivided into three groups.

The first group of methods changes the cryogenic process directionally by changing the parameters of the *external heat exchange* (Table 19.2). It includes two systematic methods; one of them regulates the relation between the radiation balance components and the other regulates the relation between the heat balance components. Each method consists of three sections, changing the elements of the external heat exchange: the integrated and the reflected shortwave radiation, the effective surface radiation of the Earth, the turbulent heat exchange with the near-surface air, the heat of moisture phase transitions (evaporation – condensation) and the heat exchange with the ground below. The main measures for changing the external heat exchange are the following: arrangement of sheds or awnings, surface covering by various diathermal films, artificial colouring of the surface, removal and planting of vegetation cover (shrubbery and trees), a modification of snow accumulation conditions, etc.

The second group combines the measures aimed at control of the processes of *heat- and mass-exchange in ground by changing composition and properties of the ground* (see Table 19.2). It is subdivided into two systematic methods: the first, involving changing the composition, properties and thermal state; the second, involving changing the properties and thermal state of the ground under amelioration. The first system of methods includes two sections. One section includes the measures facilitating the change of the composition of the organic-mineral portion of the soil, the other section is aimed at changes of ground moisture regime. The measurements for the second system allow regulation of heat- and mass-exchange in the ground by changing the properties and the thermal state of the ground without any essential transformation of its material composition.

The methods of the third group are used to change the temperature regime and the heat state of the ground under amelioration *by using additional heat sources and sinks* (see Table 19.2). It is subdivided into two systems using natural and artificial sources and sinks of heat. The first system includes two sections recognized with respect to the kind of heat carrier (water, air) being used to intensify the mass-exchange process. The second system of methods consists of three sections. Each of them uses either electric power, or thermochemical mixtures, or steam, fire, or 'thermal probes' as artificial heat sources.

The selection of control methods is fully individual in each particular case and depends on the conditions of the geological-geographical environment, on the particular features of the economic development and on the degree of efficiency of use.

When planning the amelioration measures one should bear in mind that any method, any measures being used to change a particular parameter of the environment, will inevitably result in the change of the whole complex of characteristics of the ground under amelioration to some extent. Therefore the decision to use a particular measure aimed at changing the permafrost situation, must be taken on the basis of knowing the effect of a given method on all the factors of the geological-geographical environment.

Table 19.2. *A classification of the methods of control of cryogenic processes*

Classes of methods	Sub-classes of methods	Parameters through which the natural elements being changed can be transformed	Methods of control
A group of methods regulating external heat exchange			
Modifying the radiation balance components	The integrated shortwave radiation, $Q + q$	A. Transparency of the Atmosphere B. The angle of inclination of the surface to the line of the horizon. C. Orientation of the area	Arrangement of sheds, shading by way of tree planting; shrubbery planting; change of the angle of inclination and area lay-out; screening of surface with diathermal films.
	The reflected shortwave radiation, $(Q + q)A$	A. Character and height of the surface cover B. Colour of the surface. C. Surface roughness D. Moisture content of the soil surface layer	Removal of natural covers; artificial colouring of the surface (blackening, whitening, etc.); loosening, compaction by rolling; wetting, drainage of the ground surface layer.
	The effective Earth surface radiation I_{ef}	A. Air temperature and vapour pressure. B. Colour of the surface, its roughness, character and height of cover. C. Temperature of the radiating surface	Smoke; screening of the surface with light transparent films, snow and other covers; removal of the vegetation and snow covers from the ground surface; direct control of surface temperature.
Modifying the heat balance components	Turbulent heat exchange with the near-surface air, p	A. Wind profile. B. Surface roughness measure. C. Ground surface temperature. D. Moisture content of the gound surface layer.	Arrengement of wind protective barriers (screens, tree plantations, etc.); change of the surface roughness; natural thawing of frozen ground in the course of layer by layer development; direct control of temperature and moisture content of the ground surface layer.
	Heat of moisture phase transitions, LE	A. Moisture content and condition of moisture inflow in the ground surface layer. B. Wind profile. C. Surface roughness measure. D. Temperature of ground surface.	Loosening and compacting by rolling of the ground surface layer; thawing of frozen ground in the course of layer by layer development; covering of the surface by weakly moisture permeable materials (mulch, synthetic films); impregnation of the surface ground layer by adhesives (peat glue, lignin, etc.); treatment of the upper ground layer by surface active substances; control of wind speed and ground surface temperature.

Changing composition, properties and thermal state of ground under amelioration	Heat exchange with the underlying ground, *B*	A. Ground surface temperature. B. Ground temperature regime (temperature gradients in ground). C. Ground composition, moisture content and properties.	Contol of snow and vegetation cover thickness and properties; covering of the ground by heat insulating materials (foam plastic, snow-ice, air-ice, water-ice and other artificial covers).

A group of methods modifying heat exchange by changing composition and properties of the ground under amelioration

Changing composition, properties and thermal state of the ground under amelioration	Composition of the organic-mineral portion of the ground	A. Grading, chemical-mineral ground composition and exchange cation composition. B Organic matter composition.	Complete or partial replacement of the ground under amelioration; mud injection grouting, bituminous grouting, silica enrichment, impregnation of the ground by synthetic resin; wash-out of fines from the ground, composition and content of exchangable cations in the ground; repeated change of ground thermal state; lime pretreatment of acid soils and application of gypsum to saline-like soils; soil humification and mineral and bacterial fertilizer application to soils.
	Moisture (ice) content of soil and composition of soil air	A. Grading, chemical – mineral composition of soil. B. Rock jointing and porosity. C. Conditions of ground water feeding, movement and run-off. D. Thermal state, temperature regime of ground as well as conditions of freezing and moisture migration in the soil.	Supply of water (irrigation) and dewatering (drainage) of soil; frost drainage; electroosmosis and electro-chemical ground stabilization; change of composition and content of exchange cations in the ground; mud injection, grouting, bituminous grouting, silicatization, impregnation of the ground by synthetic resin; deep loosening and artificial consolidation; change of ground water level; soil freezing and thawing and freezing rate control.
Changing properties and thermal state of ground under amelioration	Soil properties and thermal state	A. Textural-structural features of ground (including the nature of ground ice and soil cryogenic structure) B. Ground thermal state and temperature regime. C. Ground freezing-thawing conditions.	Deep soil loosening; soil consolidation with the help of explosion; soil consolidation and deconsolidation with the help of vibrators; freezing-thawing change of the ground freezing conditions and direct temperature regime control.

Table 19.2. (*Continued*)

Classes of methods	Sub-classes of methods	Parameters through which the natural elements being changed can be transformed	Methods of control
A group of methods changing temperature regime and heat state of the ground under amelioration by using additional heat sources and flows			
Using artificial heat sources and flows	Water as heat carrier	A. Filtration index of ground in thawed state. B. Nature of ground ice, soil cryogenic structures. C. Slope of the locality. D. Temperature of water used, thermal properties of soils and their temperature regime. E. Rate of flow.	Bore holes; seepage – drainage; sprinkling infiltration; conductive – seepage (thermal baths)
	Air as heat carrier	A. Composition; moisture content and properties of ground. B. Nature of ground ice, cryogenic soil structures and their temperature regime. C. Size of the cooling system surface. D. Volume (rate) of air in the cooling system and its temperature.	Natural ground freezing in the course of interaction between the atmospheric air and bare ground surface; cooling devices (ventilated cellars, ventilating conduits, ventilating pipes).
	Steam, fire, various 'heat injectors'	A. Ground composition, moisture content and properties. B. Ice content and cryogenic structure of soils and their temperature regime. C. Volume (rate) of flow of heat carrier through the ground and its parameters. D. 'Heat injection' sizes and their temperature.	Steam point thawing, methods of borehole thawing using artificially heated water, surface run-off of hot water; thawing by 'ignition' and by heaters; placement of hot rock material (quarry stone)
	Electric power	A. Ice content, cryogenic structures, unfrozen moisture content, ground temperature regime and properties. B. Voltage applied to the electrodes. C. Shape and sizes of electrodes.	Volume heating of soil by electric current drawn through the ground; electric heaters.
Using natural heat sources and flows	Thermo-chemical mixtures	A. Composition, moisture content and properties of soils and their temperature regime. B. Kind of thermo-chemical mixtures. C. Amount (volume) of mixtures used.	Coolers; cooling installations of various types; zeroters; use of heat effect of physico-chemical reactions.

The amelioration work includes as a rule not one but a number of methods complementing each other.

19.4 The basis of the rational use and protection of the geological environment in the permafrost regions

The rational siting of industrial projects, optimal development of regional infrastructure, safe and economical construction, safeguarding the region under development from dangerous geological and geocryological processes, and environmental protection must be based first of all on knowledge of the formation and development of the geocryological environment. Clearly the programme of rational development must be drawn up in accordance with the scheme of national regional planning and design. a) At the level of development of general plans for national development (All-Union, departmental and republic schemes of development and disposition of production etc.). This is when the fundamental decisions are taken on kinds and methods of construction and of extensive regional development, and on ways of protecting the environment in the course of realization of the transformation programme (the first stage). b) At the level of regional planning when the schemes (projects) of regional planning provide siting of national economic projects (mining, power-generating, industrial centres, cities and settlements, etc.) as well as the arrangement of recreation and environmental protection zones (the second stage). c) At the level of design of particular industrial and other enterprises, when the particular projects for rational use of the geological environment are worked out. This includes protective and environmental protection measures guaranteeing safe construction, necessary comforts for human activity and living conditions and protection of the geological environment against harmful technological effects (the third stage). Clearly it is essential that continuity of decisions should occur at the different stages of drawing up the plan for rational use of the geological environment.

At the first stage the integrated schemes of rational use, control and protection of the region's geological environment must be developed on the basis of existing national geological, engineering-geological and geocryological maps and other materials. Concrete definition of the plans for rational use and protection of the geological environment in the course of development of regional planning schemes (projects) calls for special-purpose engineering-geocryological and hydrogeocryological surveying of the locality under development. It is necessary that the plans for environmental protection measures should be drawn up on this basis in accordance with the plan of development of the area. Detailed development of the protective and environmental protection projects (the third stage) is carried out on the basis of engineering geocryological survey materials.

One of the main challenges in the rational use of the geological environment is to guarantee the reliability of constructions. As noted above two alternative principles are used in operating with permafrost as a foundation for construction within the permafrost regions. To select which of them it is necessary to have information not only about the engineering-geocryological conditions of the area under development but also about the structures. In the absence of the latter, the question of the selection of the construction principle cannot be solved in the majority of cases. However one can give some recommendations on the primary use of one or another principle in the context of the small-scale engineering-geocryological mapping. Thus, it is advisable to perform the construction on the first principle in the regions where low-temperature ice-rich and highly ice-rich ground occurs (with ice content higher than 0.2-0.4 on account of the ice inclusions). Preference is given to the first principle in regions of high seismicity, because use of the second principle increases the seismic design and the cost of construction. It is appropriate to use the frozen ground according to the second principle in the regions where hard rocks and ground of low thaw consolidation are developed. It is possible to use both the first and the second principles in the regions with permafrost islands.

Thus, the information on distribution and mean annual temperature of frozen ground, on the prevailing complexes of pre-Quaternary rocks, on 'genetic' types and ice content of Quaternary deposits, on the genetic type of macrotopography and seismicity of the region, taking into consideration the accumulated experience of construction within the permafrost regions, allows selection of the principle of construction (Fig. 19.1). The schematic map shows the regions where: 1) only the first principle of ground as a foundation for structures is used; 2) the first principle is mainly used, but the second one can also be used; 3) the second principle is mainly used, but the first one can also be used; 4) only the second principle is used.

Clearly the given map cannot provide an unambiguous answer on the selection of the principle of construction, by virtue of its sketchy character; however, it guides the researchers toward the advisability of selection of one or other principle within each part of the permafrost region.

Rational use of groundwater within the permafrost regions includes the solution of the problems of its protection against pollution and exhaustion. The exhaustion of the groundwater resources occurs in the course of extraction from the interior of the Earth in amounts in excess of natural (or artificial) replenishment, as well as when the groundwater recharge and

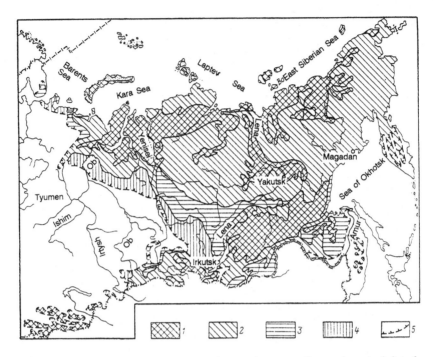

Fig. 19.1. A diagram showing the permafrost region according to the principles of usage of the ground as a foundation: 1 – principle I only; 2 – principle I mainly, however principle II may be followed; 3 – principle II mainly, however principle I may be followed; 4 – principle II only; 5 – the permafrost limit.

replenishment conditions are disrupted in the hydrogeological structure. The latter is most probable and dangerous for those permafrost regions where the water-thermal regime is most sensitive to external impacts. The exhaustion of the groundwater in the permafrost regions can be prevented if certain rules of rational use are observed and a number of measures taken. First of all it is necessary that the areas of deep groundwater recharge should be kept unfrozen. These are radiation-heat, underwater, seepage and ground-infiltration taliks. Their freezing can be caused by industrial development of a region involving vegetation reduction, snow cover removal and consolidation, or creation of artificial covers on taliks during road construction, etc. Secondly, it is necessary that discharge of deep groundwaters through mine openings (holes and shafts), left after prospecting for mineral deposits and their development, be prevented. And finally, the artificial replenishment of the groundwater must be organized, involving a series of measures aimed at increasing the amount of groundwater in the water-bearing horizons, complexes and the whole hydrogeological structures.

Protection of groundwater in the permafrost regions against pollution stipulates a series of measures aimed at prevention, retention (or improvement) of its quality and the elimination of negative effects of technological loads. First of all the assessment of the natural capacity of groundwater to protect itself against pollution should be carried out on the basis of revealing protective factors preventing the contaminants from entering the groundwater basin. The zone of aeration, the first water confining stratum below the surface and hydrodynamic isolation of water-bearing horizons must be analyzed in this context. For the permafrost regions it is necessary to analyze the character of the perennially frozen ground and distribution of taliks, because the former constitutes a complete water confining strata, while feeding, discharge and flow of the groundwater occur through taliks.

The structures situated in the zone of continuous permafrost are less sensitive to technological impacts; the structures situated within the zone of island and discontinuous permafrost with wide development of precipitation-radiation taliks are most sensitive. One can recommend creation of special-purpose water protection zones for each type of hydrogeological structure within which it is necessary that a series of measures aimed at ground water protection be performed. Protection of groundwater against exhaustion and pollution for the permafrost regions as a whole and within the water protection zones should be directed toward keeping the natural thermodynamic equilibrium in the system 'surface conditions – frozen ground – groundwater'.

Experience in many developed regions shows that it is easier to protect the groundwater against pollution than to eliminate its consequences. This means that the problem of groundwater protection arises as early as at the first stage of development and that to solve this problem it is necessary to carry out a variety of geocryological-hydrogeological investigations, including survey, forecast and special-purpose regime observations within each region of the permafrost.

In the course of the industrial/economic development of the permafrost regions vast areas are involved. Within these areas disruption of the whole complex of the natural environment takes place. This disruption of natural conditions extends as a rule not only over the site under construction but also to the whole zone of infrastructure creation (access roads, electric power lines, service lines, etc.). The restoration by nature of plant communities proceeds very slowly, with some not being restored at all. Therefore as early as before the beginning of economic activity it is necessary not only to assess the engineering-geological conditions of the area and to make a forecast of their change, but also to establish measures aimed at recultivation (restoration) of the disturbed parts.

For this purpose the methods of biological recultivation are used. Biological recultivation consists of the artificial creation of vegetation covers of various kinds, purpose and productivity. Biological recultivation within the permafrost zone can solve the following problems: a) reduction or prevention of consequences of technological disturbance of the cover, associated with sharp increase of seasonal ground thawing depth; b) stabilization of sand embankments against the action of water and wind erosion; c) creation of green landscapes in agreement with the requirements of the inhabitants so far as sanitary and aesthetic aspects are concerned; d) restoration of necessary conditions for animals to live.

Methods of biological recultivation are various for the wide range of natural-climatic conditions of the permafrost regions. The main methods used within the areas of completely disturbed soil vegetation cover or on artificial embankments, are sowing of grasses, planting of shrubs with lime pretreatment and the application of mineral and organic fertilizers. Under good moisture conditions and with a modest degree of disruption of the turf one can restrict these methods to the application of mineral fertilizers with obligatory lime pretreatment in a number of cases, to restore the vegetation cover after carrying out levelling. The kinds and sorts of plants suitable for biological recultivation must have a number of morphological features such as sufficient resistance to cold, capacity to form strong turf cover for a long time, fast growth, annual fruiting, rather high germinating capacity and be present in quantity in nature. The efficient method to prevent erosion and to establish trees in the northern settlements is planting of native willow grafts. The essential condition for the safe overwintering of plants, especially of those brought here from other regions, is a rather thick snow cover. In winter one should not consolidate or remove the snow cover within the areas under recultivation and one should institute measures promoting snow accumulation in the areas where it is blown by wind.

References

1. Budyko, M.I. Climate in the past and in the future: Gidrometeoizdat, 1980.
2. Vtyurin, B.I. Ground ice of the USSR. Moscow: Nauka, 1975.
3. Vtyurina, Ye.A. Cryogenic structure of rocks of the seasonally-thawed layer (Russian). Moscow: Nauka, 1974.
4. Vyalov, S.S. Rheological Principles of the mechanics of frozen ground (Russian). Moscow: Vysshaya Shkola, 1978.
5. Geocryology of the USSR (Russian)/Edited by E.D. Yershov. Moscow: Nedra, vols. 1–5, 1988–1989.
6. Grechishchev, S.Ye., Chistotinov, L.V., and Shur, Yu.L. Cryogenic physical and geological processes and their forecasting (Russian): Moscow: Nedra, 1980.
7. Danilov, I.D. Methods of cryolithological research (Russian). Moscow: Nedra, 1983.
8. Yershov, E.D. Moisture transport and cryogenic textures in dispersed rocks (Russian). Moscow: Moscow State University, 1979.
9. Yershov, E.D., Akimov, Yu.P. and Cheverev, V.G. Phase composition of moisture in frozen ground (Russian). Moscow: Moscow State University, 1979.
10. Yershov, E.D. Cryolithogenesis (Russian). Moscow: Nedra, 1982.
11. Yershov, E.D. Physical chemistry and mechanics of frozen ground (Russian). Moscow: Moscow State University, 1986.
12. Yershov, E.D., Danilov, I.D. and Cheverev, V.G. The petrography of frozen rock (Russian). Moscow: Moscow State University, 1987.
13. Yershov, E.D., Kuchukov, E.Z. and Komarov, I.A. Sublimation of ice in dispersed soils (Russian). Moscow: Moscow State University, 1975.
14. Laboratory methods of studying frozen ground (Russian)/edited by E.D. Yershov. Moscow: Moscow State University, 1985.
15. Methods of ice surveying (Russian). Moscow: Moscow State University.
16. Microstructure of frozen ground (Russian)/edited by E.D. Yershov. Moscow: Moscow State University, 1988.
17. General permafrost studies (Russian)/edited by V.A. Kudryavtsev. Moscow: Moscow State University, 1976.
18. Principles of frozen ground forecasting during engineering and geocryological investigations (Russian)/Edited by V.A. Kudryavtsev. Moscow: Moscow

State University.
19. Savel'yev, B.A. Physics, chemistry and structure of natural ice and frozen rocks (Russian). Moscow: Moscow State University, 1971.
20. Tsytovich, N.A. Mechanics of frozen ground (Russian). Moscow: Vysshaya shkola, 1973.
21. Shilo, N.A. Principles of studying deposits (Russian). Moscow: Nauka, 1981.

Index

Printed in the United States
By Bookmasters